Subcellular Biochemistry

Volume 76

Series editor

J. Robin Harris, Mainz, Germany

More information about this series at http://www.springer.com/series/6515

Caroline Fonta · László Négyessy
Editors

Neuronal Tissue-Nonspecific Alkaline Phosphatase (TNAP)

Springer

Editors
Caroline Fonta
National Centre for Scientific Research (CNRS)
University of Toulouse-UPS
Toulouse
France

László Négyessy
Wigner Research Centre for Physics
Hungarian Academy of Sciences
Budapest
Hungary

and

Department of Anatomy, Histology and Embryology
Semmelweis University Medical School
Budapest
Hungary

ISSN 0306-0225
Subcellular Biochemistry
ISBN 978-94-017-7196-2 ISBN 978-94-017-7197-9 (eBook)
DOI 10.1007/978-94-017-7197-9

Library of Congress Control Number: 2015941876

Springer Dordrecht Heidelberg New York London

Printed on acid-free paper

Springer Science+Business Media B.V. Dordrecht is part of Springer Science+Business Media (www.springer.com)

Preface

Serum alkaline phosphatase (AP) is perhaps the most widely used biomarker since it was discovered in bone tissue by Robison (*Biochem. J.* 17: 286, 1923). The work by Kay (*Brit. J. exp. Path.* 10: 253, 1929) and Roberts (*Brit. J. exp. Path.* 11: 90, 1930) demonstrating elevation of this enzyme in bone and hepatobiliary diseases helped establish the usefulness of this simple enzymatic assay for clinical chemistry. However, most of what we know about the function(s) of this enzyme comes not from the study of clinical conditions displaying increased levels of the enzyme but rather from studying hypophosphatasia (HPP), a life-threatening inborn-error-of-metabolism described by Rathbun (*Am J Dis Child* 75:822–831, 1943) caused by a deficiency in the expression of the tissue-nonspecific AP (TNAP, a.k.a liver-bone-kidney type AP) isozyme. Fast forward to 2015, we now understand that TNAP plays a fundamental role in establishing the phosphate/pyrophosphate ratio conducive to proper skeletal and dental mineralization, and that increasing concentration of pyrophosphate alters this ratio causing the rickets (in children) or osteomalacia (in adults) characteristic of HPP as well as the lack of acellular cementum, dentin and tooth enamel mineralization in patients and mouse models of this disease. Enzyme replacement has proven efficacious in correcting these mineralization defects in HPP mice and clinical trials are now underway in infants and children with severe disease where it appears to preserve life and provide benefit to their skeletal condition. Little is known about the role(s) of TNAP in the kidney and even less on its function(s) in the liver, the brain and other soft tissues. Now that HPP patients under treatment are expected to survive their skeletal disease we must fill this void in knowledge to help us understand other manifestations of HPP, not associated with the skeleton or teeth.

This book by Caroline Fonta and László Négyessy is a timely contribution to the literature focusing on what we know and do not know about TNAP in the nervous system that will undoubtedly help focus more research on this important area. The book is organized into four parts. The first part predictably starts with a recapitulation of what we have learned about TNAP function through the study of hypophosphatasia patients. Dr. Salles (Chap. 1) provides a comprehensive review of the pathophysiology and symptomatology of the various forms of HPP, while

Dr. Mornet (Chap. 2) summarizes the structure of the *ALPL* gene and mutation analyses that have led to our the current knowledge regarding genotype–phenotype correlations. Much of this important work has been driven by Dr. Mornet and his group who also maintain a publicly available database of all known HPP mutations (http://www.sesep.uvsq.fr/03_hypo_mutations.php). This first part ends with a chapter by Dr. Narisawa (Chap. 3) who has contributed to the generation and characterization of many of the currently available animal models examining TNAP function, including transgenic models that overexpress TNAP, global and conditional ablation models of *Alpl* function and knockin models of dominant HPP-causing *Alpl* mutations. These models are very important research tools to probe the metabolic pathways affected in HPP and also for the pre-clinical evaluation of possible treatments for HPP and other diseases of TNAP dysfunction.

The second part of the book (Chaps. 4–7) summarizes what is known about TNAP expression in the central nervous system (CNS). Chapter 4, by Drs. Zimmermann and Langer, reviews the patterns of expression of TNAP during embryonic and post-natal neurogenesis and summarizes *in vitro* experiments that point to a role of TNAP in axonal growth and neuronal cell differentiation. Chapter 5, by Drs. Fonta, Négyessy and collaborators delve into the methods that have been used to document expression of TNAP in a cell-specific manner in CNS, and describe this groups extensive comparative studies of TNAP expression in the rodent and primate cerebral cortex. Dr. Kántor and collaborators continue in Chap. 6 with comparative studies throughout speciation, including zebrafish, frogs, rats, mice, squirrel monkeys, ferrets, chickens, golden hamsters, cats, rabbits, guinea pigs and humans, but focusing on the pattern of expression of TNAP in the retina of the eye, a subject where little prior information existed and to which this group has recently contributed. In Chap. 7, Dr. Deracinois and collaborators review published data as well as their own results using a proteomics approach, demonstrating the expression of TNAP in brain capillary endothelial cells and discussing the possible implication of this expression for the function of the blood–brain barrier.

The third part of the book, Chaps. 8–13, presents some of the new active areas of research regarding TNAP function in the brain. Some are tentative explorations of hypothetical new ideas while some document well established lines of investigation. My own chapter (Chap. 8) summarizes what we know about TNAP function mainly using mouse models and how that information can help us design further studies to understand the role(s) of TNAP in the brain. In Chap. 9, Dr. Ermonval and collaborators point to the possibility that TNAP may exert pathophysiological roles as part of a dynamic versatile lipid rafts population present in neuronal cells. Dr. Négyessy and collaborators (Chap. 10) build on this theme but using bioinformatic tools to come to the conclusion that TNAP is at the crossroad of numerous biochemical pathways. Not unexpectedly, many of those connectivity diagrams support a role for TNAP in neurotransmission and purinergic signaling in good agreement with currently available data. In fact, in Chap. 11, Dr. Coburn, a world-renowned investigator in the vitamin B6 field, summarizes the vast amount of data linking TNAP function with the metabolism of pyridoxal-5'phosphate, the major chemical form of the vitamin, and its essential role in the synthesis of neurotransmitters such as gamma-aminobutyric acid (GABA). Indeed, GABA insufficiency appears to largely explain the seizures experienced by infants as well as mice with life-threatening HPP. In Chap. 12, Dr. Nowak and collaborators share a cautionary note on the widespread use of levamisole and tetramisole as inhibitors of TNAP. These compounds are known to cause many off-target effects and are therefore not an optimal choice for *in vivo* studies, especially now that better pharmacological inhibitors have been developed. Chapter 13, by Drs. Street and Sowa, present truly exciting new information regarding the role of TNAP in the production of the anti-nociceptive adenosine in dorsal root ganglia neurons, at synapsis that transmit pain signals. These studies not only point to TNAP as a possible new target for pain management but also may help explain the mechanism of alleviation of pain in HPP patients undergoing enzyme replacement therapy.

The fourth and last part of the book (Chaps. 14–18) takes us back to the clinical issues that these new directions of research should help us understand. Thus, in Chap. 14, Dr. Taketani, points us beyond the seizures, whose pathophysiology are better understood, towards encephalopathy, intracranial hypertension, mental retardation, deafness and growth hormone deficiency as conditions associated with HPP that we do not yet understand. Dr. Hofmann and collaborators review the clinical experience with the use of enzyme replacement therapy with asfotase alfa, what works well and what is yet unknown regarding efficacy with this therapeutic approach. Drs Cole and Thompson (Chap. 15) summarize current information on Marbry Syndrome, a condition with hyperphosphatasemia caused by defective synthesis of the glycosyl phosphatidylinositol (GPI) anchor needed for TNAP to be tethered to the membrane of cells and vesicles. It is unclear if the neurological deficits in Marbry patients are caused by the persistent hyperphosphatasemia or by another pathway that depends on GPI anchoring for its function. In Chap. 17, Drs Kellett and Hooper discuss clinical data related to elevated plasma and brain TNAP level in Alzheimer's Disease patients and the possibility that neuronal toxicity and cell death might be associated with ability of unregulated TNAP to dephosphorylate Tau protein. To conclude this volume, Dr. Diaz-Hernandez and collaborators (Chap. 18) recount their studies that demonstrate that dephosphorylated tau protein behaves as an agonist of muscarinic M1 and M3 receptors, provoking a robust and sustained intracellular calcium increase that triggers neuronal death while also increasing TNAP expression. These authors have demonstrated that the promoting effect of TNAP on axonal growth is due to its ability to hydrolyze extracellular ATP thus preventing P2X7 receptor activation.

To close on these introductory remarks, I think that after finishing this book the readers will be left with the impression that while much needs to be learned about the role of TNAP in the CNS, there are some good solid leads that researchers have started to pursue and that much progress and new understanding of the pathophysiology of HPP and other diseases associated with TNAP dysfunction is to be expected in the not-too-distant future.

José Luis Millán

Contents

Abbreviations

A1-4R	Adenosine receptors 1-4
AADC	Aromatic L-amino acid decarboxylase
Aβ	Amyloid−β peptide
ABCA7	ATP binding cassette transporter 7 gene
ACh	Acetylcholine
ACSF	Artificial cerebro-spinal fluid
AD	Alzheimer's disease
ADA	Adenosine deaminase
ADO	Adenosine
AGXT2L1	Alanine-glyoxylate aminotransferase 2-like 1
AMP, ADP, ATP	Adenosine mono-, di-, triphosphate
$Akp2^{-/-}$ mouse	TNAP-homozygous null mouse
ALPL	Tissue non-specific alkaline phosphatase gene
AMPAR	α-amino-3-hydroxy-5-methyl-4-isoxazolepropionic acid receptor
ANK	Ankylosis protein
ANKH	Progressive ankylosis protein homolog
AR/AD	Autosomal recessive/dominant
AP	Alkaline phosphatase
APP	Amyloid precursor protein
APS	Antidromic population spikes
BBB	Blood brain barrier
BCEC	Brain capillary endothelial cell
BCIP	5-bromo-4-chloro-3'-indolyphosphate p-toluidine salt
BIC	Bayesian information criteria
BIN1	Bridging integrator 1 gene
BMP	Bone morphogenetic protein
CAMPGEFII	cAMP-regulated guanine nucleotide exchange factor II
CD	Convergence degree
CD33	Myeloid cell surface antigen CD33 gene
CD2AP	CD2 associated protein gene

CFP	Cyan fluorescent protein
CGRP	Calcitonin gene-related protein
ChAT	Choline acetyltransferase
CHO	Chinese hamster ovary
CLU	Clusterin gene
CNM	Clauset-Newman-Moore clustering
CNS	Central nervous system
CPAP	Continuous positive airway pressure
CPPD, CPPDC	Calcium PPi dihydrate crystals
CR	Calretinin
CR1	Complement receptor 1 gene
CREB	cAMP response element-binding protein
CSF	Cerebrospinal fluid
CSPG	Chondroitin proteoglycans
CT scan	Computed tomography scan
CW	Community walktrap clustering
CWD	Chronic wasting disease
DA	Dopamine
DAG	Diacyglycerol
DCX	Doublecortin
DLX	Distal-less homeobox
DMSO	Dimethyl sulfoxide
LDOPA	LDopamine
DRG	Dorsal root ganglia
D1R	Dopamine receptor D1
E 'n'	Embryonic day 'n'
EAE	Experimental allergic encephalomyelitis
EAP	Embryonic isoform of alkaline phosphatase
EB	Edge betweenness
EC	Endothelial cells
ECM	Extracellular matrix
EEG	Electroencephalogram
EGFR	Epidermal growth factor receptor
EGFP	Enhanced green fluorescent protein
EM	Electron microscopy
ENPP1	Ectonucleotide pyrophosphatase/phosphodiesterase 1
ENTPD/CD39	Ectonucleotide triphosphate diphosphohydrolase
ENU	*N*-ethyl-*N*-nitrosourea
EPHA1	Ephrin receptor A1 gene
EPI	Epinephrine
EPSP	Excitatory postsynaptic potential
ER	Endoplasmic reticulum
ERT	Enzyme replacement therapy
ES	Embryonic stem

ESTs	Expressed sequence tag, a short sub-sequence of a cDNA sequence
EV	Extracellular vesicles
FIVEHT2BR	Serotonin 2B receptor
FSCV	Fast-scan cyclic voltammetry
GABA	γ-Aminobutyric acid, gamma-aminobutyrate
$GABA_AR/GABA_BR$	γ-Aminobutyric acid receptor type A/type B
GABA-T	GABA transaminase
GACI	Generalized arterial calcification in infants
GAD	Glutamic acid decarboxylase
GCAP	Germ-cell alkaline phosphatase
GCL	Ganglion cell layer
G1-n	1st-nth generation
GFP	Green fluorescent protein
GH	Growth hormone
GHD	Growth hormone deficiency
GLAST	Astrocyte-specific glial glutamate transporter
GlcNAc	*N*-acetylglucosamine
G-protein	Guanine nucleotide-binding protein
GPI	Glycosyl phosphatidylinositol
GPCR	G-protein coupled receptor
GWAS	Genome-wide association studies
HA	Hydroxyapatite
HDAC	Histone deacetylase
5-HIAA	5-hydroxyindole acetic acid
HPP	Hypophosphatasia
HPLC	High-performance liquid chromatography
HSPG	Heparin sulfate proteoglycans
5-HTTP	Hydroxytryptophan
5 HTT	Serotonin (5HT) transporter
HVA	Homovanillic acid
IAP	Intestinal alkaline phosphatase
IC_{50}	Half maximal inhibitory concentration
IGF1	Insulin growth factor 1
INL	Inner nuclear layer
IPL	Inner plexiform layer
IP_3	Inositol triphosphate
dKO	Gene knock out, double
LFP	Local field potential
LPA	Lysophosphatidic acid
LPS	Lipopolysaccharide
LSD	Lysosomal storage disease
M1R, 3R, M1, 3	Muscarinic acetylcholine receptor M1R, M3R
MAP2	Microtubule-associated protein 2
MAPK	Mitogen-activated protein kinase

MCI	Mild cognitive impairment
MCL	Markov clustering
3-*O*-MD	3-*O*-methyldopa
MeV	Medium vesicles
MHPG	3-methoxy-4 hydroxyphenylglycol
MLR	Membrane lipid raft
MLS 0038949	TNAP inhibitor
MRI	Magnetic resonance imaging
MS4A	Membrane-spanning 4-domains subfamily A gene
MSC	Mesenchymal stem cell
MV	Matrix vesicles
MVB	Multi vesicular body
NA	Noradrenaline
NB	Node betweenness
NBT	Nitro-blue tetrazolium chloride
NFTs	Neurofibrillary tangles
NIK	NF-kappa-*B*-inducing kinase also known as mitogen-activated protein kinase kinase kinase 14 (MAP3K14)
NMDA	N-methyl-D-aspartate
pNPP	p nitrophenyl phosphate
NPP1	Nucleotide pyrophosphatase/phosphodiesterase 1
NSAID	Non-steroidal anti-inflammatory drug
NT5E/CD73	Ecto-5′-nucleotidase
NTPs	Nucleotide triphosphates
dNTP	Desoxynucleoside triphosphate
NTPD	Nucleoside triphosphate diphosphohydrolase
OI	Osteogenesis imperfecta
OPL	Outer plexiform layer
OPN	Osteopontin
OPS	Orthodromic population spikes
P1 receptors	A_1, A_{2A}, A_{2B}, A_3 adenosine receptors
P, PND	Postnatal day
P2X, P2Y	Purinergic receptors
PAP	Prostatic acid phosphatase
PC1	Plasma cell membrane glycoprotein-1
PCho	Phosphocholine
PCR	Polymerase chain reaction
PEA	Phosphoethanolamine
PGAP	Post-GPI attachment to proteins
PHEX	Phosphate regulating endopeptidase homolog, X-linked

PHOSPHO1	Phosphatase orphan
Pi	Inorganic phosphate
PICALM	Phosphatidylinositol clathrin associated lymphoid myeloid gene
P_iT	Phosphate transporter 1
PIGV	Phosphatidylinositol Glycan Anchor Biosynthesis, Class V
PIP_2	Phosphatidylinositol 4,5-bisphosphate
PIPL	Phosphatidylinositol-phospholipases
PKCα	Protein kinase C alpha subunit
PL	Pyridoxal
PLAP	Placental alkaline phosphatase
PLC	Phospholipase C
PLP	Pyridoxal 5′phosphate
PN	Pyridoxine
PNPO	Gene for pyridox(am)ine 5′-phosphate oxidase
PPi	Inorganic pyrophosphate (i=intracellular, e=extracellular)
PrP^C	Cellular prion protein
PRS	Pyridoxine responsive seizure
fPSP	Field post-synaptic potential
PTH	Parathyroid hormone
RARE	Retinoic acid response element
RIM	Regulating synaptic membrane exocytosis protein
RMS	Rostral migratory stream
shRNA	Small hairpin RNA
RT-PCR	Reverse transcription polymerase chain reaction
RUNX	Runt-related transcription factor
SAGE	Serial analysis of gene expression
s.c.	Sub cutaneous
SD	Standard deviation
SERT	See at 5HTT
SEM	Standard error of the mean
SMAD	Mothers against decapentaplegic homolog
SNP	Single nucleotide polymorphism
SSH	Suppression subtractive hybridization
STAT	Signal transducer and activator of transcription
SVZ	Subventricular zone
TCA	Tricarboxylic acid
TEER	Transendothelial electrical resistance
TEM	Transmission electron microscopy
TH	Tyrosine hydroxylase
TNAP	Tissue non-specific alkaline phosphatase
TRPV1	Transient receptor potential cation channel subfamily V member 1

TTX	Tetrodotoxine
UDP	Uridine diphosphate
UTP	Uridine triphosphate
uPAR	Urokinase plasminogen activator receptor
VDRE	Vitamin-D response element
VMA	Vanillylmandelic acid
VSMC	Vascular smooth muscle cells
VZ	Ventricular zone

Contributors

Jesús Avila Centro de Biología Molecular "Severo Ochoa", CSIC-UAM, Madrid, Spain; Centro de Investigación Biomédica en Red Sobre Enfermedades Neurodegenerativas (CIBERNED), Madrid, Spain

Mihály Bányai Computational Systems Neuroscience Lab, Wigner Research Center for Physics, Hungarian Academy of Sciences, Budapest, Hungary

Fülöp Bazsó Theoretical Neuroscience and Complex Systems Research Group, Wigner Research Center for Physics, Hungarian Academy of Sciences, Budapest, Hungary; SU-Tech College of Applied Sciences, Subotica, Serbia

Pascal Barone Centre de Recherche Cerveau et Cognition (CerCo), Université de Toulouse UPS; CNRS UMR 5549, Toulouse, France

Florence Baychelier Département de Virologie, Institut Pasteur, Unité Des Stratégies Antivirales, Paris, France

Stephen P. Coburn Department of Chemistry, Indiana University—Purdue University, Fort Wayne, IN, USA

David E.C. Cole Department of Laboratory Medicine and Pathobiology, University of Toronto, Toronto, ON, Canada; Department of Clinical Pathology, Sunnybrook Health Sciences Center, Toronto, ON, Canada

Dorottya Cserpán Theoretical Neuroscience and Complex Systems Research Group, Wigner Research Center for Physics, Hungarian Academy of Sciences, Budapest, Hungary

Dávid Czégé Department of Physiology and Neurobiology, Eötvös Loránd University, Budapest, Hungary

Marie-Pierre Dehouck University of Lille Nord de France, Lille, France; University of Artois, LBHE, Lens, France; IMPRT-IFR114, Lille, France

Barbara Deracinois University of Lille Nord de France, Lille, France; University of Artois, LBHE, Lens, France; IMPRT-IFR114, Lille, France

Miguel Diaz-Hernandez Departamento de Bioquímica y Biología Molecular, Facultad de Veterinaria, Universidad Complutense de Madrid, Madrid, Spain; Instituto de Investigación Sanitaria San Carlos IDISSC, Madrid, Spain

Myriam Ermonval Département de Virologie, Institut Pasteur, Unité Des Stratégies Antivirales, Paris, France

Christophe Flahaut University of Lille Nord de France, Lille, France; University of Artois, LBHE, Lens, France; IMPRT-IFR114, Lille, France

Caroline Fonta Centre de Recherche Cerveau et Cognition (CerCo), Université de Toulouse UPS; CNRS UMR 5549, Toulouse, France

Hermann J. Girschick Children's Hospital, Vivantes Hospital im Friedrichshain, Berlin, Germany

Stephanie Graser Orthopedic Department, Orthopedic Center for Musculoskeletal Research, University of Würzburg, Würzburg, Germany

Balázs Györffy Laboratory of Proteomics and MTA-ELTE NAP B Neuroimmunology Research Group, Institute of Biology, Eötvös Loránd University, Budapest, Hungary

János Hanics Department of Anatomy, Histology and Embryology, Semmelweis University Medical School, Budapest, Hungary

Felix Hernandez Centro de Biología Molecular "Severo Ochoa", CSIC-UAM, Madrid, Spain; Centro de Investigación Biomédica en Red Sobre Enfermedades Neurodegenerativas (CIBERNED), Madrid, Spain

Christine Hofmann Children's Hospital, University of Würzburg, Würzburg, Germany

Nigel M. Hooper Institute of Brain, Behaviour and Mental Health, Faculty of Medical and Human Sciences, University of Manchester, Manchester, UK

Franz Jakob Orthopedic Department, Orthopedic Center for Musculoskeletal Research, University of Würzburg, Würzburg, Germany

Orsolya Kántor Department of Anatomy, Histology and Embryology, Semmelweis University, Budapest, Hungary

Katherine A. B. Kellett Institute of Brain, Behaviour and Mental Health, Faculty of Medical and Human Sciences, University of Manchester, Manchester, UK

David Langer Institute for Cell Biology and Neuroscience, Molecular and Cellular Neurobiology, Goethe University, Frankfurt am Main, Germany

Anne-Marie Lenfant University of Lille Nord de France, Lille, France; University of Artois, LBHE, Lens, France; IMPRT-IFR114, Lille, France

Johannes Liese Children's Hospital, University of Würzburg, Würzburg, Germany

Ákos Lukáts Department of Human Morphology and Developmental Biology, Semmelweis University, Budapest, Hungary

Birgit Mentrup Orthopedic Department, Orthopedic Center for Musculoskeletal Research, University of Würzburg, Würzburg, Germany

José Luis Millán Sanford-Burnham Medical Research Institute, La Jolla, CA, USA

Maria Teresa Miras-Portugal Departamento de Bioquímica y Biología Molecular, Facultad de Veterinaria, Universidad Complutense de Madrid, Madrid, Spain; Instituto de Investigación Sanitaria San Carlos IDISSC, Madrid, Spain

Etienne Mornet Unité de Génétique Constitutionnelle, Centre Hospitalier de Versailles, Le Chesnay, France

Sonoko Narisawa Sanford-Burnham Medical Research Institute, La Jolla, CA, USA

László Négyessy Theoretical Neuroscience and Complex Systems Research Group, Wigner Research Center for Physics, Hungarian Academy of Sciences, Budapest, Hungary; Department of Anatomy, Histology and Embryology, Semmelweis University Medical School, Budapest, Hungary

Lionel G. Nowak Centre de Recherche Cerveau et Cognition (CerCo), Université de Toulouse UPS; CNRS UMR 5549, Toulouse, France

Horacio Plotkin University of Nebraska Medical Center, Omaha, NE, USA

Laia Rodriguez Martinez Centre de Recherche Cerveau et Cognition (CerCo), Université de Toulouse UPS; CNRS UMR 5549, Toulouse, France

Benoît Rosay Centre de Recherche Cerveau et Cognition (CerCo), Université de Toulouse UPS; CNRS UMR 5549, Toulouse, France

Jean Pierre Salles Unité d'Endocrinologie Maladies Osseuses, Hôpital Des Enfants, CHU de Toulouse; Inserm UMR 1043 UPS, Toulouse Cedex, France

Lothar Seefried Orthopedic Department, Orthopedic Center for Musculoskeletal Research, University of Würzburg, Würzburg, Germany

Zoltán Somogyvári Theoretical Neuroscience and Complex Systems Research Group, Wigner Research Center for Physics, Hungarian Academy of Sciences, Budapest, Hungary

Nathaniel A. Sowa Department of Psychiatry, University of North Carolina at Chapel Hill and UNC Hospitals, Chapel Hill, NC, USA

Sarah E. Street Department of Cell Biology and Physiology, University of North Carolina at Chapel Hill, Chapel Hill, NC, USA

Takeshi Taketani Department of Pediatrics, Shimane University Faculty of Medicine, Izumo, Shimane, Japan

Miles D. Thompson Department of Laboratory Medicine and Pathobiology, University of Toronto, Toronto, ON, Canada

Béla Völgyi Department of Experimental Zoology and Neurobiology, University of Pécs, Pécs, Hungary; János Szentágothai Research Center, Pécs, Hungary

Herbert Zimmermann Institute for Cell Biology and Neuroscience, Molecular and Cellular Neurobiology, Goethe University, Frankfurt am Main, Germany

Part I
Disease Forms and Animal Models of Hypophosphatasia

Chapter 1
Clinical Forms and Animal Models of Hypophosphatasia

Jean Pierre Salles

Abstract Hypophosphatasia (HPP) is due to mutations of the tissue non-specific alkaline phosphatase (TNAP) gene expressed in the liver, kidney, and bone. TNAP substrates include inorganic pyrophosphate cleaved into inorganic phosphate (Pi) in bone, pyridoxal-5'-phosphate (PLP), the circulating form of vitamin B6, and phosphoethanolamine (PEA). As an autosomal recessive or dominant disease, HPP results in a range of clinical forms. Its hallmarks are low alkaline phosphatase (AP) and elevated PLP and PEA levels. Perinatal HPP may cause early death with respiratory insufficiency and hypomineralization resulting in deformed limbs and sometimes near-absence of bones and skull. Infantile HPP is diagnosed before 6 months of life. Respiratory failure, rib fractures and seizures due to vitamin B6deficiency in the brain indicate poor prognosis. Craniosynostosis is frequent. Unlike in other forms of rickets, calcium and phosphorus are not decreased, resulting in hypercalciuria and nephrocalcinosis. Hypercalcemic crisis may occur. Failure to thrive and growth retardation are concerns. In infantile and adult forms of HPP, non-traumatic fractures may be the prominent manifestation, with otherwise unexplained chronic pain. Progressive myopathy has been described. Dental manifestations with early loss of teeth are usual in HPP and in a specific form, odontohypophosphatasia. HPP has been studied in knock-out mice models which mimic its severe form. Animal models have made a major contribution to the development of an original enzyme therapy for human infantile HPP, which is however essentially targeted at mineralized tissues. Better knowledge of its

---

Endocrinology, Bone Diseases, Obesity, Genetics and Gynecology Unit, Children's Hospital, Toulouse University Hospital, University of Toulouse, France.

INSERM UMR 1043, Molecular Mechanisms of Growth, Osteogenesis and Osteolysis, Biotherapies, Paul Sabatier University, Toulouse, France.

National Reference Center for Diseases of the Metabolism of Calcium and Phosphate, Toulouse, France.

J.P. Salles (✉)
Unité d'Endocrinologie Maladies Osseuses, Hôpital Des Enfants, CHU de Toulouse and Inserm UMR 1043 UPS, Toulouse Cedex, France
e-mail: Salles.jp@chu-toulouse.fr

C. Fonta and L. Négyessy (eds.), *Neuronal Tissue-Nonspecific Alkaline Phosphatase (TNAP)*, Subcellular Biochemistry 76,
DOI 10.1007/978-94-017-7197-9_1

extraskeletal manifestations, including pain and neurological symptoms, is therefore required.

Keywords Hypophosphatasia · Bone · Rickets · Calcium · Phosphate

1.1 Introduction

Hypophosphatasia (HPP, OMIM 146300, 241500, 241510) is an inborn error of metabolism. Its biological hallmark, hypophosphatasemia, results from the defective activity of alkaline phosphatase (orthophosphoric-monoester phosphohydrolase, alkaline optimum, EC 3.1.3.1) (AP). HPP is due to inactivating mutations within one of the four genes responsible for alkaline phosphatase activity (Whyte 1994; Mornet 2008; Whyte 2010, 2012; Rockman-Greenberg 2013) (see Chap. 2). Three of these genes produce tissue-specific AP isoenzymes (intestinal, placental and germ-cell). The fourth gene, responsible for HPP, produces an ubiquitous AP isoenzyme abundant in liver, bone and kidney and is known as the tissue non-specific alkaline phosphatase gene (ALPL: OMIM 171760) (TNAP in the present paper). While the human gene symbol for TNAP is ALPL (Alkaline Phosphatase-Liver), TNAP is more appropriate to the gene and its products, given its pathophysiological role as demonstrated in HPP (Whyte 2012). The TNAP products from liver, bone and kidney, which undergo post-translational modifications, represent a family of isoenzymes with distinctive physicochemical properties (Whyte 2010, 2012; Buchet et al. 2013). The TNAP gene is located on chromosome 1 (1p36.1–34). HPP usually presents as rickets or osteomalacia. The disease is due to loss-of-function mutations in the TNAP gene. Depending on the age of clinical onset, the disorder has been classified into perinatal, infantile, childhood and adult forms, and odontohypophosphatasia. Pediatricians encounter various forms with impaired skeletal mineralization, rickets, bone deformity, fractures, growth delay and dental abnormalities and also, in the most severe cases, respiratory and neurological complications. The adult form often begins with the loss of adult teeth followed by recurrent bone pain and fractures. We will briefly examine the pathophysiology of HPP before describing its clinical forms and their diagnosis. We will also analyze the manifestations of HPP in animal models, in rodents and primates. These models have opened new perspectives to a better understanding of HPP pathophysiology and paved the way for original treatments of the disease.

1.2 Pathophysiology of HPP

More than 260 mutations of the *TNAP* gene are known to be associated with HPP, most of them being missense mutations (see also Chap. 2). Most HPP cases result from compound heterozygous mutations transmitted in an autosomal recessive manner (Mornet 2000, 2007; Rockman-Greenberg 2013), but HPP may also follow an autosomal dominant pattern (Mornet 2007). The phenotypic expression of HPP is varied and the genotype-phenotype correlation is sometimes difficult to establish. Modifying genes may be involved, responsible for modulating the expression of a given mutation (Rockman-Greenberg 2013).

The observed mutations of *TNAP* in HPP have confirmed an early hypothesis regarding the essential role of AP in skeletal mineralization (Robison 1932; Whyte 1994, 2010). TNAP is now known to be necessary for the development of bone and primary teeth. The AP isoenzymes encoded by TNAP are not only expressed in mineralized tissues like bone and teeth. Other organs, first of all the liver, which also expresses TNAP, are apparently unaffected in HPP. The significance of TNAP expression in non-mineralized tissues thus remains unclear (Whyte 2010). For example, although the expression of TNAP in the brain has been extensively studied, its true function has not been clarified (Fonta et al. 2004, 2005; Negyessy et al. 2011; Brun-Heath et al. 2011; Hanics et al. 2012). The bone isoforms account for approximately 57 % of total AP activity detected in plasma. Together with the activity of the liver isoform, they account for 95 % of AP activity. The bone-specific TNAP isoforms can be detected in plasma by immunoanalytical techniques. Circulating AP activity is indicative of the AP activity present in tissues. However, "physiological" AP is bound to the surface of plasma membranes as an ectoenzyme, linked to the polar group of a phosphatidylinositol glycan (Seetharam et al. 1987). Although incompletely understood, deficient AP release from cell membranes results in hyperphosphatasemia, as observed in Mabry syndrome (OMIM 2393000), due to a defect of PIGV mannosyltransferase (Krawitz et al. 2010) (see also Chap. 16).

The visible consequence of loss-of-function mutations of *TNAP* is extracellular accumulation of TNAP substrates. In the current view, the clearly understood physiological role of TNAP is related to two major substrates, whose levels are elevated in HPP patients and in animal models: inorganicpyrophosphate (PPi) and pyridoxal 5'- phosphate (PLP), also known as vitamin B6 (Whyte 2010). Accumulated PPi inhibits bone mineralization by blocking hydroxyapatite crystal formation. In addition, TNAP deficiency leads to decreased inorganic phosphate (Pi) availability and reduced hydroxyapatite formation. There is strong evidence from observation in humans and animal models that mineralization is controlled by TNAP, which is responsible for PPi hydrolysis and Pi production, and by the ectonucleotide nucleotide pyrophosphatase/phosphodiesterase 1 (NPP1), previously called plasma cell membrane glycoprotein-1 (PC-1), which releases PPi by desoxynucleoside triphosphate (dNTP) hydrolysis (Hessle et al. 2002). TNAP is a positive factor of mineralization and NPP1 a negative one. Osteopontin and the cell surface protein ANKH (Ank in the mouse), which channels PPi to the extracellular medium, are other inhibitors of mineralization (Harmey et al. 2006; Narisawa et al. 2007). Observations thus converge to conclude that disruption of TNAP activity results in insufficient mineralization of calcified tissues, bone and teeth.

Defective mineralization in the growing skeleton of children impairs endochondral bone formation and leads to rickets, while in adults, after closure of growth plates, it is responsible forosteomalacia. Abnormal calcification in growth plates results in growth retardation and bone deformity. In most cases of rickets, phosphate and calcium levels are in fact decreased. Because of inability to utilize calciumand phosphate for mineralization, it is noteworthy that circulating levels of calcium or phosphate are normal or elevated In HPP.

On the other hand, high PLP levels reflect defective PLP hydrolysis by TNAP and decreased availability of pyridoxal (PL)to cross the blood-brain barrier and feed PLP production in the brain. PLP is a coenzyme involved in glutamic acid decarboxylase activity and gamma-aminobutyric (GABA) synthesis (see Chap. 11). There is strong evidence thatdeficiency of vitamin B6 in the central nervous system is responsible for seizures in HPP children (Waymire et al. 1995; Whyte et al. 1985, 1988; Baumgartner-Sigl et al. 2007).

The principal symptoms of HPP can be thus explained by defective TNAP activity for hydrolysis of PPi and PLP. More systematic analysis of the metabolic effects of TNAP deficiency would increase our knowledge of less obvious mechanisms involved in HPP pathophysiology.

1.3 Clinical Manifestations of HPP

The first clinical case introducing the term "hypophosphatasia" was described by Rathbun (1948). To date, more than 350 patients with HPP have been reported with a wide range of disease severity. In Europe, based on the TNAP mutation database, the prevalence of HPP has been estimated at approximately 1/300,000 and 1/6000 for the severe and moderate forms, respectively (Mornet et al. 2011). HPP is highly prevalent in Mennonite families, 1/25 individuals being carriers of a specific mutation (Greenberg et al. 1993). There is a correlation between clinical forms of the disease and the findings of functional and structural studies, which make it possible to distinguish moderate alleles from severe alleles (Zurutuza et al. 1999). This suggests that the phenotypic heterogeneity observed in HPP is due mainly to the variable enzymatic activity arising from missense mutations (see Chap. 2). However, disease severity may also vary within a given sibship (Whyte et al. 1982a; Moore et al. 1990; Macfarlane et al. 1992). In addition, individuals exhibiting biochemical abnormalities due to a defective *TNAP* allele may be asymptomatic and be considered carriers while other relatives display typical HPP (Whyte et al. 1982a; Whyte 2012). The description of HPP thus remains primarily clinical. Its presentations are classically divided into perinatal, infantile, childhood, adult HPP, and odonto HPP. It is noteworthy that prenatal forms are heterogeneous with spontaneous improvement in some cases, which has led to description of a "benign prenatal" form of HPP.

1.3.1 Perinatal Hypophosphatasia

Perinatal HPP manifests with profound hypomineralization in utero. Nevertheless, the so-called perinatal form of HPP with fatal prognosis and the severe infantile forms are now often detected early during pregnancy, which raises specific concerns regarding diagnosis and prognosis. Some skeletal features may make it possible to distinguish HPP from other severe forms of hypomineralization, first of all osteogenesis imperfecta (OI) (Cormier-Daire et al. 2004; Rockman-Greenberg 2013). In some cases bones are nearly devoid of mineral or have severe rachitic changes, with the skull bones showing only central calcification. Metaphyseal spurs are characteristic of HPP. Prenatal HPP may be difficult to differentiate from osteogenesis imperfecta or other skeletal dysplasias that result in bowing of the long bones and hypomineralization (Cormier-Daire et al. 2004). The prognosis of prenatal HPP is nearly as difficult to establish as that of osteogenesis imperfecta. If HPP is strongly suspected and genetic analysis is performed, it may help to provide a genotype-phenotype correlation. However, it is still difficult to establish an accurate prognosis of HPP. For instance, some forms may improve during the third trimester (Whyte et al. 2006; Stevenson et al. 2008) corresponding to a so-called "benign prenatal" form. Marked hypomineralization, especially of the skull, predicts a very poor prognosis. A more positive prognosis may be established by the result of genetic analysis, if known. Comparison of prenatal and postnatal findings has also suggested that fetal crowding with relatively normal fetal mineralization and chest size, together with heterozygous TNAP mutations, may predict benign forms. Conversely, recessive HPP with some specific mutations suggests lethality, although not with complete reliability (Wenkert et al. 2011; Leung et al. 2013). Prenatal counseling must always be very cautious in view of the new perspectives offered by potential treatments of HPP (see below). Therefore the distinction between lethal and benign prenatal forms is challenging (Taketani et al. 2014). Figure 1.1 illustrates examples of antenatal diagnosis of HPP, with distinctive outcomes.

In classic perinatal HHP, major hypomineralization usually leads to death in utero or soon after birth. Affected neonates may live a few days with respiratory compromise resulting from rachitic defects in the chest. Half of infants die after birth from serious complications, hypoplastic lungs and respiratory failure. In the less severe forms, on a spectrum with the infantile form of HPP, rickets remains the hallmark with shortened and deformed limbs. Respiratory failure and pyridoxine-responsive seizures (see below) are major concerns.

1.3.2 Infantile HPP

If not diagnosed prenatally, infantile HPP becomes manifest before 6 months of age with progressive failure to thrive and rickets. Metaphyseal spurs may markedly

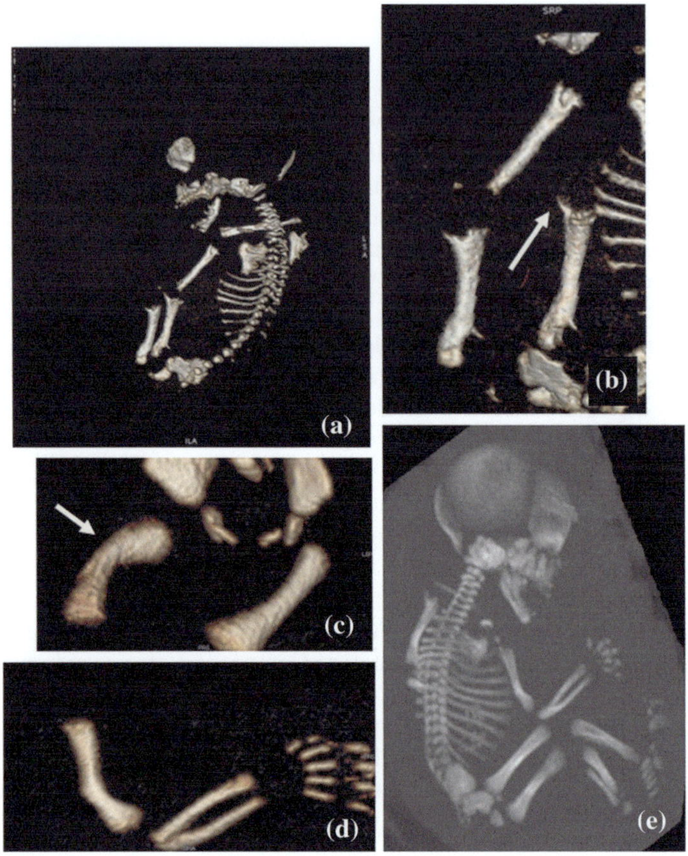

Fig. 1.1 Prenatal diagnosis of HPP **a** and **b**: diagnosis of a lethal form of HPP; 3rd trimester CT scan showed severe hypomineralization, almost no mineralization of the skull and typical metaphyseal abnormalities with spurs (*arrow*) suggesting severe rickets; **c**-**e**: diagnosis of an infantile form of HPP, with hypomineralization (see the skull in E) and femur bowing (*arrow*) but no metaphyseal spurs. Personal data

protrude from the ulna or fibula. The chest deformity and fractures are a contributory factor in respiratory infections. Ventilatory support is often required (Rockman-Greenberg 2013), and a recent study has reported that respiratory work in complicated HPP is mainly done by abdominal displacement (Rodriguez et al. 2012). Increasing demineralization and fractures usually predict a negative outcome but spontaneous, unexplained clinical improvement may occur, making it hazardous to establish a prognosis.

Craniostenosis is frequent. On radiographs the fontanels appear to be open, due to hypomineralized calvarium. The premature fusion of cranial sutures often causes cranial synostosis with a characteristic appearance due to abnormal sutures and skull osteomalacia (so-called "functional craniostenosis") (Fig. 1.2a, b) with

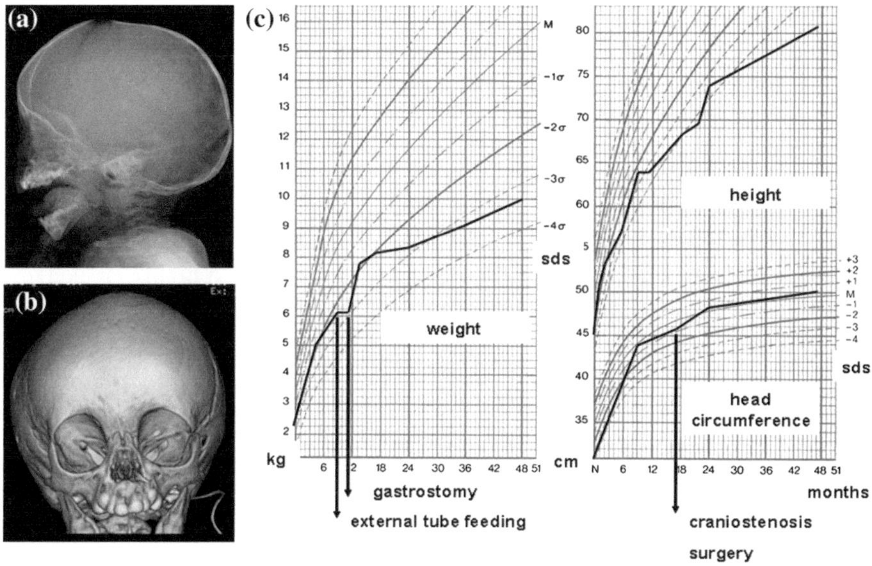

Fig. 1.2 Course of an infantile form of HPP **a** and **b**: craniostenosis with secondary intracranial hypertension occurred by two years of age and was successfully operated. Skull radiography showed the typical appearance due to "functional" craniostenosis. **c**: stagnation of weight gain led to tube feeding and gastrostomy for nocturnal nutrition. Severe growth retardation is a hallmark of the course of infantile hypophosphatasia. Personal data. sds: standard deviation score

subsequent intracranial pressure and potential brain damage. An isolated widened fontanel may be a feature that reveals HPP (Mohn et al. 2011).

Hypercalcemia, due to inability to utilize calcium and phosphate for mineralization, is common in infantile HPP and may trigger the diagnosis. It may explain recurrent vomiting as well as nephrocalcinosis with renal impairment. Severe hypercalcemia with hyperphosphatemia may appear around the age of 3 months, when mineral uptake by the poorly growing skeleton becomes insufficient. Clinically, hypercalcemia results in irritability, anorexia, vomiting and colitis, polyuria, and hypotonia. It may later cause nephrocalcinosis (Auron and Alon 2005). Conventional approaches to managing HPP-linked hypercalcemia include low calcium diet, vitamin D withdrawal, calcitonin and hydrochlorothiazide (Utsch et al. 2009). In our experience, in this particular situation timely administration of bisphosphonate may be useful by decreasing bone resorption, normalizing the calcium level and decreasing hypercalciuria without detectable skeletal damage. It may even be followed by a temporary improvement of growth (Mimoun et al. 2012), bisphosphonate therapy being then associated with a freeing of parathyroid hormone (PTH) secretion and a slight increase of AP level. It should however be noted that bisphosphonates are not recommended for long-term use in HPP (see below).

Pyridoxine-responsive seizures, refractory to standard anticonvulsive therapy and stopped by pyridoxine, may be the first symptom of infantile HPP. Such seizures occurring in the newborn predict a poor prognosis. Preventive therapy with pyridoxine may also be recommended in infantile HPP (Nunes et al. 2002; Baumgartner-Sigl et al. 2007). Other symptoms may be present in the course of HPP such as anemia (interpreted as a consequence of relative medullary insufficiency), unexplained fever, intracranial hemorrhage, irritability and high-pitched cry (Rockman-Greenberg 2013). Failure to thrive and poor growth are inevitable in surviving children with infantile HPP. Failure to thrive is often highly challenging, leading to nutritional supplementation by tube-feeding or gastrostomy in many cases (Fig. 1.2c).

Until recently, the treatment of infantile HPP has been highly disappointing (Deeb et al. 2000). Transplantation therapy and various enzyme replacement therapies have been attempted with no or marginal effect (Whyte et al. 1982b, 2003; Cahill et al. 2007). Recently a bone-targeted recombinant enzyme replacement therapy in infants with life-threatening HPP has yielded highly promising results, improving skeletal radiographs and pulmonary and physical function (Rodriguez et al. 2012; Whyte et al. 2012) (see also Chap. 15). As yet there are no published data regarding the long-term effect and the impact on growth retardation of such enzyme replacement therapy.

1.3.3 Childhood Hypophosphatasia

This form is very variable in its expression and course. Classically, the diagnosis is made after age 6 months. Premature loss of deciduous teeth before 5 years of age is a major feature. It results from abnormal cementum with weak connection between the tooth root and the periodontal ligament. The incisors are lost first. The entire primary dentition may be affected. On the other hand, rickets causes short stature with bowed legs and enlarged wrists, knees and ankles, and beading of the costochondral junctions. Radiographs show characteristic focal bony defects near the ends of major long bones, "tongues" of radiolucency that project from the rachitic growth plate into the metaphysis and can be mistaken for infection or leukemia (Fig. 1.3b). This feature, together with laboratory findings, makes it possible to distinguish hypophosphatasia from other forms of rickets or metaphyseal dysplasia. Growth plates are irregular with widened metaphyses but the epiphyses are preserved. Frontal bossing may be present. Craniostenosis may occur as in infantile HPP. The skull usually displays a "beaten copper" appearance which may persist in adulthood (Fig. 1.3b).

Walking is delayed with a characteristic "waddling" gait reflecting the severity of skeletal disease. Patients also describe bone pain, joint discomfort (stiffness, swelling) and muscle weakness of the limbs (especially the thighs) suggestive of myopathy (Seshia et al. 1990) (see below). Bone marrow edema with periarticular bone pain and soft tissue swelling may complicate HPP and may be confused with

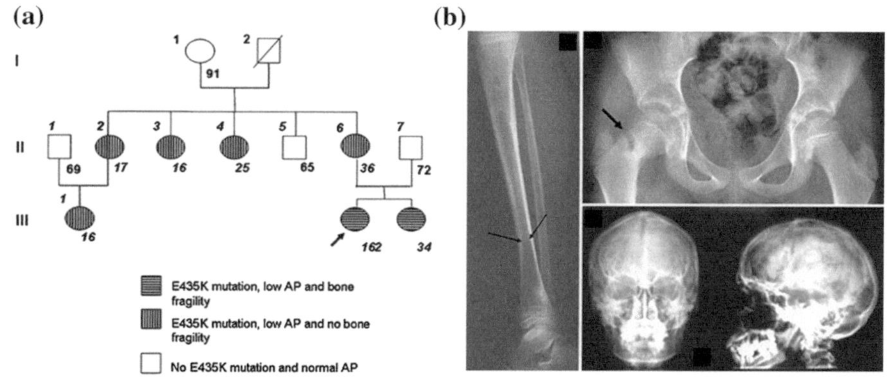

Fig. 1.3 Fractures in childhood HPP **a**: pedigree analysis shows the variety of expression of a given mutation within a family. Most carriers have low levels of blood AP without clinical manifestations. Two individuals (the proband and her sister) demonstrated long bone fractures. **b**: radiographs of the proband show reduced cortical bone and typical radiolucent "tongue" areas (*left, arrows*) and femoral neck fracture, quite unusual in children (*right, upper, arrow*); skull radiograph of the mother (who had no symptoms) (*right, lower*) shows a beaten copper appearance typical of HPP (adapted from Moulin et al. 2009)

chronic recurrent multifocal osteomyelitis (Girschick et al. 2007; Whyte et al. 2009). Bone marrow edema can be detected by magnetic resonance imaging (MRI).

The course of childhood HPP is highly variable. Fractures occur, especially metatarsal stress fractures, pseudofractures or poorly healing fractures which may require surgery. They can also reveal HPP at a later stage. We have described the occurrence of multiple fractures of the hip and femurs revealing HPP in two young sisters, with no other bone manifestation in any family members (Moulin et al. 2009), except for the mother whose skull had a beaten copper appearance (Fig. 1.3a, b). Levels of AP activity were heterogeneous in affected patients and carriers. These forms revealed during childhood are on a spectrum with the adult form. They illustrate the relatively unpredictable course and outcome of HPP.

There is no effective treatment of childhood HPP. Anti-inflammatory drugs or non-steroidal anti-inflammatory agents may reduce the pain (Girschick et al. 2006), which usually resolves after growth plate closure. Ongoing studies ofrecombinant enzyme replacement therapy suggest improvement of childhood HPP (Rockman-Greenberg 2013) (see also Chap. 15).

Growth deficiency is usually observed and may be significant in childhood HPP. It is often associated with low insulin growth factor 1 (IGF-1) level and decreased response to provocative tests of growth hormone (GH) secretion, suggesting GH deficiency (GHD) (Moulin et al. 2009; Taketani et al. 2014). In our experience low IGF-1 levels, with or without proven GHD, are actually common in HPP. GH therapy may be proposed in order to decrease the growth retardation of HPP patients without adverse effects (Moulin et al. 2012; Taketani et al. 2014). The Fig. 1.4 shows an example of GH treatment in an HPP patient.

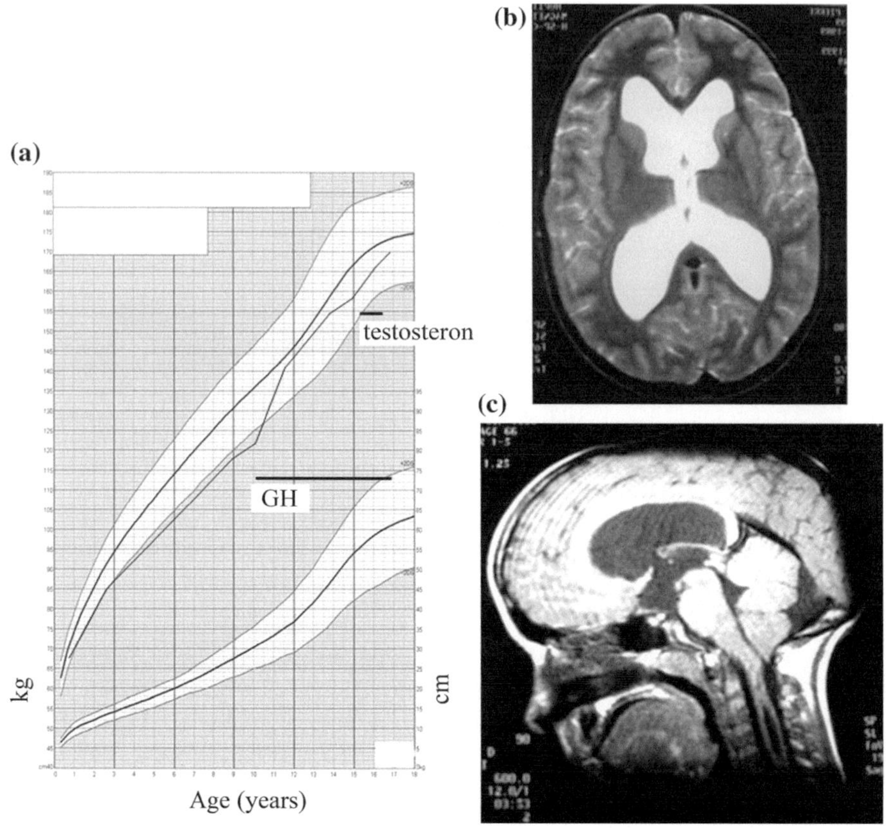

Fig. 1.4 Growth hormone treatment in childhood hypophosphatasia Growth retardation is observed in childhood HPP. Growth hormone testing was carried out because of delayed growth and GH deficiency was proven by two GH provocation tests. Growth chart is provided in **a**. Treatment (35 μ/kg/day) was given almost until adult height. Testosterone was added due to delayed puberty. Final height was close to the target height deduced from parental heights. No complication was noted. Of note, uncomplicated ventricle dilatation due to Arnold-Chiari malformation was present in this patient (shown in **b** and **c**); hypophysis hypoplasia was documented by CT scan (**c**). Unpublished personal data

Delayed fracture healing is frequent in HPP and may be a major concern. Surgical fixation may be required. All interventions and follow-up must be done by experienced pediatric orthopedic surgeons. Appliances including wheelchairs are frequently needed in cases of prolonged immobilization (Fig. 1.5). The specific subtrochanteric femoral fractures observed in HPP mimic those observed in adults treated chronically with bisphosphonates (Deeb et al. 2000; Shane et al. 2010; Sutton et al. 2012). The risk of fracture in adults treated with bisphosphonates increases with the duration of exposure, which leads to low remodeling with adynamic bone. So while this treatment is effective in other cases of bone fragility in children such as osteogenesis imperfecta (Marini 2009), bisphosphonates are not

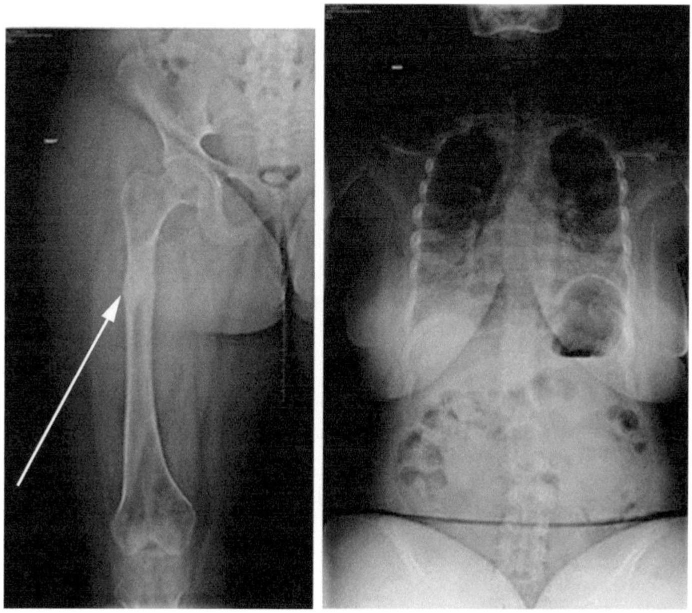

Fig. 1.5 Course of an infantile form of HPP After relative improvement during childhood, an infantile form of HPP led during adolescence to a subtrochanteric fracture of the femur that is typical of HPP (*left, arrow*) and is also encountered in long-term bisphosphonate-treated adult patients. Chronic pain, very delayed healing of femoral fracture and long-term impaired mobility contributed to massive weight gain and wheelchair confinement. The narrowed chest is still visible on *right*

recommended in HPP (Deeb et al. 2000; Sutton et al. 2012). Treatment with therecombinant analog of PTH, teriperatide, which has been proposed in adult HPP (see below), is not indicated during growth in childhood in view of the occurrence of osteosarcomas in rodent models treated with PTH.

1.3.4 Adult Hypophosphatasia

The adult form of HPP is characterized by short stature, osteomalacia, pseudo-fractures and pathological fractures after minimal trauma. Muscle and joint pain is present (Barvencik et al. 2011). The diagnosis is based on clinical and radiological findings, bone mineral density measurements and biochemical assays. The disease usually presents in middle age. Osteomalacia results in recurrent, poorly healing, metatarsal stress fractures. In advanced disease, aching in the thighs results from incompletely healed femoral pseudofractures. Pseudofractures are noted most often in the proximal femora, rather than medially as in most other types of osteomalacia. As mentioned above, subtrochanteric femoral pseudofractures resemble those due to long-term use of bisphosphonatesand are suggested to result from low bone turnover. Loss of secondary dentition may occur in adult HPP and may be the first manifestation. Adult HPP patients often have a history of dental problems during childhood. Premature loss of deciduous teeth followed by relatively good health is often observed. Calcium PPi dihydrate crystals (CPPDC) accumulate in tissues and joints and induce inflammation (attacks of pseudogout). HPP patients suffer from pain in their lower extremities. The mechanisms involved are imperfectly understood. A chronic inflammatory process secondary to the metabolic effect, involving activation of Toll-like receptors and interleukin-1 receptor, has been suggested in HPP-linked CPPDC (Beck et al. 2009). Sometimes degeneration of articular cartilage and ossified ligaments mimics spinal hyperostosis.

Treatment of adult hypophosphatasia by teriperatide (see above) has been proposed, with inconsistent results (Whyte et al. 2007; Camacho et al. 2008; Doshi et al. 2009; Gagnon et al. 2010; Schalin-Jantti et al. 2010; Laroche 2012). In our experience, teriperatide treatment of an adolescent girl with HPP after growth plate closure did not relieve metaphyseal pain (unpublished results). There is currently no treatment for adult HPP, but bone-targeted recombinant enzyme replacement therapy seems promising (see Chap. 15).

1.3.5 Odontohypophosphatasia

This form consists of dental manifestations without evidence of rickets or osteomalacia. Premature loss of deciduous teeth is the only symptom. Serum AP levels that are low or at the lower limit of normal are highly suggestive of the diagnosis, which can easily be confirmed by genetic analysis. Dominant negative mutations are frequent.

Dental manifestations are otherwise common in HPP, with premature loss of primary teeth, sometimes an early sign of the disease. Reduction of alveolar bone height of the mandible, cementum hypoplasia, and enlargement of the pulp chamber and root canals are also found (Reibel et al. 2009). Murine models have increased our knowledge of the pathophysiology of dental manifestations of HPP and their potential treatments (Millán et al. 2008; McKee et al. 2011).

1.3.6 Neurological Manifestations of HPP and Pain

Neurological complications of HPP include seizures in the newborn and craniostenosis, as mentioned earlier. Japanese series report mental retardation with acquired deafness, which does not clearly result from hypomineralization of the ear ossicles (Taketani et al. 2014). GHD, observed in 50 % of these patients, has also been suggested to result from neuron dysfunction (Taketani et al. 2014) (see also Chap. 14). In addition, HPP patients often experience more or less generalized pain, which may be a major challenge in management of the disease (see also Chap. 13). Pain may partly be explained by chronic inflammatory processes secondary to CPPDC accumulation. However, anti-inflammatory drugs are rarely effective. Nucleosides and nucleotides play important roles in pain signaling. According to recent studies, TNAP is expressed in thedorsal root ganglia neurons and spinal cord and can dephosphorylate AMP in these tissues, suggesting a possible anti-nociceptive effect of TNAP (Street et al. 2013).

A static myopathy with weakness is also present, which some authors have interpreted as a proximal myopathy, an early sign of HPP (Seshia et al. 1990). Elevated pyrophosphate level may inhibit muscle function. As a result, delayed walking and waddling gait are characteristics of HPP.

1.4 Diagnosis

Diagnosis of HPP is strongly suggested by the age of onset of symptoms and the clinical manifestations that are characteristic of the different forms described above. We have already mentioned some initial presentations that can reveal HPP and may confuse diagnosis (Table 1.1). All symptoms and biological and radiological data must always be carefully analyzed to avoid an incorrect diagnosis.

In children, other causes of rickets must be excluded, even when low AP level has suggested the diagnosis. Calcium, phosphorus, 25-hydroxyvitamin D and PTH levels must be documented. As we have said, HPP is an unusual cause of rickets associated with normal or high levels of calcium and phosphorus, and a resulting low level of PTH. At all ages a diagnosis of HPP can easily be missed (Moulin et al. 2009; Mohn et al. 2011). AP level must be accurately checked. It is variable with age, therefore accurate normal and abnormal levels for the age of the patient must be available from the laboratory, which is not always a simple task, especially in neonates (Tinnion and Embleton 2012). Of note, reduced AP levels are often unnoticed by clinicians. Elevated levels of TNAP substrates in plasma and urine (PPi, PLP and PEA) may contribute to the diagnosis. Nevertheless, PPi measurement is essentially carried out for research purposes and is not readily available. PLP/vitamin B6 level is more commonly performed but the results are not always rapidly delivered. PLP may be influenced by pyridoxine levels. Urine PEA levels lack the sensitivity required for a diagnostic marker, and reliable reference levels are often unavailable, especially for very young children or neonates (Imbard et al. 2012). At the present time, repeated accurate measurement of AP and PLP levels is the usual laboratory test to support a diagnosis of HPP when it is suggested by the clinical and radiological symptoms, with or without a family history. It should be noted that pseudohypophosphatasia seems a very rare presentation of HPP in which AP displays normal activity levels in the plasma, while PEA, PPi and PLP accumulate abnormally as in HPP. Pseudohypophosphatasia is thought to be due to special TNAP mutations affecting AP activity only in physiological conditions (Whyte 2012). Dental abnormalities are highly suggestive of HPP if associated with

Table 1.1 Clinical manifestations of HPP

Perinatal HPP	Infantile HPP	Childhood HPP	Adult HPP
Hypomineralization	Rickets	Rickets	Osteomalacia
Bones/chest deformity	Bones/chest deformity	Bones/chest deformity	Fractures/pseudofractures
Fractures	Fractures/delayed healing	Fractures/delayed healing	Chondrocalcinosis
Metaphyseal spurs	Bone radiolucencies	Craniosynostosis	Arthropathy
Bone radiolucencies	Craniosynostosis	Bone radiolucencies	Pseudogout
Seizures	Tooth loss	Low bone density	Chronic pain (muscle/bone)
Respiratory failure	Seizures	Tooth loss	Tooth loss
Neonatal death	Respiratory failure	Respiratory failure	Dental abnormalities
Hypercalcemia	Failure to thrive	Failure to thrive	Muscle weakness
	Hypotonia	Growth retardation	
	Hypercalcemia crisis	Muscle weakness	
		Chronic pain (muscle/bone)	
		Delayed walking	
		Hypercalcemia	

bone abnormalities. Low bone-specific TNAP levels may be documented but are of no real help in diagnosis. HPP diagnosis may be confirmed by appropriate genetic analysis leading to identification of the causative mutation(s) (see Chap. 2).

Perinatal/prenatal forms deserve special discussion. As we have previously said, some cases of HPP with prenatal diagnosis may spontaneously improve ("benign perinatal form"). This is an important point, especially in view of the recent introduction of recombinant enzyme therapy for HPP (Whyte et al. 2012). Moreover, while OI is the most common confounding diagnosis with HPP prenatally and in neonates, bisphosphonates, the usual treatment of OI, are not indicated in HPP. Type V OI in particular, with metaphyseal flaring and radiolucent tongues, may be confused with mild perinatal forms of HPP (Arundel et al. 2011). Laboratory tests for calcium, phosphate, and, above all, AP and PLP are usually discriminatory, but genetic analysis may sometimes be recommended before starting the OI treatment with bisphosphonates. Other causes of hypomineralization such as Wïedemann syndrome are more easily distinguishable (Akawi et al. 2012). However, it is noteworthy that a major mineralization defect such as cleido-cranial dysplasia may be responsible for a biological phenotype of HPP (Unger et al. 2002).

Laboratory tests of kidney, liver and general functions are usually normal in HPP.

1.5 Animal Models of HPP

Animal models have provided useful information for understanding the clinical manifestations of hypophosphatasia and identifying possible treatments. Disruption of the TNAP gene has been performed by homologous recombination in embryonic stem ES cells (MacGregor et al. 1995; Waymire et al. 1995). Mutants had no skeletal abnormalities but displayed seizures caused by abnormal PLP metabolism and decreased levels of GABA in the brain. Rescued mice further exhibited defective enamel formation (Waymire et al. 1995). An independent generation of TNAP-negative knock-out mice demonstrated a different effect (Narisawa et al. 1997) (see also Chap. 3). These mice also developed spontaneous epileptic seizures but were only temporarily rescued and showed significant demineralization at death. They displayed morphological abnormalities in the spinal nerve roots along the spinal cord. It was therefore suggested that disturbance ofvitamin B6 metabolism could not explain all the symptoms of the TNAP knock-out mice. Strikingly, affected mice showed obvious bone abnormalities with low mineralization, fractures, shorter growth plates with growth delay, and abnormal morphology of osteoblasts (Narisawa et al. 1997). The phenotypic differences with the previous report of Waymire et al. (1995) were attributed to possible differences in genetic background. Another important finding was that disruption of embryonic TNAP, expressed during mouse early embryogenesis and further reexpressed in adult mice in the thymus, intestine and testis, had no effect, supporting evidence for a functional role of TNAP during the development of the neural tube (Narisawa et al. 1997) (see also Chaps. 4 and 18). Further studies have clearly demonstrated that TNAP knock-out in mice recapitulates the metabolic and skeletal abnormalities of human HPP (Fedde et al. 1999, Anderson et al. 2004). The levels of TNAP substrates, PEA, Pi and PLP, were increased endogenously in the knock-out mice while skeletal manifestations of the disease were radiographically apparent by 10 days post-natal age with worsening rickets and fractures. Arrested chondrocyte differentiation was observed in the epiphyses and growthplates with reduced hypertrophic zones (Fedde et al. 1999). Progressive osteoidosis was noted due to defective mineralization, but, unlike the natural infantile form of HPP in humans, calciumand phosphate levels remained unremarkable in plasma and urine. Of note, there were also differences in the severity of the HPP phenotype between the two different strains of knock-out mice (Fedde et al. 1999). Further detailed characterization of ultrastructural mineralization suggested that hypomineralization in HPP mice results from inability of the initial mineral crystals to nucleate and to proliferate, a failure of the second stage of mineral formation possibly caused by the excess of local PPi (Anderson et al. 2004). This was in good agreement with a previous study showing that TNAP and the membrane glycoprotein-1/NPP1, an enzyme responsible for PPi generation, are antagonistic regulators of bone mineralization (Hessle et al. 2002). Abnormal PPi levels and mineralization deficits of the calvarium and spine in TNAP knock-out mice can be rescued by the double-knock out of TNAP and NPP1. The NPP1 defect results in hypermineralized phenotype of soft tissues and certain sites in the skeleton. The human syndrome of generalized arterial calcification in infants (GACI) has been reported in recessive inactivating mutations in ENPP1, while recent findings have also linked hypophosphatemia to a lack of NPP1 function (Mackenzie et al. 2012).

Therefore, animal models have largely supported TNAP as a major actor of bone mineralization and bone development and growth, mainly through its action on PPi hydrolysis which allows the normal process of mineralization to occur. These models have provided clear data regarding the metabolic consequences of TNAP deficiency in different organs. In addition, the disparities observed between different strains of TNAP-deficient mice are in good agreement with the clinical observation of various expressions of the disease in humans (Whyte 2012) (see also Chap. 3). Animal models have also been used to test proof of the concept and efficacy of original treatments. The most impressive results have been obtained by enzyme replacement directed at bone manifestations of HPP. Initial trials of enzyme therapy described in case reports used intravenous infusions of AP-rich plasma (purified AP) from Paget's disease patients, however they failed to fully rescue patients (Whyte et al. 1982b, 1984). Nevertheless, a favorable outcome was obtained in girls with infantile HPP after transplantation of mesenchymal-derived cells, with a small number of TNAP-positive cells detectable in osteoblasts and chondrocytes after transplantation (Whyte et al. 2003; Cahill et al. 2007). The failure of therapy using enriched fractions of natural TNAP was considered to be due to insufficient concentrations of TNAP, or, more likely, to an inability of TNAP to reach the sites of mineralization. A soluble form of human TNAP specially designed to have a high affinity for hydroxyapatite crystals due to the presence of a repetitive C-terminal extension of 10 Asp residues was shown to be highly effective in treating the bone manifestations and epilepsy of TNAP mice (Millán et al. 2008). Enzyme replacement also prevented enamel defects in these mice (Yadav et al. 2012). Of note, the results of similar enzyme replacement therapy in infantile HPP were published less than 5 years later, with highly promising findings regarding the outcome of these severely affected children (Whyte et al. 2012) (see also Chap. 15).

Alternative strategies have been explored in order to compensate the TNAP defect in genetically deprived animals. Gene therapy has been tested by means of a single injection of a lentiviral vector expressing bone–targeted TNAP into neonatal HPP mice, which resulted in sustained levels of AP and phenotypic correction (Yamamoto et al. 2011). Therapy with adeno-associated virus serotype 8 (AAV8) resulted in the rescue of TNAP knock-out mice (Matsumoto et al. 2011). More recently, AAV9 gene therapy in utero was successfully applied to HPP mice (Sugano et al. 2012). Gene therapy would present the advantage of a potentially one-shot treatment. In parallel, the potential of cell therapy combined with gene therapy has been explored by the use of mesenchymal stem cell (MSC) transplantation. A proof of concept has been provided by transducing promoter-driven TNAP gene by means of a retroviral vector in MSCs. These modified cells expressed significant AP activity, displayed mineralization, and provided newly formed bone in rats (Katsube et al. 2010). So far, new bone formation has been obtained only from allogeneic MSCs which were infused in a patient with severe perinatal HPP after bone marrow transplantation and previous differentiation of MSCs in osteogenic conditions (Tadokoro et al. 2009). Such complex strategies aiming to counteract the bone AP defect are obviously highly challenged by bone-directed enzyme therapy. Therefore, a clinically applicable therapy now exists to treat and prevent the skeletal manifestations of infantile HPP, while other potential strategies, in particular using gene therapy, are still under investigation.

Previous approaches have been directed at decreasing the consequences of abnormal AP in mineralized tissues. Enzyme therapy targeted to bone clearly improves not only the skeletal manifestations of HPP but also the general course of the disease, preventing the fatal outcome in many clinical cases (Millán et al. 2008; Whyte et al. 2012). Seizures, generally attributed to decreased hydrolysis of PLP in HPP patients and untreated TNAP-deficient mice, with increased PLP levels and decreased PL availability, were also improved with normalization of PL levels. This suggests that the infused bone-targeted enzyme may substitute for the deficient metabolism of PPi and PLP and prevent the most spectacular neurological manifestations of the disease. However, TNAP expression has also been directly observed in neural tissue. An initial report by Millán et al. described aberrant development of the lumbar nerve roots in TNAP deficient mice (Millán et al. 2008). Seminal works in the primate by Fonta et al. demonstrated localization of TNAP in the cerebral cortex, with evidence for a role in neurotransmission (Fonta et al. 2004). and correlation with the maturation of neurotransmission (Fonta et al. 2005). A distinctive role of TNAP in neurological disorders due to its dysfunctions in the human brain was further suggested by the observation of its expression in the thalamo-recipient granular layer of the sensory cortex (Negyessy et al. 2011). AP activity was attributed to TNAP expression in primates (humans and marmosets) and in rodents (rats and mice) (Brun-Heath et al. 2011). Neuronal and endothelial cells exclusively expressed the bone TNAP transcript in all species tested, except in mouse neurons where liver TNAP transcripts were also detected, with developmental regulation of TNAP expression (Brun-Heath et al. 2011). More recently, myelin abnormalities and synaptic dysfunction observed in TNAP knock-out mice suggested a contribution to neurological symptoms, including the epilepsy found in severe HPP (Hanics et al. 2012). These works open new perspectives for the role of TNAP in the brain. The role of brain TNAP should considered with regard to specific clinical manifestations of HPP, especially the pathophysiology of pain.

1.6 Brief Overview of the Clinical Manifestations of HPP and Future Perspectives

The clinical manifestations of HPP are summarized in the Table 1.1. To sum up, the clinical presentation of the disease varies highly in terms of severity and age at onset of symptoms. In some cases genotype and phenotype correlations may be made. However the actual prognosis of forms detected prenatally often remains difficult to establish. True perinatal HPP is frequently fatal but other forms detected prenatally or early after birth may have a less deleterious outcome.

Managing HPP is a difficult challenge. In children, failure to thrive and skeletal manifestations with their consequences on pulmonary function may require ventilatory support and prolonged tube feeding. Enzyme replacement with bone-targeted TNAP is highly promising (Whyte et al. 2012), but muscle or neurological manifestations of HPP may not be significantly concerned by this therapy. Therefore, the long-term effect of this treatment on the extra-skeletal manifestations and growth prognosis of HPP is a question that is still outstanding. In adults, the major clinical concerns result from the history of HPP in childhood, including abnormal dentition, and from fractures and pseudofractures, osteoarthropathy with pseudogout and chondrocalcinosis, which mainly develop during adult life. Most of these symptoms may be explained by the role of TNAP in phosphate metabolism and its consequences on defective mineralization. Nevertheless, the chronic muscle and bone pain, on which the usual antalgic treatment is largely ineffective, are not well explained by abnormal mineralization. Our knowledge of the extraskeletal manifestations of the disease will benefit from careful analysis of the animal models of HPP.

Conflict of Interest The author declares no conflict of interest.

References

Akawi NA, Ali BR, Al-Gazali L (2012) Stuve-wiedemann syndrome and related bent bone dysplasias. Clin Genet 82:12–21

Anderson HC, Sipe JB, Hessle L, Dhanyamraju R, Atti E, Camacho NP, Millán JL (2004) Impaired calcification around matrix vesicles of growth plate and bone in alkaline phosphatase-deficient mice. Am J Pathol 164:841–847

Arundel P, Offiah A, Bishop NJ (2011) Evolution of the radiographic appearance of the metaphyses over the first year of life in type V osteogenesis imperfecta: clues to pathogenesis. J Bone Miner Res 26:894–898

Auron A, Alon US (2005) Resolution of medullary nephrocalcinosis in children with metabolic bone disorders. Pediatr Nephrol 20:1143–1145

Barvencik F, Beil FT, Gebauer M, Busse B, Koehne T, Seitz S, Zustin J, Pogoda P, Schinke T, Amling M (2011) Skeletal mineralization defects in adult hypophosphatasia–a clinical and histological analysis. Osteoporos Int 22:2667–2675

Baumgartner-Sigl S, Haberlandt E, Mumm S, Scholl-Burgi S, Sergi C, Ryan L, Ericson KL, Whyte MP, Hogler W (2007) Pyridoxine-responsive seizures as the first symptom of infantile hypophosphatasia caused by two novel missense mutations (c.677T > C, p. M226T; c.1112C > T, p.T371I) of the tissue-nonspecific alkaline phosphatase gene. Bone 40:1655–1661

Beck C, Morbach H, Richl P, Stenzel M, Girschick HJ (2009) How can calcium pyrophosphate crystals induce inflammation in hypophosphatasia or chronic inflammatory joint diseases? Rheumatol Int 29:229–238

Brun-Heath I, Ermonval M, Chabrol E, Xiao J, Palkovits M, Lyck R, Miller F, Couraud PO, Mornet E, Fonta C (2011) Differential expression of the bone and the liver tissue non-specific alkaline phosphatase isoforms in brain tissues. Cell Tissue Res 343:521–536

Buchet R, Millán JL, Magne D (2013) Multisystemic functions of alkaline phosphatases. Methods Mol Biol 1053:27–51

Cahill RA, Wenkert D, Perlman SA, Steele A, Coburn SP, McAlister WH, Mumm S, Whyte MP (2007) Infantile hypophosphatasia: transplantation therapy trial using bone fragments and cultured osteoblasts. J Clin Endocrinol Metab 92:2923–2930

Camacho PM, Painter S, Kadanoff R (2008) Treatment of adult hypophosphatasia with teriparatide. Endocr Pract 14:204–208

Cormier-Daire V, Genevieve D, Munnich A, Le Merrer M (2004) New insights in congenital bowing of the femora. Clin Genet 66:169–176

Deeb AA, Bruce SN, Morris AA, Cheetham TD (2000) Infantile hypophosphatasia: disappointing results of treatment. Acta Paediatr 89:730–733

Doshi KB, Hamrahian AH, Licata AA (2009) Teriparatide treatment in adult hypophosphatasia in a patient exposed to bisphosphonate: a case report. Clin Cases Miner Bone Metab 6:266–269

Fedde KN, Blair L, Silverstein J, Coburn SP, Ryan LM, Weinstein RS, Waymire K, Narisawa S, Millán JL, MacGregor GR, Whyte MP (1999) Alkaline phosphatase knock-out mice recapitulate the metabolic and skeletal defects of infantile hypophosphatasia. J Bone Miner Res 14:2015–2026

Fonta C, Negyessy L, Renaud L, Barone P (2004) Areal and subcellular localization of the ubiquitous alkaline phosphatase in the primate cerebral cortex: evidence for a role in neurotransmission. Cereb Cortex 14:595–609

Fonta C, Negyessy L, Renaud L, Barone P (2005) Postnatal development of alkaline phosphatase activity correlates with the maturation of neurotransmission in the cerebral cortex. J Comp Neurol 486:179–196

Gagnon C, Sims NA, Mumm S, McAuley SA, Jung C, Poulton IJ, Ng KW, Ebeling PR (2010) Lack of sustained response to teriparatide in a patient with adult hypophosphatasia. J Clin Endocrinol Metab 95:1007–1012

Girschick HJ, Schneider P, Haubitz I, Hiort O, Collmann H, Beer M, Shin YS, Seyberth HW (2006) Effective NSAID treatment indicates that hyperprostaglandinism is affecting the clinical severity of childhood hypophosphatasia. Orphanet J Rare Dis 1:24

Girschick HJ, Mornet E, Beer M, Warmuth-Metz M, Schneider P (2007) Chronic multifocal non-bacterial osteomyelitis in hypophosphatasia mimicking malignancy. BMC Pediatr 7:3

Greenberg CR, Taylor CL, Haworth JC, Seargeant LE, Philipps S, Triggs-Raine B, Chodirker BN (1993) A homoallelic Gly317- > Asp mutation in ALPL causes the perinatal (lethal) form of hypophosphatasia in Canadian mennonites. Genomics 17:215–217

Hanics J, Barna J, Xiao J, Millán JL, Fonta C, Negyessy L (2012) Ablation of TNAP function compromises myelination and synaptogenesis in the mouse brain. Cell Tissue Res 349:459–471

Harmey D, Johnson KA, Zelken J, Camacho NP, Hoylaerts MF, Noda M, Terkeltaub R, Millán JL (2006) Elevated skeletal osteopontin levels contribute to the hypophosphatasia phenotype in Akp2(-/-) mice. J Bone Miner Res 21:1377–1386

Hessle L, Johnson KA, Anderson HC, Narisawa S, Sali A, Goding JW, Terkeltaub R, Millán JL (2002) Tissue-nonspecific alkaline phosphatase and plasma cell membrane glycoprotein-1 are central antagonistic regulators of bone mineralization. Proc Natl Acad Sci U S A 99:9445–9449

Imbard A, Alberti C, Armoogum-Boizeau P, Ottolenghi C, Josserand E, Rogal O, Benoist JF (2012) Phosphoethanolamine normal range in pediatric urines for hypophosphatasia screening. Clin Chem Lab Med 50:2231–2233

Katsube Y, Kotobuki N, Tadokoro M, Kanai R, Taketani T, Yamaguchi S, Ohgushi H (2010) Restoration of cellular function of mesenchymal stem cells from a hypophosphatasia patient. Gene Ther 17:494–502

Krawitz PM, Schweiger MR, Rodelsperger C, Marcelis C, Kolsch U, Meisel C, Stephani F, Kinoshita T, Murakami Y, Bauer S, Isau M, Fischer A, Dahl A, Kerick M, Hecht J, Kohler S, Jager M, Grunhagen J, de Condor BJ, Doelken S, Brunner HG, Meinecke P, Passarge E, Thompson MD, Cole DE, Horn D, Roscioli T, Mundlos S, Robinson PN (2010)

Identity-by-descent filtering of exome sequence data identifies PIGV mutations in hyperphosphatasia mental retardation syndrome. Nat Genet 42:827–829

Laroche M (2012) Failure of teriparatide in treatment of bone complications of adult hypophosphatasia. Calcif Tissue Int 90:250

Leung EC, Mhanni AA, Reed M, Whyte MP, Landy H, Greenberg CR (2013) Outcome of perinatal hypophosphatasia in manitoba mennonites: a retrospective cohort analysis. JIMD Rep 11:73–78

Macfarlane JD, Kroon HM, van der Harten JJ (1992) Phenotypically dissimilar hypophosphatasia in two sibships. Am J Med Genet 42:117–121

MacGregor GR, Zambrowicz BP, Soriano P (1995) Tissue non-specific alkaline phosphatase is expressed in both embryonic and extraembryonic lineages during mouse embryogenesis but is not required for migration of primordial germ cells. Development 121:1487–1496

Mackenzie NC, Huesa C, Rutsch F, MacRae VE (2012) New insights into NPP1 function: lessons from clinical and animal studies. Bone 51:961–968

Marini JC (2009) Bone: use of bisphosphonates in children-proceed with caution. Nat Rev Endocrinol 5:241–243

Matsumoto T, Miyake K, Yamamoto S, Orimo H, Miyake N, Odagaki Y, Adachi K, Iijima O, Narisawa S, Millán JL, Fukunaga Y, Shimada T (2011) Rescue of severe infantile hypophosphatasia mice by AAV-mediated sustained expression of soluble alkaline phosphatase. Hum Gene Ther 22:1355–1364

McKee MD, Nakano Y, Masica DL, Gray JJ, Lemire I, Heft R, Whyte MP, Crine P, Millán JL (2011) Enzyme replacement therapy prevents dental defects in a model of hypophosphatasia. J Dent Res 90:470–476

Millán JL, Narisawa S, Lemire I, Loisel TP, Boileau G, Leonard P, Gramatikova S, Terkeltaub R, Camacho NP, McKee MD, Crine P, Whyte MP (2008) Enzyme replacement therapy for murine hypophosphatasia. J Bone Miner Res 23:777–787

Mimoun E, Moulin P, Sales de Gauzy J, Mornet E, Salles J (2012) Biphosphonates in hypophosphatasia: not the evil? Bull Group Int Rech Sci Stomatol Odontol 51:e41

Mohn A, De Leonibus C, de Giorgis T, Mornet E, Chiarelli F (2011) Hypophosphatasia in a child with widened anterior fontanelle: lessons learned from late diagnosis and incorrect treatment. Acta Paediatr 100:e43–e46

Moore CA, Ward JC, Rivas ML, Magill HL, Whyte MP (1990) Infantile hypophosphatasia: autosomal recessive transmission to two related sibships. Am J Med Genet 36:15–22

Mornet E (2000) Hypophosphatasia: the mutations in the tissue-nonspecific alkaline phosphatase gene. Hum Mutat 15:309–315

Mornet E (2007) Hypophosphatasia. Orphanet J Rare Dis 2:40

Mornet E (2008) Hypophosphatasia. Best Pract Res Clin Rheumatol 22:113–1277

Mornet E, Yvard A, Taillandier A, Fauvert D, Simon-Bouy B (2011) A molecular-based estimation of the prevalence of hypophosphatasia in the European population. Ann Hum Genet 75:439–445

Moulin P, Vaysse F, Bieth E, Mornet E, Gennero I, Dalicieux-Laurencin S, Baunin C, Tauber MT, De Gauzy JS, Salles JP (2009) Hypophosphatasia may lead to bone fragility: don't miss it. Eur J Pediatr 168:783–788

Moulin P, Sales de Gauzy J, Vaysse Y, Mornet E, Salles JP (2012) Growth hormone increases growth velocity and alkaline phosphatase level in children with hypophosphatasia. Bull Group Int Rech Sci Stomatol Odontol 51:e40

Narisawa S, Frohlander N, Millán JL (1997) Inactivation of two mouse alkaline phosphatase genes and establishment of a model of infantile hypophosphatasia. Dev Dyn 208:432–446

Narisawa S, Harmey D, Yadav MC, O'Neill WC, Hoylaerts MF, Millán JL (2007) Novel inhibitors of alkaline phosphatase suppress vascular smooth muscle cell calcification. J Bone Miner Res 22:1700–1710

Negyessy L, Xiao J, Kantor O, Kovacs GG, Palkovits M, Doczi TP, Renaud L, Baksa G, Glasz T, Ashaber M, Barone P, Fonta C (2011) Layer-specific activity of tissue non-specific alkaline phosphatase in the human neocortex. Neuroscience 172:406–418

Nunes ML, Mugnol F, Bica I, Fiori RM (2002) Pyridoxine-dependent seizures associated with hypophosphatasia in a newborn. J Child Neurol 17:222–224

Rathbun JC (1948) Hypophosphatasia; a new developmental anomaly. Am J Dis Child 75:822–831

Reibel A, Maniere MC, Clauss F, Droz D, Alembik Y, Mornet E, Bloch-Zupan A (2009) Orodental phenotype and genotype findings in all subtypes of hypophosphatasia. Orphanet J Rare Dis 4:6

Robison R (1932) The significance of phosphoric esters in metabolism. New York Academic Press, New York

Rockman-Greenberg C (2013) Hypophosphatasia. Pediatr Endocrinol Rev 10(Suppl 2):380–388

Rodriguez E, Bober MB, Davey L, Zamora A, Li Puma AB, Chidekel A, Shaffer TH (2012) Respiratory mechanics in an infant with perinatal lethal hypophosphatasia treated with human recombinant enzyme replacement therapy. Pediatr Pulmonol 47:917–922

Schalin-Jantti C, Mornet E, Lamminen A, Valimaki MJ (2010) Parathyroid hormone treatment improves pain and fracture healing in adult hypophosphatasia. J Clin Endocrinol Metab 95:5174–5179

Seetharam B, Tiruppathi C, Alpers DH (1987) Hydrophobic interactions of brush border alkaline phosphatases: the role of phosphatidyl inositol. Arch Biochem Biophys 253:189–198

Seshia SS, Derbyshire G, Haworth JC, Hoogstraten J (1990) Myopathy with hypophosphatasia. Arch Dis Child 65:130–131

Shane E, Burr D, Abrahamsen B, Adler RA, Brown TD, Cheung AM, Cosman F, Curtis JR, Dell R, Dempster DW, Ebeling PR, Einhorn TA, Genant HK, Geusens P, Klaushofer K, Lane JM, McKiernan F, McKinney R, Ng A, Nieves J, O'Keefe R, Papapoulos S, Howe TS, van der Meulen MC, Weinstein RS, Whyte MP (2010) Atypical subtrochanteric and diaphyseal femoral fractures: second report of a task force of the American society for bone and mineral research. J Bone Miner Res 29:1–23

Stevenson DA, Carey JC, Coburn SP, Ericson KL, Byrne JL, Mumm S, Whyte MP (2008) Autosomal recessive hypophosphatasia manifesting in utero with long bone deformity but showing spontaneous postnatal improvement. J Clin Endocrinol Metab 93:3443–3448

Street SE, Kramer NJ, Walsh PL, Taylor-Blake B, Yadav MC, King IF, Vihko P, Wightman RM, Millán JL, Zylka MJ (2013) Tissue-nonspecific alkaline phosphatase acts redundantly with PAP and NT5E to generate adenosine in the dorsal spinal cord. J Neurosci 33:11314–11322

Sugano H, Matsumoto T, Miyake K, Watanabe A, Iijima O, Migita M, Narisawa S, Millán JL, Fukunaga Y, Shimada T (2012) Successful gene therapy in utero for lethal murine hypophosphatasia. Hum Gene Ther 23:399–406

Sutton RA, Mumm S, Coburn SP, Ericson KL, Whyte MP (2012) "Atypical femoral fractures" during bisphosphonate exposure in adult hypophosphatasia. J Bone Miner Res 27:987–994

Tadokoro M, Kanai R, Taketani T, Uchio Y, Yamaguchi S, Ohgushi H (2009) New bone formation by allogeneic mesenchymal stem cell transplantation in a patient with perinatal hypophosphatasia. J Pediatr 154:924–930

Taketani T, Onigata K, Kobayashi H, Mushimoto Y, Fukuda S, Yamaguchi S (2014) Clinical and genetic aspects of hypophosphatasia in Japanese patients. Arch Dis Child 99:211–215

Tinnion RJ, Embleton ND (2012) How to use… alkaline phosphatase in neonatology. Arch Dis Cjild Educ Pract Ed 97:157–163

Unger S, Mornet E, Mundlos S, Blaser S, Cole DE (2002) Severe cleidocranial dysplasia can mimic hypophosphatasia. Eur J Pediatr 161:623–626

Utsch B, Brun-Heath I, Staatz G, Gravou-Apostolatou C, Karle S, Jacobs U, Ludwig M, Zenker M, Dorr HG, Rascher W, Mornet E, Dotsch J (2009) Infantile hypophosphatasia due to a new compound heterozygous TNSALP mutation—functional evidence for a hydrophobic side-chain? Exp Clin Endocrinol Diab 117:28–33

Waymire KG, Mahuren JD, Jaje JM, Guilarte TR, Coburn SP, MacGregor GR (1995) Mice lacking tissue non-specific alkaline phosphatase die from seizures due to defective metabolism of vitamin B-6. Nat Genet 11:45–51

Wenkert D, McAlister WH, Coburn SP, Zerega JA, Ryan LM, Ericson KL, Hersh JH, Mumm S, Whyte MP (2011) Hypophosphatasia: nonlethal disease despite skeletal presentation in utero (17 new cases and literature review). J Bone Miner Res 26:2389–2398

Whyte MP (1994) Hypophosphatasia and the role of alkaline phosphatase in skeletal mineralization. Endocr Rev 15:439–461

Whyte MP (2010) Physiological role of alkaline phosphatase explored in hypophosphatasia. Ann N Y Acad Sci 1192:190–200

Whyte MP (2012) Hypophosphastasia. In: Glorieux FH, Pettifor JM, Jüppner H (eds) Pediatric bone biology and diseases, 2nd edn. Elsevier, Amsterdam, pp 771–794

Whyte MP, Essmyer K, Geimer M, Mumm S (2006) Homozygosity for TNSALP mutation 1348c > T (Arg433Cys) causes infantile hypophosphatasia manifesting transient disease correction and variably lethal outcome in a kindred of black ancestry. J Pediatr 148:753–758

Whyte MP, Greenberg CR, Salman NJ, Bober MB, McAlister WH, Wenkert D, Van Sickle BJ, Simmons JH, Edgar TS, Bauer ML, Hamdan MA, Bishop N, Lutz RE, McGinn M, Craig S, Moore JN, Taylor JW, Cleveland RH, Cranley WR, Lim R, Thacher TD, Mayhew JE, Downs M, Millán JL, Skrinar AM, Crine P, Landy H (2012) Enzyme-replacement therapy in life-threatening hypophosphatasia. N Engl J Med 366:904–913

Whyte MP, Kurtzberg J, McAlister WH, Mumm S, Podgornik MN, Coburn SP, Ryan LM, Miller CR, Gottesman GS, Smith AK, Douville J, Waters-Pick B, Armstrong RD, Martin PL (2003) Marrow cell transplantation for infantile hypophosphatasia. J Bone Miner Res 18:624–636

Whyte MP, Mahuren JD, Fedde KN, Cole FS, McCabe ER, Coburn SP (1988) Perinatal hypophosphatasia: tissue levels of vitamin B6 are unremarkable despite markedly increased circulating concentrations of pyridoxal-5'-phosphate. Evidence for an ectoenzyme role for tissue-nonspecific alkaline phosphatase. J Clin Invest 81:1234–1239

Whyte MP, Mahuren JD, Vrabel LA, Coburn SP (1985) Markedly increased circulating pyridoxal-5'-phosphate levels in hypophosphatasia. Alkaline phosphatase acts in vitamin B6 metabolism. J Clin Invest 76:752–756

Whyte MP, McAlister WH, Patton LS, Magill HL, Fallon MD, Lorentz WB Jr, Herrod HG (1984) Enzyme replacement therapy for infantile hypophosphatasia attempted by intravenous infusions of alkaline phosphatase-rich Paget plasma: results in three additional patients. J Pediatr 105:926–933

Whyte MP, Mumm S, Deal C (2007) Adult hypophosphatasia treated with teriparatide. J Clin Endocrinol Metab 92:1203–1208

Whyte MP, Murphy WA, Fallon MD (1982a) Adult hypophosphatasia with chondrocalcinosis and arthropathy. Variable penetrance of hypophosphatasemia in a large Oklahoma kindred. Am J Med 72:631–641

Whyte MP, Valdes R Jr, Ryan LM, McAlister WH (1982b) Infantile hypophosphatasia: enzyme replacement therapy by intravenous infusion of alkaline phosphatase-rich plasma from patients with Paget bone disease. J Pediatr 101:379–386

Whyte MP, Wenkert D, McAlister WH, Mughal MZ, Freemont AJ, Whitehouse R, Baildam EM, Coburn SP, Ryan LM, Mumm S (2009) Chronic recurrent multifocal osteomyelitis mimicked in childhood hypophosphatasia. J Bone Miner Res 24:1493–1505

Yadav MC, de Oliveira RC, Foster BL, Fong H, Cory E, Narisawa S, Sah RL, Somerman M, Whyte MP, Millán JL (2012) Enzyme replacement prevents enamel defects in hypophosphatasia mice. J Bone Miner Res 27:1722–1734

Yamamoto S, Orimo H, Matsumoto T, Iijima O, Narisawa S, Maeda T, Millán JL, Shimada T (2011) Prolonged survival and phenotypic correction of Akp2(-/-) hypophosphatasia mice by lentiviral gene therapy. J Bone Miner Res 26:135–142

Zurutuza L, Muller F, Gibrat JF, Taillandier A, Simon-Bouy B, Serre JL, Mornet E (1999) Correlations of genotype and phenotype in hypophosphatasia. Hum Mol Genet 8:1039–1046

Chapter 2
Molecular Genetics of Hypophosphatasia and Phenotype-Genotype Correlations

Etienne Mornet

Abstract Hypophosphatasia (HPP) is due to deficient activity of the tissue-nonspecific isoenzyme of alkaline phosphatase (TNAP). This enzyme cleaves extracellular substrates inorganic pyrophosphates (PPi), pyridoxal-5′-phosphate (PLP), phosphoethanolamine (PEA) and nucleotides, and probably other substrates not yet identified. During the last 15 years the role of TNAP in mineralization, and to a less degree in brain, has been investigated, providing hypotheses and explanations for both bone and neuronal HPP phenotypes. *ALPL*, the gene encoding TNAP, is subject to many mutations, mostly missense mutations. A few number of mutations are recurrently found and may be quite frequent in particular populations. This reflects founder effects. The great variety of mutations results in a great number of compound heterozygous genotypes and in highly variable clinical expressivity. A good correlation was observed between the severity of the disease and in vitro enzymatic activity of the mutant protein measured after site-directed mutagenesis. Many missense mutations found in severe hypophosphatasia produced a mutant protein that failed to reach the cell membrane, was accumulated in the cis-Golgi and was subsequently degraded in the proteasome. Missense mutations located in the catalytic site or in the homodimer interface were often shown by site-directed mutagenesis to have a dominant negative effect. Currently molecular diagnosis of HPP is based on the sequencing of the coding sequence of *ALPL* that allows detection of approximately 95 % of mutations in severe cases. In addition, other genes, especially genes encoding proteins involved in the regulation of extracellular PPi concentration, could modify the phenotype (modifier genes).

Keywords Hypophosphatasia · Genotype-phenotype correlation · Mutation · Tissue-nonspecific alkaline hypophosphatasia

E. Mornet (✉)
Unité de Génétique Constitutionnelle, Centre Hospitalier de Versailles, Le Chesnay, France
e-mail: emornet@ch-versailles.fr

C. Fonta and L. Négyessy (eds.), *Neuronal Tissue-Nonspecific Alkaline Phosphatase (TNAP)*, Subcellular Biochemistry 76,
DOI 10.1007/978-94-017-7197-9_2

2.1 Introduction

For geneticists dealing with genotype-phenotype correlation, Hypophosphatasia (HPP) is a challenging model of inherited disorder because it cumulates various deviations from Mendelian transmission : variable expressivity, even in patients sharing the same genotype, dominant or recessive inheritance of moderate forms, incomplete penetrance of dominant forms, and parental bias of transmission in the so-called prenatal benign form of HPP. The study of the *ALPL* mutations by various experimental approaches including site-directed mutagenesis, 3D modeling and immunofluorescence have been very informative to progress in understanding the particular genetic aspects of HPP.

2.2 The Tissue-Nonspecific Alkaline Phosphatase (TNAP)

2.2.1 Alkaline Phosphatases

Alkaline phosphatases (E.C.3.1.3.1) (APs) form a large family of enzymes common to all organisms. The core fold and the active-site architecture of the enzyme have been shown to be similar in bacteria, archea and eukaryotes (Wende et al. 2010). APs catalyze the hydrolysis of phosphomonoesters with release of inorganic phosphate (Schwartz and Lipmann 1961). The catalytic activity of APs depends on a multimeric configuration of identical monomers. Each identical subunit possesses one active site and contains two Zn^{2+} and one Mg^{2+} ions that stabilize the tertiary structure (Kim and Wyckoff 1991). APs are believed to be homodimeric in serum and in membranes, but the tissue-nonspecific isoform could exist as homodimeric and/or homotetrameric structure in membranes (Hawrylak and Stinson 1987).

In humans, three out of four AP isozymes are tissue-specific : one is placental (PLAP), the second appears in germ cells (GCAP), and the third in the intestine (IAP). They are 90–98 % homologous, and their genes are tandemly arranged on chromosome 2q37.1, likely arising from relatively recent ancestral duplications. The fourth, TNAP, is 50 % identical to the other three, suggesting that it arised from a more ancient duplication. It is ubiquitous but strongly expressed in liver, bone and kidney (Stigbrand 1984; Millan 1992), in cerebral cortex (Fonta et al. 2004) and in developing spinal cord (Narisawa et al. 1994). Its gene is located on chromosome 1p34–36 (Greenberg et al. 1990).

2.2.2 Structure of the TNAP According to Modelling from PLAP

For many years the only source of structural information on APs was the crystallographic coordinates of the *E. coli* AP (Kim and Wyckoff 1991). Building a

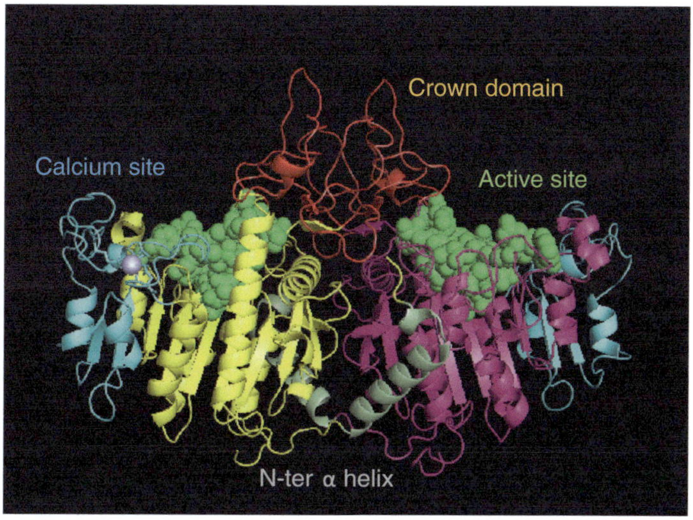

Fig. 2.1 Three-dimensional model of TNAP based on the crystal structure of the placental isoform (Le Du et al. 2001). The two identical monomers forming the homodimer are displayed in *yellow* and *magenta*. The functional domains are identified by distinct *colours*. The active site is represented in *green* spacefill, the crown domain in *red*, the calcium binding site in *blue* (calcium atom is shown in *blue* spacefill), and the N-ter α helix in *green* ribbon. The homodimer interface also includes residues from the crown domain and the N-ter alpha helix

three-dimensional model of human TNAP using alignment to the structure of *E. coli* AP as a template evidenced a number of regions that could not be modelled (Zurutuza et al. 1999), suggesting that human APs have functions that do not exist in *E. coli* AP. When the three-dimensional structure of the first mammalian AP, i.e. PLAP, was elucidated (Le Du et al. 2001), these regions were modelled and evidence was provided for their conservation in mammals (Le Du and Millan 2002) and for their functional role (Le Du et al. 2001; Mornet et al. 2001). Now, the accumulation of molecular and structural data suggests that APs progressively acquired specialization and new functions during evolution.

Five functional domains have been identified in TNAP (Fig. 2.1). The active site is the centre of the catalytic activity. It is conserved in all species including *E. coli*. The crown domain is a key factor of uncompetitive inhibition (Kozlenkov et al. 2004), heat-stability (Bossi et al. 1993) and allosteric behaviour (Hoylaerts et al. 1997). Together with the N-terminal arm, the crown domain contributes to stabilize the dimeric structure. It may be also involved in the binding of TNAP to collagen (Hoylaerts and Millan 1991; Bossi et al. 1993), corroborating previous studies suggesting this property of TNAP (Vittur et al. 1984; Wu et al. 1991). The crown domain is highly conserved in mammals. The homodimer interface is crucial because dimerization is indispensable for allostery and because APs are active only in dimeric form (Bossi et al. 1993). Monomer-monomer interactions are made of noncovalent hydrogen bonds and of van der Waals and hydrophobic contacts, and involve about 90 residues and 25 % of the overall surface of the protein (Millan 2006). Interestingly, whereas the distribution of hydrophobic/hydrophilic amino acids significantly differ from an isoform to another one, it has been suggested that TNAP monomer could dimerize with the intestinal isoform (IAP) monomer (Vergnes et al. 2000). The exact role of the calcium site remains to be elucidated. However this site is conserved among vertebrates, suggesting a possible vertebrate-specific function (Brun-Heath et al. 2005). The N-terminal α helix is essential for stability and allosteric properties of the enzyme (Hoylaerts et al. 2006).

2.2.3 TNAP Substrates and Functions

TNAP cleaves extracellular substrates such as inorganic pyrophosphates (PPi), pyridoxal-5′-phosphate (PLP), phosphoethanolamine (PEA) and nucleotides. However, the demonstration that PEA is also a natural substrate of TNAP in vivo remains to be confirmed (Millan 2006), and it remains possible that all substrates of TNAP are not yet identified. The exact function of TNAP in bone and dental mineralization has now been greatly deciphered and involves hydrolysis of PPi, and perhaps mammalian-specific activities such as collagen (Hoylaerts et al. 1997) and calcium (Mornet et al. 2001) binding.

2.2.3.1 Inorganic Pyrophosphate (PPi)

TNAP is present in matrix vesicles and activity of TNAP is required to generate the inorganic phosphate (Pi) needed for hydroxyapatite crystallization (Fallon et al. 1980). Because PPi suppresses the formation and growth of hydroxyapatite crystals (Meyer 1984), the ability of TNAP to hydrolyze PPi promotes osteoblastic mineralization (Rezende et al. 1994), resulting in antagonistic roles of the plasma cell membrane glycoprotein-1 (PC1) that products PPi from dNTPs, and TNAP that hydrolyzes PPi to generate Pi (Hessle et al. 2002) (Fig. 2.2). Finally five proteins have been identified as central regulators of extracellular PPi and Pi levels: TNAP, which hydrolyzes PPi, PC-1 which generates PPi, the multiple-pass transmembrane protein ANKH, which mediates intracellular to extracellular channeling of PPi (Ho et al. 2000; Harmey et al. 2004), osteopontin (OPN), another mineralization inhibitor, and PHOSPHO1, a phosphatase involved in the first step of mineralization in matrix vesicles (Yadav et al. 2011) (Fig. 2.2).

2.2.3.2 Pyridoxal 5′-Phosphate (PLP)

It is known that pyridoxine-responsive seizures are an indicator of hypophosphatasia severity and lethal prognosis (Litmanovitz et al. 2002; Nunes et al. 2002; Baumgartner-Sigl et al. 2007). PLP is the phosphorylated version of pyridoxine

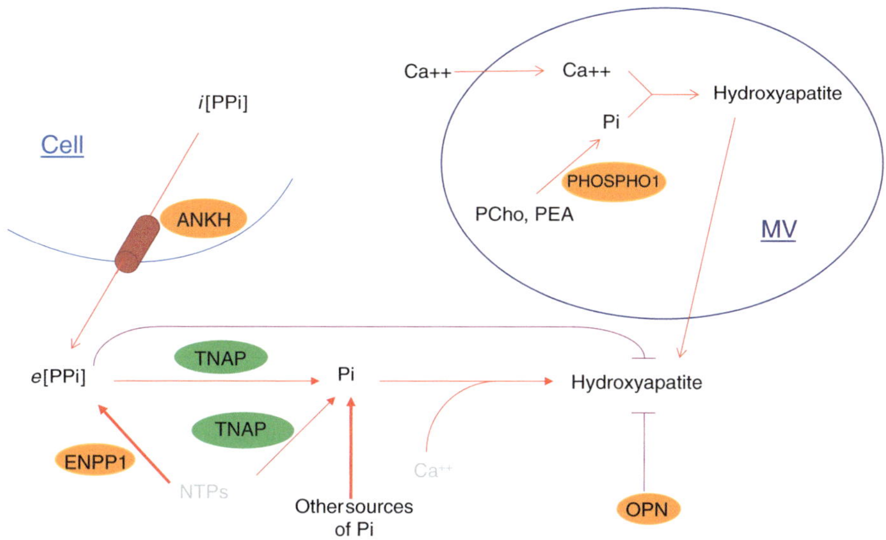

Fig. 2.2 Schematic representation of the key players in bone mineralization and its regulation [adapted from (Hessle et al. 2002; Harmey et al. 2004; Orimo 2010; Yadav et al. 2011)]. MV = matrix vesicle; PEA = Phosphoethanolamine, PPi = inorganic pyrophosphate (i = intracellular, e = extracellular), Pi = inorganic phosphate; NTPs = nucleotide triphosphates. ENPP1 = ectonucleotide pyrophosphatase/phosphodiesterase 1; OPN = osteopontin; TNAP = tissue-nonspecific alkaline phosphatase; ANKH = progressive ankylosis protein homolog; PEA = phosphoethanolamine; PCho = phosphocholine; PHOSPHO1 = phosphatase orphan 1

(vitamin B6) (see Chap. 11). TNAP hydrolyzes PLP, and the unphosphorylated vitamer PL crosses the blood-brain barrier, to be re-generated as PLP into the cell (Whyte 2001) where it plays a crucial role as co-enzyme for the glutamic acid decarboxylase, an enzyme responsible for gamma-aminobutyric acid synthesis (Baxter 2003). Consequently, in patients with hypophosphatasia, inability to cleave PLP may result in a central nervous system-localized vitamin B6 deficiency and seizures (Waymire et al. 1995) (see also Chap. 14). Interestingly, in patients affected with Mabry's Disease, a particular form of hyperphosphatasia characterized by neurologic defects and seizures, pyridoxine responsiveness of the electroencephalographic abnormalities suggests that TNAP could be also involved in the intraneuronal balance between PL and PLP (Thompson et al. 2006). Mabry's disease has been shown in at least a part of cases, to be due to mutations in PIG (Phosphatidylinositol Glycan anchor biosynthesis) genes resulting in absence of membrane anchoring and secretion of TNAP (Krawitz et al. 2010; Horn et al. 2011; Krawitz et al. 2012; Murakami et al. 2012; Thompson et al. 2012; Krawitz et al. 2013) (see also Chap. 16).

Studies of alkaline phosphatase activity in primates brain supports the important role of this enzyme in neurotransmission (Fonta et al. 2004; Fonta et al. 2005), and the bone isoform of TNAP has been shown to be the only AP expressed in human brain (Brun-Heath et al. 2011). In addition, kinetic characterization of TNAP mutations evidenced that some mutations were efficient in hydrolyzing PPi but inefficient in hydrolyzing PLP (Di Mauro et al. 2002), which could explain why some patients manifest more severe seizures than others.

2.2.3.3 Nucleotides

In rat, TNAP is able to hydrolyze ATP, ADP and AMP (Say et al. 1991) (see Chap. 4). In mouse TNAP has been recently highlighted as one of the three phosphatases (along with prostatic acid phosphatase [PAP] and ecto-5'-nucleotidase [NT5E]) generating adenosine from AMP in dorsal root ganglia (DRG) neurons (Street et al. 2013). Adenosine shows antinociceptive properties and, as other nucleosides, plays a key role in pain signaling and sensory biology, suggesting that mutations in *ALPL* could result in a lower concentration of adenosine in DRG neurons. This could explain bone, joint, and muscle pain suffered by a proportion of HPP patients (see Chap. 13).

2.3 The Liver/Bone/Kidney Alkaline Phosphatase (*ALPL*) Gene and the Mutations Responsible for Hypophosphatasia

2.3.1 The ALPL *Gene*

ALPL consists of 12 exons distributed over 65 kb. Two transcripts have notably been identified, driven by two alternative exons 1 in the 5'-untranslated region (UTR) (Weiss et al. 1988b) and responsible for differential transcription and subsequent co-translational and post-translational modifications (Matsuura et al. 1990; Terao et al. 1990; Studer et al. 1991). The transcription of the upstream exon 1A is preferentially driven in osteoblasts, whereas transcription is preferentially initiated with exon 1B in liver and kidney (Toh et al. 1989; Matsuura et al. 1990; Studer et al. 1991). AP activity detected in the parenchyma and in endothelial cells of brain from human and other species results from the expression of the ALPL gene driven by the bone promoter (Brun-Heath et al. 2011). However, a remarkable species specificity was highlighted by the finding of an additional transcript starting with exon 1B in mouse neurons. The mechanism of tissue-specific regulation of TNAP is not yet elucidated. In addition three transcript variants of the bone ALPL are thus far identified (http://www.ncbi.nlm.nih.gov/gene/249), the longest variant NM_00478.4 harboring the 12 exons, and two shorter variants lacking a portion of the 5' coding region, NM_001177520.1. and NM_001127501.2. The latter was also reported in a patient with adult HPP (Mentrup et al. 2011).

2.3.2 The Hypophosphatasia Mutations

Since the first mutation was identified in the *ALPL* gene (Weiss et al. 1988a), around 300 distinct mutations have been reported. A constantly updated list of mutations is available online (http://www.sesep.uvsq.fr/database_hypo/Mutation.html). Most of them (74 %) are missense mutations. The remaining reported mutations are deletions (11 %), splicing mutations (6 %), nonsense mutations (4 %), small insertions (2 %), a complex deletion + insertion and a nucleotide substitution affecting the major transcription initiation site. Two gross deletions (Spentchian et al. 2006), 2 de novo mutations (Taillandier et al. 2005; Zhang et al. 2012) and a homozygous mutation resulting from paternal isodisomy of chromosome 1 (Watanabe et al. 2012) were also reported.

A small number of mutations are recurrently found and may be quite frequent in particular populations. This reflects founder effects, i.e. ancient ancestral mutations that occurred on single chromosomes and spread throughout populations of small sizes in limited area. For instance, the mutations c.1559delT and p.F327L represent 40.9 and 13.6 % of HPP chromosomes, respectively, in the Japanese population (Michigami et al. 2005) (see also Chap. 14), the mutation c.571G > A (p.E191 K) is found in half of patients from European ancestry affected with moderate HPP (Herasse et al. 2002), and the mutation c.1001G > A (p.G334D) represents most of HPP chromosomes in the Canadian Mennonite community (Greenberg et al. 1993). In USA, the missense mutation c.1133A > T (p.D378 V), probably originating from North-West Europe where it is rare represents 14 % of HPP chromosomes and is a common cause of dominant prenatal benign HPP (Mumm et al. 2007).

Except one case (Whyte et al. 2006), HPP was never reported in populations from black African ancestry.

2.3.3 Genotype-Phenotype Correlations

The great variety of *ALPL* mutations results in a great number of compound heterozygous genotypes and in highly variable clinical expressivity. This makes difficult to assess to a mutation a degree of severity. However, attempts to assess the relative importance of missense mutations and the genotype-phenotype relationship were performed for a number of mutations by using site-directed mutagenesis. These experiments allowed to study alkaline phosphatase activity, cell localization, and degradation of mutant proteins.

A correlation was observed between the severity of the disease and in vitro enzymatic activity of the mutant protein (Zurutuza et al. 1999). Patients with recessive moderate HPP carry at least one mutation that, when tested, exhibits significant residual enzymatic activity, while patients with severe HPP carry mutations that, when tested, mostly do not product enzymatic activity. The results of in vitro assays of over 80 mutations are listed in the Tissue Nonspecific Alkaline

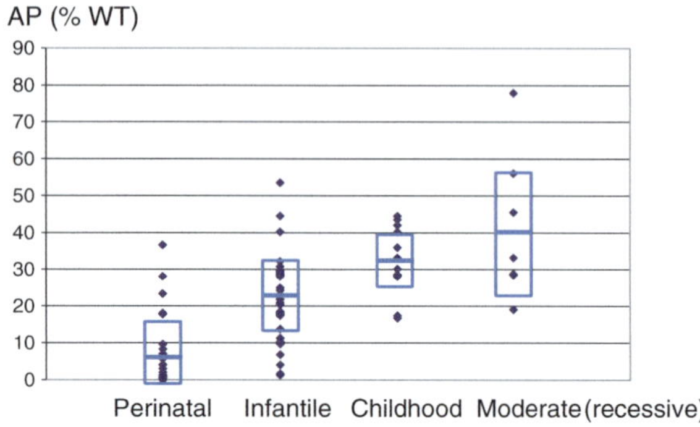

Fig. 2.3 Expected mean AP calculated from the 2 mutations composing the genotype (n = 81). Expected mean alkaline phosphatase relative level calculated from in vitro assays of the two missense mutations composing a genotype. Eighty-one distinct unrelated genotypes were tested. Heretozygous patients, i.e. with dominant HPP, are not shown since the model assumes that the mean activity reflects absence of dominant negative effect

Phosphatase Gene Mutations Database (http://www.sesep.uvsq.fr/03_hypo_mutations.php). Moreover, similarly to the level of serum alkaline phosphatase that correlates with the severity (Whyte 1994), the expected mean AP activity of recessive genotypes with missense mutations tested by site-directed mutagenesis also correlates with the observed phenotype, suggesting that in vitro assessment of mutations is relevant to evaluate the severity of a mutation (Fig. 2.3).

By using GFP fusion proteins or immunofluorescence and biochemical treatments such as cell-surface biotinylation, digestion with phosphatidylinositol-specific phospholipase C, various mutations were characterized for their cell localization and their ability to undergo post-translational processes or to be degraded (Cai et al. 1998; Fukushi et al. 1998; Shibata et al. 1998; Fukushi-Irie et al. 2000; Ito et al. 2002; Ishida et al. 2003; Komaru et al. 2005; Watanabe et al. 2005; Nasu et al. 2006; Brun-Heath et al. 2007; Numa et al. 2008; Ishida et al. 2011; Mentrup et al. 2011; Satou et al. 2012). These studies showed that most of the missense mutations found in severe HPP product a mutant protein that fails to reach the cell membrane, is accumulated in the cis-Golgi and is subsequently degraded in the proteasome. By contrast, the missense mutations responsible for moderate HPP were found to be at least in part correctly localized to the cell membrane.

Analysis of the 3D model of TNAP showed that severe missense mutations mostly affect residues localized in crucial domains of the protein while mutations found in mild forms affect residues more randomly dispatched (Zurutuza et al. 1999; Mornet et al. 2001; Mornet 2008).

As previously indicated neurologic symptoms of HPP (seizures) are observed in the most severe forms of the disease and are assumed to be related to the deficiency of TNAP to dephosphorylate PLP at the surface of neuronal cells (see Chap. 14). Most of the patients presenting with seizures have no detectable serum ALP activity, suggesting that neuronal forms of HPP result from absence of TNAP and that dephosphorylation of PLP may be fulfilled by only very residual ALP activity.

Alternatively, the ability of mutant TNAP to metabolize substrates may differ from one substrate to another, depending on the mutation, as previously shown in vitro for PLP and PPi (Di Mauro et al. 2002). Consequently neuronal phenotype of HPP could depend on the substrat-specificity allowed by the mutation together with its too low ALP activity. Interestingly missense mutations observed in patients with seizures seem to be located in particular regions of the TNAP protein, especially the active site and the calcium site (our unpublished data).

2.3.4 The Dominant Effect of TNAP Mutations

Dominant transmission of HPP has been suggested on the basis of pedigree and laboratory data (Whyte et al. 1979; Whyte et al. 1982; Eastman and Bixler 1983; Eberic et al. 1984; Moore et al. 1999; Hu et al. 2000). Molecular diagnosis, now routinely performed, confirms that dominant inheritance is common in moderate forms of HPP, and that milder is the disease more the inheritance is autosomal dominant (Fauvert et al. 2009). Thus far, at least 22 missense mutations were shown to have a dominant negative effect by site-directed mutagenesis (Table 2.1). Corroborating the idea that the dominant negative effect is due to negative interactions between wild-type and mutant monomers, most of the mutations with a dominant negative effect affect residues located in the active site and the homodimer interface, especially the crown domain which constitute the top of the

Table 2.1 The mutations with a dominant negative effect

Mutation	Residual activity (%WT)	Dominant negative effect (%WT)	Mutation	Residual activity (%WT)	Dominant negative effect (%WT)
p.G456 W	0.8	14.5	p.R71H	1.0	30.5
p.G420A	4.2	19.7	p.E429 K	1.3	31.0
p.D378 V	1.2	19.9	p.G339R	1.1	33.0
p.L414 M	0.4	23.5	p.R71C	2.5	35.0
p.P108L	1.9	24.0	p.S445P	2.1	35.1
p.G63 V	0.8	26.0	p.R184 W	0.6	36.7
p.N417S	3.0	26.5	p.N478I	1.5	38.2
p.E452 K	1.4	27.5	p.G334R	5.0	39.0
p.T100 M	1.3	28.2	p.E476A	0.4	39.3
p.R391H	0	29.1	p.G334D	1.7	40.0
p.V382I	0[a]	30[a]	p.A116T	0.6	40.0

The inhibition of the activity of the mutated monomer onto the wild type (WT) monomer was measured by co-transfecting WT and mutant cDNAs in COS cells and assaying the AP activity. Absence of dominant negative effect by inhibition is expected to produce 50 % of WT activity

[a]As reported by Takinami et al. 2004 (Takinami et al. 2004)

homodimer interface (Fauvert et al. 2009). However, it has been also suggested that dominant negative effects could arise from negative interactions between monomers, homodimers, or multimers, forming aggregates (Lia-Baldini et al. 2008; Sultana et al. 2013).

While the penetrance of the disease (the proportion of affected persons among the persons carrying the affected genotype) in recessive HPP is assumed complete, in dominant forms the penetrance may vary from a mutation to another one, and even with the same mutation from a patient to another one. Thus, dominance is sometimes difficult to demonstrate by using familial analysis. The dominant negative effect was often demonstrated by using site-directed mutagenesis. Immunofluoresence studies with fusions GFP-TNAP or CFP-TNAP proteins showed that the mutant protein may sequestrate the heterodimer in the cytoplasm, explaining the dominant negative effect (Lia-Baldini et al. 2008). AP activity assays showed that the mutant protein inhibited the normal monomer in the heterodimer made of mutant and normal proteins, resulting in decreased levels of alkaline phosphatase activity. Instead of the 50 % expected in heterozygotes, alkaline phosphatase activities were found to range from 20 to 40 % of wild-type (Lia-Baldini et al. 2001). The most strong in vitro inhibition was found with mutation p.D378 V, a mutation found in patients with the perinatal benign form of HPP. Interestingly, parents of these patients express only very mild symptoms (mostly premature loss of teeth) or even, may be completely unaffected (Moore et al. 1999; Pauli et al. 1999; Muller et al. 2000). This is also the case of families with mild HPP due to other dominant missense mutations. This may be attributable both to the progressive improvement of affected patients from infancy to adulthood and to genetic or epigenetic factors involved in the variable expression of the disease. It is possible that in particular stages of development alkaline phosphatase requirements are beyond the capacity of the heterozygous cell, resulting in HPP symptoms. Then, AP requirements may be less important and filled by the heterozygous cell, which may explain the improvement in adult patients. It is also possible that the maternal alkaline phosphatase plays a role via fetal-maternal exchanges, as suggested by the dominant prenatal benign form that is mostly observed when the mutation is inherited from the mother (Moore et al. 1999; Pauli et al. 1999; Wenkert et al. 2011).

2.3.5 Prevalence of HPP

The overall incidence of HPP is unknown. However, as a consequence of the existence of two modes of inheritance, it is expected that the incidence of severe forms strongly differ from moderate forms. The incidence of severe forms was estimated at 1/100 000 in Canada (Fraser 1957) and more recently 1/300 000 in Europe (Mornet et al. 2011). The higher frequency observed in Fraser's study may be due to the founder effect observed in the Canadian Menonnite population (Greenberg et al. 1993) in the Toronto's area where this study was performed.

The incidence of milder forms remains difficult to estimate because of incomplete penetrance and possible dominant inheritance. However, it is likely that these milder forms are more common than severe forms since heterozygotes may be affected (Mornet et al. 2011).

2.4 Molecular Diagnosis and Genetic Counselling

In addition to clinical and radiographic examinations (see Chap. 1), hypophosphatasia diagnosis is based on laboratory assays, and from 1990s, molecular biology.

2.4.1 Laboratory Assays

Total serum alkaline phosphatase activity is markedly reduced in HPP. So, the diagnosis can be suggested in patients where serum AP activity is clearly and consistently subnormal. In general, the more severe the disease, the lower the serum AP activity level (Whyte 1994). However, AP activity is only a helpful diagnostic indicator because other conditions may also show this finding: early pregnancy, drugs administration, hypothyroidism, anemia etc. It must be also noticed that serum AP dramatically varies with age and sex, so that patients and controls must be matched for these two parameters. Serum PLP and urinary PEA may be also used for diagnosis of HPP, however PEA is not pathognomonic since it is observed in a variety of other conditions, including several metabolic bone diseases, and some hypophosphatasia patients may have normal PEA excretion. PLP may be a sensitive marker for HPP (Whyte 1994) and there is a correlation between the severity of the disease and the level of serum PLP.

Heterozygous carriers for severe forms are usually clinically normal but often show modestly reduced serum AP activity and increased urinary PEA (Rasmussen 1968).

2.4.2 Molecular Biology

Screening for mutations in the *ALPL* gene is essential to confirm the hypophosphatasia diagnosis and to offer molecular prenatal diagnosis to families affected with severe forms of the disease. Mutation screening is usually performed by sequencing the *ALPL* coding sequence in genomic DNA. The exons are small and few in number, making the analysis relatively easy. But some mutations remain undetectable because they lie in intronic or regulatory sequences. In our experience, by using sequencing, approximately 95 % of mutations are detected in severe (perinatal and infantile) hypophosphatasia, while patients with mild forms often carry only one detected mutated allele. This may be due to failure to detect the second allelic mutation or to expression of the disease at the heterozygous state.

Prenatal assessment of severe hypophosphatasia by mutation analysis of chorionic villus DNA is now well documented (Henthorn and Whyte 1995; Orimo et al. 1996; Mornet et al. 1999; Watanabe et al. 2007). It seems that mutation analysis is more reliable than alkaline phosphatase assay of chorionic villus at least for heterozygote detection where low AP values may be misinterpreted (Mornet et al. 1999).

Genetic counselling of HPP is complicated by the inheritance that may be autosomal dominant or autosomal recessive and by the existence of the uncommon prenatal benign form. The risk of recurrence of severe forms is 25 %. In moderate forms, it may be 25 % (recessive transmission), 50 % (dominant transmission) or still different (less than 50 %) due to the variable expressivity of dominant forms (Lia-Baldini et al. 2001; Herasse et al. 2003). The mutations detected in dominant forms and responsible for moderate HPP are also found in severe recessive HPP, associated to other mutations (Henthorn et al. 1992; Hu et al. 2000; Lia-Baldini et al. 2001; Herasse et al. 2003; Mumm et al. 2007; Fauvert et al. 2009). Testing patient's relatives is useful since heterozygotes may express a mild form of the disease. In regard to the frequency of the disease (1/300 000 in Europe), testing spouses of carriers is not relevant unless there is a history of consanguinity.

In prenatal context it may be difficult to distinguish prenatal benign HPP and perinatal severe HPP. In fact prenatal benign HPP was previously described with mild outcome (Moore et al. 1999; Pauli et al. 1999), but several years of experience suggest that it may result in variable outcome, ranging from mild HPP to infantile non lethal HPP (Wenkert et al. 2011). Combined with clinical examination, the identification of the mutations and the prediction of their degree of severity could be very useful to distinguish severe and benign prenatal HPP. However, testing new mutations by in vitro functional tests remain time-consuming and existing in silico predictive tools still show relatively low predictive powers.

2.5 Is HPP an Oligogenic Disease?

In a few proportion of patients, estimated to 5 %, no mutation in the *ALPL* gene was found, which may suggest intronic mutations or mutations in the regulatory sequence, but also possible mutations in other genes. In addition there is evidence of patients with the same genotype but significant different phenotypes although these differences mostly correspond to adjacent classes of severity (childhood and infantile for instance) (Whyte et al. 2006; Petkovic Ramadza et al. 2009; Wenkert et al. 2011). This may be due to environmental or epigenetic factors but also to modifiers genes. Since the mechanism of mineralization has been greatly deciphered (Millan 2006; Yadav et al. 2011), a series of genes involved in Pi and PPi metabolism has been suggested as possible modifiers of the HPP phenotype (Fig. 2.2), especially the genes encoding ENPP1 (Ectonucleotide pyrophosphatase/phosphodiesterase), an antagonist to TNAP that produces PPi (Hessle et al. 2002; Rodrigues et al. 2011), ANK (ankylosis protein), a transmembrane protein exporting intracellular PPi (Harmey et al. 2004), OPN, an inhibitor of mineralization that binds hydroxyapatite (Harmey et al. 2004) and PHOSPHO1 a phosphatase involved in the initiation of mineralization (Yadav et al. 2011). Although there is no yet direct proof of implication of these genes in the modulation of the HPP phenotype, exhaustive sequencing of these genes should efficiently answer the question.

Conflict of Interest The author states the absence of conflict of interest.

References

Baumgartner-Sigl S, Haberlandt E, Mumm S, Scholl-Burgi S, Sergi C, Ryan L, Ericson KL, Whyte MP, Hogler W (2007) Pyridoxine-responsive seizures as the first symptom of infantile hypophosphatasia caused by two novel missense mutations (c.677T > C, p. M226T; c.1112C > T, p.T371I) of the tissue-nonspecific alkaline phosphatase gene. Bone 40:1655–1661

Baxter P (2003) Pyridoxine-dependent seizures: a clinical and biochemical conundrum. Biochim Biophys Acta 1647:36–41

Bossi M, Hoylaerts MF, Millan JL (1993) Modifications in a flexible surface loop modulate the isozyme-specific properties of mammalian alkaline phosphatases. J Biol Chem 268:25409–25416

Brun-Heath I, Ermonval M, Chabrol E, Xiao J, Palkovits M, Lyck R, Miller F, Couraud PO, Mornet E, Fonta C (2011) Differential expression of the bone and the liver tissue non-specific alkaline phosphatase isoforms in brain tissues. Cell Tissue Res 343:521–536

Brun-Heath I, Lia-Baldini AS, Maillard S, Taillandier A, Utsch B, Nunes ME, Serre JL, Mornet E (2007) Delayed transport of tissue-nonspecific alkaline phosphatase with missense mutations causing hypophosphatasia. Eur J Med Genet 50:367–378

Brun-Heath I, Taillandier A, Serre JL, Mornet E (2005) Characterization of 11 novel mutations in the tissue non-specific alkaline phosphatase gene responsible for hypophosphatasia and genotype-phenotype correlations. Mol Genet Metab 84:273–277

Cai G, Michigami T, Yamamoto T, Yasui N, Satomura K, Yamagata M, Shima M, Nakajima S, Mushiake S, Okada S, Ozono K (1998) Analysis of localization of mutated tissue-nonspecific alkaline phosphatase proteins associated with neonatal hypophosphatasia using green fluorescent protein chimeras. J Clin Endocrinol Metab 83:3936–3942

Di Mauro S, Manes T, Hessle L, Kozlenkov A, Pizauro JM, Hoylaerts MF, Millan JL (2002) Kinetic characterization of hypophosphatasia mutations with physiological substrates. J Bone Miner Res 17:1383–1391

Eastman JR, Bixler D (1983) Clinical, laboratory, and genetic investigations of hypophosphatasia: support for autosomal dominant inheritance with homozygous lethality. J Craniofac Genet Dev Biol 3:213–234

Eberic FHS, Pralle H, Kabish A (1984) Adult hypophosphatasia without apparent skeletal disease: "ondotohypophosphatasia" in four heterozygote members of a family. Klin Wochenschr 62:371

Fallon MD, Whyte MP, Teitelbaum SL (1980) Stereospecific inhibition of alkaline phosphatase by L-tetramisole prevents in vitro cartilage calcification. Lab Invest 43:489–494

Fauvert D, Brun-Heath I, Lia-Baldini AS, Bellazi L, Taillandier A, Serre JL, de Mazancourt P, Mornet E (2009) Mild forms of hypophosphatasia mostly result from dominant negative effect of severe alleles or from compound heterozygosity for severe and moderate alleles. BMC Med Genet 10:51

Fonta C, Negyessy L, Renaud L, Barone P (2004) Areal and subcellular localization of the ubiquitous alkaline phosphatase in the primate cerebral cortex: evidence for a role in neurotransmission. Cereb Cortex 14:595–609

Fonta C, Negyessy L, Renaud L, Barone P (2005) Postnatal development of alkaline phosphatase activity correlates with the maturation of neurotransmission in the cerebral cortex. J Comp Neurol 486:179–196

Fraser D (1957) Hypophosphatasia. Am J Med 22:730–746

Fukushi-Irie M, Ito M, Amaya Y, Amizuka N, Ozawa H, Omura S, Ikehara Y, Oda K (2000) Possible interference between tissue-non-specific alkaline phosphatase with an Arg 54– > Cys substitution and a counter part with an Asp 277– > Ala substitution found in a compound heterozygote associated with severe hypophosphatasia. Biochem J 348(Pt 3):633–642

Fukushi M, Amizuka N, Hoshi K, Ozawa H, Kumagai H, Omura S, Misumi Y, Ikehara Y, Oda K (1998) Intracellular retention and degradation of tissue-nonspecific alkaline phosphatase with a Gly 317– > Asp substitution associated with lethal hypophosphatasia. Biochem Biophys Res Commun 246:613–618

Greenberg CR, Evans JA, McKendry-Smith S, Redekopp S, Haworth JC, Mulivor R, Chodirker BN (1990) Infantile hypophosphatasia: localization within chromosome region 1 p 36.1–34 and prenatal diagnosis using linked DNA markers. Am J Hum Genet 46:286–292

Greenberg CR, Taylor CL, Haworth JC, Seargeant LE, Philipps S, Triggs-Raine B, Chodirker BN (1993) A homoallelic Gly317– > Asp mutation in ALPL causes the perinatal (lethal) form of hypophosphatasia in Canadian mennonites. Genomics 17:215–217

Harmey D, Hessle L, Narisawa S, Johnson KA, Terkeltaub R, Millan JL (2004) Concerted regulation of inorganic pyrophosphate and osteopontin by akp2, enpp1, and ank: an integrated model of the pathogenesis of mineralization disorders. Am J Pathol 164:1199–1209

Hawrylak K, Stinson RA (1987) Tetrameric alkaline phosphatase from human liver is converted to dimers by phosphatidylinositol phospholipase C. FEBS Lett 212:289–291

Henthorn PS, Raducha M, Fedde KN, Lafferty MA, Whyte MP (1992) Different missense mutations at the tissue-nonspecific alkaline phosphatase gene locus in autosomal recessively inherited forms of mild and severe hypophosphatasia. Proc Natl Acad Sci U S A 89:9924–9928

Henthorn PS, Whyte MP (1995) Infantile hypophosphatasia: successful prenatal assessment by testing for tissue-non-specific alkaline phosphatase isoenzyme gene mutations. Prenat Diagn 15:1001–1006

Herasse M, Spentchian M, Taillandier A, Keppler-Noreuil K, Fliorito AN, Bergoffen J, Wallerstein R, Muti C, Simon-Bouy B, Mornet E (2003) Molecular study of three cases of odontohypophosphatasia resulting from heterozygosity for mutations in the tissue non-specific alkaline phosphatase gene. J Med Genet 40:605–609

Herasse M, Spentchian M, Taillandier A, Mornet E (2002) Evidence of a founder effect for the tissue-nonspecific alkaline phosphatase (TNSALP) gene E174 K mutation in hypophosphatasia patients. Eur J Hum Genet 10:666–668

Hessle L, Johnson KA, Anderson HC, Narisawa S, Sali A, Goding JW, Terkeltaub R, Millan JL (2002) Tissue-nonspecific alkaline phosphatase and plasma cell membrane glycoprotein-1 are central antagonistic regulators of bone mineralization. Proc Natl Acad Sci U S A 99:9445–9449

Ho AM, Johnson MD, Kingsley DM (2000) Role of the mouse ank gene in control of tissue calcification and arthritis. Science 289:265–270

Horn D, Krawitz P, Mannhardt A, Korenke GC, Meinecke P (2011) Hyperphosphatasia-mental retardation syndrome due to PIGV mutations: expanded clinical spectrum. Am J Med Genet A 155A:1917–1922

Hoylaerts MF, Ding L, Narisawa S, Van Kerckhoven S, Millan JL (2006) Mammalian alkaline phosphatase catalysis requires active site structure stabilization via the N-terminal amino acid microenvironment. Biochemistry 45:9756–9766

Hoylaerts MF, Manes T, Millan JL (1997) Mammalian alkaline phosphatases are allosteric enzymes. J Biol Chem 272:22781–22787

Hoylaerts MF, Millan JL (1991) Site-directed mutagenesis and epitope-mapped monoclonal antibodies define a catalytically important conformational difference between human placental and germ cell alkaline phosphatase. Eur J Biochem 202:605–616

Hu JC, Plaetke R, Mornet E, Zhang C, Sun X, Thomas HF, Simmer JP (2000) Characterization of a family with dominant hypophosphatasia. Eur J Oral Sci 108:189–194

Ishida Y, Komaru K, Ito M, Amaya Y, Kohno S, Oda K (2003) Tissue-nonspecific alkaline phosphatase with an Asp (289)– > Val mutation fails to reach the cell surface and undergoes proteasome-mediated degradation. J Biochem (Tokyo) 134:63–70

Ishida Y, Komaru K, Oda K (2011) Molecular characterization of tissue-nonspecific alkaline phosphatase with an Ala to Thr substitution at position 116 associated with dominantly inherited hypophosphatasia. Biochim Biophys Acta 1812:326–332

Ito M, Amizuka N, Ozawa H, Oda K (2002) Retention at the cis-Golgi and delayed degradation of tissue-non-specific alkaline phosphatase with an Asn 153– > Asp substitution, a cause of perinatal hypophosphatasia. Biochem J 361:473–480

Kim EE, Wyckoff HW (1991) Reaction mechanism of alkaline phosphatase based on crystal structures two-metal ion catalysis. J Mol Biol 218:449–464

Komaru K, Ishida Y, Amaya Y, Goseki-Sone M, Orimo H, Oda K (2005) Novel aggregate formation of a frame-shift mutant protein of tissue-nonspecific alkaline phosphatase is ascribed to three cysteine residues in the C-terminal extension. Retarded secretion and proteasomal degradation. FEBS J 272:1704–1717

Kozlenkov A, Le Du MH, Cuniasse P, Ny T, Hoylaerts MF, Millan JL (2004) Residues determining the binding specificity of uncompetitive inhibitors to tissue-nonspecific alkaline phosphatase. J Bone Miner Res 19:1862–1872

Krawitz PM, Murakami Y, Hecht J, Kruger U, Holder SE, Mortier GR, Delle Chiaie B, De Baere E, Thompson MD, Roscioli T, Kielbasa S, Kinoshita T, Mundlos S, Robinson PN, Horn D (2012) Mutations in PIGO, a member of the GPI-anchor-synthesis pathway, cause hyperphosphatasia with mental retardation. Am J Hum Genet 91:146–151

Krawitz PM, Murakami Y, Riess A, Hietala M, Kruger U, Zhu N, Kinoshita T, Mundlos S, Hecht J, Robinson PN, Horn D (2013) PGAP2 mutations, affecting the GPI-anchor-synthesis pathway, cause hyperphosphatasia with mental retardation syndrome. Am J Hum Genet 92:584–589

Krawitz PM, Schweiger MR, Rodelsperger C, Marcelis C, Kolsch U, Meisel C, Stephani F et al (2010) Identity-by-descent filtering of exome sequence data identifies PIGV mutations in hyperphosphatasia mental retardation syndrome. Nat Genet 42:827–829

Le Du MH, Millan JL (2002) Structural evidence of functional divergence in human alkaline phosphatases. J Biol Chem 277:49808–49814

Le Du MH, Stigbrand T, Taussig MJ, Menez A, Stura EA (2001) Crystal structure of alkaline phosphatase from human placenta at 1.8 A resolution. Implication for a substrate specificity. J Biol Chem 276:9158–9165

Lia-Baldini AS, Brun-Heath I, Carrion C, Simon-Bouy B, Serre JL, Nunes ME, Mornet E (2008) A new mechanism of dominance in hypophosphatasia: the mutated protein can disturb the cell localization of the wild-type protein. Hum Genet 123:429–432

Lia-Baldini AS, Muller F, Taillandier A, Gibrat JF, Mouchard M, Robin B, Simon-Bouy B, Serre JL, Aylsworth AS, Bieth E, Delanote S, Freisinger P, Hu JC, Krohn HP, Nunes ME, Mornet E (2001) A molecular approach to dominance in hypophosphatasia. Hum Genet 109:99–108

Litmanovitz Reish O, Dolfin T, Arnon S, Regev R, Grinshpan G, Yamazaki M, Ozono K (2002) Glu274Lys/Gly309Arg mutation of the tissue-nonspecific alkaline phosphatase gene in neonatal hypophosphatasia associated with convulsions. J Inherit Metab Dis 25:35–40

Matsuura S, Kishi F, Kajii T (1990) Characterization of a 5'-flanking region of the human liver/bone/kidney alkaline phosphatase gene: two kinds of mRNA from a single gene. Biochem Biophys Res Commun 168:993–1000

Mentrup B, Marschall C, Barvencik F, Amling M, Plendl H, Jakob F, Beck C (2011) Functional characterization of a novel mutation localized in the start codon of the tissue-nonspecific alkaline phosphatase gene. Bone 48:1401–1408
Meyer JL (1984) Can biological calcification occur in the presence of pyrophosphate? Arch Biochem Biophys 231:1–8
Michigami T, Uchihashi T, Suzuki A, Tachikawa K, Nakajima S, Ozono K (2005) Common mutations F310L and T1559del in the tissue-nonspecific alkaline phosphatase gene are related to distinct phenotypes in Japanese patients with hypophosphatasia. Eur J Pediatr 164:277–282
Millan J (2006) Mammalian alkaline phosphatases: from biology to applications in medicine and biotechnology. Wiley-VCH Verlag GmbH, Weinheim
Millan JL (1992) Alkaline phosphatase as a reporter of cancerous transformation. Clin Chim Acta 209:123–129
Moore CA, Curry CJ, Henthorn PS, Smith JA, Smith JC, O'Lague P, Coburn SP, Weaver DD, Whyte MP (1999) Mild autosomal dominant hypophosphatasia: in utero presentation in two families. Am J Med Genet 86:410–415
Mornet E (2008) Hypophosphatasia. Best Pract Res Clin Rheumatol 22:113–127
Mornet E, Muller F, Ngo S, Taillandier A, Simon-Bouy B, Maire I, Oury JF (1999) Correlation of alkaline phosphatase (ALP) determination and analysis of the tissue non-specific ALP gene in prenatal diagnosis of severe hypophosphatasia. Prenat Diagn 19:755–757
Mornet E, Stura E, Lia-Baldini AS, Stigbrand T, Menez A, Le Du MH (2001) Structural evidence for a functional role of human tissue nonspecific alkaline phosphatase in bone mineralization. J Biol Chem 276:31171–31178
Mornet E, Yvard A, Taillandier A, Fauvert D, Simon-Bouy B (2011) A molecular-based estimation of the prevalence of hypophosphatasia in the European population. Ann Hum Genet 75:439–445
Muller HL, Yamazaki M, Michigami T, Kageyama T, Schonau E, Schneider P, Ozono K (2000) Asp361Val Mutant of alkaline phosphatase found in patients with dominantly inherited hypophosphatasia inhibits the activity of the wild-type enzyme. J Clin Endocrinol Metab 85:743–747
Mumm S, Wenkert D, Zhang X, Geimer M, Zerega J, MPW (2007) Hypophosphatasia: the c.1133A > T, D378 V transversion is the most common American TNSALP mutation. In: Paper presented at Fifth International Alkaline Phosphatase Symposium: Understanding alkaline phosphatase function—Pathophysiology and treatment of Hypophosphatasia and other AP-related diseases. Huningue, France
Murakami Y, Kanzawa N, Saito K, Krawitz PM, Mundlos S, Robinson PN, Karadimitris A, Maeda Y, Kinoshita T (2012) Mechanism for release of alkaline phosphatase caused by glycosylphosphatidylinositol deficiency in patients with hyperphosphatasia mental retardation syndrome. J Biol Chem 287:6318–6325
Narisawa S, Hasegawa H, Watanabe K, Millan JL (1994) Stage-specific expression of alkaline phosphatase during neural development in the mouse. Dev Dyn 201:227–235
Nasu M, Ito M, Ishida Y, Numa N, Komaru K, Nomura S, Oda K (2006) Aberrant interchain disulfide bridge of tissue-nonspecific alkaline phosphatase with an Arg 433– > Cys substitution associated with severe hypophosphatasia. FEBS J 273:5612–5624
Numa N, Ishida Y, Nasu M, Sohda M, Misumi Y, Noda T, Oda K (2008) Molecular basis of perinatal hypophosphatasia with tissue-nonspecific alkaline phosphatase bearing a conservative replacement of valine by alanine at position 406. Structural importance of the crown domain. FEBS J 275:2727–2737
Nunes ML, Mugnol F, Bica I, Fiori RM (2002) Pyridoxine-dependent seizures associated with hypophosphatasia in a newborn. J Child Neurol 17:222–224
Orimo H (2010) The mechanism of mineralization and the role of alkaline phosphatase in health and disease. J Nihon Med Sch 77:4–12
Orimo H, Nakajima E, Hayashi Z, Kijima K, Watanabe A, Tenjin H, Araki T, Shimada T (1996) First-trimester prenatal molecular diagnosis of infantile hypophosphatasia in a Japanese family. Prenat Diagn 16:559–563

Pauli RM, Modaff P, Sipes SL, Whyte MP (1999) Mild hypophosphatasia mimicking severe osteogenesis imperfecta in utero: bent but not broken. Am J Med Genet 86:434–438

Petkovic Ramadza D, Stipoljev F, Sarnavka V, Begovic D, Potocki K, Fumic K, Mornet E, Baric I (2009) Hypophosphatasia: phenotypic variability and possible Croatian origin of the c.1402 g > A mutation of TNSALP gene. Coll Antropol 33:1255–1258

Rasmussen K (1968) Phosphorylethanolamine and hypophosphatasia. Dan Med Bull 15(Suppl 2):1–112

Rezende AA, Pizauro JM, Ciancaglini P, Leone FA (1994) Phosphodiesterase activity is a novel property of alkaline phosphatase from osseous plate. Biochem J 301(Pt 2):517–522

Rodrigues TL, Foster BL, Silverio KG, Martins L, Casati MZ, Sallum EA, Somerman MJ, Nociti FH (2011) Correction of Hypophosphatasia (Hpp) Associated Mineralization Deficiencies In vitro by Phosphate/Pyrophosphate Modulation in Periodontal Ligament Cells. J Periodontol

Satou Y, Al-Shawafi HA, Sultana S, Makita S, Sohda M, Oda K (2012) Disulfide bonds are critical for tissue-nonspecific alkaline phosphatase function revealed by analysis of mutant proteins bearing a C(201)-Y or C(489)-S substitution associated with severe hypophosphatasia. Biochim Biophys Acta 1822:581–588

Say JC, Ciuffi K, Furriel RP, Ciancaglini P, Leone FA (1991) Alkaline phosphatase from rat osseous plates: purification and biochemical characterization of a soluble form. Biochim Biophys Acta 1074:256–262

Schwartz JH, Lipmann F (1961) Phosphate incorporation into alkaline phosphatase of E. coli. Proc Natl Acad Sci U S A 47:1996–2005

Shibata H, Fukushi M, Igarashi A, Misumi Y, Ikehara Y, Ohashi Y, Oda K (1998) Defective intracellular transport of tissue-nonspecific alkaline phosphatase with an Ala 162– > Thr mutation associated with lethal hypophosphatasia. J Biochem (Tokyo) 123:968–977

Spentchian M, Brun-Heath I, Taillandier A, Fauvert D, Serre JL, Simon-Bouy B, Carvalho F, Grochova I, Mehta SG, Muller G, Oberstein SL, Ogur G, Sharif S, Mornet E (2006) Characterization of Missense Mutations and Large Deletions in the ALPL Gene by Sequencing and Quantitative Multiplex PCR of Short Fragments. Genet Test 10:252–257

Stigbrand T (1984) Present status and future trends of human alkaline phosphatases. Prog Clin Biol Res 166:3–14

Street SE, Kramer NJ, Walsh PL, Taylor-Blake B, Yadav MC, King IF, Vihko P, Wightman RM, Millan JL, Zylka MJ (2013) Tissue-Nonspecific Alkaline Phosphatase Acts Redundantly with PAP and NT5E to Generate Adenosine in the Dorsal Spinal Cord. J Neurosci 33:11314–11322

Studer M, Terao M, Gianni M, Garattini E (1991) Characterization of a second promoter for the mouse liver/bone/kidney-type alkaline phosphatase gene: cell and tissue specific expression. Biochem Biophys Res Commun 179:1352–1360

Sultana S, Al-Shawafi HA, Makita S, Sohda M, Amizuka N, Takagi R, Oda K (2013) An asparagine at position 417 of tissue-nonspecific alkaline phosphatase is essential for its structure and function as revealed by analysis of the N417S mutation associated with severe hypophosphatasia. Mol Genet Metab 109:282–288

Taillandier A, Sallinen SL, Brun-Heath I, De Mazancourt P, Serre JL, Mornet E (2005) Childhood hypophosphatasia due to a de novo missense mutation in the tissue-nonspecific alkaline phosphatase gene. J Clin Endocrinol Metab 90:2436–2439

Takinami H, Goseki-Sone M, Watanabe H, Orimo H, Hamatani R, Fukushi-Irie M, Ishikawa I (2004) The mutant (F310L and V365I) tissue-nonspecific alkaline phosphatase gene from hypophosphatasia. J Med Dent Sci 51:67–74

Terao M, Studer M, Gianni M, Garattini E (1990) Isolation and characterization of the mouse liver/bone/kidney-type alkaline phosphatase gene. Biochem J 268:641–648

Thompson MD, Killoran A, Percy ME, Nezarati M, Cole DE, Hwang PA (2006) Hyperphosphatasia with neurologic deficit: a pyridoxine-responsive seizure disorder? Pediatr Neurol 34:303–307

Thompson MD, Roscioli T, Marcelis C, Nezarati MM, Stolte-Dijkstra I, Sharom FJ, Lu P, Phillips JA, Sweeney E, Robinson PN, Krawitz P, Yntema HG, Andrade DM, Brunner HG,

Cole DE (2012) Phenotypic variability in hyperphosphatasia with seizures and neurologic deficit (Mabry syndrome). Am J Med Genet A 158A:553–558

Toh Y, Yamamoto M, Endo H, Misumi Y, Ikehara Y (1989) Isolation and characterization of a rat liver alkaline phosphatase gene. A single gene with two promoters. Eur J Biochem 182:231–237

Vergnes H, Grozdea J, Denier C, Bourrouillou G, Calvas P, De La Farge F, Valdiguie P, Calot M (2000) Lower alkaline phosphatase activity and occurrence of an abnormal hybrid intestinal/tissue non-specific isoform in Down's syndrome amniotic fluids. Early Hum Dev 58:17–24

Vittur F, Stagni N, Moro L, de Bernard B (1984) Alkaline phosphatase binds to collagen; a hypothesis on the mechanism of extravesicular mineralization in epiphyseal cartilage. Experientia 40:836–837

Watanabe A, Satoh, S, Fujita, A, Naing, B.T, Orimo, H, Shimada, T (2012) Perinatal (lethal) type of hypophosphatasia resulting from paternal isodisomy of chromosome 1. In: Paper presented at 6th Alkaline Phosphatase and Hypophosphatasia Symposium, May pp 16–19. Huningue, France

Watanabe A, Yamamasu S, Shinagawa T, Suzuki Y, Miyake H, Takeshita T, Orimo H, Shimada T (2007) Prenatal genetic diagnosis of severe perinatal (lethal) hypophosphatasia. J Nippon Med Sch 74:65–69

Watanabe H, Takinami H, Goseki-Sone M, Orimo H, Hamatani R, Ishikawa I (2005) Characterization of the mutant (A115 V) tissue-nonspecific alkaline phosphatase gene from adult-type hypophosphatasia. Biochem Biophys Res Commun 327:124–129

Waymire KG, Mahuren JD, Jaje JM, Guilarte TR, Coburn SP, MacGregor GR (1995) Mice lacking tissue non-specific alkaline phosphatase die from seizures due to defective metabolism of vitamin B-6. Nat Genet 11:45–51

Weiss MJ, Cole DE, Ray K, Whyte MP, Lafferty MA, Mulivor RA, Harris H (1988a) A missense mutation in the human liver/bone/kidney alkaline phosphatase gene causing a lethal form of hypophosphatasia. Proc Natl Acad Sci U S A 85:7666–7669

Weiss MJ, Ray K, Henthorn PS, Lamb B, Kadesch T, Harris H (1988b) Structure of the human liver/bone/kidney alkaline phosphatase gene. J Biol Chem 263:12002–12010

Wende A, Johansson P, Vollrath R, Dyall-Smith M, Oesterhelt D, Grininger M (2010) Structural and biochemical characterization of a halophilic archaeal alkaline phosphatase. J Mol Biol 400:52–62

Wenkert D, McAlister WH, Coburn SP, Zerega JA, Ryan LM, Ericson KL, Hersh JH, Mumm S, Whyte MP (2011) Hypophosphatasia: nonlethal disease despite skeletal presentation in utero (17 new cases and literature review). J Bone Miner Res 26:2389–2398

Whyte M (2001) Hypophosphatasia, vol IV. McGraw-Hill, New-York

Whyte MP (1994) Hypophosphatasia and the role of alkaline phosphatase in skeletal mineralization. Endocr Rev 15:439–461

Whyte MP, Essmyer K, Geimer M, Mumm S (2006) Homozygosity for TNSALP mutation 1348c > T (Arg433Cys) causes infantile hypophosphatasia manifesting transient disease correction and variably lethal outcome in a kindred of black ancestry. J Pediatr 148:753–758

Whyte MP, Teitelbaum SL, Murphy WA, Bergfeld MA, Avioli LV (1979) Adult hypophosphatasia. Clinical, laboratory, and genetic investigation of a large kindred with review of the literature. Medicine (Baltimore) 58:329–347

Whyte MP, Vrabel LA, Schwartz TD (1982) Adult hypophosphatasia: generalized deficiency of alkaline phosphatase activity demonstrated with cultured skin fibroblasts. Trans Assoc Am Physicians 95:253–263

Wu LN, Genge BR, Lloyd GC, Wuthier RE (1991) Collagen-binding proteins in collagenase-released matrix vesicles from cartilage. Interaction between matrix vesicle proteins and different types of collagen. J Biol Chem 266:1195–1203

Yadav MC, Simao AM, Narisawa S, Huesa C, McKee MD, Farquharson C, Millan JL (2011) Loss of skeletal mineralization by the simultaneous ablation of PHOSPHO1 and alkaline phosphatase function: a unified model of the mechanisms of initiation of skeletal calcification. J Bone Miner Res 26:286–297

Zhang H, Ke YH, Wang C, Yue H, Hu WW, Gu JM, Zhang ZL (2012) Identification of the mutations in the tissue-nonspecific alkaline phosphatase Gene in two Chinese families with hypophosphatasia. Arch Med Res 43:21–30

Zurutuza L, Muller F, Gibrat JF, Taillandier A, Simon-Bouy B, Serre JL, Mornet E (1999) Correlations of genotype and phenotype in hypophosphatasia. Hum Mol Genet 8:1039–1046

Chapter 3
Genetically Modified Mice for Studying TNAP Function

Sonoko Narisawa

Abstract Genetically modified mice are powerful tools for understanding the functions of genes and proteins and often serve as models of human disease. Here, several knockout and transgenic mouse lines related to tissue-nonspecific alkaline phosphatase (TNAP) are described. Conventional TNAP knockout mice die before weaning and show vitamin B6 dependent epilepsy and impaired bone mineralization, mimicking infantile hypophosphatasia. Administration of recombinant human TNAP rescues the lethal phenotype and improves bone mineralization in the null knockout mice, and this enzyme replacement therapy has been successfully applied to the treatment of human patients. Transgenic expression of human TNAP also rescues the TNAP knockout mice. Studies of the TNAP knockout mice and their double knockouts with ectonucleotide pyrophosphatase/phosphodiesterase 1 or progressive ankylosis protein revealed that pyridoxal phosphate and inorganic pyrophosphate are natural substrates of TNAP. Bone osteopontin from TNAP knockout mice is highly phosphorylated, whereas osteopontin from TNAP knockout mice expressing human TNAP is de-phosphorylated, similar to that in wild type mice, indicating that osteopontin is also a natural substrate of TNAP and that phosphorylated osteopontin contributes the impaired bone mineralization in TNAP knockout mice. Conditional TNAP knockout mice and TNAP mutants produced by ENU (*N*-ethyl-*N*-nitrosourea) mutagenesis show milder hypophosphatasia and are expected to be useful models of adult hypophosphatasia.

Keywords Alkaline phosphatase · Conditional knockout · Hypophosphatasia · Knockout mice · Transgenic mice

S. Narisawa (✉)
Sanford-Burnham Medical Research Institute, La Jolla, CA, USA
e-mail: sonokon@sanfordburnham.org

C. Fonta and L. Négyessy (eds.), *Neuronal Tissue-Nonspecific Alkaline Phosphatase (TNAP)*, Subcellular Biochemistry 76,
DOI 10.1007/978-94-017-7197-9_3

3.1 Introduction

A range of mutations in *ALPL*, the gene that encodes tissue-nonspecific alkaline phosphatase (TNAP), can result in hypophosphatasia (HPP) (cf. Chap. 2). The clinical manifestations of HPP include rickets and osteomalacia, as well as seizures associated with vitamin B6 deficiency (cf. Chap. 1). Manipulation of TNAP gene in mice can provide insights into the function of TNAP and the pathophysiology of HPP. Two common manipulations are to remove function through gene knockouts and to add function through the introduction of a transgene. This chapter will describe how both approaches have been used in mouse models to understand TNAP function.

3.2 Conventional and Conditional TNAP Knockout Mice

3.2.1 Brief Overview of Methods to Generate Knockout Mice

The classic method for generating knockout mice is to inactivate a gene in embryonic stem (ES) cells via homologous recombination using a targeting construct. The basic structure of the construct consists of a left/upstream homology arm [several kilobases (kb) of genomic DNA], a cassette for a drug-resistance gene (*Neo*) inserted into an exon or replacing an area between inside of an exon and inside of other exon, and a right/downstream homology arm (several kb of genomic DNA) (Fig. 3.1b). ES cell clones transfected with the targeting vector are screened by Southern blot hybridization and/or polymerase chain reaction (PCR) to identify homologous recombination events. A targeted ES cell clone is then injected into donor blastocysts to produce chimeric mice, which will be bred to wild type animals. Germline transmission will produce some heterozygous animals (in which one allele of the gene carries the targeted structure) among the offspring of the chimeric founder. The heterozygotes can then be bred to produce homozygous knockout animals (in which both alleles of the gene are replaced by the targeting construct).

Conditional knockouts allow the inactivation of the gene of interest to be limited to specific tissues or cell types. To create a conditional knockout mouse, the drug-resistance gene is placed within an intron of the gene, and two LoxP sites are inserted, one in the left arm and one in the right arm (Fig. 3.1c). This targeted allele is called "floxed" (flanked by LoxP), and the floxed genomic region will be excised when Cre recombinase is expressed in the same cell. By crossing a floxed knockout mouse to a transgenic mouse that expresses Cre recombinase in a tissue- or cell-specific manner, a conditional knockout mouse line can be established in which the targeted gene is removed only in the specific tissue or cell where Cre recombinase is expressed.

Instead of gene disruption, homologous recombination can also be used to generate knockin mice in which a mutation can be introduced in the genome (Fig. 3.1d). Knockin mice often mimic genetic diseases in humans.

(a)
(b)
(c)
(d)
(e)
(f)

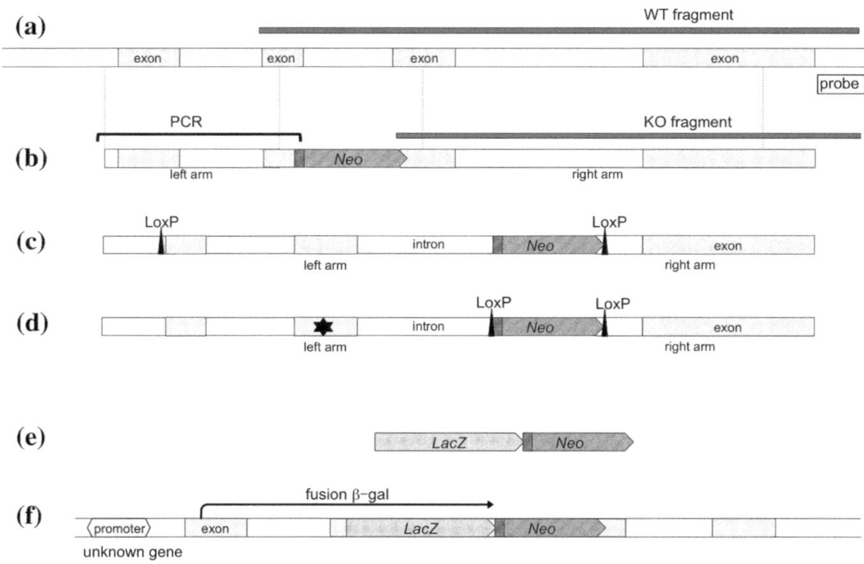

Fig. 3.1 Strategies to modify mouse genome. **a** Genomic gene. **b** Classic targeting vector. Neo gene is driven by a promoter such as PGK (Phosphoglycerate kinase) promoter. Homologous recombination event can be identified by analyzing DNA fragments digested by a restriction enzyme which site is created in the vector by using Southern blot with a probe from outside of the construct. PCR amplification with a primer from outside of the construct and a primer from Neo cassette can also prove the homologous recombination. **c** Conditional KO vector. A Neo cassette flanked by a LoxP sequence is placed in an intron. Another LoxP site is placed in an intron. The region between the two LoxP sites containing exon(s) will be excised by Cre recombinase. **d** Knock-in vector. A mutation is created in an exon and Neo cassette is flanked by LoxP sites. When mice carrying this structure are mated to a transgenic mouse expressing Cre recombinase in their germ cells, the Neo cassette is removed in their offspring's genome. **e** Promoter trap vector consists of a promoter-less LacZ gene and a Neo cassette. **f** A promoter trap vector is driven by an unknown gene expressing a fusion β-galactosidase

In addition to the classic gene targeting strategy that requires specific vectors for each gene, high-throughput approaches such as gene traps are also widely used to obtain knockout mice. The DNA construct for a gene trap normally consists of a promoterless reporter gene (*LacZ*) flanked by an upstream 3′ splice site (splice acceptor), drug-resistance gene (*Neo*), and a downstream polyA sequence (Fig. 3.1e). When this construct is integrated into an intron of an active gene, a fusion protein with the reporter protein will be produced by the endogenous promoter of the gene, and production of the normal protein will be inhibited (Fig. 3.1f). The disrupted gene in each ES cell clone is identified by sequencing. Information of the gene trapped ES clones or mice can be found in the database of International Mouse Strain Resource (http://www.findmice.org). ES clones, frozen embryos/sperm or live mice with a disrupted gene of interest may be available from the depositories; however, one should select clones that have an insertion site in early (upstream) introns, because insertion into a downstream intron can result in a fusion protein that retains some function.

Another high-throughput approach to obtaining knockout mice is *N*-ethyl-*N*-nitrosourea (ENU)-induced mutagenesis. ENU is a potent mutagen and causes point mutations in the genomic DNA in spermatogonia when injected into male mice. Offspring from those ENU-treated male mice exhibit a wide variety of abnormal phenotypes, including behaviors, physiological factors, and morphology. The first generation (G1) produced from the treated males can be screened for dominant mutations, and G3 animals produced from the mating of G2 females with a G1 male can be analyzed for recessive mutations. To identify any mutations in a gene of interest or in a specific chromosomal region, sperm samples from the G1 offspring are screened by sequencing.

3.2.2 Conventional TNAP Knockout Mice

Two conventional knockout lines, $Alpl^{tm1Sor}$ and $Alpl^{tm1Jlm}$, were independently developed by the Soriano (MacGregor et al. 1995) and Millan (Narisawa et al. 1997) groups, respectively, using the classic gene targeting strategy.

In $Alpl^{tm1Sor}$ mice, a part of the genomic gene of TNAP (from the middle of exon 2 to the middle of exon 6) was replaced with a LacZ-Neo cassette via homologous recombination, which allowed the expression of β-galactosidase under the control of the endogenous TNAP promoter. Despite the inactivation of TNAP in the primordial germ cells, the development and migration of the germ cells were not affected (MacGregor et al. 1995). Waymire and colleagues reported that $Alpl^{tm1Sor}$ mice exhibited elevated pyridoxal phosphate (PLP) and died from seizures. Administration of pyridoxal (PL) reduced the seizures and prolonged the lifespan and dental dysplasia was observed in the surviving animals (Waymire et al. 1995).

The $Alpl^{tm1Jlm}$ line was generated by inserting a *Neo* cassette into exon 6 via homologous recombination, and the homozygous animals exhibited vitamin B6 dependent epilepsy and impaired bone mineralization (Narisawa et al. 1997). The average lifespan of the null mice was 8.8 ± 2.3 days in the 129 J background (n = 57) and 10.6 ± 3.4 days in the 129 J; C57Bl/6 J hybrid background (n = 28). Plasma of $Alpl^{-/-}$ mice contains very low levels of AP activity derived from two intestinal APs, duodenum intestinal AP and global intestinal AP, and embryonic AP; AP activity in $Alpl^{+/-}$ plasma was reduced to near 50 % of that in $Alpl^{+/+}$ littermates (Linder et al. 2013). $Alpl^{+/-}$ animals are healthy and fertile, and no abnormality has been observed under normal conditions. Other obvious abnormalities seen in $Alpl^{-/-}$ animals include apoptotic thymus, thinning descending nerve roots, accumulation of gas in the small intestines, and reduced numbers of white blood cells (Narisawa et al. 1997). We have shown that the epilepsy, abnormal thymus, and thinning nerve roots in $Alpl^{-/-}$ mice are not distinguishable from symptoms seen in vitamin B6 depleted wild type mice, but the bone phenotype is independent of vitamin B6 deficiency (Narisawa et al. 2001).

The two TNAP knockout lines, $Alpl^{tm1Sor}$ and $Alpl^{tm1Jlm}$, were compared in detail by Whyte and colleagues (Fedde et al. 1999). Putative natural substrates of TNAP, phosphoethanolamine, inorganic pyrophosphate (PP_i), and PLP, were increased in both lines of TNAP knockout mice. Radiographical bone abnormalities were first seen at around 10 days of age, and osteopenia and fractures worsened at later ages. They concluded that both lines faithfully mimic infantile HPP.

The $Alpl^{tm1Jlm}$ line was crossed to $Enpp1^{tm1Gdg}$, a knockout mouse line of ectonucleotide pyrophosphatase/phosphodiesterase 1 (*Enpp1*) (Sali et al. 1999), to produce double knockout [*Alpl*$^{-/-}$; *Enpp1*$^{-/-}$] animals. *Enpp1* knockout mice show excess mineralization in the vasculature and arthritic joints due to decreased PP_i, a potent inhibitor of bone mineralization, because the *Enpp1* protein (NPP1, also known as PC1) produces PP_i by hydrolyzing ATP in the extracellular milieu. The impaired bone mineralization in the $Alpl^{tm1Jlm}$ knockout was partially corrected in the double knockout animals; the increased PP_i in *Alpl*$^{-/-}$ osteoblasts and decreased PP_i in *Enpp1*$^{-/-}$ osteoblasts were normalized in the [*Alpl*$^{-/-}$; *Enpp1*$^{-/-}$] osteoblasts, indicating that TNAP and NPP1 function antagonistically in bone mineralization by regulating extracellular PP_i (Hessle et al. 2002). Ank^{ank} is an over mineralizing mutant mouse that is caused by a spontaneous point mutation in the progressive ankylosis (Ank) gene (Ho et al. 2000). This mutant also exhibits decreased extracellular PP_i levels because the *Ank* product transports PP_i from the cytoplasm to the extracellular space. Double mutant [*Alpl*$^{-/-}$; *ank/ank*] animals showed limited normalization of bone mineralization and extracellular PP_i levels (Harmey et al. 2004).

In addition to PP_i, both the RNA and protein levels of osteopontin (OPN) are upregulated in the $Alpl^{tm1Jlm}$ knockout mouse, whereas OPN is decreased in *Enpp1*$^{-/-}$ and ank/ank mice (Harmey et al. 2004). OPN is a major bone matrix protein, and phosphorylated OPN inhibits bone mineralization (Jono et al. 2000; Addison et al. 2010). The increased plasma OPN in *Alpl*$^{-/-}$ mice was normalized in the [*Alpl*$^{-/-}$; *Enpp1*$^{-/-}$] and [*Alpl*$^{-/-}$; *ank/ank*] double mutants, suggesting that the impaired mineralization in *Alpl*$^{-/-}$ mice is caused by both PP_i and OPN (Harmey et al. 2004).

The $Alpl^{tm1Jlm}$ line was also crossed to $Spp1^{tm1Rit}$, a knockout mouse lacking OPN, to produce a double knockout mouse [*Alpl*$^{-/-}$; *Spp1*$^{-/-}$]. OPN knockout mice are healthy and have normal bone structure, but show altered osteoclastogenesis (Rittling et al. 1998) and have an abnormal immune phenotype (Chabas et al. 2001; Diao et al. 2008). However, these mice are resistant to ovariectomy induced osteoporosis (Yoshitake et al. 1999) and increased bone mineralization (Harmey et al. 2006). The impaired bone mineralization in *Alpl*$^{-/-}$ mice was improved in the double knockout [*Alpl*$^{-/-}$; *Spp1*$^{-/-}$] animals and osteoblasts despite uncorrected PP_i levels, suggesting that increased OPN also contributes to the HPP phenotype in *Alpl*$^{-/-}$ mice (Harmey et al. 2006).

3.2.3 Chemically Induced TNAP Knockout Mice

The Gena 328 mouse, $Alpl^{Hpp}$, is a mutant mouse generated by ENU mutagenesis that exhibits autosomal semidominant adult HPP (Hough et al. 2007). These mice show low plasma AP and late-onset skeletal abnormalities, but have a normal life span and show no epilepsy, unlike $Alpl^{tm1Sor}$ and $Alpl^{tm1Jlm}$ mice. A point mutation at the splice site for exon 8 produces a truncated, inactive TNAP consisting of 276 amino acids (aa) rather than the normal TNAP peptide of 525 aa; however, a small amount of normal splicing is detected in the homozygous mutants, suggesting that the mutation is hypomorphic.

The $Alpl^{ALPLD1}$ mutant was also produced by ENU mutagenesis and carries a mutation of A to G at nucleotide (nt) position 326 in exon 5, resulting in an Asp to Gly change at aa 109 (Aigner et al. 2009). Sabrautzki and colleagues reported new mouse models for metabolic bone diseases produced by the Munich ENU Mutagenesis Project (Sabrautzki et al. 2012). The following lines have phenotypes of lowered plasma AP due to mutations in *Alpl.* $Alpl^{BAP023}$ carries a missense mutation of T to G in exon 7 at nt 755 (Leu to Pro at aa 251). $Alpl^{BAP026}$ has splice-site mutation in intron 9. $Alpl^{BAP027}$ carries a missense mutation of T to A in exon 10 at nt 1194 (Ile to Asn at aa 395). $Alpl^{BAP032}$ has a missense mutation of A to G in exon 11 at nt 1217 (Asp to Gly at aa 406). The authors pointed that the same mutation was found in a HPP patient. $Alpl^{SAP007}$ has an A to G point mutation in exon 12 at nt 1357 (Thr to Ala at aa 453). Mutants produced by ENU mutagenesis including these reported lines may serve as models of adult HPP and are possibly important tools for understanding the variations in severity of HPP.

3.2.4 Conditional TNAP Knockout Mice

We have generated a floxed mouse line, $Alpl^{flox/flox}$, by using a construct consisting of a 6 kb left homology arm (introns 1–4), a pGK *Neo* cassette, and a 6 kb right homology arm (introns 4–7). One LoxP site was placed in intron 2 and the second one in intron 4 to allow deletion of exons 3 and 4 as well as a frame shift of exon 5 when Cre recombinase is expressed. The collagen type 1 α 1 chain (col1a1) is a bone-specific molecule, and a transgenic mouse expressing *Cre* under the promoter of the col1a1 gene, Tg(col1a1-cre)1Kry, (here referred to as "*Col1a1-Cre*") was used to generate a bone-specific conditional TNAP knockout line [$Alpl^{flox/flox}$; *Col1a1-Cre*]. Plasma AP levels were approximately 30 % of those in the wild-type controls, and the mutant bone phenotype is currently under investigation (unpublished data). The $Alpl^{flox/flox}$ line is expected to be a useful tool for understanding the function of TNAP in non-bone cells such as neuronal cells, endothelial cells, and renal tubules, if crossed to appropriate *Cre* transgenic lines.

3.3 Transgenic Mice with Bone-Specific and Vasculature-Specific TNAP Overexpression

3.3.1 *Brief Overview of Methods to Generate Transgenic Mice*

Transgenic mice express foreign or additional genes, such as human or viral genes, in the background of the mouse genome. They are classically produced by pronuclear microinjection of a DNA fragment into fertilized embryos at the single-cell stage. Animals derived from the injected embryo are screened by Southern blot or PCR for the integration of the transgene into the genome and transmission to the next generation. The integration site of the injected DNA is random, and the copy number of the DNA varies from one to $\sim$100 tandem repeats. The expression and stability of the transgene could be affected by neighboring genes and the chromosomal location. Transgenic mice are still useful approaches to study function of genes by overexpression under tissue-specific promoters, and to test promoter regions connected with a reporter gene (such as *GFP* and *LacZ*). In recent years, tissue/cell-specific *Cre* transgenic lines have become very important, because a successful study with conditional knockout lines depends on the specificity and expression levels of *Cre* in the transgenic mice used.

To overcome the problems caused by the randomness of the integration site and copy number, new methods of producing transgenic animals have been developed. One is to introduce a transgene by gene targeting into a permissive locus such as *Hprt*. This knockin approach makes insertion of a single copy next to the *Hprt* gene possible but requires the same amount of time and effort as producing a knockout mouse.

3.3.2 *Transgenic Mice Expressing TNAP*

We have generated a transgenic mouse that expresses human TNAP cDNA under control of the *col1a1* promoter [*Col1a1-Tnap*]. These animals overexpress human TNAP in osteoblasts and cortical/calvarial bones, and develop normally except they showed increased bone mineralization in vertebrae bones compared to wild type controls (Narisawa et al. 2013). A transgenic mouse line expressing human TNAP cDNA under control of the apolipoprotein E promoter, [*ApoE-Tnap*], was originally produced to overexpress human TNAP in the liver (Murshed et al. 2005) but also expresses the transgene in the endochondral bones. Increased bone mineralization is also found in the femur and vertebrae bones of *ApoE-Tnap* mice (Narisawa et al. 2013). These data suggest that overexpressed TNAP increases mineralization without harmful effects.

Our laboratory is interested in TNAP's role in medial calcification, which often develops in the arterial tunica of patients with diabetes and end-stage renal disease and increases the risk for lower-extremity amputation (Vattikuti and Towler 2004). Medial calcification is caused by transdifferentiation of vascular smooth muscle cells (VSMCs) to an osteogenic lineage, and mineralization of cultured VSMCs expressing TNAP can be inhibited by inhibitors of TNAP (Narisawa et al. 2007). Recently, we generated a "targeted transgenic mouse" via homologous recombination with a targeting vector to insert an inducible human TNAP cDNA in the *Hprt* locus (Hprt$^{\text{tm1(ALPL) Millan}}$). The TNAP cDNA is controlled by a strong CAG promoter, but a floxed stop sequence was placed between the promoter and the cDNA, so that the human TNAP can be expressed when the stop sequence is removed by Cre recombinase. The Hprt$^{\text{tm1(ALPL) Millan}}$ line was crossed to a smooth-muscle-specific *Cre* transgenic mouse, the *SM22-Cre* transgenic line Tg (Tagln-cre)1Her. Male [*Hprt*$^{ALPL/Y}$; *SM22-Cre*$^{+/-}$] mice, whose single X chromosome carries the targeted Hprt allele, showed massive calcification in the arteries. The phenotype is quite severe, and the majority of these males die within several weeks after birth, whereas female [*Hprt*$^{ALPL/X}$; *SM22-Cre*$^{+/-}$] animals show a milder phenotype and live longer because X inactivation lowers the expression levels (Sheen et al. 2015). We are planning to utilize this transgenic model to test treatment methods for medial calcification.

3.3.3 Other Transgenic Mice Related to TNAP

A targeted *Cre* transgenic mouse (Alpl$^{\text{tm1(cre)Nagy}}$) was generated by inserting the Cre gene under control of the *Alpl* promoter (Lomeli et al. 2000). Because heterozygous TNAP knockout mice show no detectable phenotype, this line can be used to generate a TNAP-specific knockout line when crossed to a conditional knockout line carrying LoxP sites.

Placental AP (PLAP) is exclusively expressed in human term placenta and certain cancer cells, and the PLAP gene (*ALPP*) is often utilized as an alternative reporter to *LacZ* in genetically modified mice due to its stability as an enzyme and the availability of PLAP-specific antibodies. Also, overexpression of PLAP does not cause any apparent phenotype (Lobe et al. 1999; Skynner et al. 1999). However, the in vivo function of PLAP is not known, and APs can dephosphorylate a wide range of substrates in vitro, including phosphorylated proteins, DNA, PP_i, PEA, PLP, ATP, and lipopolysaccharides. Therefore, the possible side effects of overexpressing an AP should always be considered as a reporter in genetically modified animals.

3.4 Rescue of TNAP Knockout Mice by Enzyme Replacement Therapy and Transgene Expression

A recombinant TNAP (sALP-FcD10) that consists of the entire human TNAP peptide, the human IgG1 Fc domain, and a deca-aspartate motif (D10) was administered subcutaneously to the Alpl$^{\text{tm1Jlm}}$ mice to test its use as a potential strategy for treating human HPP. sALP-FcD10 was produced in mammalian cells (CHO DG44), purified on a large scale with a Protein A column, and injected into the knockout mice every 24 h beginning on postnatal day 1. Epilepsy was prevented, and their lifespan was dramatically extended. The HPP model mice receiving high-dose sALP-FcD10 (8.2 mg/kg body weight) grew normally and appeared well without epilepsy, showing correction of skeletal and dental abnormality. Plasma PP_i and PLP concentrations remained in their normal ranges (Millan et al. 2008; Yadav et al. 2012; Foster et al. 2013). This enzyme replacement therapy in the $Alpl^{tm1Jlm}$ mouse model was successfully adapted to clinical treatment of human HPP patients (Whyte et al. 2012) (cf. Chap. 15).

In another strategy, a lentiviral vector to express human TNAP containing the D10 motif was injected into the jugular vein of 1-day-old $Alpl^{tm1Jlm}$ mice. The injected *Alpl*$^{-/-}$ mice exhibited no epilepsy and survived more than 10 months with an improved bone phenotype (Yamamoto et al. 2011). This result suggests that a gene therapy approach also has potential for treating HPP.

We crossed the $Alpl^{tm1Jlm}$ line to transgenic mice overexpressing TNAP to obtain two lines of transgenic mice in *Alpl*$^{-/-}$ background, [*Col1a1-Tnap*; *Alpl*$^{-/-}$] and [*ApoE-Tnap*; *Alp*$^{-/-}$]. Expression of both transgenes rescued the lethal phenotype of *Alpl*$^{-/-}$ mice (Narisawa et al. 2013), prevented epilepsy, and improved bone mineralization (Fig. 3.2). As mentioned earlier, OPN is upregulated in *Alpl*$^{-/-}$ mice, but the phosphorylation status of the increased OPN was not known. By comparing the phosphorylation status of OPN from *Alpl*$^{-/-}$, wild type, and [*Col1a1-Tnap*; *Alpl*$^{-/-}$] mice with a proteomics approach, we have been able to show that OPN from *Alpl*$^{-/-}$ mice is highly phosphorylated, and the expressed human TNAP dephosphorylates OPN, indicating that OPN is a natural substrate of TNAP (Narisawa et al. 2013). Accumulation of phosphorylated OPN contributes to the reduced mineralization in *Alpl*$^{-/-}$ mice since phosphorylated OPN inhibits mineralization in vitro. Because expression of *Col1a1-Tnap* is restricted to osteoblasts, the chondrocytes in [*Col1a1-Tnap*; *Alpl*$^{-/-}$] mice express no TNAP, resulting in compromised mineralization in the adult growth plates, although these mice can live a normal life span in the standard sedentary condition.

3.5 Future Directions

The two major phenotypes in both the null knockout lines, $Alpl^{tm1Sor}$ and $Alpl^{tm1Jlm}$, are vitamin B6 deficiency and reduced bone mineralization seen postnatally despite high TNAP expression in the inner cell mass of blastocysts, the developing neural tubes, primordial germ cells, and placenta during embryogenesis (Narisawa et al. 1994) (Fig. 3.3). This implies that the functions of TNAP in the early developmental stages may be protected by redundant mechanisms involving still-unidentified molecules. One example of such a mechanism is the PHOSPHO1 expressed in mineralizing bone. PHOSPHO1 is a member of the large haloacid dehalogenase superfamily of hydrolases and preferentially cleaves

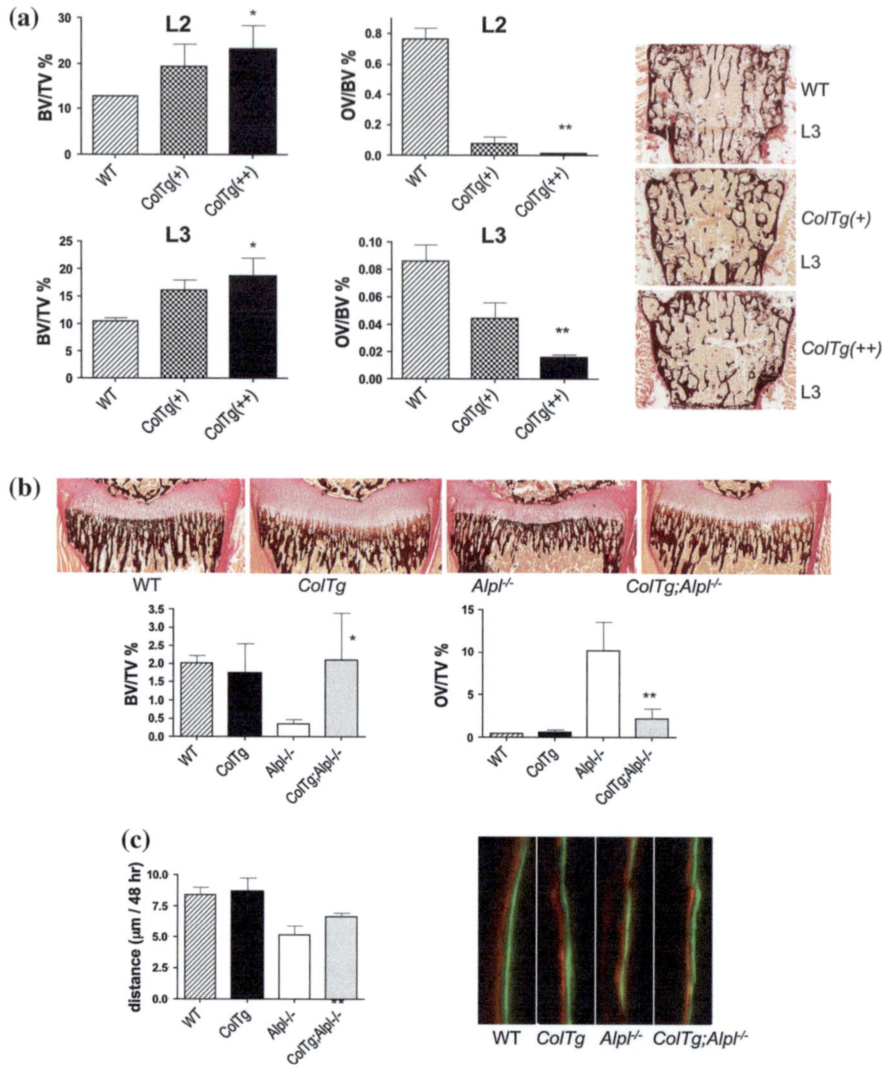

Fig. 3.2 Bone histomorphometric analysis. **a** Comparison of area of mineralized bone (BV/TV) and unmineralized osteoind (OV/BV) in L2 and L3 vertebrae bones from 4-month old WT, *Col1a1-Tnap*$^{+/-}$, and *Col1a1-Tnap*$^{+/+}$ mice (n = 3). **b** Comparison of area of mineralized bone (BV/TV) and unmineralized osteoind (OV/BV) in tibia from 16-day-old WT, *Col1a1-Tnap*$^{+/-}$, *Alpl*$^{-/-}$ and [*Col1a1-Tnap*$^{+/-}$; *Alpl*$^{-/-}$] mice (n = 3). **c** Mineral apposition rates in parietal bones from 11-day-old WT, *Col1a1-Tnap*$^{+/-}$, *Alpl*$^{-/-}$ and [*Col1a1-Tnap*$^{+/-}$; *Alpl*$^{-/-}$] mice (n = 3). Histological staining on **a**, **b**: Von Kossa and Van Gieson staining and **c** in vivo labeling with alizarin (*red*) and calcein (*green*). Taken from Narisawa et al. 2013 with permission from Journal of Bone Mineral Research

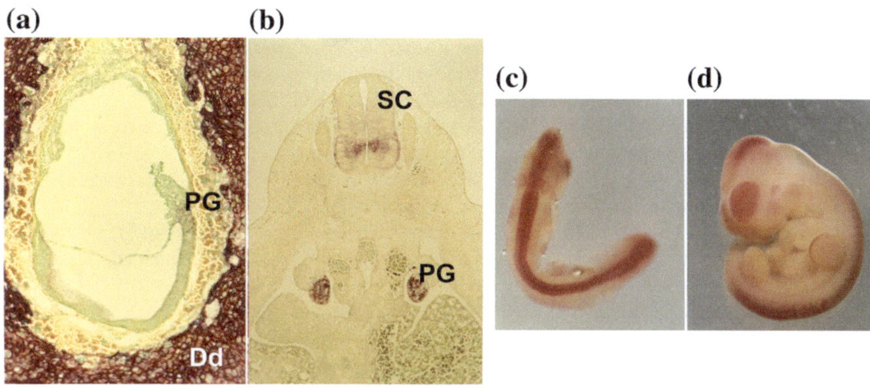

Fig. 3.3 Expression of TNAP during embryogenesis (AP activity staining). **a** Sagittal section of E7.5 embryo (*Dd* decidua, *PG* primordial germ cells near allantois). **b** Transverse section of E11.5 embryo (*SC* spinal cord, *PG* primordial germ cells entered in gonad). **c** E9.5 embryo. **d** E11.5 embryo

phosphoethanolamine and phosphocholine in vitro. *Phospho*$^{-/-}$ knockout (Phospho1^{m1Jlm}) mice show osteomalacia and thoracic scoliosis but are able to live normal lifespans, whereas double-knockout [*Alpl*$^{-/-}$; *Phospho*$^{-/-}$] mice are embryonic lethal with no bone mineralization (Yadav et al. 2011). This result indicates that PHOSPHO1 is one of the molecules that compensates for the lack of TNAP in *Alpl*$^{-/-}$ skeletal tissue during early postnatal stages. In the future, similar redundancy mechanisms might be revealed for the roles of TNAP in inner cell mass, embryonic neural tube, primordial germ cell, and placenta. The conditional *Alpl*$^{flox/flox}$ line will be a useful tool to investigate TNAP's function by crossing it to specific *Cre* transgenic mice. No significant phenotype due to lack of TNAP has been observed in the kidney, endothelial cells, or neuronal cells in postnatal *Alpl*$^{-/-}$ mice; however, it is possible that TNAP in those cell types may play more significant roles in the adult stages. The availability of suitable *Cre* transgenic lines for each cell type is crucial for these future investigations.

Conflict of Interest The author has no conflicts of interest.

References

Addison WN, Masica DL, Gray JJ et al (2010) Phosphorylation-dependent inhibition of mineralization by osteopontin ASARM peptides is regulated by PHEX cleavage. J Bone Miner Res 25:695–705

Aigner B, Rathkolb B, Klaften M et al (2009) Generation of N-ethyl-N-nitrosourea-induced mouse mutants with deviations in plasma enzyme activities as novel organ-specific disease models. Exp Physiol 94:412–421

Chabas D, Baranzini SE, Mitchell D et al (2001) The influence of the proinflammatory cytokine, osteopontin, on autoimmune demyelinating disease. Science 294:1731–1735

Diao H, Iwabuchi K, Li L et al (2008) Osteopontin regulates development and function of invariant natural killer T cells. Proc Natl Acad Sci USA 105:15884–15889

Fedde KN, Blair L, Silverstein J et al (1999) Alkaline phosphatase knock-out mice recapitulate the metabolic and skeletal defects of infantile hypophosphatasia. J Bone Miner Res 14:2015–2026

Foster BL, Nagatomo KJ, Tso HW et al (2013) Tooth root dentin mineralization defects in a mouse model of hypophosphatasia. J Bone Miner Res 28:271–282

Harmey D, Hessle L, Narisawa S et al (2004) Concerted regulation of inorganic pyrophosphate and osteopontin by Akp2, Enpp1, and Ank: an integrated model of the pathogenesis of mineralization disorders. Am J Pathol 164:1199–1209

Harmey D, Johnson KA, Zelken J et al (2006) Elevated skeletal osteopontin levels contribute to the hypophosphatasia phenotype in $Akp2^{-/-}$ mice. J Bone Miner Res 21:1377–1386

Hessle L, Johnson KA, Anderson HC et al (2002) Tissue-nonspecific alkaline phosphatase and plasma cell membrane glycoprotein-1 are central antagonistic regulators of bone mineralization. Proc Natl Acad Sci USA 99:9445–9449

Ho AM, Johnson MD, Kingsley DM (2000) Role of the mouse ank gene in control of tissue calcification and arthritis. Science 289:265–270

Hough TA, Polewski M, Johnson K et al (2007) Novel mouse model of autosomal semidominant adult hypophosphatasia has a splice site mutation in the tissue nonspecific alkaline phosphatase gene Akp2. J Bone Miner Res 22:1397–1407

Jono S, Peinado C, Giachelli CM (2000) Phosphorylation of osteopontin is required for inhibition of vascular smooth muscle cell calcification. J Biol Chem 275:20197–20203

Linder CH, Englund UH, Narisawa S et al (2013) Isozyme profile and tissue-origin of alkaline phosphatases in mouse serum. Bone 53:399–408

Lobe CG, Koop KE, Kreppner W et al (1999) Z/AP, a double reporter for cre-mediated recombination. Dev Biol 208:281–292

Lomeli H, Ramos-Mejia V, Gertsenstein M et al (2000) Targeted insertion of Cre recombinase into the TNAP gene: excision in primordial germ cells. Genesis 26:116–117

MacGregor GR, Zambrowicz BP, Soriano P (1995) Tissue non-specific alkaline phosphatase is expressed in both embryonic and extraembryonic lineages during mouse embryogenesis but is not required for migration of primordial germ cells. Development 121:1487–1496

Millan JL, Narisawa S, Lemire I et al (2008) Enzyme replacement therapy for murine hypophosphatasia. J Bone Miner Res 23:777–787

Murshed M, Harmey D, Millan JL et al (2005) Unique coexpression in osteoblasts of broadly expressed genes accounts for the spatial restriction of ECM mineralization to bone. Genes Dev 19:1093–1104

Narisawa S, Frohlander N, Millan JL (1997) Inactivation of two mouse alkaline phosphatase genes and establishment of a model of infantile hypophosphatasia. Dev Dyn 208:432–446

Narisawa S, Harmey D, Yadav MC et al (2007) Novel inhibitors of alkaline phosphatase suppress vascular smooth muscle cell calcification. J Bone Miner Res 22:1700–1710

Narisawa S, Hasegawa H, Watanabe K et al (1994) Stage-specific expression of alkaline phosphatase during neural development in the mouse. Dev Dyn 201:227–235

Narisawa S, Wennberg C, Millan JL (2001) Abnormal vitamin B6 metabolism in alkaline phosphatase knock-out mice causes multiple abnormalities, but not the impaired bone mineralization. J Pathol 193:125–133

Narisawa S, Yadav MC, Millan JL (2013) In vivo overexpression of tissue-nonspecific alkaline phosphatase increases skeletal mineralization and affects the phosphorylation status of osteopontin. J Bone Miner Res 28:1587–1598

Rittling SR, Matsumoto HN, McKee MD et al (1998) Mice lacking osteopontin show normal development and bone structure but display altered osteoclast formation in vitro. J Bone Miner Res 13:1101–1111

Sabrautzki S, Rubio-Aliaga I, Hans W et al (2012) New mouse models for metabolic bone diseases generated by genome-wide ENU mutagenesis. Mamm Genome 23:416–430

Sali A, Favaloro J, Terkeltaub R et al (1999) Germline deletion of the nucleoside triphosphate pyrophosphohydrolase (NTPPPH) plasma cell glycoprotein (PC-1) produces abnormal calcification of periarticular tissues. In: Vanduffel L, Lemmens R (eds) Ecto-ATPases and related ectonucleotidases: proceedings of the second international workshop on Ecto-ATPases and related ecotonucleotidases, 1st edn. Shaker Publishing B.V., Diepenbeek, pp 267–282

Skynner MJ, Drage DJ, Dean WL et al (1999) Transgenic mice ubiquitously expressing human placental alkaline phosphatase (PLAP): an additional reporter gene for use in tandem with beta-galactosidase. Int J Dev Biol 43:85–90

Sheen CR, Kuss P, Narisawa S et al (2015) Pathophysiological role of vascular smooth muscle alkaline phosphatase in medial artery calcification. J Bone Miner Res 30:824–836

Vattikuti R, Towler DA (2004) Osteogenic regulation of vascular calcification: an early perspective. Am J Physiol Endocrinol Metab 286:E686–E696

Waymire KG, Mahuren JD, Jaje JM et al (1995) Mice lacking tissue non-specific alkaline phosphatase die from seizures due to defective metabolism of vitamin B-6. Nat Genet 11:45–51

Whyte MP, Greenberg CR, Salman NJ et al (2012) Enzyme-replacement therapy in life-threatening hypophosphatasia. N Engl J Med 366:904–913

Yadav MC, de Oliveira RC, Foster BL et al (2012) Enzyme replacement prevents enamel defects in hypophosphatasia mice. J Bone Miner Res 27:1722–1734

Yadav MC, Simao AM, Narisawa S et al (2011) Loss of skeletal mineralization by the simultaneous ablation of PHOSPHO1 and alkaline phosphatase function: a unified model of the mechanisms of initiation of skeletal calcification. J Bone Miner Res 26:286–297

Yamamoto S, Orimo H, Matsumoto T et al (2011) Prolonged survival and phenotypic correction of $Akp2^{-/-}$ hypophosphatasia mice by lentiviral gene therapy. J Bone Miner Res 26:135–142

Yoshitake H, Rittling SR, Denhardt DT et al (1999) Osteopontin-deficient mice are resistant to ovariectomy-induced bone resorption. Proc Natl Acad Sci USA 96:8156–8160

Part II
Expression and Regulation of TNAP in the Central Nervous System

Chapter 4
Tissue-Nonspecific Alkaline Phosphatase in the Developing Brain and in Adult Neurogenesis

Herbert Zimmermann and David Langer

Abstract Tissue-nonspecific alkaline phosphatase is expressed both in the developing and adult mammalian brain. Expression is particularly strong in the developing nervous system and associated with neurogenic activity. It ceases during later development but high expression remains in the rodent adult subventricular zone of the lateral ventricles, a neurogenic niche generating new neurons for the olfactory bulb. Lower activity is maintained in specific brain regions. In spite of the wide expression of the enzyme very little is known concerning its regulation of expression and physiological function in nervous tissue. This may be one of the reasons why the expression of TNAP in the nervous system to date has received relatively little attention. The strong and in part transient expression of the enzyme in the developing brain implies, however, a significant role in the control of neural development. Interestingly, several factors that stimulate TNAP expression or activity have a strong impact on neural development. The chapter summarizes major findings regarding the cellular distribution of TNAP in the developing brain and the neurogenic niches of the adult brain and in vitro evidence for a functional role of TNAP in axon growth and progenitor cell proliferation and differentiation. Potential mechanisms of TNAP function are discussed.

Keywords Alkaline phosphatase · Brain · Neural development · Neurogenesis · TNAP

H. Zimmermann (✉) · D. Langer
Institute for Cell Biology and Neuroscience, Molecular and Cellular Neurobiology, Goethe University, Max-von-Laue-Str. 13, 60438, Frankfurt am Main, Germany
e-mail: h.zimmermann@bio.uni-frankfurt.de

D. Langer
e-mail: david_langer@web.de

C. Fonta and L. Négyessy (eds.), *Neuronal Tissue-Nonspecific Alkaline Phosphatase (TNAP)*, Subcellular Biochemistry 76,
DOI 10.1007/978-94-017-7197-9_4

4.1 Two Alkaline Phosphatases in Early Mammalian Embryos

Alkaline phosphatase is expressed very early on. Catalytic activity was studied by means of enzyme histochemical staining protocols in a variety of mammalian embryos such as rat, mouse and hamster. In mouse embryos the enzyme histochemical reaction becomes apparent at the 2–4-cell stage (Ishiyama and Izquierdo 1977) and is strong in the 8-cells embryo (Solter et al. 1973). Following differentiation into the trophectoderm and inner cell mass of the blastocyst alkaline phosphatase activity becomes restricted to the inner cell mass and primary ectoderm of pregastrulation embryos (Mulnard and Huygens 1978; Izquierdo et al. 1980). During early development a switch in isoenzyme expression takes place. Whereas the 2 to 8-cell stage mouse embryos express the embryonic isoform (EAP) of the enzyme, blastocysts coexpress both the EAP and the tissue-nonspecific (TNAP) isoenzyme (Lepire and Ziomek 1989; Hahnel et al. 1990). TNAP remains the major isoenzyme of 7.5–14 day embryos and is expressed at restricted stages of postimplantation development (Hahnel et al. 1990; Narisawa et al.1994) and also in the adult (Brun-Heath et al. 2011).

4.2 TNAP Expression During Neural Development

The distribution of alkaline phosphatase activity in the developing nervous system has been investigated in a variety of vertebrate species by enzyme histochemical methods. In the mouse it was analyzed in detail using sodium-β-glycerophosphate as a substrate (Chiquoine 1954), an azo-dye technique (Kwong and Tam 1984; Tam and Kwong 1987; Narisawa et al. 1994) or the combination of NBT (nitro-blue tetrazolium chloride) and BCIP (5-bromo-4-chloro-3′-indolyphosphate p-toluidine salt) and immunostaining (Langer et al. 2007). The isoform responsible for this catalytic activity has further been identified as TNAP using Northern blot analysis and reverse-transcriptase polymerase chain reaction amplification of RNA (Hahnel et al. 1990; Narisawa et al. 1994).

It should be noted that the enzyme histochemical procedures provide no information regarding the actual catalytic activity since the experimenter typically applies the substrate solution to the tissue section until satisfactory staining intensity is obtained. Enzyme histochemical procedures can, however, provide information regarding the relative distribution of catalytic activity between brain regions within a given tissue section. If sections from different tissues are to be compared it is essential to keep substrates, reaction times and reaction temperatures identical. Thus staining intensities obtained in individual reports are difficult to compare as they depend on varying substrate solutions and substrate incubation times.

In general, the earliest stages of morphological differentiation of the neural plate and of the central nervous system reveal high levels of TNAP catalytic activity whereas enzyme activity falls as differentiation progresses. For this reason it was assumed earlier on that alkaline phosphatase activity is not present (except blood vessels and the choroid plexus) in the adult brain. The following summarizes data from Chiquoine (1954), Kwong and Tam (1984), Tam and Kwong (1987), and Narisawa et al. (1994). At E8.5 cell surface-located TNAP activity becomes prominent in the neural fold and is associated with the neuroepithelial cells of the closed spinal cord region of the neuroepithelium. The differentiation of the neural plate into the five brain vesicles and spinal cord in E9 embryos is accompanied by a strong increase in TNAP activity. Strongly TNAP-positive cells appear in the brain and spinal cord at E9.5 and at E10.5 when the development of the brain tube advances. The telencephalon exhibits a uniform and maximal activity at stages E9 and E10. Strong activity is also observed in the diencephalon, mesencephalon and spinal cord. The lateral portions of the brain vesicle and the diencephalon and mesencephalon display a ventrodorsal gradient in enzyme staining from maximal in the ventral portion to reduced activity of the roof plate. At that stage also cranial nerves emerging from the myelencephalon express the enzyme. During further development TNAP-positive fiber-like structures become apparent that run in longitudinal and transverse orientations in the myelencephalon and also in the metencephalon and spinal cord. The telencephalon displays diffuse, weak staining. In the spinal cord strong activity is associated with the ventral horn gray matter, around motoneurons and their processes projecting to the ventral roots. At E13.5 TNAP-positive fibers are prominent in tegmentum, pons and in the marginal layer but from E14.5 onwards such fibers can only rarely be observed. In the spinal cord TNAP-positive processes traversing from the luminal side to the marginal layer become prominent. Cells with high enzymatic activity were especially densely packed near the luminal side.

In general, the later differentiation of the neural tube in embryos of stages E11 to E14 is characterized by a progressive decrease in TNAP activity the extent of which varies between brain regions that in turn vary regarding their state of neural differentiation. In the wall of the cerebral vesicles of E11 and E12 embryos staining intensity is, however, enhanced at the ventricular layer. At E13 and E14 the marginal and the mantle layers reveal only moderate activity while the activity at the ventricular layer is further increased. This distribution of staining intensity is similar in corpus striatum and pallium. Increasing enzyme staining is obtained in E11 to E14 embryos in the cranial and spinal ganglia.

4.3 Expression of TNAP Relates to Neurogenesis

In the mammalian nervous system, neurogenesis occurs predominantly during embryogenesis whereas glial cells are prevalently generated after birth (Kriegstein and Alavarez-Buylla 2009). The neuroepithelial cells of the neural tube function as pluripotent neural stem cells for the generation of neurons and macroglia of brain and spinal cord. Neurons and glia of the peripheral nervous system are derived from the neural crest cells. A major focus has been on the development of the forebrain with its laminar structure. The cortical projection neurons are generated from neural precursors situated in an embryonic proliferative zone at the ventricle surface. Young neurons born in this region migrate into the overlaying cortical tissue. Of central importance in this process is the radial glia. Neuroepithelial cells transform into cortical radial glia that function as neural stem cells in both the prenatal and postnatal telencephalon. At the same time they serve to guide young neurons migrating out of the proliferative zone (Kriegstein and Götz 2003; Malatesta et al. 2003; Gal et al. 2006). In the mouse forebrain radial glial cells appear at E9/10. The early progenitors are positioned at the ventricular zone (VZ) that lines the ventricular cavity. At the onset of cortical neurogenesis (E12) (Guillemot and Parras 2005; Attardo et al. 2008), a second proliferating precursor population appears at the subjacent subventricular zone (SVZ). At the end of neurogenesis (E18), cortical radial glia becomes exclusively gliogenic and eventually radial glia retracts their processes giving rise to multipolar astrocytes (Pinto and Götz 2007). The peak of astrocyte formation occurs early postnatally and oligodendrocyte formation peaks around the second postnatal week (Levison et al. 1993; Zerlin et al. 1995; Parnavelas 1999; Sauvageot and Stiles 2002).

It is obvious that regions of increasing TNAP activity during E11 and E14 correspond to parts of the brain undergoing early morphological differentiation at this time, such as the neuroepithelial cells, cells of the ventricular zone, the basal plates of the telencephalon and the alar plates of the metencephalon. In contrast other brain regions that have passed through the earlier stages of differentiation exhibit decreasing enzyme activity. This becomes even more apparent when later prenatal stages and postnatal stages of cortical development are analyzed.

4.4 TNAP During Late Embryonic and Postnatal Neurogenesis

In contrast to previous belief it has become clear within the last two decades that neurogenesis is not terminated at embryonic age but continues during postnatal life and in the adult. During prenatal stages neurogenesis takes place in two actively proliferating zones, the apical VZ and the basal SVZ, two relatively thin layers lining the primitive ventricular cavities (Peretto et al. 1999). Early basal progenitors arise from the mitosis of neuroepithelial and radial glial cells at the VZ and later form the SVZ (Götz and Huttner 2005). Late borne neurons originate from an intermediate precursor pool that forms a large reservoir of undifferentiated neuroblasts in the SVZ (Smart and McSherry 1982). The VZ is the source of deep layer neurons whereas neurons arising from an intermediate neural precursor pool of the SVZ generate the upper layer cortical neurons. During embryonic development, the VZ eventually disappears whereas the SVZ achieves prominence, peaks in size during early postnatal development and becomes a major source of macroglia (Brazel et al. 2003).

After birth, these proliferating layers progressively disappear, leaving the mature nervous tissue in contact with a continuous, ventricle-lining cell monolayer, the ependyma. In the adult, neurogenesis (and to a minor extent gliogenesis) is restricted to specific brain regions, the neurogenic niches. Neurogenesis persists in the adult SVZ (situated below the ependyma and also referred to as subependymal zone) of the lateral ventricles (Alvarez-Buylla and García-Verdugo 2002) and in the subgranular layer of the hippocampal dentate gyrus (Kempermann et al. 2004). Both regions house radial glia-derived astrocyte-like precursors (Zhao et al. 2008). Neural stem cells of the adult SVZ generate neurons for the olfactory bulb. Young neurons (neuroblasts) derived from the stem cells in the SVZ migrate via the rostral migratory stream (RMS) to the olfactory bulb, where they either undergo apoptosis or differentiate into interneurons. Neural precursors of the hippocampal dentate gyrus produce granule cells that—after short distance migration—can integrate into the granule cell layer (Ming and Song 2005).

A detailed analysis of the distribution of TNAP in the developing mouse brain clearly depicts this transition from the embryonic to postnatal and adult neurogenesis. It shows that TNAP can serve as a marker for neural progenitor cells of the VZ and SVZ throughout cortical neurogenesis in the mouse. The distribution of TNAP activity was investigated by enzyme histochemical staining (Langer et al. 2007) from stage E14 onwards, with particular emphasis on the VZ and SVZ of the lateral walls of the lateral ventricles bordering the striatum that is derived from the lateral ganglionic eminence (Sturrock and Smart 1980; Wichterle et al. 2001). This was compared to the immunostaining for doublecortin (DCX), a microtubule-associated protein that marks the simultaneous appearance and tissue distribution of immature neurons. At E14, a developmental state with ongoing cortical neurogenesis in mice (Götz and Huttner 2005), the entire cerebral wall stains for alkaline phosphatase activity (Fig. 4.1). However, enzyme staining is enhanced at the VZ and gradually reduced toward the pial surface. Catalytic activity is associated with essentially all cells of the telencephalic vesicle including radially oriented cells presumably representing radial glial cells. In contrast, the young neuron marker DCX is restricted to cells that had migrated out of the VZ into the periphery. Thus, the distribution of TNAP activity is essentially complementary to that of DCX. This general distribution pattern of the two markers is maintained during further embryonic development and the thickening of the cortical and striatal tissues at stages E16 to E18. Strong TNAP activity remains associated with cells of the VZ and SVZ whereas it becomes reduced in the outer layers. The intense subventricular enzyme activity corresponds to the regions of increased proliferative activity previously identified in autoradiographs following [^{3}H]thymidine injection in the developing and adult mouse and rat brain (Smart 1961; Altman and Bayer 1990). Fibrous TNAP staining proceeding at right angles with the ventricular surface (E18) could represent chains of migrating young neurons. Immunostaining for DCX at E16 is still strongest at the pial side but is also increased toward the ventricular region. During further embryonic and early postnatal development DCX immunostaining is excluded from the VZ but otherwise becomes widely associated with cells of the developing striatum and cortex, indicating progressive neuronal maturation. At postnatal stage P2, cells at both the ventricular surface and the SVZ are intensely stained for TNAP and DCX-positive cells are now bordering the SVZ.

This situation dramatically changes during progressive postnatal development. On continued maturation of the brain tissue and development of the ependyma (Sturrock and Smart 1980), the TNAP-positive cell layer becomes considerably thinner. Moreover, TNAP-positive cells begin to form subependymal clusters. This becomes apparent already at P10 and is depicted in Fig. 4.1 for P14 and P21. At P21, the ventricular distribution of TNAP-positive cells essentially corresponded to that of the adult stage. Retraction of TNAP from the cell layer lining the ventricular surface is very obvious at P10, corresponding to the development of the ciliated ependyma within the first 15 postnatal days (Sturrock and Smart 1980). The distribution of DCX also changes. While DCX-positive cells are essentially absent from the ventricular surface, DCX-positive cells now populate the SVZ and begin to form subependymal clusters in coronal sections of the SVZ—typical for the adult SVZ. These clustered DCX-positive cells represent the neuroblasts that migrate from the SVZ toward the olfactory bulb.

Notably, immunostaining for TNAP in the developing brain revealed the same cellular distribution as enzyme histochemical staining (Langer et al. 2007). Double immunostaining for TNAP and DCX showed that DCX-positive cells of the SVZ are immunopositive also for TNAP. Additional closely associated TNAP-positive cells were, however, DCX-negative. The contribution of DCX-positive cells to the total of TNAP-positive cells increased from 24 % at P2 to 60 % in the adult, paralleled by a concomitant reduction in width of the SVZ. Importantly TNAP was absent from DCX-positive cells that had migrated out into the adjacent striatum and had started to extend neuritic processes (Fig. 4.1, P14, P21).

These data suggest that TNAP is expressed in neural progenitors in the developing brain and that the change in the distribution of TNAP at the lateral ventricles reflects the rearrangement of the neurogenic layers from embryonic to adult. TNAP is associated both with the early ventricle-lining progenitors and with the progenitors of the SVZ. With progressing maturation TNAP is also associated with DCX-positive young neurons as long as these are associated with the neurogenic niche.

During further development, TNAP activity is transiently increased in additional brain regions such as the striatum and the septum. Alkaline phosphatase–positive microvessels are absent during early development and appear from P12 onwards (Vorbrodt et al. 1986; Langer et al. 2007). A detailed study of the postnatal development of the marmoset brain (Fonta et al. 2005) revealed that at birth white matter throughout the brain is intensely stained. This is followed by a subsequent decrease whereby TNAP activity and myelin content display a complementary pattern. Depending on brain region neuronal TNAP activity is already present at birth or develops during the following months in various sensory cortical areas, presumably reflecting differential levels of maturation. The electron microscopic analysis of this primate brain revealed that TNAP activity is localized at the surface

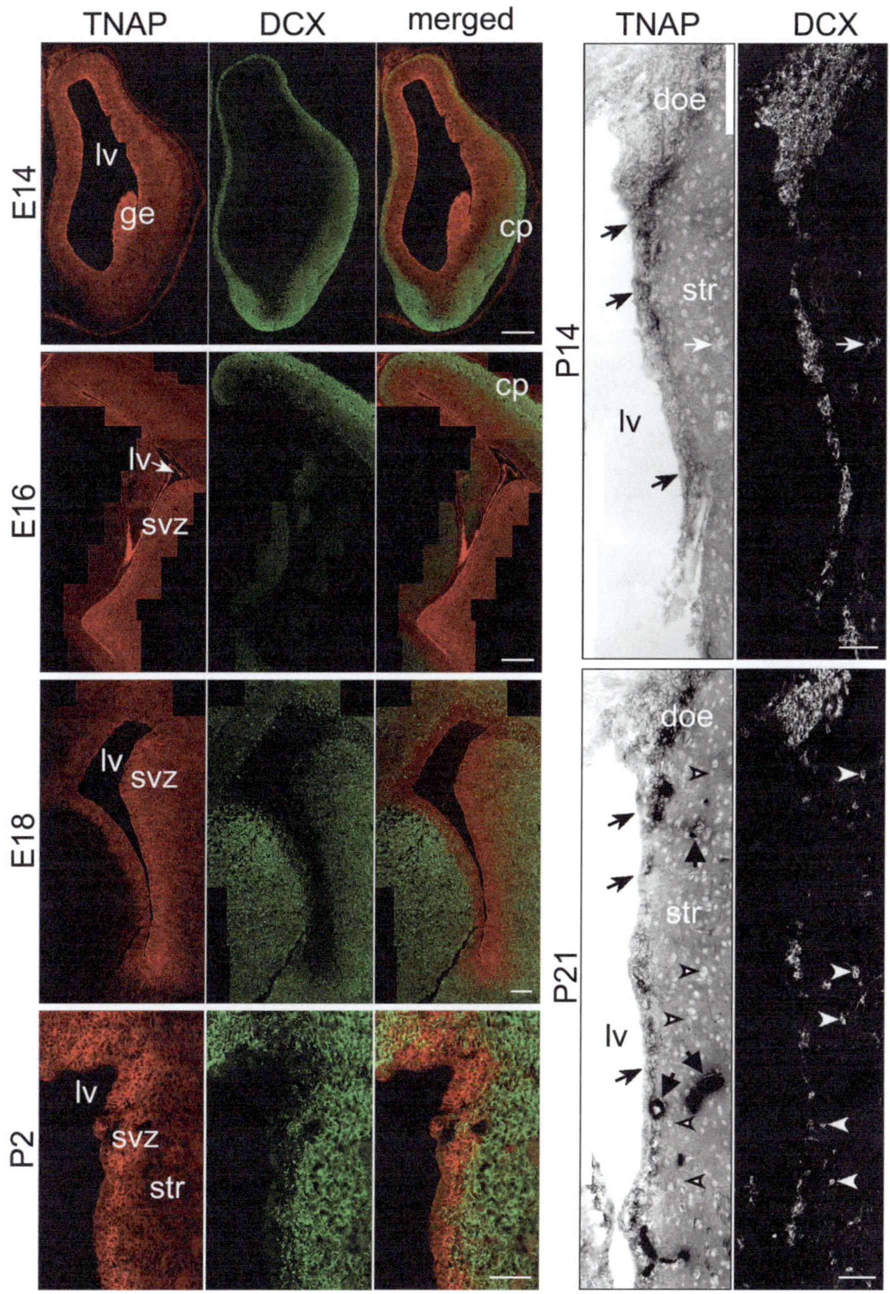
TNAP
DCX
merged
TNAP
DCX
E14
lv
ge
cp
E16
cp
lv
svz
E18
lv
svz
P2
lv
svz
str
P14
doe
str
lv
P21
doe
str
lv

◀ **Fig. 4.1** Distribution of TNAP catalytic activity and immunostaining for doublecortin (DCX) in the embryonic and postnatal mouse brain at the level of the lateral ventricles (coronal sections). Stages E14 to P2 are presented in *false colors* (TNAP, *red*; DCX, *green*), stages P14 and P21 in *black* and *white*. E14, staining of the cerebral vesicle. E16 and E18 (compound pictures), intense staining at the side facing the developing striatum but also along the septal and dorsal ventricular walls. Thin radially oriented cellular elements are stained. P2, dorsal aspect of the lateral ventricle wall. Intense staining is restricted to the SVZ. P14, TNAP staining (dark staining) is reduced to a small region at the SVZ and becomes patchy (*small black arrows*). P21, The extent of ventricular TNAP staining is considerably reduced (in parallel to a reduction in thickness of the SVZ) and is associated with patches of cells (*small black arrows*). At that stage TNAP staining of blood vessels becomes prominent (*thick black arrows*). Immunostaining for DCX is initially (E14, E16) most prominent toward the pial surface, including the cortical plate. At E18 DCX is more homogeneously distributed but absent from the SVZ. At P2 DCX is absent from the ventricular surface but associated with cells of the SVZ and the adjacent striatum. At P14 and P21 DCX (*white*) has largely disappeared from the striatum and is now selectively associated with cell clusters of the SVZ. *White arrows* and *arrowheads* at P14 and P21 depict DCX-positive cells in the striatum that are no longer immunopositive for TNAP. *cp* Cortical plate, *doe* dorsal extension of the SVZ, *ge* ganglionic eminence, *lv* lateral ventricle, *str* striatum, *svz* subventricular zone. *Scale bars* E14, 250 μm; E16, 200 μm; E18, 100 μm; P2–P21, 50 μm (Methods as described by Langer et al. 2007)

of the bare axons prior to myelination and restricted to nodes of Ranvier in the mature brain (Fonta et al. 2005). Furthermore it revealed that the development of TNAP activity coincides with the time course of high-rate synaptogenesis. Whereas profiles of immature synapses do not express the enzyme, TNAP activity becomes associated with synaptic contacts suggesting that TNAP may contribute to the function of mature synapses.

4.5 TNAP in Adult Neurogenesis

In the adult mouse brain elevated TNAP activity is associated with cells of the SVZ and the RMS but it is absent from the hippocampal neurogenic niche (Langer et al. 2007). The SVZ houses heterogeneous populations of precursor cells. At least three closely adjacent cell types situated within or below the ependyma are distinguished morphologically and functionally. B cells represent astrocyte-like stem cells derived from the embryonic radial glia. They give rise to the transit-amplifying C cells, which, in turn, generate (DCX-positive) migrating neuroblasts (A cells). Chains of A cells migrate towards the olfactory bulb. At least two types of B cells have been distinguished (Chojnacki et al. 2009). One population (B1 cells), assumed to represent the actual neural stem cells, is interspersed between ependymal cells, contacts the ventricle lumen with a ciliated protrusion and approaches blood vessels with its basal process. Other B cells located in the subependyma mainly wrap around the chains of migrating neuroblasts, forming the gliotubes including those within the rostral migratory stream (Zhao et al. 2008; Ming and Song 2011).

In the adult SVZ there is selective and enhanced TNAP staining of subependymal cells along the lateral walls of the lateral ventricles and within the entire

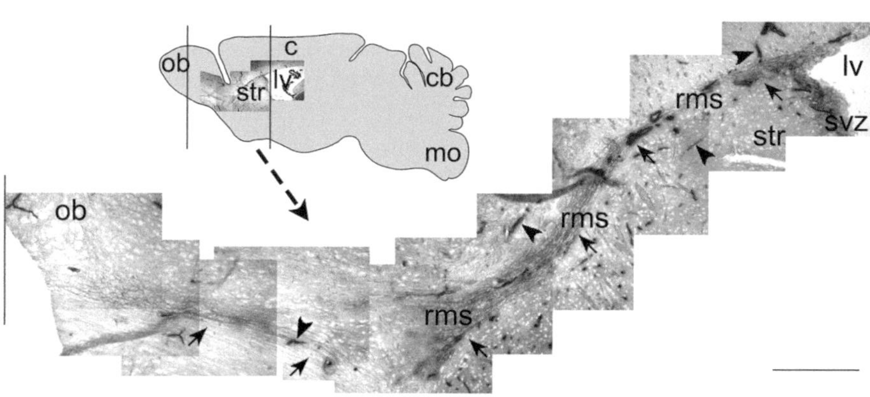

Fig. 4.2 Distribution of TNAP activity in the adult mouse subventricular zone (SVZ) and rostral migratory stream (RMS) as revealed by enzyme histochemistry. Parasagittal section revealing dark staining of the SVZ and the RMS (*arrows*) and of blood vessels (*arrow heads*). Staining in the RMS reveals a longitudinal arrangement of the stained cells. The RMS can be traced into the olfactory bulb (ob) where enzyme histochemical staining fades. The cartoon above illustrates the position of SVZ and RMS in an adult mouse brain. *c* Cerebral cortex, *cb* cerebellum, *lv* lateral ventricle, *mo* medulla oblongata, *str* striatum. *Scale bar* 250 μm

RMS. Staining ceases when the RMS has entered the olfactory bulb (Fig. 4.2). Thus, TNAP becomes downregulated as the progenitors arrive in the olfactory bulb and differentiate into interneurons (Langer et al. 2007). The ependyma is free of TNAP activity. TNAP activity at the plasma membrane of subependymal cells of the adult rat lateral ventricle could be detected also by electron microscopy (Mori and Nagano 1985). No activity was associated with ependymal cells but occasional alkaline phosphatase–positive processes were observed protruding between ependymal cells that most likely correspond to processes of B1 cells (Doetsch et al. 1999). Clusters of subependymal cells immunopositive for DCX (representing A cells) reveal TNAP catalytic activity and are immunopositive for TNAP (Langer et al. 2007). Similarly immunostaining for βIII tubulin, a marker for early neuronal differentiation, colocalizes with TNAP. Moreover, there is partial co-localization between TNAP and the nucleus-located transcription factor oligodendrocyte transcription factor-2 (Olig2) (Hack et al. 2005; Ligon et al. 2006; Menn et al. 2006), suggesting that TNAP is also expressed by C cells. The astrocyte-like B cells that form the glial tubes around clusters of migrating neuroblasts (A cells) are equally immunopositive for TNAP. This is exemplified by co-localization of TNAP with both nucleoside triphosphate diphosphohydrolase 2 (NTPDase2), an ectonucleotidase strongly expressed by B cells (Braun et al. 2003) and the astrocyte-specific glial glutamate transporter (GLAST) (Langer et al. 2007). The same results are obtained when cells from the SVZ are isolated and immunostained following adhesion to coverslips. This and additional evidence suggests that all progenitor cell types of the adult SVZ express TNAP.

While TNAP activity in the mouse brain (besides blood vessels and the choroid plexus) is by far highest in the SVZ neurogenic niche, enhanced background staining for alkaline phosphatase can be observed in the parenchyma of several brain regions, such as parts of the cortex, the septum or the external plexiform layer of the OB (Langer et al. 2007). This staining could not be assigned to individual cells. This observation is in accord with other studies that observed TNAP activity in various brain regions of several mammalian species (Brun-Heath et al. 2011). Using electron microscopy TNAP has been allocated to myelin-free axonal membranes, including the Node of Ranvier and synaptic contacts (Mori and Nagano 1985; Fonta et al. 2004, 2005). Similarly, TNAP appears to be absent from the axon cylinder of peripheral nerves but it remains associated with parts of the myelin sheath (nodes of Ranvier, incisures of Schmidt-Lanterman) (Pinner et al. 1964; Scherer 1996). Accordingly, ablation of TNAP compromises myelination and synaptogenesis in mouse brain (Hanics et al. 2012). These data suggest that TNAP continues to be expressed by select neuronal populations in the adult even though catalytic activity is strongly reduced as compared to the developmental stages of the nervous system.

4.6 TNAP is Essential for Progenitor Cell Proliferation and Differentiation in Vitro

Enzyme histochemical and immunohistochemical experiments had shown that TNAP is associated with all types of progenitor cells in the adult SVZ of the mouse (Langer et al. 2007). TNAP may thus potentially impact cell proliferation or also cell differentiation within the neurogenic niche. To gain insight into the functional role of TNAP a knockdown protocol using RNA interference with shRNA and retroviral infection was applied to cultured TNAP-expressing multipotent neural stem cells isolated from the adult mouse SVZ (Kermer et al. 2010). The experiments revealed that knockdown of TNAP resulted in a strong decrease in progenitor cell proliferation. When the cellular differentiation of the progenitors was induced by switching to differentiation medium young neurons and oligodendrocytes as identified by immunostaining for DCX and the oligodendrocyte marker O4, respectively were generated. TNAP shRNA strongly reduced the formation of both neurons and oligodendrocytes. The effects of shRNA on either proliferation or differentiation were completely abrogated when soluble alkaline phosphatase was simultaneously added to the culture medium. These in vitro studies imply that TNAP is essential for principal functions of neural stem cells such as proliferation and differentiation. They show that the mode of action of TNAP must be cell surface-located as the addition of native alkaline phosphatase protein (but not of heat inactivated alkaline phosphatase) abolished the effect of TNAP RNA interference.

4.7 TNAP Controls Axon Development in Vitro

In the embryonic stages between 8.5 and 13.5 TNAP was found to be expressed in nerve roots in the pons, medulla oblongata, cranial nerves and spinal nerves. Furthermore immunostaining for neurofilament protein was associated with the emerging nerves (Narisawa et al. 1994; Brun-Heath et al. 2011). In TNAP knockout mice nerve roots from the lumbar spinal cord were thinner than those from heterozygous control mice, presumably a consequence of the lack of TNAP in the developing neural tube between stages E8.5 and E13.5 (Narisawa et al. 1997, 2001). This suggests that TNAP also impacts axonal development.

A recent study shows that TNAP promotes axonal growth of hippocampal neurons (Díez-Zaera et al. 2011). Hippocampal neurons cultured to grow extensive microtubule-associated protein 2 (MAP2)-positive neurites and an axon identified with anti-Tau-1 immunostaining were found to stain for TNAP catalytic activity (Fig. 4.3a). Immunostaining for TNAP revealed that the enzyme is prominent on the growth cones. TNAP reveals a dotted distribution and is colocalized with the $P2X_7$ receptor, an ATP-gated, nonselective ion channel permeable to Na^+, K^+ and Ca^{2+} (Khakh and North 2006) (Fig. 4.3b, c). The experiments further showed that TNAP effectively reduces the extracellular ATP levels in the culture supernatant. TNAP inhibition with levamisole or TNAP shRNA knockdown (Fig. 4.3d, e) induced shortening of the principal axon and its ramifications. Since the authors had previously shown that extracellular ATP reduces axon growth and branching in cultured hippocampal neurons via the $P2X_7$ receptor (Díaz-Hernández et al. 2008) and inhibits neuritogenesis in Neuro 2a neuroblastoma cells (Gomez-Villafuertes et al. 2009) they analyzed the interrelation of ATP, the $P2X_7$ receptor and TNAP in the cultured hippocampal neurons. Overexpression of the $P2X_7$ receptor impaired axon growth whereas knockdown of $P2X_7$ increased axonal length (Fig. 4.3f, g). When TNAP was added to $P2X_7$-overexpressing neurons axon length was increased (Fig. 4.3h). Parallel neuronal knockdown of $P2X_7$ receptors and TNAP permitted normal neuronal axon development and reversed the axon growth impairment induced by TNAP shRNA (Fig. 4.3i). While alkaline phosphatase hydrolyzes the receptor agonist ATP to adenosine the authors excluded a role of adenosine in the control of axon growth. To the contrary, application of inhibitors of $P2X_7$ receptors induced axon growth. In a previous study with cultured hippocampal neurons the authors had shown that ATP induces $P2X_7$-dependent Ca^{2+} transients in the distal domain of the axon. These Ca^{2+} transients could mediate local actin polymerization and repress axon growth. Regarding the underlying mechanisms, $P2X_7$-receptor inhibition was found to decrease Ca^{2+}-calmodulin dependent protein kinase II phosphorylation whereas it increased the activity of focal adhesion kinase, the activation of phosphatidylinositide 3-kinase and the modification of its downstream targets, Akt/protein kinase B and glycogen synthase kinase-3. This could lead to actin depolymerization and permit axon growth, as well as the formation of multiple axon terminals (Díaz-Hernández et al. 2008) (see Chap. 18).

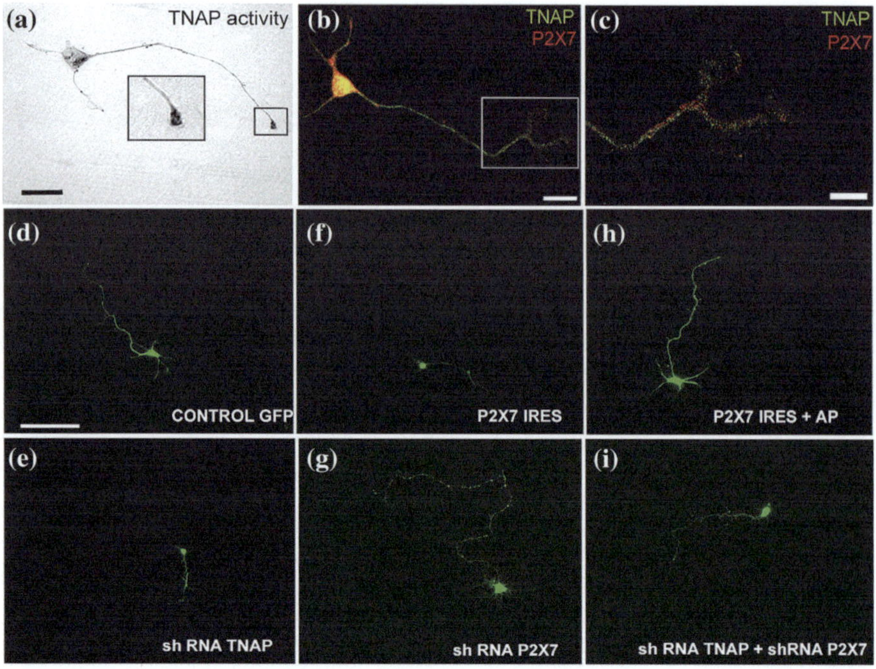

Fig. 4.3 Enzyme histochemical staining and immunostaining of cultured hippocampal neurons for TNAP and the $P2X_7$ receptor and effect of shRNA knockdown of TNAP and the $P2X_7$ receptor on axonal growth. **a** TNAP catalytic activity in a neuron as revealed by enzyme histochemistry. Note the intense staining of the axonal growth cone (2× magnification of the *boxed* region). **b** Double immunofluorescence of a hippocampal neuron fixed after 3 days in vitro and stained with antibodies against TNAP (*green*) and $P2X_7$ (*red*). **c** Blowup of box in (**b**) of axon terminal region revealing dotted immunofluorescence and colocalization of TNAP and $P2X_7$ immunoreactivity. **d**–**i** Modulation of axon growth by expression of TNAP and $P2X_7$ and respective knockdowns. Hippocampal neurons (*green*) were transfected at 1 DIV with GFP alone or with the $P2X_7$-IRES-GFP plasmid, which independently encodes the $P2X_7$ receptor and GFP and fixed after 3 DIV. **d** GFP-transfected control neuron. **e** shRNA knockdown of TNAP impairs axon growth. **f** $P2X_7$ IRES-transfected neuron. **g** shRNA knockdown of $P2X_7$ IRES-transfected cell stimulates axon growth. **h** Addition of alkaline phosphatase (AP) to $P2X_7$ IRES-transfected cells stimulates axon growth. **i** Double knockdown of TNAP and $P2X_7$ permits normal neuronal axon development and reverses the axon growth impairment induced by TNAP shRNA. *Scale bars* **a**, **b** 50 μm, **c** 20 μm, **d**–**i** 50 μm. **a**–**c** is from Díez-Zaera et al. 2011, Copyright (2011) American Society for Cell Biology (ASCB), USA. **d**–**i** has kindly been provided by Miguel Díaz-Hernández, Madrid

These and other experiments reported by Díez-Zaera et al. (2011) show that endogenous TNAP expressed by outgrowing axons reduces extracellular ATP levels and thus controls ligand availability at ATP receptors. By eliminating extracellular ATP, TNAP impairs the activation of the growth inhibiting axonal $P2X_7$ receptors. The authors suggest that this tandem mechanism of TNAP and the $P2X_7$ receptor could be critically involved in the control of axon growth during development and possibly also during axon regeneration in the adult brain.

4.8 Regulation of TNAP Expression

TNAP expression is obviously regulated during development of the nervous system. It can be transiently associated with cells but it can also be constitutively expressed as for example by specific cell types of the adult nervous system. This raises the question of the mechanisms of transcriptional or posttranscriptional regulation of TNAP. The situation is complicated by the fact that expression of TNAP can alternatively be driven by two different promoters.

Of the 13 exons of the TNAP gene the first two exons (Ia, Ib) are noncoding. Each exon has its own promoter sequence and thus the presence of exons Ia and Ib in the mRNAs is mutually exclusive. The mRNAs have different 5′-untranslated sequences but the resulting protein sequences are identical (Kishi et al. 1989; Toh et al. 1989a, b). In humans and rats the upstream promoter Ia is preferentially utilized by osteoblasts (bone type) and the downstream promoter Ib in kidney/liver (liver type) (Zernik et al. 1991). A recent analysis of TNAP transcripts of human, marmoset monkey, rats and mice provides interesting new information regarding the control of TNAP expression in nervous tissue (Brun-Heath et al. 2011). Bone TNAP was expressed in the adult brain of all species investigated. However mouse (but not rat) brain contained in addition the liver TNAP transcript. This dual expression of bone and liver TNAP isoforms was also observed in the E15 mouse embryonic brain and spinal cord. As in the adult, the bone TNAP transcript appeared to be preferentially expressed. Freshly isolated endothelial cells from brain vessels contained the bone TNAP transcript. Taken together the data obtained by Brun-Heath et al. (2011) show that in the brains of various mammalian species including humans TNAP expression in cells of neuronal and endothelial origin is driven by the bone promoter. In neuronal cells of mouse brain TNAP expression can in addition be driven by the liver TNAP promoter.

Very little information is available regarding the control of TNAP expression in neural cells. However, the cell-specific regulation of TNAP gene expression has been a matter of considerable interest with an emphasis on osteogenetic pathways—that presumably is similar to that in the mammalian brain. The promoters of the human and murine genes have been analyzed and 5′-upstream elements responding to a considerable variety of transcription factors have been identified and the puzzle of regulation of AP expression is being pieced together (ref in Millán 2006a). Only few examples are selected here. Interestingly, a variety of factors that stimulate TNAP expression or catalytic activity simultaneously have an important impact on neural development.

Bone TNAP is typically expressed during osteoblast differentiation. Analysis of the basal promoter of human TNAP involving transfections of a series of 5′ deletion mutants indicate that promoter activity is dependent on multiple motifs including several putative Spl binding sites and a TATA box (Kiledjian and Kadesch 1990). All-*trans*-retinoic acid, an important modulator of neural development, promotes development of neural progenitors from mouse embryonic stem cells (Engberg et al. 2010). It is has been identified in a considerable variety of cellular systems to enhance TNAP activity (Gianni et al. 1991). It stimulates cellular differentiation via retinoic acid receptor (RAR)–retinoid X receptor (RXR) heterodimers, which bind to the specific *cis*-element retinoic acid response element (RARE). In the murine TNAP gene (*Akp2*) regulation by retinoic acid is via a RARE in the upstream promoter region of exon Ia (Escalante-Alcalde et al. 1996. Also in the human TNAP gene a RARE was identified in this promoter region. Supershift assays revealed that RARs and a RXR bound to the RARE, suggesting that the complex enhances TNAP expression (Orimo and Shimada 2005). Additional investigations suggest that independent of gene transcription retinoic acid can enhance the cytoplasmic stability of nascent TNAP mRNA chains, thus enhancing enzyme activity (Zhou et al. 1994; Müller et al. 2000). Retinoic acid also affects the degree of TNAP N-glycosylation (Müller et al. 2000). Similarly, experiments investigating the upregulation of human TNAP expression resulting from co-cultivation of endothelial cells and human primary osteoblasts revealed that co-cultivation elevated the half-life of osteoblastic TNAP mRNA without affecting its promoter activity. Rather it was posttranscriptionally regulated via p38 mitogen–activated protein kinase-dependent mRNA turnover (Hager et al. 2009).

Furthermore vitamin D that exhibits a trophic role in differentiation and maturation of neurons (Taniura et al. 2006), stimulates expression of TNAP in bone cells, presumably by binding to vitamin-D-regulating elements (VDRE) (Kyeyune-Nyombi et al. 1991; Johnson-Pais and Leach 1996). RARE and VDRE were identified in the brain TNAP promoter of the marmoset (Brun-Heath et al. 2011). Also phosphate can induce TNAP expression (Orimo and Shimada 2008). However, luciferase assays failed to detect any transcriptional activation of the promoter region of the human TNAP gene by β-glycerophosphate or NaH_2PO_4, suggesting that the effects of these phosphates may be indirect. Transcription repressors frequently interact with histone deacetylases (HDACs) to down-regulate target gene expression. Suppression of HDAC activity with the HDAC inhibitor sodium butyrate (Lee et al. 2006) or MS-275 (Kim et al. 2011) induced osteoblasts marker genes including TNAP.

Bone morphogenetic proteins (BMPs), that are members of the transforming growth factor-β superfamily, regulate the differentiation of osteoprogenitor cells. They also represent a group of powerful morphogens that play an important role in the development of the nervous system (Bond et al. 2012). Amongst others, BMP signaling is critical for progenitor fate specification in both neurogenesis and later in astrogliogenesis. During neural fate specification, BMP signaling initially promotes neurogenesis by pushing progenitors toward a neuronal fate and suppressing the oligodendroglial fate. Towards the end of neurogenesis BMP signaling switches from a pro-neuronal to proastroglial cue. Adult SVZ cells express BMB and BMP signaling blocks the production of neurons by directing glial differentiation (Lim et al. 2000). BMP-6 and BMP-7 induce alkaline phosphatase activity through SMAD1 and SMAD2 in murine cell lines (Aoki et al. 2001). BMP-2 and the glucocorticoid dexamethasone can regulate TNAP transcription by regulating osteogenic transcription factors, such as Runx2, Osterix (OSX), and the homeobox genes Dlx5 and Msx2, via Smad- or GR-dependent signaling pathways (Kim et al.

2004; Ito et al. 2007). Treatment with both BMP-2 and dexamethasone was found to synergistically increase TNAP levels in C3H10T1/2 cell cultures independent of the effects of Runx2, OSX, Dlx5, or Msx2, and to act, at least in part, through a STAT3-dependent mechanism (Mikami et al. 2010). It remains to be investigated whether these mechanisms of transcriptional regulation also apply to neural cells.

The liver type promoter that is also relevant for nerve cells in the mouse may be activated by dibutyryl cAMP (Studer et al. 1991) or dexamethasone (Zernik et al. 1991). cAMP response element-binding protein (CREB) is involved in multiple aspects of neuronal development and plasticity. It also regulates specific phases of adult neurogenesis in the subventricular zone/olfactory bulb (Giachino et al. 2005). A detailed analysis of the regulation of TNAP expression in neural progenitor and neural cells merits further investigation.

4.9 Functional Role of TNAP in the Developing Brain

One of the major questions concerns the mechanism(s) of action of TNAP. Alkaline phosphatases display several functional properties. On the one hand they act as enzymes and catalyze the hydrolysis of monoesters of phosphoric acid, releasing phosphate (Millán 2006b). Under conditions of TNAP deficiency three phospho-compounds accumulate endogenously: phosphoethanolamine, PP_i, and pyridoxal 5′-phosphate, indicating that these are natural substrates for TNAP (Whyte 1996; Fedde et al. 1999). TNAP converts extracellular pyridoxal 5′-phosphate into pyridoxal that is taken up into cells and rephosphorylated by intracellular kinases. Pyridoxal 5′-phosphate is a cofactor of several enzymes including glutamate decarboxylase, the final catalytic step in the synthesis of the inhibitory neurotransmitter γ-aminobutyric acid (GABA). Mice lacking TNAP are characterized by elevated serum pyridoxal 5′-phosphate levels and reduced levels in the inhibitory neurotransmitter GABA. They develop fatal seizures at approximately 2 weeks after birth (Waymire et al. 1995). Importantly, the mutant seizure phenotype can be rescued by the administration of pyridoxal. It also rescues the development of lumbar spinal nerve roots (Narisawa et al. 2001) that were found to be thinner in the TNAP knockout mice (Narisawa et al. 1997). For this reason TNAP may not only have local but also systemic effects on cellular metabolism. Interestingly, the neurotransmitter GABA plays critical roles on the different steps of adult neurogenesis. GABA decreases the proliferation of neural stem cells and neuroblasts while also reducing neuroblast migration. GABA thus exerts a negative influence on neuroblast production (Platel et al. 2007). Pyridoxal 5′-phosphate is also a cofactor of amino acid decarboxylase, an enzyme relevant for the synthesis of biogenic amines. TNAP inhibition was found to promote a decrease in the levels of the neurotransmitter serotonin (5-hydroxytryptamine) or dopamine in a neuronal cell model (Ermonval et al. 2009). Serotonin is a regulator of morphogenetic activities during early brain development and it increases neurogenesis, including cell proliferation, migration, differentiation, and synaptogenesis (Schmitt et al.

2007). Similarly, dopamine stimulates endogenous adult neurogenesis in the SVZ by activating D2-like receptors on transit-amplifying progenitor cells (Borta and Höglinger 2007; Young et al. 2011). High expression of TNAP in the neurogenic regions may thus be relevant for maintaining activity of these transmitter pathways.

Physiologically relevant substrates of TNAP include phosphoproteins of the cell surface. For example TNAP can dephosphorylate the extracellular matrix protein laminin (Ermonval et al. 2009) or the hyperphosphorylated microtubule-associated tau protein once it is released upon neuronal death (Díaz-Hernández et al. 2010) (see Chap. 18). To date this catalytic activity of TNAP has not yet been investigated in relation to neural development.

TNAP may also exert its function by interacting with other proteins of the plasma membrane as has been suggested for the interaction with the equally GPI-anchored cellular prion protein. This interaction takes place in lipid rafts of cultured neuroepithelial cell lines (Ermonval et al. 2009) (see Chap. 9). Furthermore, collagen type I, II, and X have been identified as binding partners of TNAP (Vittur et al. 1984; Wu et al. 1991, 1992). A putative protein-protein interaction domain has been identified that is in part responsible for the binding of TNAP to collagen (Bossi et al. 1993) and located within a top flexible extracellular loop of the enzyme, the crown domain (Le Du et al. 2001). This domain is located at the very tip of the crown and highly accessible. Thus, the interaction of progenitor cells with the extracellular matrix may be regulated by TNAP.

4.10 A Role in Purinergic Signaling?

Alkaline phosphatases are the only ectonucleotidases that can sequentially dephosphorylate nucleoside triphosphates to the respective nucleoside. Depending on the nucleotide involved extracellular nucleotides such as ATP, ADP, UTP and UDP act via ionotropic P2X or metabotropic P2Y receptors. Adenosine activates P1 receptors. Many cellular systems express multiple subtypes of these receptors (Abbracchio et al. 2009). For example, TNAP hydrolyzes extracellular ATP via ADP and AMP to adenosine. It thus has the potential of inactivating agonists of P2 receptors and to produce the P1 receptor agonist adenosine and to serve purine salvage through cellular reuptake of the nucleoside (Zimmermann et al. 2012). To date perhaps the best characterized example for an ectonucleotidase function of TNAP in the development of the nervous system concerns the role of TNAP in reducing extracellular ATP (nucleotide) concentrations at the surface of cultured hippocampal neurons (Díez-Zaera et al. 2011).

There are multiple examples that nucleotides increase neural precursor cell proliferation, migration and also differentiation by nucleotide receptor-specific mechanisms (Zimmermann 2011). ATP was found to be released from cultured SVZ-derived neural progenitors (Lin et al. 2007) and from B1 cells in acute brain slices (Lacar et al. 2012). Several studies provided evidence that ATP or UTP stimulate proliferation or also migration of embryonic or adult SVZ-derived neural progenitor cells in vitro (Scemes et al. 2003; Tran et al. 2004; Mishra et al. 2006; Lin et al. 2007, Striedinger et al. 2007; Grimm et al. 2010). Work on acute rat and mouse embryonic brain slices further revealed that neural precursors in situ can release ATP via hemichannels. Released ATP induces P2 receptor-mediated Ca^{2+} fluctuations in ventricular zone precursors and facilitates progenitor cell proliferation and migration within the developing cortex (Weissman et al. 2004; Liu et al. 2008, 2010). Additional studies on organotypic cultures of the SVZ (Khodosevich et al. 2012) or of the adult SVZ in situ (Suyama et al. 2012) provide evidence that ATP promotes progenitor cell proliferation also in the adult. Thus, hydrolysis of released nucleotides would be an important regulatory step in the control of both embryonic and adult neurogenesis.

Alkaline phosphatases have a pH optimum in the high alkaline range (>9). But obviously their catalytic activity is sufficient to constitutively dephosphorylate extracellular nucleotides also at physiological pH (Ohkubo et al. 2000; Díez-Zaera et al. 2011; Street et al. 2013). It is of interest that in the adult murine subventricular zone and in the rostral migratory stream another ATP- and ADP-hydrolyzing enzyme is expressed, NTPDase2 (Braun et al. 2003). In contrast to TNAP the expression of this enzyme is restricted to the neural stem cell population, the B cells. At physiological pH TNAP catalytic activity for the hydrolysis of ATP and ADP in tissue sections of the adult SVZ is negligible as compared to that of NTPDase catalytic activity (Langer et al. 2008). It has been suggested that NTPDase2 is the predominant enzyme controlling the availability of purinergic ligands of P2 receptors in the neurogenic niche, thereby acting as a homeostatic regulator of nucleotide-mediated neural progenitor cell proliferation and expansion in the adult SVZ (Gampe et al. 2015). Yet, as it cannot hydrolyze AMP, NTPDase2 requires TNAP to produce the P1 receptor agonist adenosine. Moreover, whereas TNAP is expressed from the onset of embryonic neurogenesis, NTPDase2 is first expressed in the murine VZ and SVZ during late embryonic stages (Braun et al. 2003). Thus TNAP represents the predominant nucleotide-hydrolyzing enzyme of the VZ and SVZ during earlier development. Both enzymes may impact neurogenesis via hydrolysis of nucleotides.

The notion of a functional role of TNAP in purinergic signaling in the developing brain is further supported by the parallel expression of nucleotide receptors (Zimmermann 2006). This is best exemplified by the expression pattern of the P2X5 receptor. This receptor appears as early as E8 specifically in the neural tube and at E9 in the spinal cord and dorsal root ganglia and then becomes gradually confined to new-born neurons in the cortical plate and ventral horn (Guo et al. 2013). Also the P2X3 receptor is already expressed in the neural tube (Cheung and Burnstock 2002) and in the developing peripheral ganglia (Boldogköi et al. 2002; Ruan et al. 2004). P2X5 and P2X3 receptors may form heteromeric multimers with other P2X receptors forming ATP-gated ion channels permeable to Na^+, K^+ and Ca^{2+} (North 2002). Similar to TNAP these receptors are transiently expressed during the early stages of neural development. Also a variety of P2Y receptors are transiently expressed in developing brain structures beginning with the ADP- and

ATP-sensitive P2Y1 receptor in the floor plate of the neural tube (Cheung et al. 2003). Immunocytochemical double-labeling would help to further define the co-localization of TNAP and specific P2 receptors.

4.11 Synopsis

TNAP is highly and transiently expressed in the neurogenic regions of the developing brain and spinal cord as well as at outgrowing axons. During later development TNAP expression ceases but high activity is maintained in the neurogenic niche at the lateral ventricles that, in the adult, provides new neurons for the olfactory bulb (adult neurogenesis). There TNAP is associated with all neural progenitor cell types. Low activity of TNAP remains associated with nerve cells in specific regions of the adult mammalian brain. Little is known concerning the regulation of expression and physiological function of TNAP in nervous tissue. The strong and transient expression of the enzyme during brain development and in adult neurogenesis implies, however a significant role in the control of neural development. Specifying the functional role of TNAP at the various stages of neural development therefore remains an important challenge.

Acknowledgements We would like to thank Miguel Díaz-Hernández, Madrid, for providing Fig. 4.3.

Disclosure of Potential Conflict of Interest The authors indicate no potential conflicts of interest.

References

Abbracchio MP, Burnstock G, Verkhratsky A, Zimmermann H (2009) Purinergic signalling in the nervous system: an overview. Trends Neurosci 32:19–29

Altman J, Bayer SA (1990) Horizontal compartimentation in the germinal matrices and intermediate zone of the embryonic rat cerebral cortex. Eptl Neurol 107:36–47

Alvarez-Buylla A, García-Verdugo JM (2002) Neurogenesis in the adult ventricular zone. J Neurosci 22:629–634

Aoki H, Fujii M, Imamura T, Yagi K, Takehara K, Kato M, Miyazono K (2001) Synergistic effects of different bone morphogenetic protein type I receptors on alkaline phosphatase induction. J Cell Sci 114:1483–1489

Attardo A, Calegari F, Haubensak W, Wilsch-Brauninger M, Huttner WB (2008) Live imaging at the onset of cortical neurogenesis reveals differential appearance of the neuronal phenotype in apical versus basal progenitor progeny. PLoS One 3:e2388

Boldogköi Z, Schutz B, Sallach J, Zimmer A (2002) P2X3 receptor expression at early stage of mouse embryogenesis. Mech Dev 118:255–260

Bond AM, Bhalala OG, Kessler JA (2012) The dynamic role of morphogenetic proteins in neural stem cell fate and maturation. Dev Neurobiol 72:1068–1084

Borta A, Höglinger GU (2007) Dopamine and adult neurogenesis. J Neurochem 100:587–595

Bossi M, Hoylaerts MF, Millán JL (1993) Modifications in a flexible surface loop modulate the isozyme-specific properties of mammalian alkaline phosphatases. J Biol Chem 268:25409–25416

Braun N, Sévigny J, Mishra S, Robson SC, Barth SW, Gerstberger R, Hammer K, Zimmermann H (2003) Expression of the ecto-ATPase NTPDase2 in the germinal zones of the developing and adult rat brain. Eur J Neurosci 17:1355–1364

Brazel CY, Romanko MJ, Rothstein RP, Levison SW (2003) Roles of the mammalian subventricular zone in barin development. Prog Neurobiol 69:49–69

Brun-Heath I, Ermonval M, Chabrol E, Xiao J, Palkovits M, Lyck R, Miller F, Couraud PO, Mornet E, Fonta C (2011) Differential expression of the bone and the liver tissue non-specific alkaline phosphatase isoforms in brain tissues. Cell Tissue Res 343:521–536

Cheung KK, Burnstock G (2002) Localization of P2X3 receptors and coexpression with P2X2 receptors during rat embryonic neurogenesis. J Comp Neurol 443:368–382

Cheung KK, Ryten M, Burnstock G (2003) Abundant and dynamic expression of G protein-coupled P2Y receptors in mammalian development. Dev Dyn 228:254–266

Chiquoine AD (1954) Distribution of alkaline phosphomonoesterase in the central nervous system of the mouse embryo. J Comp Neurol 100:415–439

Chojnacki AK, Mak GK, Weiss S (2009) Identity crisis for adult periventricular neural stem cells: subventricular zone astrocytes, ependymal cells or both? Nat Rev Neurosci 10:153–163

Díaz-Hernández M, del Puerto A, Díaz-Hernández JI, Díez-Zaera M, Lucas JJ, Garrido JJ, Miras-Portugal MT (2008) Inhibition of the ATP-gated $P2X_7$ receptor promotes axonal growth and branching in cultured hippocampal neurons. J Cell Sci 121:3717–3728

Díaz-Hernández M, Gomez-Ramos A, Rubio A, Gomez-Villafuertes R, Naranjo JR, Miras-Portugal MT, Avila J (2010) Tissue-nonspecific alkaline phosphatase promotes the neurotoxicity effect of extracellular tau. J Biol Chem 285:32539–32548

Díez-Zaera M, Díaz-Hernández JI, Hernández-Alvarez E, Zimmermann H, Díaz-Hernández M, Miras-Portugal MT (2011) Tissue-nonspecific alkaline phosphatase promotes axonal growth of hippocampal neurons. Mol Biol Cell 22:1014–1024

Doetsch F, Caillé I, Lim DA, García-Verdugo JM, Alvarez-Buylla A (1999) Subventricular zone astrocytes are neural stem cells in the adult mammalian brain. Cell 97:703–716

Engberg N, Kahn M, Petersen DR, Hansson M, Serup P (2010) Retinoic acid synthesis promotes development of neural progenitors from mouse embryonic stem cells by suppressing endogenous, Wnt-dependent nodal signaling. Stem Cells 28:1498–1509

Ermonval M, Baudry A, Baychelier F, Pradines E, Pietri M, Oda K, Schneider B, Mouillet-Richard S, Launay JM, Kellermann O (2009) The cellular prion protein interacts with the tissue non-specific alkaline phosphatase in membrane microdomains of bioaminergic neuronal cells. PLoS One 4:e6497

Escalante-Alcalde D, Recillas-Targa F, Hernández-Garcia D, Castro-Obregon S, Terao M, Garattini E, Covarrubias L (1996) Retinoic acid and methylation cis-regulatory elements control the mouse tissue non-specific alkaline phosphatase gene expression. Mech Dev 57:21–32

Fedde KN, Blair L, Silverstein J, Coburn SP, Ryan LM, Weinstein RS, Waymire K, Narisawa S, Millán JL, MacGregor GR, Whyte MP (1999) Alkaline phosphatase knock-out mice recapitulate the metabolic and skeletal defects of infantile hypophosphatasia. J Bone Miner Res 14:2015–2026

Fonta C, Negyessy L, Renaud L, Barone P (2004) Areal and subcellular localization of the ubiquitous alkaline phosphatase in the primate cerebral cortex: Evidence for a role in neurotransmission. Cereb Cortex 14:595–609

Fonta C, Negyessy L, Renaud L, Barone P (2005) Postnatal development of alkaline phosphatase activity correlates with the maturation of neurotransmission in the cerebral cortex. J Comp Neurol 486:179–196

Gal JS, Morozov YM, Ayoub AE, Chatterjee M, Rakic P, Haydar TF (2006) Molecular and morphological heterogeneity of neural precursors in the mouse neocortical proliferative zones. J Neurosci 26:1045–1056

Gampe K, Stefani J, Hammer K, Brendel P, Pötzsch A, Enikolopov G, Enjyoji K, Acker-Palmer A, Robson SC, Zimmermann H (2015) NTPDase2 and purinergic signaling control progenitor cell proliferation in neurogenic niches of the adult mouse brain. Stem Cells 33:253–264

Giachino C, De Marchis S, Giampietro C, Parlato R, Perroteau I, Schütz G, Fasolo A, Peretto P (2005) cAMP response element-binding protein regulates differentiation and survival of newborn neurons in the olfactory bulb. J Neurosci 25:10105–10118

Gianni M, Studer M, Carpani G, Terao M, Garattini E (1991) Retinoic acid induces liver/bone/kidney-type alkaline phosphatase gene expression in F9 teratocarcinoma cells. Biochem J 274:673–678

Gomez-Villafuertes R, del Puerto A, Díaz-Hernández M, Bustillo D, Díaz-Hernández JI, Huerta PG, Artalejo AR, Garrido JJ, Miras-Portugal MT (2009) Ca^{2+}/calmodulin-dependent kinase II signalling cascade mediates $P2X_7$ receptor-dependent inhibition of neuritogenesis in neuroblastoma cells. FEBS J 276:5307–5325

Götz M, Huttner WB (2005) The cell biology of neurogenesis. Nature Rev Mol Cell Biol 6:777–788

Grimm I, Ullsperger SN, Zimmermann H (2010) Nucleotides and epidermal growth factor induce parallel cytoskeletal rearrangements and migration in cultured adult murine neural stem cells. Acta Physiol 199:181–189

Guillemot F, Parras C (2005) Adult neurogenesis: a tale of two precursors. Nat Neurosci 8:846–848

Guo W, Zhang Z, Liu X, Burnstock G, Xiang Z, He C (2013) Developmental expression of P2X5 receptors in the mouse prenatal central and peripheral nervous systems. Purinergic Signal 9:239–248

Hack MA, Saghatelyan A, deChevigny A, Pfeifer A, Ashery-Padan R, Lledo PM, Götz M (2005) Neuronal fate determinants of adult olfactory bulb neurogenesis. Nat Neurosci 8:865–872

Hager S, Lampert FM, Orimo H, Stark GB, Finkenzeller G (2009) Up-regulation of alkaline phosphatase expression in human primary osteoblasts by cocultivation with primary endothelial cells is mediated by p38 mitogen-activated protein kinase-dependent mRNA stabilization. Tissue Eng Part A 15:3437–3447

Hahnel AC, Rappolee DA, Millán JL, Manes T, Ziomek CA, Theodosiu NG, Werb Z, Pedersen RA, Schultz GA (1990) Two alkaline phosphatase genes are expressed during early development in the mouse embryo. Development 110:555–564

Hanics J, Barna J, Xiao J, Millán JL, Fonta C, Negyessy L (2012) Ablation of TNAP function compromises myelination and synaptogenesis in the mouse brain. Cell Tissue Res 349:459–471

Ishiyama V, Izquierdo L (1977) The onset of phosphatase activity in the early mammalian embryos. J Embryol Exp Morph 42:305–308

Ito S, Suzuki N, Kato S, Takahashi T, Takagi M (2007) Glucocorticoids induce the differentiation of a mesenchymal progenitor cell line, ROB-C26 into adipocytes and osteoblasts, but fail to induce terminal osteoblast differentiation. Bone 40:84–92

Izquierdo L, Lopez T, Marticorena P (1980) Cell membrane regions in preimplantation mouse embryos. J Embryol Exp Morphol 59:89–102

Johnson-Pais TL, Leach RJ (1996) 1,25-Dihydroxyvitamin D3 and transforming growth factor-beta act synergistically to override extinction of liver/bone/kidney alkaline phosphatase in osteosarcoma hybrid cells. Exp Cell Res 226:67–74

Kempermann G, Jessberger S, Steiner B, Kronenberg G (2004) Milestones of neuronal development in the adult hippocampus. Trends Neurosci 27:447–452

Kermer V, Ritter M, Albuquerque B, Leib C, Stanke M, Zimmermann H (2010) Knockdown of tissue nonspecific alkaline phosphatase impairs neural stem cell proliferation and differentiation. Neurosci Lett 485:208–211

Khakh BS, North RA (2006) P2X receptors as cell-surface ATP sensors in health and disease. Nature 442:527–532

Khodosevich K, Zuccotti A, Kreuzberg MM, Le MC, Frank M, Willecke K, Monyer H (2012) Connexin45 modulates the proliferation of transit-amplifying precursor cells in the mouse subventricular zone. Proc Natl Acad Sci USA 109:20107–20112

Kiledjian M, Kadesch T (1990) Analysis of the human liver/bone/kidney alkaline phosphatase promoter in vivo and in vitro. Nucleic Acids Res 18:957–961

Kim HN, Lee JH, Bae SC, Ryoo HM, Kim HH, Ha H, Lee ZH (2011) Histone deacetylase inhibitor MS-275 stimulates bone formation in part by enhancing Dhx36-mediated TNAP transcription. J Bone Miner Res 26:2161–2173

Kim YJ, Lee MH, Wozney JM, Cho JY, Ryoo HM (2004) Bone morphogenetic protein-2-induced alkaline phosphatase expression is stimulated by Dlx5 and repressed by Msx2. J Biol Chem 279:50773–50780

Kishi F, Matsuura S, Kajii T (1989) Nucleotide sequence of the human liver-type alkaline phosphatase cDNA. Nucleic Acids Res 17:2129

Kriegstein A, Alvarez-Buylla A (2009) The glial nature of embryonic and adult neural stem cells. Annu Rev Neurosci 32(149–184):149–184

Kriegstein A, Götz M (2003) Radial glia diversity: a matter of cell fate. Glia 43:37–43

Kwong WH, Tam PP (1984) The pattern of alkaline phosphatase activity in the developing mouse spinal cord. J Embryol Exp Morphol 82:241–251

Kyeyune-Nyombi E, Lau KH, Baylink DJ, Strong DD (1991) 1,25-Dihydroxyvitamin D3 stimulates both alkaline phosphatase gene transcription and mRNA stability in human bone cells. Arch Biochem Biophys 291:316–325

Lacar B, Herman P, Hartman NW, Hyder F, Bordey A (2012) S phase entry of neural progenitor cells correlates with increased blood flow in the young subventricular zone. PLoS One 7: e31960

Langer D, Ikehara Y, Takebayashi H, Hawkes R, Zimmermann H (2007) The ectonucleotidases alkaline phosphatase and nucleoside triphosphate diphosphohydrolase 2 are associated with subsets of progenitor cell populations in the mouse embryonic, postnatal and adult neurogenic zones. Neuroscience 150:863–879

Langer D, Hammer K, Koszalka P, Schrader J, Robson S, Zimmermann H (2008) Distribution of ectonucleotidases in the rodent brain revisited. Cell Tissue Res 334:199–217

LeDu MH, Stigbrand T, Taussig MJ, Menez A, Stura EA (2001) Crystal structure of alkaline phosphatase from human placenta at 1.8 A resolution: implication for a substrate specificity. J Biol Chem 276:9158–9165

Lee HW, Suh JH, Kim AY, Lee YS, Park SY, Kim JB (2006) Histone deacetylase 1-mediated histone modification regulates osteoblast differentiation. Mol Endocrinol 20:2432–2443

Lepire ML, Ziomek CA (1989) Preimplantation mouse embryos express a heat-stable alkaline phosphatase. Biol Reprod 41:464–473

Levison SW, Chuang C, Abramson BJ, Goldman JE (1993) The migrational patterns and developmental fates of glial precursors in the rat subventricular zone are temporarily regulated. Development 119:611–622

Ligon KL, Fancy SPJ, Franklin RJP, Rowitch DH (2006) *Olig* gene function in CNS development and disease. Glia 54:1–10

Lim DA, Tramontin AD, Tervejo JM, Herrera DG, Garcia-Verdugo JM, Alvarez-Buylla A (2000) Noggin antagonizes BMP signaling to create a niche for adult neurogenesis. Neuron 28:713–726

Lin JHC, Takano T, Arcuino G, Wang XH, Hu FR, Darzynkiewicz Z, Nunes M, Goldman SA, Nedergaard M (2007) Purinergic signaling regulates neural progenitor cell expansion and neurogenesis. Dev Biol 302:356–366

Liu X, Hashimoto-Torii K, Torii M, Haydar TF, Rakic P (2008) The role of ATP signaling in the migration of intermediate neuronal progenitors to the neocortical subventricular zone. Proc Natl Acad Sci USA 105:11802–11807

Liu X, Hashimoto-Torii K, Torii M, Ding C, Rakic P (2010) Gap junctions/hemichannels modulate interkinetic nuclear migration in the forebrain precursors. J Neurosci 30:4197–4209

Malatesta P, Hack MA, Hartfuss E, Kettenmann H, Klinkert W, Kirchhoff F, Götz M (2003) Neuronal or glial progeny: regional differences in radial glia fate. Neuron 37:751–764
Menn B, Garcia-Verdugo JM, Yaschine C, Gonzalez-Perez O, Rowitch D, Alvarez-Buylla A (2006) Origin of oligdendrocytes in the subventricular zone of the adult brain. J Neurosci 26:7907–7918
Mikami Y, Asano M, Honda MJ, Takagi M (2010) Bone morphogenetic protein 2 and dexamethasone synergistically increase alkaline phosphatase levels through JAK/STAT signaling in C3H10T1/2 cells. J Cell Physiol 223:123–133
Millán JL (2006a) Mammalian alkaline phosphatase. From biology to applications in medicine and biotechnology. Wiley Verlag GmbH, Weinheim
Millán JL (2006b) Structure, substrate specificity and functional relatedness to other members of a large superfamily of enzymes. Purinergic Signal 2:335–341
Ming GL, Song HJ (2005) Adult neurogenesis in the mammalian central nervous system. Annu Rev Neurosci 28:223–250
Ming GL, Song HJ (2011) Adult neurogenesis in the mammalian brain: significant answers and significant questions. Neuron 70:687–702
Mishra SK, Braun N, Shukla V, Füllgrabe M, Schomerus C, Korf H-W, Gachet C, Ikehara Y, Sévigny J, Robson SC, Zimmermann H (2006) Extracellular nucleotide signaling in adult neural stem cells: Synergism with growth factor-mediated cellular proliferation. Development 133:675–684
Mori S, Nagano M (1985) Ultracytochemical demonstration of alkaline phosphatase activity in astrocytes and subependymal cells in the rat brain. Arch Hist Jap 48:511–517
Müller WH, Kleefeld D, Khattab B, Meissner JD, Scheibe RJ (2000) Effects of retinoic acid on N-glycosylation and mRNA stability of the liver/bone/kidney alkaline phosphatase in neuronal cells. J Cell Physiol 182:50–61
Mulnard J, Huygens R (1978) Ultrastructural localization of non-specific alkaline phosphatase during cleavage and blastocyst formation in the mouse. J Embryol Exp Morphol 44:121–131
Narisawa S, Wennberg C, Millán JL (2001) Abnormal vitamin B6 metabolism in alkaline phosphatase knock-out mice causes multiple abnormalities, but not the impaired bone mineralization. J Pathol 193:125–133
Narisawa S, Hasegawa H, Watanabe K, Millán JL (1994) Stage-specific expression of alkaline phosphatase during neural development of the mouse. Dev Dynam 201:227–235
Narisawa S, Fröhlander N, Millán JL (1997) Inactivation of two mouse alkaline phosphatase genes and establishment of a model of infantile hypophosphatasia. Dev Dynam 208:432–446
North RA (2002) Molecular physiology of P2X receptors. Physiol Rev 82:1013–1067
Ohkubo S, Kimura J, Matsuoka I (2000) Ecto-alkaline phosphatase in NG108-15 cells: a key enzyme mediating P1 antagonist-sensitive ATP response. Br J Pharmacol 131:1667–1672
Orimo H, Shimada T (2005) Regulation of the human tissue-nonspecific alkaline phosphatase gene expression by all-trans-retinoic acid in SaOS-2 osteosarcoma cell line. Bone 36:866–876
Orimo H, Shimada T (2008) The role of tissue-nonspecific alkaline phosphatase in the phosphate-induced activation of alkaline phosphatase and mineralization in SaOS-2 human osteoblast-like cells. Mol Cell Biochem 315:51–60
Parnavelas JG (1999) Glial cell lineages in the rat cerebral cortex. Exp Neurol 156:418–429
Peretto P, Merighi A, Fasolo A, Bonfanti L (1999) The subependymal layer in rodents: a site of structural plasticity and cell migration in the adult mammalian brain. Brain Res Bull 49:221–243
Pinner B, Davison JF, Campbell JB (1964) Alkaline phosphatase in peripheral nerves. Science 145:936–938
Pinto L, Götz M (2007) Radial glial cell heterogeneity: the source of diverse progeny in the CNS. Progr Neurobiol 83:2–23
Platel JC, Lacar B, Bordey A (2007) GABA and glutamate signaling: homeostatic control of adult forebrain neurogenesis. J Mol Histol 38:303–311
Ruan HZ, Moules E, Burnstock G (2004) Changes in P2X3 purinoceptors in sensory ganglia of the mouse during embryonic and postnatal development. Histochem Cell Biol 122:539–551

Sauvageot CM, Stiles CD (2002) Molecular mechanisms controlling cortical gliogenesis. Curr Opin Neurobiol 12:244–249

Scemes E, Duval N, Meda P (2003) Reduced expression of $P2Y_1$ receptors in connexin43-null mice alters calcium signaling and migration of neural progenitor cells. J Neurosci 23:11444–11452

Scherer SS (1996) Molecular specializations at nodes and paranodes in peripheral nerve. Microsci Res Tech 34:452–461

Schmitt A, Benninghoff J, Moessner R, Rizzi M, Paizanis E, Doenitz C, Gross S, Hermann M, Gritti A, Lanfumey L, Fritzen S, Reif A, Hamon M, Murphy DL, Vescovi A, Lesch KP (2007) Adult neurogenesis in serotonin transporter deficient mice. J Neural Transm 114:1107–1119

Smart I (1961) The subependymal layer of the mouse brain and its cell production as shown by autoradiography after thymidine-H3 injection. J Comp Neurol 116:325–347

Smart IH, McSherry GM (1982) Growth patterns in the lateral wall of the mouse telencephalon. II. Histological changes during and subsequent to the period of isocortical neuron production. J Anat 134:415–442

Solter D, Damjanov I, Skreb N (1973) Distribution of hydrolytic enzymes in early rat and mouse embryos–a reappraisal. Z Anat Entwicklungsgesch 139:119–126

Street SE, Kramer NJ, Walsh PL, Taylor-Blake B, Yadav MC, King IF, Vihko P, Wightman RM, Millán JL, Zylka MJ (2013) Tissue-nonspecific alkaline phosphatase acts redundantly with PAP and NT5E to generate adenosine in the dorsal spinal cord. J Neurosci 33:11314–11322

Striedinger K, Meda P, Scemes E (2007) Exocytosis of ATP from astrocyte progenitors modulates spontaneous Ca^{2+} oscillations and cell migration. Glia 55:652–662

Studer M, Terao M, Gianni M, Garattini E (1991) Characterization of a second promoter for the mouse liver/bone/kidney-type alkaline phosphatase gene: cell and tissue specific expression. Biochem Biophys Res Commun 179:1352–1360

Sturrock RR, Smart IHM (1980) A morphological study of the mouse subependymal layer from embryonic life to old age. J Anat 130:391–415

Suyama S, Sunabori T, Kanki H, Sawamoto K, Gachet C, Koizumi S, Okano H (2012) Purinergic signaling promotes proliferation of adult mouse subventricular zone cells. J Neurosci 32:9238–9247

Tam PP, Kwong WH (1987) A study on the pattern of alkaline phosphatase activity correlated with observations on silver-impregnated structures in the developing mouse brain. J Anat 150:169–180

Taniura H, Ito M, Sanada N, Kuramoto N, Ohno Y, Nakamichi N, Yoneda Y (2006) Chronic vitamin D3 treatment protects against neurotoxicity by glutamate in association with upregulation of vitamin D receptor mRNA expression in cultured rat cortical neurons. J Neurosci Res 83:1179–1189

Toh Y, Yamamoto M, Endo H, Fujita A, Misumi Y, Ikehara Y (1989a) Sequence divergence of 5' extremities in rat liver alkaline phosphatase mRNAs. J Biochem 105:61–65

Toh Y, Yamamoto M, Endo H, Misumi Y, Ikehara Y (1989b) Isolation and characterization of a rat liver alkaline phosphatase gene. A single gene with two promoters. Eur J Biochem 182:231–237

Tran PB, Ren DJ, Veldhouse TJ, Miller RJ (2004) Chemokine receptors are expressed widely by embryonic and adult neural progenitor cells. J Neurosci Res 76:20–34

Vittur F, Stagni N, Moro L, de Bernhard B (1984) Alkaline phosphatase binds to collagen; a hypothesis on the mechanism of extravesicular mineralization in epiphyseal cartilage. Experientia 40:836–837

Vorbrodt AW, Lossinsky AS, Wisniewski HM (1986) Localization of alkaline phosphatase activity in endothelia of developing and mature mouse blood-brain barrier. Dev Neurosci 8:1–13

Waymire JC, Mahuren JD, Jaje JM, Guilarte T, Coburn SP, MacGregor GR (1995) Mice lacking tissue non-specific alkaline phosphatase die from seizures due to defective metabolism of vitamin B-6. Nat Genet 11:45–51

Weissman TA, Riquelme PA, Ivic L, Flint AC, Kriegstein AR (2004) Calcium waves propagate through radial glial cells and modulate proliferation in the developing neocortex. Neuron 43:647–661

Whyte PP (1996) Hypophosphatasia: natures window on alkaline phosphatase function in man. In: Bilezkian J, Raisz L, Rodan G (eds) Principles of bone biology. Academic Press, San Diego, pp 951–968

Wichterle H, Turnbull DH, Nery S, Fishell G, Alvarez-Buylla A (2001) In utero fate mapping reveals distinct migratory pathways and fates of neurons born in the mammalian forebrain. Development 128:3759–3771

Wu LN, Genge BR, Lloyd GC, Wuthier RE (1991) Collagen-binding proteins in collagenase-released matrix vesicles from cartilage. Interaction between matrix vesicle proteins and different types of collagen. J Biol Chem 266:1195–1203

Wu LN, Genge BR, Wuthier RE (1992) Evidence for specific interaction between matrix vesicle proteins and the connective tissue matrix. Bone Miner 17:247–252

Young SZ, Taylor MM, Bordey A (2011) Neurotransmitters couple brain activity to subventricular zone neurogenesis. Eur J Neurosci 33:1123–1132

Zerlin M, Levison SW, Goldman JE (1995) Early patterns of migration, morphogenesis, and intermediate filament expression of subventricular zone cells in the postnatal rat forebrain. J Neurosci 15:7238–7249

Zernik J, Kream B, Twarog K (1991) Tissue-specific and dexamethasone-inducible expression of alkaline phosphatase from alternative promoters of the rat bone/liver/kidney/placenta gene. Biochem Biophys Res Commun 176:1149–1156

Zhao CM, Deng W, Gage FH (2008) Mechanisms and functional implications of adult neurogenesis. Cell 132:645–660

Zhou H, Manji SS, Findlay DM, Martin TJ, Heath JK, Ng KW (1994) Novel action of retinoic acid: stabilization of newly synthesized alkaline phosphatase transcripts. J Biol Chem 269:22433–22439

Zimmermann H (2006) Nucleotide signaling in nervous system development. Pflugers Arch Eur J Physiol 452:573–588

Zimmermann H (2011) Purinergic signaling in neural development. Semin Cell Dev Biol 22:194–204

Zimmermann H, Zebisch M, Sträter N (2012) Cellular function and molecular structure of ecto-nucleotidases. Purinergic Signal 8:437–502

Chapter 5
Rediscovering TNAP in the Brain: A Major Role in Regulating the Function and Development of the Cerebral Cortex

Caroline Fonta, Pascal Barone, Laia Rodriguez Martinez and László Négyessy

Abstract The presence of alkaline phosphatase (AP) activity in the neural tissue has been described decades ago. However, only recent studies clarified the isotype, regional distribution and subcellular localization of the AP expressed in the cerebral cortex of diverse mammalian species including the human. In the primate brain the discovery that the bone AP isotype (TNAP) is expressed provided the opportunity of a deeper understanding of the role of this enzyme in neuronal functions based on the knowledge acquired by studying the role of the enzyme in hypophosphatasia, mostly in bone mineralization. TNAP exhibits widespread substrate specificity and, in the brain, it is potentially involved in the regulation of molecules which play fundamental roles in signal transmission and development. In light of these observations, the localization of TNAP in the human cerebral cortex is of high significance when considering that epilepsy is often diagnosed in hypophosphatasia. Here we overview our results on the identification of TNAP in the primate cerebral cortex: TNAP exhibits a noticeably high activity in the synapses and nodes of Ranvier, is specifically present in layer 4 of the sensory cortices and additionally in layer 5 of prefrontal, temporal and other associational areas in human. Our studies also indicate that bone AP activity depends on the level of sensory input and

---

C. Fonta (✉) · P. Barone · L. Rodriguez Martinez
Centre de Recherche Cerveau et Cognition (CerCo), Université de Toulouse UPS; CNRS UMR 5549, Toulouse, France
e-mail: caroline.fonta@cerco.ups-tlse.fr

L. Négyessy
Theoretical Neuroscience and Complex Systems Research Group, Wigner Research Center for Physics, Budapest 1121, Hungary

L. Négyessy
Department of Anatomy, Histology and Embryology, Semmelweis University Medical School, Budapest 1094, Hungary

C. Fonta and L. Négyessy (eds.), *Neuronal Tissue-Nonspecific Alkaline Phosphatase (TNAP)*, Subcellular Biochemistry 76,
DOI 10.1007/978-94-017-7197-9_5

that its developmental time-course exhibits characteristic regional differences. The relevance of our findings regarding human cortical physiology and brain disorders are discussed.

Keywords Synaptic cleft · Node of Ranvier · Neurotransmission · Cortical layer · Enzyme histochemistry

5.1 Rediscovering Alkaline Phosphatase in the Brain

The non-specific phosphomonoesterase (alkaline phosphatase -AP-) is the first enzyme whose demonstration in tissues was performed in the 1920s. Since then the detection of AP enzymatic activity gave rise to a substantial number of publications to propose modifications or alternatives to limit artefacts that might be generated by the initial methods. The histochemical methods were applied to different non osseous tissues, both normal and pathological. However these enthusiastic explorations, revealing the wide distribution of AP in many organs had to face the paucity of data on the physiological role of the enzyme (except the bone tissue, see Chap. 8). A growing interest in this enzyme has become apparent in the last few years, especially in regard to the nervous system (Buchet et al. 2013).

Shimizu (1950) initially demonstrated AP activity in the vertebrate brain using modified procedures of Takamatsu and Gömöri (Gomori 1939; Takamatsu 1939). He showed alkaline phosphatase activity in the choroid plexus, the endothelium and noted weaker activity in the parenchyma of the brain and spinal cord of mice, rats, guinea pigs and cats. Interestingly observations were different in the rabbit brain where AP staining was strong in the parenchyma and absent from the vessels (Shimizu 1950). No AP activity was detected either in the vessel walls of the frog central nervous system (Latker et al. 1987). In the chicken brain, a weak vessel reaction was found in some parts of the brain but the parenchyma was reported as strongly reactive (Shimizu 1950). In the Old World monkey brain Friede (1966) described a regional distribution of alkaline phosphatase activity in the neuropil and in blood vessels. In all neocortical areas the enzymatic activity was the highest in the superficial part of the first layer and in the mid-layers. Robins et al. (1956) had previously reported that alkaline phosphatase activity was higher in the upper and middle layers of the visual and motor cortex of the macaque. Such a pattern was also observed in New World monkeys (Fonta and Imbert 2002; Fonta et al. 2004). In the human brain, alkaline phosphatase activity has been mainly used to study intracerebral microvascular patterns (Bell and Scarrow 1984; Norman and O'Kusky 1986; Anstrom et al. 2002). However, we recently discovered a specific pattern of AP activity in the neuropil of different parts of the human neocortex including primary sensory and higher order associative cortical areas (Négyessy et al. 2011) (Fig. 5.1). In rodents, an exhaustive description of AP activity in the mouse brain

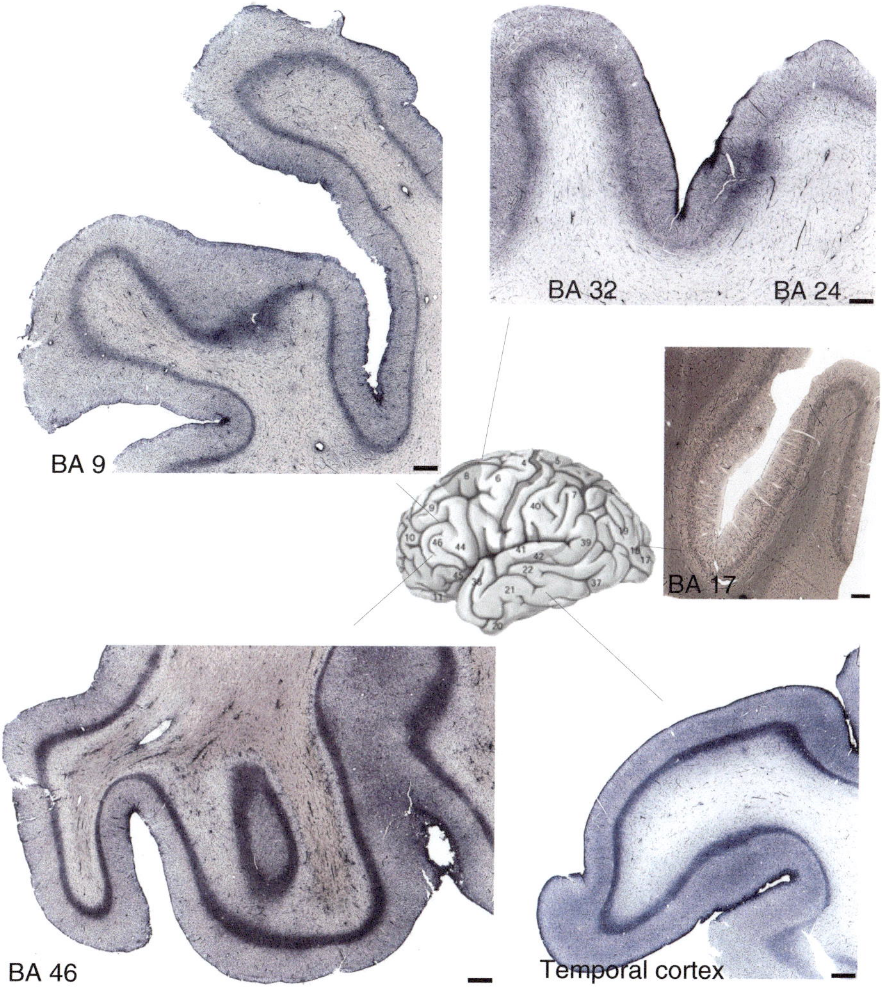

Fig. 5.1 Histological detection of TNAP enzymatic activity in coronal sections of the human cortex. Beside the strong activity in the vessels, a TNAP-positive stripe, located in the innermost half of the cortex, runs parallel to the cortical surface in different samples. This stripe was localised infragranulary in layer 5 of the temporal and frontal cortices, specifically in the cingular cortex (areas 24 and 32) and the dorso-lateral prefrontal cortex (areas 9 and 46). In sensory areas, such as the primary visual cortex (area 17) of the occipital lobe, TNAP activity was localized in the granular layer 4, the main target of visual thalamic inputs. *Scale bar* = 500 μm

revealed that this enzyme is associated with the parenchyma of many cerebral regions, more particularly the cerebral cortex, olfactory bulb, tectum, hypothalamus and the medulla (Langer et al. 2008). However in the cortex AP activity exhibits different patterns in rodents and primates. In rodents there is a relatively high, homogenous AP activity throughout the different cortical layers with a bit more

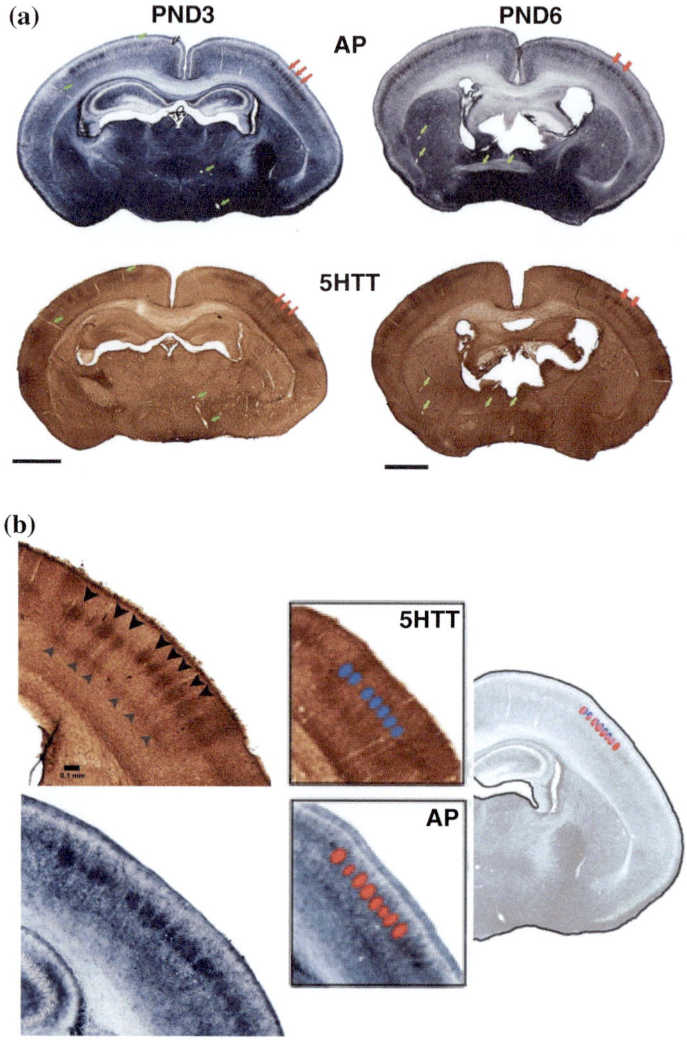

Fig. 5.2 Histological detection of TNAP enzymatic activity in coronal sections of the mouse brain, in the somato-sensory cortex. **a** Serial coronal sections. (40 μm thick) were stained to detect TNAP activity and the serotonin transporter (5HTT) in the barrel field 3 (PND3) or 6 (PND6) days after birth. Green arrows show landmarks (vessels). Both markers were found in the middle layers of the barrel field cortex (*red arrows*) as soon as PND3. **b** TNAP activity (*red spots*) was found in the barrels (*blue spots*) receiving the whisker inputs identified by 5HTT immunohistochemical staining (*black arrowheads*). *Scale bar* = 1 mm

elevated activity in the upper and middle layers. Thus, granular layer 4 of the barrel cortex exhibit high AP activity localized into the barrels while septal regions exhibit lower AP enzymatic activity (Fonta et al. 2004; Brun-Heath et al. 2011) (Fig. 5.2). In contrast, a characteristic laminar pattern is present in the cerebral cortex of primates including human (Friede 1966; Fonta et al. 2004; Négyessy et al. 2011) (Fig. 5.1). The localization of AP activity into a particular layer of different areas, especially in the human, suggests that this phosphatase is specialized for some distinct functions in the cerebral cortex.

5.2 Converging Evidence for the Expression of TNAP in the Neural Tissue

The identification of the Tissue Non-specific Alkaline Phosphatase (TNAP) in brain was proposed by Goldstein and Harris (1981) who based their results on the possibility to discriminate TNAP from the other alkaline phosphatases by testing different **inhibitors** (amino acids or small peptides e.g. phenylalanine, homoarginine, phenylalanylglycylglycine) and thermostability on brain extracts. They thus showed TNAP activity in a large range of species (guinea pig, rat, mouse, hamster, squirrel, sheep, cow, baboon, tamarin and human), although in a much lower level in the brain than in other organs (e.g. kidney and liver).

Till recently the validation of TNAP activity in neural tissue was obtained by incubation of brain tissue or neuronal cells with specific inhibitors. TNAP has been shown to be inhibited by some compounds among which levamisole found to be highly specific (Van Belle 1976; Kozlenkov et al. 2004). In studies analysing TNAP functions at the neuronal level, levamisole (or tetramisole which is the racemic mixture of levamisole and dexamisole (Vanhoutte et al. 1977; Przegalinski et al. 1980; Casale et al. 2012) has been mainly used to test whether the AP activity that was under investigation corresponded to the TNAP activity. Accordingly, concentrations ranging from 2 to 10 mM have been used in histological sections in primates (Fonta et al. 2004; Négyessy et al. 2011; Vardy et al. 2012) and in rodents (Langer et al. 2007, 2008; Street et al. 2013), from 500 μM to 5 mM in cell cultures (Ermonval et al. 2009; Diaz-Hernandez et al. 2010; Diez-Zaera et al. 2011) or 100 μM for human blood plasma (Vardy et al. 2012).

However, recent studies show that levamisole interferes neuronal activity by inhibiting not only TNAP but apparently some ion channels as well, suggesting that the specificity of levamisole for the TNAP isoenzyme is questionable (see Chap. 12). Novel, highly selective molecules have been identified that are potent TNAP inhibitors in luminescent and colorimetric TNAP assays (Dahl et al. 2009; Sergienko et al. 2009). One of them, 2,5-Dimethoxy-N-(quinolin-3-yl) benzenesulfonamide, available as MLS-0038949, was recently used to test TNAP ectonucleotidase function at 50 μM in TNAP transfected HEK293 cells (Street et al. 2013) and in mice hippocampal slices at 10 μM (Zhang et al. 2012a) (see also Chaps. 12 and 13).

Antibodies directed against TNAP were also used to detect TNAP localization in cell culture or brain sections. Polyclonal antibodies either available commercially (Diaz-Hernandez et al. 2010; Diez-Zaera et al. 2011) or provided by other groups working on the role of TNAP in bone mineralization (Miki et al. 1986; Oda et al. 1999) showed distribution of the enzyme at the surface of rodent neuronal cells (Langer et al. 2007; Ermonval et al. 2009) and human neuroblastoma cells (Diaz-Hernandez et al. 2010). Unfortunately, immunohistochemical studies are painfully lacking in regard to the subcellular localization of the enzyme TNAP in the adult brain.

TNAP was also identified in lipid rafts of 1C11 cells differentiated into bioaminergic neurons by western blot and **mass spectrometry** (Ermonval et al. 2009).

At the level of TNAP gene transcription**, in situ hybridization** revealed TNAP transcripts in primary sensory neurons of the dorsal root ganglia (DRG) and in neurons of the spinal cord gray matter in the mouse, using a 1190 bp riboprobe (Street et al. 2013) (see also Chap. 13). Confirmation of TNAP expression in the DRG was obtained by **RT-PCR** with primers spanning the ninth intron of TNAP gene (Street et al. 2013). Quantitative PCR was also performed to detect TNAP expression in mice hippocampal slices (Zhang et al. 2012a) and in the human brain (Diaz-Hernandez et al. 2010) with probes targeting a common domain of bone and liver TNAP transcripts (in the respective murine and human TNAP gene). TNAP isoenzyme was also identified in the neural tube of developing mouse embryos by PCR with primers directed to nucleotides 1283 through 1302 (exon 11) and nucleotides 1591 through 1612 (exon 12) (Hahnel et al. 1990; Narisawa et al. 1994).

Finally, the inhibitory effects of knocking down TNAP on cell proliferation and differentiation with **small interfering RNA** (antisense shRNA oligonucleotides) of TNAP were abrogated by adding active TNAP enzyme in mouse neuronal stem cell culture medium (Kermer et al. 2010). This approach, used recently in humans as enzyme replacement therapy (see Chaps. 8 and 15), provides some evidence for functional role of TNAP in neural progenitors (see Chaps. 3 and 4).

Considering all the approaches developed, it seems sensible to consider that alkaline phosphatase activity observed in the neuronal tissue or at the level of neuronal cells can be attributed to the tissue non-specific alkaline phosphatase subtype. However two different TNAP transcripts have been identified. The TNAP gene contains twelve exons, the first one being part of a region of the TNAP mRNA that is not translated (Weiss et al. 1988). This region consists either in exon 1A preferentially driven in osteoblasts or in exon 1B driven preferentially in liver and kidney (Toh et al. 1989; Matsuura et al. 1990; Studer et al. 1991). The two isoforms, so-called the **bone and the liver TNAP isoforms**, have the same amino acid sequence but differ at the level of their mRNAs. Their presence in the neuronal tissue has been investigated recently (Brun-Heath et al. 2011). Notably, expression of tissue-specific alkaline phosphatases and a neuronal-specific promoter for TNAP were discarded in the brain of rodents and primates (Brun-Heath et al. 2011). Instead, the bone-type TNAP has been identified in human, monkey, rat and mouse brains and in rodent neuronal cells. Interestingly, the liver-type transcript was additionally found but only in the mouse brain and neuronal cells. Also the bone-type TNAP transcript was identified in the endothelium of the human and mouse brains (Brun-Heath et al. 2011). These results suggest that in addition to its expression in the bone tissue, TNAP exhibits organ and cell-type specific regulation in the brain.

All these studies using diverse techniques indicate that TNAP is expressed in the neural tissue.

5.3 Cell Types Expressing TNAP in the Nervous System

Both neuronal and endothelial cells have been shown positive for TNAP activity in the brain.

Endothelial cells exhibit TNAP activity both at the luminal and abluminal plasma membranes (Vorbrodt et al. 1986; Fonta et al. 2004). However, it should be noted that capillaries and arteries stain strongly while the endothelium of venules lack this AP activity after using histochemical labelling (Vorbrodt et al. 1986) (see Chap. 7).

TNAP functions as an ectoenzyme, bound to the cell membrane on its extracellular surface. Therefore, the localization of TNAP activity in the narrow extracellular space of the brain parenchyma renders the assignment of the staining to individual **neuronal cells** (either neuron or glia) difficult. Thus cellular identification of TNAP expression in cerebral tissue is scarce.

In the **adult central nervous system**, as already mentioned, using in situ hybridization Street et al. (2013) discovered expression of TNAP in neurons of dorsal root ganglia and of spinal cord gray matter and suggested that all subtypes of sensory neurons express TNAP.

In the neurogenic zones of adult mouse brain, TNAP is highly active in clusters of subependymal cells along the walls of the lateral ventricles and within the rostral migratory stream (Langer et al. 2007). However, the neurogenic dentate gyrus lacks TNAP activity (Langer et al. 2007). Most, if not all, TNAP positive cells are either neuronal or glial progenitors in these proliferating neurogenic zones (see Chaps. 4 and 18).

Few works have been published **on embryonic tissues.** They used northern blot (Brun-Heath et al. 2011), western blot (Kermer et al. 2010) or multiple labelling techniques including histochemistry and cell type specific immunohistochemistry (Narisawa et al. 1994; Langer et al. 2007; Kermer et al. 2010). TNAP transcript is expressed both in neuronal and glial cell types in cortical preparations of the rat embryo (E15) (Brun-Heath et al. 2011). In mice activity of TNAP is strong in the neuroepithelium at early embryonic development (E8.5) and overlaps with neurofilament expression at E9.5 (Narisawa et al. 1994). Latter at E14 it is widely expressed by cells of the cerebral vesicles. TNAP characterizes the ventricular and subventricular zones (Langer et al. 2007; Kermer et al. 2010) that are zones of proliferative activity during prenatal and early postnatal stages. An exhaustive description of the spatio-temporal pattern of TNAP activity during brain development shows that this enzyme is a marker of immature neurons but may not be restricted to a neuronal lineage (Langer et al. 2007) (see also Chap. 4). Results obtained by **in vitro preparations**, by using different methodologies (mass spectrometry, western blot, RT-PCR analyses) convincingly converged to demonstrate that TNAP is expressed in differentiated serotonergic and noradrenergic 1C11 derived-cells and not in their 1C11 neuroepithelial cell line precursors in vitro (Ermonval et al. 2009). Studies of Kermer et al. (2010) who manipulated TNAP expression in cultured multipotent stem cells, suggest that TNAP plays role both in proliferation and differentiation of neurones and oligodendrocytes.

In conclusion, TNAP is expressed by several cell types in the cerebral tissue and is not restricted to neurons, which is supported by ultrastructural analyses (Table 5.1). Moreover TNAP is found at different stages of neuronal differentiation and maturation.

5.4 Sub-Cellular Localization of TNAP

TNAP activity is generally higher in the grey matter than in the white matter in the adult brain. At the light microscopic level the parenchyma between the cell bodies ('the neuropil') is TNAP positive in the gray matter. TNAP is localised at the cell surface, both around cell bodies and neuronal processes (neurites) (Ermonval et al. 2009; Fonta et al. 2004). Ermonval et al. (2009) identified TNAP in lipid microdomains of 1C11 cells differentiated to bioaminergic neurons (cf. Chap. 9).

By using transmission electron microscopy (TEM), some authors have also reported enzymatic activity within the neuronal somata, on the circumference of the nucleolus or even inside the nucleolus and the nucleus, and also in the cytoplasm (Shimizu 1950) (Table 5.1). TNAP activity was found in association with tubulin preparation from the bovine cerebral cortex (Larsson et al. 1979). On the contrary no activity was observed in cell bodies in the cortex of adult macaque (Friede 1966) and marmoset (Fonta et al. 2004) monkeys, or in adult mouse brain (Langer et al. 2008).

Evidence for TNAP in synapses was brought by TEM (see references in Table 5.1) and biochemical investigations. Thus spectrophotometric measurements of TNAP activity in native and lysed synaptic vesicles extracted from bovine cerebral cortex were reported at the level of the interior surface of the synaptic vesicle membrane (Zisapel and Haklai 1980; Zisapel and Levi 1980).

Interestingly, TNAP activity is known as a marker of the nodes of Ranvier along the axons, that are myelin-free domains in adult brain (Pinner and Campbell 1965; Scherer 1996). Interestingly a temporal correlation had been put forward between decrease of TNAP activity and the extent of myelination on axons in rodents and primates (Ng and Tam 1986; Vorbrodt et al. 1986; Fonta et al. 2005).

Table 5.1 Cellular and subcellular localisation of alkaline phosphatase activity

Species	Structure	Vessels	Neurons	Glial cells	Pericytes	References
Adult rat	Cerebrum	Plasma membrane	*Membrane of some cellular processes	*Glial membrane in contact with the basement lamina of capillaries	0	Mayahara et al. (1967)
			* Mitochondria (cell processes and bodies)	* Membrane covering glial cell bodies		
Adult rat	Cerebrum		*Postsynaptic membrane			Sugimura and Mizutani (1979)
			*Presynaptic membrane			
			*Synaptic vesicles membrane			
Adult rat	Cerebral cortex and corpus callosum	Plasma membrane (luminal and abluminal) cytoplasm	Surface membrane of synaptic bouton	*Plasma membrane of oligodendrocyte processes	Plasma membrane, nucleus, cytoplasm	Mori and Nagano (1985a)
				*Cytoplasm of immature oligodendrocytes		
				*Mitochondria (oligodendrocytes)		
Adult rat	Cerebral cortex		*Plasma membrane of nerve cell bodies, dendrites processes, terminals, axons			Mori and Nagano (1985b)
			*Outer membrane of some mitochondria			
			*Node of Ranvier (nodal ≫paranodal)			

(continued)

Table 5.1 (continued)

Species	Structure	Vessels	Neurons	Glial cells	Pericytes	References
	Optic and sciatic nerves		*Axonal plasma membrane *Node of Ranvier (nodal ≫paranodal)			Mori and Nagano (1985b)
Newborn rat	Cerebral cortex	Basement membrane	Nuclear membrane of nerve cells	*Glial endfeet and cytoplasm *Nuclear membrane of glial cells	Vesicles	Ovtscharoff (1973)
Chicken	Retina		*Plasma membrane of dendritic processes (horizontal and bipolar cells) *Pre-synaptic membrane (photoreceptor)	Muller cells		Araki and Saito (1983)
Chick embryo	Cultured spinal cord		Membrane of nerve cell bodies and axons if in contact with glial cells	Membrane of glial cells and processes		Ciani et al. (1973)
Marmoset monkey	Cerebral cortex	Plasma membrane (luminal and abluminal)	*Neuron membrane, synaptic membrane *Node of Ranvier			Fonta et al. (2004, 2005)

5.5 Regulation of TNAP by Neural Activity

Based on the findings that TNAP activity is observed in neuronal micro-domains (synapses and nodes of Ranvier) tightly associated with signal transmission (respectively synaptic transmission and axonal conduction), it could be hypothesized that neuronal and TNAP activities are tightly linked. This possibility was tested in two situations: 1, postnatal brain development and 2, manipulating neuronal activity in the adult brain. The prenatal evolution of TNAP activity is exhaustively described in Chap. 4.

5.5.1 TNAP Activity of the Grey Matter During Postnatal Development

In the marmoset **monkey**, a longitudinal analysis was realized from birth to several months after birth. This period of life is essential for the production and maturation of synapses, and similarly for the myelination of fibers in the cerebral hemispheres. Interestingly TNAP activity showed a differential developmental timecourse at different scales. For example at birth, visual cortex is devoid of enzymatic activity, analysed at the light and electron microscopy levels, except in the endothelial cells. Meanwhile the parenchyma of the auditory cortex and the somatosensory cortex display TNAP activity. TNAP activity clearly appears in the visual cortex one week later and then increases during the three first postnatal months (Fig. 5.3). The delayed appearance of TNAP activity in the visual system reflects the different maturation time courses of the different sensory modalities during marmoset development (Missler et al. 1992; Barone et al. 2003; Fonta et al. 2005).

In the primary visual cortex, TNAP activity appears in the first postnatal week in the part of the visual cortex devoted to the integration of signals originating from the peripheral retina. The operculum, receiving its inputs from the central retina, becomes TNAP positive later, during the second week of postnatal life (Fonta et al. 2005). Moreover, in this same cortex, localization of TNAP activity reveals the two sublayers targeted by the visual thalamic afferents. These two sub-layers correspond respectively to the segregation of the magno- and parvo-cellular systems. Notably TNAP activity is first observed in the magno-cellular system which is known to mature earlier (Kennedy et al. 1985; Spatz et al. 1993; Fonta et al. 1997). Therefore unlike classical neuronal activity dependent markers used in histology such as the cytochrome oxidase activity, detection of TNAP activity enables the evaluation of the areal, regional and sublaminar functional maturation of the cerebral cortex at a much better temporal scale (a few days). In parallel EM analyses reveal that TNAP activity is present only in fully-formed synapses, suggesting that this enzyme can be considered as a marker of synaptic maturation (Fonta et al. 2005). Interestingly both synaptogenesis and synaptic maturation on the one hand

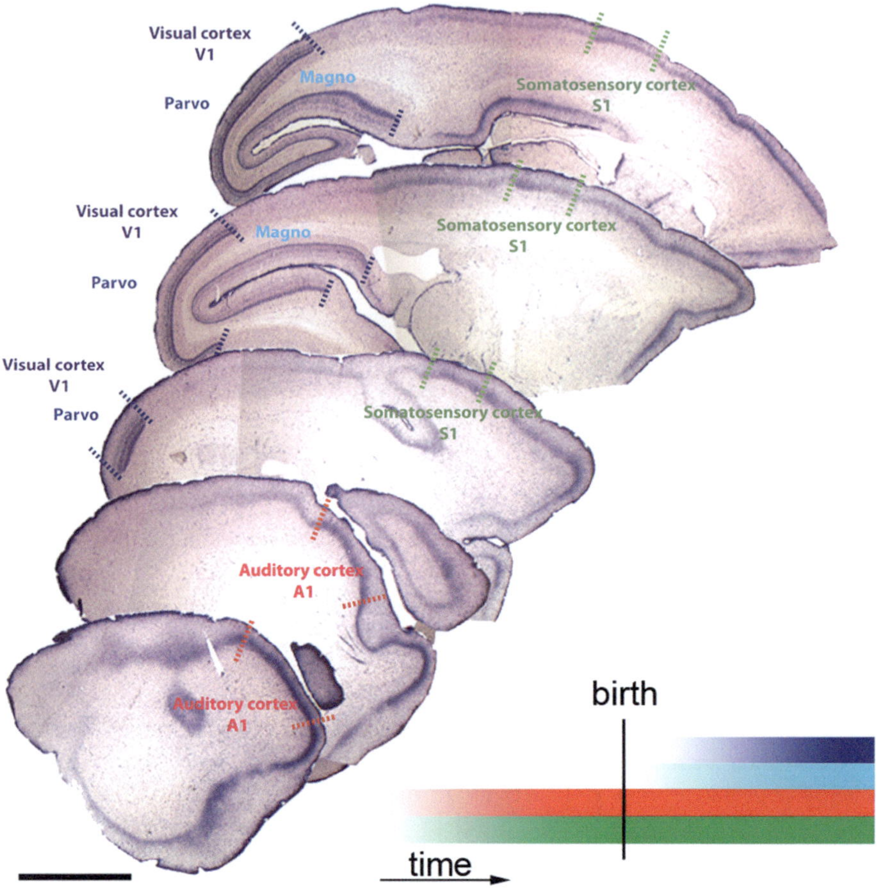

Fig. 5.3 TNAP enzyme histochemical activity exhibits different developmental time course in the sensory areas of the cerebral cortex. Samples of sections from series cut in the sagittal plan of the marmoset brain show the localization of the different areas investigated. Primary somatosensory cortex (S1) is shown in green, auditory cortex (A1) is marked by red and parvo- and magnocellular parts of the primary visual cortex (V1) are marked by dark and light blue, respectively. Grading in color bars indicate the emergence of TNAP activity in the perinatal period. Colors correspond to those used for delineating the areas on the sections. While in somatosensory and auditory cortex TNAP activity reaches adult level by birth, TNAP activity increases in the visual cortex after birth with a delay of the parvocellular system relative to the magnocellular system, in agreement with their sequential functional maturation. Rostral is on right, dorsal is up. Sections arranged in the medio-lateral order with the most lateral section at the bottom. *Scale bar* = 3 mm

and TNAP activity level on the other hand match during the first three postnatal months, supporting the possible function of TNAP in neurotransmission (Missler et al. 1993a, b; Fonta et al. 2005).

In primary cortices of **rodents**, as in primates, TNAP activity is particularly strong in layer 4, targeted by thalamic afferent fibers (Fonta et al. 2004). In the adult murine vibrissal somatosensory cortex, TNAP activity clearly reveals the barrels, that receive the thalamic afferents conveying facial whiskers related information (Fonta et al. 2004). The emergence of these cortical modules analysed with AP histochemistry was identified at the third postnatal day by correlated immunohistochemical staining of 5HTT (or SERT, the serotonin transporter) that is present in the thalamo-cortical axons (Lebrand et al. 1996). Figure 5.2 shows the total correspondence between 5 HTT and TNAP labelling at the time of the emergence of the barrels between the third and the seventh postnatal days (Rebsam et al. 2002).

The deletion of the TNAP gene does not appear to modify the macroscopic organization of the barrel field in the 6 first postnatal days of life (our own observations). However, the maturation of the synaptic connections, evaluated by a quantitative ultrastructural analysis, was shown to be delayed in the parietal cortex of one–week-old TNAP-null mice (Hanics et al. 2012).

These findings together suggest that TNAP may be involved both in synaptogenesis and the proper functioning of the mature synaptic contacts.

5.5.2 TNAP Activity of the White Matter During Postnatal Development

Opposite to the developmental increase of the intensity of AP activity in the grey matter, TNAP activity drastically decreases in the white matter during postnatal development in primates. Different phenomena are observed in mice (see Chap. 4). Histological observations clearly capture a post-developmental timecourse of TNAP activity in the white matter of the marmoset (Fonta et al. 2005). The cerebral white matter (including the corpus callosum and the commissures), strongly TNAP-active at birth and during the first postnatal weeks, progressively loses this activity. The weakening in TNAP activity is not homogeneous along the caudo-rostral axis of the corpus callosum: it is observed first in the occipital, then in the frontal pole, while myelination occurs according to an opposite, rostro-caudal gradient (Fonta et al. 2005), as classically described in humans (Paus et al. 2001; Levitt 2003). TNAP activity also decreases in the commissures during the second postnatal month. Amazingly, ultrastructural analyses revealed that TNAP activity covers the neuronal fibers that were not myelinated in neonates and juvenile monkeys and that this activity is maintained at the nodes of Ranvier latter on (Fonta et al. 2005). This inverse relationship between TNAP activity and myelin sheath around the neuronal fibers would support an important role of TNAP in maintaining the structural and functional integrity of the developing axonal process until it is enwrapped by myelin.

In order to approach the relationship between TNAP and myelination, electron microscopic studies were performed in Akp2$^{-/-}$ mice invalidated for the TNAP gene (Hanics et al. 2012). This study was made on pups up to the 8th postnatal day (PND8), as long as these animals are viable (see Chap. 3). Hanics et al. (2012)

focused on the spinal cord that myelinates earlier than the cerebral hemispheres. TNAP gene deletion impaired white matter growth from the fourth postnatal day leading to a decrease in the relative number of myelinated fibers, thinning of myelin sheath and decrease in axon diameter. Qualitatively, myelin degradation was observed in the nodal regions concurring to suggest a delay or inhibition of myelination in TNAP-null animals. This was confirmed by the practically absent or degraded myelin profiles in the cortex of $Akp2^{-/-}$ mice compared to wild-type mice. These ultrastructural analyses furthermore suggest that beside a role in the development of axonal processes, TNAP could also be involved in their myelination.

Noteworthy, TNAP activity in brain vessels of the mouse is delayed, compared to the TNAP activity in the neurons. A discreet vascular activity is found around the 10th postnatal day. However in rats and primates, vessels are TNAP-positive well before birth at E15 (Brun-Heath et al. 2011) and 28 weeks of human gestation (Anstrom et al. 2002), respectively. Such animal-species characteristics might be linked to differences in the developmental time course of the blood brain barrier (see Chap. 7).

5.5.3 TNAP Activity Is Dependent on Neuronal Activity in the Adult

The impact of modulating the activity level of sensory afferences on TNAP activity in the brain has been tested in non-human primates. After maturation of the visual system was completed monocular deprivation paradigms were developed either to lower neuronal activity in the visual system by a two weeks long eye lid suture in marmoset monkeys or to totally suppress retinal afferent activity by intra-ocular injection of the Na^+ channels blocker (TTX) in macaques (Fonta et al. 2004). Also retinal scotomas were made by laser pulses in baboons. All these protocols induced a decrease in TNAP activity in the cortical columns of the primary visual cortex receiving the inputs from the deprived eye (Fonta et al. 2004). Interestingly, the complete blocking of afferent activity (TTX and laser exposure) resulted in the parallel decrease of TNAP activity and cortical metabolism evaluated by cytochrome oxidase activity in the tissue. However, in case of the short, moderate deprivation protocol (lid suture) that preserves spontaneous retinal activity, cytochrome oxidase activity was not affected, suggesting that TNAP is a more sensitive marker of afferent activity in the adult cortex (Fonta et al. 2004).

The molecular mechanisms of regulation of TNAP expression are poorly examined in neuronal cells. Studies on this topic were mainly focused on the processes of TNAP control in bone formation during development (see Chap. 8). For a comprehensive review about the factors of TNAP regulation involved in neuronal development and differentiation, the reader is suggested to look up Chap. 4. Notably, the regulation of TNAP in the brain has recently been studied also in neurodegeneration (Diaz-Hernandez et al. 2010): extracellular hyperphosphorylated tau becomes an agonist of the muscarinic cholinergic receptors after dephosphorylation by TNAP. Activation of M1 triggers an increase of the intracellular level of calcium and the expression of TNAP. Accordingly, TNAP activity and TNAP expression are higher in the temporal gyrus and hippocampus in patients with Alzheimer's disease when compared to healthy subjects (see Chaps. 17 and 18, see also Chap. 10).

5.6 Significance of the Pattern of TNAP Distribution in the Human Cerebral Cortex

5.6.1 Clinical Relevance

The above descriptions strongly support the hypothesis that TNAP deficiency in humans, clearly identified as producing poor bone mineralization, may also directly affect neural functions and behaviour.

Recent works have successfully tested the contribution of TNAP in somatosensory circuits in the mouse spinal cord, suggesting that TNAP is involved in pain control (see Chap. 13). Pain is indeed reported in clinical cases of mild hypophosphatasia in teenagers and adults (Whyte et al. 2009; Iida et al. 2012; Zhang et al. 2012b).

However the only published information about neurological disorders in patients with hypophosphatasia refers to epileptic seizures in the cases of the severe form of the disease (Whyte et al. 1988; Litmanovitz et al. 2002; Baumgartner-Sigl et al. 2007; Leung et al. 2013). Till recently this neurological disorder was considered as a consequence of cranial deformities, intracerebral haemorrhage and hypoxia. Respiratory insufficiency observed in these babies is also explained by a reduced thoracic volume or insufficient mineralization of the ribs. However some authors had hypothesized that severe forms of hypophosphatasia may be related in alkaline phosphatase deficit in vital organs where they act (Bethenod et al. 1967): babies with the severe form of hypophosphatasia would die because of fatal encephalopathy rather than skeleton problems. Interestingly, patients with infantile hypophosphatasia suffer from lethal seizures, either with or without bone abnormalities (Baumgartner-Sigl et al. 2007; de Roo et al. 2014; Taketani et al. 2014) (see Chaps. 14 and 15). Importantly, mice invalidated for the TNAP gene also exhibit serious epileptic seizures (Waymire et al. 1995; Narisawa et al. 1997). From their observations on the abnormal nerve roots of the spinal cord, Narisawa et al. (1997) hypothesized a primary effect of TNAP depletion on neural tube during embryonic development, and suggested consequent disorders in the autonomic nervous system of the TNAP-knock out mice. All together these human and rodent observations suggest that neurological consequences of TNAP gene deficits can be dissociated from TNAP defects in other organs and tissues.

In all cases of severe hypophosphatasia, which were accompanied with seizures, epilepsy was successfully treated with pyridoxine, but only transiently (Litmanovitz et al. 2002; Nunes et al. 2002; Yamamoto et al. 2004; Balasubramaniam et al. 2010). Several lines of evidence suggest that TNAP functions as an ectoenzyme that metabolizes the B6-vitamer PLP, the essential co-factor of the GABA-synthesizing enzyme, glutamic acid decarboxylase (GAD). Thus, lack of TNAP activity could produce imbalance between excitation and inhibition in the neural networks affected, which finally can result in epilepsy as hypothesized by Waymire et al. (1995). However PLP is the co-factor of numerous enzymes, the so called B6-enzymes, involved among others in the metabolism of neurotransmitters and biogenic amines (see Chap. 11). Accordingly, in two cases on neonatal hypophosphatasia, it was shown a functional deficiency of AADC (Balasubramaniam et al. 2010), an enzyme required for the production of serotonin and dopamine, which neurotransmitters are heavily implicated in neuropsychiatric diseases (Rogers 2011; Arnsten et al. 2012). Serotonin and dopamine metabolism was also impaired in the cases reported by the groups of Hofmann and Baumgartner (Baumgartner-Sigl et al. 2007; Hofmann et al. 2013). Also, as detailed above, TNAP plays important role in neural development and transmission. Therefore, several mechanisms other than GABA deficit could be involved in the development of epilepsy in hypophosphatasia. Accordingly PLP is also important in the metabolism of sphingolipids (Bourquin et al. 2011), which are essential structural components of the cell membrane and are involved in numerous cellular processes including excitability. Therefore troubles in PLP availability may have multiple consequences, added to the other possible functions of TNAP such as nucleotide metabolism or protein dephosphorylation (see Chaps. 4, 9 and 10). Besides, phosphoethanolamine, a natural substrate candidate of TNAP (Fraser 1957), and eventually epileptogenic, might be involved in the epileptic phenotype (Whyte et al. 1988) (see also Chap. 11).

As a consequence of TNAP malfunction, brain structure appears greatly impacted in patients with severe hypophosphatasia. Cerebral imaging has revealed several impairments in the nervous tissue of the affected new-borns, such as hypodensity of the white matter, dilated ventricles, polycystic encephalopathy, atrophy of the hemispheres, cortical laminar necrosis and anomalies in several parts of the cerebral hemispheres (Nunes et al. 2002; Demirbilek et al. 2012; Hofmann et al. 2013; de Roo et al. 2014). Again, these observations show the important role of TNAP in the development both of the grey and white matters of the central nervous system and add further data supporting the fundamental role of TNAP in the human brain.

5.6.2 Relevance in Cortical Physiology

Considering cortical physiological processes, an important observation was that TNAP activity in non-human primate cortex is particularly strong at the Ranvier nodes and synapses, the critical sites of neurotransmission (Fonta et al. 2004, 2005). Also, TEM studies suggest that TNAP is expressed in inhibitory and in excitatory synapses. Moreover, TNAP activity exhibits specific laminar distribution in the human cerebral cortex with layer 4 in the primary sensory areas and layer 5 in other, higher order association areas (Négyessy et al. 2011). This is an intriguing finding considering the different functions of the cortical layers, which are determined in part by their afferents (for a review see Ahissar and Staiger, 2010). EM observations show strong TNAP activity in apparently all synaptic contacts of the TNAP positive layer (Fonta et al. 2004). Therefore, TNAP is apparently not input specific and it can be questioned if its expression is determined by the postsynaptic neuron(s) of the particular cortical layer considering that some dendrites, especially the apical ones, extend over that layer where the soma is localized. TNAP could play important role in controlling the output of that particular layer, where it is expressed at high level. Accordingly, in layer 4, where TNAP activity is not reduced to thalamo-cortical synapses, TNAP activity can be important in regulating the transmission of thalamo-cortical input to the upper layers by modulating synaptic inputs of layer 4 neurons. On the other hand, layer 5 is thought as the origin of one of the major efferent pathways of the cerebral cortex to subcortical structures including the thalamus, tectum, spinal cord, striatum and other brainstem structures.

An important aspect of cortical physiology is activity dependent redistribution of synaptic weights and connectivity. Interestingly, TNAP activity in the visual cortex apparently depends on neural activity as shown in visual deprivation and retinal neuronal inhibition experiments. The activity dependent function of TNAP seems to be a very important feature when considering its strategic position to modulate neural transmission. Therefore these observations suggest that TNAP can be an important player in regulating synaptic plasticity in the layers with high TNAP activity. This suggestion is in line with the results of network analyses showing the close relationship of TNAP and secondary signal transduction machineries in the neuronal signal transduction network (see Chap. 10).

Another notable characteristics of TNAP in regard to its cortical function, is that TNAP is expressed both in excitatory and inhibitory synapses. This subcellular localization and the selective laminar distribution raise questions about the mechanisms leading to serious seizures in hypophosphatasia: can layer specific disturbance of synaptic transmission produce unbalanced excitability leading to seizures?

It is also worth notig that TNAP contributes to the synthesis of different neurochemicals (GABA, serotonin, dopamine, adenosine), whose interactions subserve specific cognitive demands (e.g. attention, executive functions, control of learning and memory, social behaviour) or regulation of mood (e.g. in anxiety and depression…). Therefore dysregulation of neural processing in these circuits may be implicated, through TNAP deficiency, in specific neuropsychiatric disorders.

Acknowledgements We acknowledge Professor Jose Luis Millan and Professor Etienne Mornet for their perpetual support, from the beginning of the studies, and Dr Myriam Ermonval for fruitful discussions and precious advices. The support of Hypophosphatasia Europe is highly appreciated. We are grateful to Jinsong Xiao and Kevin Richetin for their contribution to the histological studies. We thank Luc Renaud, Gregory Marsal, and the staff of the animal rearing facilities for their technical assistance, Catherine Marlot for helping with bibliography.

These studies were supported by PHC Egide (Balaton 17341UE), CNRS (PICS 4331), Hypophosphatasie Europe, University of Toulouse (ASUPS) (C.F.), Erasmus Student placement European Programme between Toulouse University and Universidad Politecnica de Valencia (LRM) and the joint funding by the National Office for Research and Technology (NKTH) and the Hungarian Science and Technology Foundation (TETALAP) (FR-16/2007) (L.N).

Conflict of Interest The authors declare no conflict of interest.

References

Ahissar E, Staiger J (2010) S1 laminar specialization. Scholarpedia 5(8):7457

Anstrom JA, Brown WR, Moody DM, Thore CR, Challa VR, Block SM (2002) Temporal expression pattern of cerebrovascular endothelial cell alkaline phosphatase during human gestation. J Neuropathol Exp Neurol 61:76–84

Araki M, Saito T (1983) Electron microscopic histochemistry of alkaline phosphatase activity in the developing chick retina. Exp Eye Res 36:187–198

Arnsten AF, Wang MJ, Paspalas CD (2012) Neuromodulation of thought: flexibilities and vulnerabilities in prefrontal cortical network synapses. Neuron 76:223–239

Balasubramaniam S, Bowling F, Carpenter K, Earl J, Chaitow J, Pitt J, Mornet E, Sillence D, Ellaway C (2010) Perinatal hypophosphatasia presenting as neonatal epileptic encephalopathy with abnormal neurotransmitter metabolism secondary to reduced co-factor pyridoxal-5'-phosphate availability. J Inherit Metab Dis 33(Suppl3):25–33

Barone P, Morau S, Fonta C (2003) Postnatal angiogenesis in the auditory system of the monkey: comparison with the visual system. In: Budinger E, Gaschler-Markefski B (eds) International Conference on Auditory Cortex - Towards a synthesis of human and animal research, Shaker Verlag, Aachen, Magdeburg, Germany, Abst 31.

Baumgartner-Sigl S, Haberlandt E, Mumm S, Scholl-Burgi S, Sergi C, Ryan L, Ericson KL, Whyte MP, Hogler W (2007) Pyridoxine-responsive seizures as the first symptom of infantile hypophosphatasia caused by two novel missense mutations (c.677T > C, p. M226T; c.1112C > T, p.T371I) of the tissue-nonspecific alkaline phosphatase gene. Bone 40:1655–1661

Bell MA, Scarrow WG (1984) Staining for microvascular alkaline phosphatase in thick celloidin sections of nervous tissue: morphometric and pathological applications. Microvasc Res 27:189–203

Bethenod M, Cotte MF, Collombel C, Frederich A, Cotte J (1967) Neonatal discovery of hypophosphatasia. Bone improvement. Fatal convulsant encephalopathy. Ann Pediatr (Paris) 14:835–841

Bourquin F, Capitani G, Grutter MG (2011) PLP-dependent enzymes as entry and exit gates of sphingolipid metabolism. Protein Sci 20:1492–1508

Brun-Heath I, Ermonval M, Chabrol E, Xiao J, Palkovits M, Lyck R, Miller F, Couraud PO, Mornet E, Fonta C (2011) Differential expression of the bone and the liver tissue non-specific alkaline phosphatase isoforms in brain tissues. Cell Tissue Res 343:521–536

Buchet R, Millan JL, Magne D (2013) Multisystemic functions of alkaline phosphatases. Methods Mol Biol 1053:27–51

Casale JF, Colley VL, Legatt DF (2012) Determination of phenyltetrahydroimidazothiazole enantiomers (Levamisole/Dexamisole) in illicit cocaine seizures and in the urine of cocaine abusers via chiral capillary gas chromatography-flame-ionization detection: clinical and forensic perspectives. J Anal Toxicol 36:130–135

Ciani F, Contestabile A, Minelli G, Quaglia A (1973) Ultrastructural localization of alkaline phosphatase in cultures of nervous tissue in vitro. J Neurocytol 2:105–116

Dahl R, Sergienko EA, Su Y, Mostofi YS, Yang L, Simao AM, Narisawa S, Brown B, Mangravita-Novo A, Vicchiarelli M, Smith LH, O'Neill WC, Millan JL, Cosford ND (2009) Discovery and validation of a series of aryl sulfonamides as selective inhibitors of tissue-nonspecific alkaline phosphatase (TNAP). J Med Chem 52:6919–6925

de Roo MG, Abeling NG, Majoie CB, Bosch AM, Koelman JH, Cobben JM, Duran M, Poll-The BT (2014) Infantile hypophosphatasia without bone deformities presenting with severe pyridoxine-resistant seizures. Mol Genet Metab 111:404–407

Demirbilek H, Alanay Y, Alikasifoglu A, Topcu M, Mornet E, Gonc N, Ozon A, Kandemir N (2012) Hypophosphatasia presenting with pyridoxine-responsive seizures, hypercalcemia, and pseudotumor cerebri: case report. J Clin Res Pediatr Endocrinol 4:34–38

Diaz-Hernandez M, Gomez-Ramos A, Rubio A, Gomez-Villafuertes R, Naranjo JR, Miras-Portugal MT, Avila J (2010) Tissue-nonspecific alkaline phosphatase promotes the neurotoxicity effect of extracellular tau. J Biol Chem 285:32539–32548

Diez-Zaera M, Diaz-Hernandez JI, Hernandez-Alvarez E, Zimmermann H, Diaz-Hernandez M, Miras-Portugal MT (2011) Tissue-nonspecific alkaline phosphatase promotes axonal growth of hippocampal neurons. Mol Biol Cell 22:1014–1024

Ermonval M, Baudry A, Baychelier F, Pradines E, Pietri M, Oda K, Schneider B, Mouillet-Richard S, Launay JM, Kellermann O (2009) The cellular prion protein interacts with the tissue non-specific alkaline phosphatase in membrane microdomains of bioaminergic neuronal cells. PLoS ONE 4:e6497

Fonta C, Imbert M (2002) Vascularization in the primate visual cortex during development. Cereb Cortex 12:199–211

Fonta C, Chappert C, Imbert M (1997) N-Methyl-D-Aspartate subunit R1 involvement in the postnatal organization of the primary visual cortex of *Callithrix jacchus*. J Comp Neurol 386:260–276

Fonta C, Négyessy L, Renaud L, Barone P (2004) Areal and subcellular localization of the ubiquitous alkaline phosphatase in the primate cerebral cortex: evidence for a role in neurotransmission. Cereb Cortex 14:595–609

Fonta C, Négyessy L, Renaud L, Barone P (2005) Postnatal development of alkaline phosphatase activity correlates with the maturation of neurotransmission in the cerebral cortex. J Comp Neurol 486:179–196

Fraser D (1957) Hypophosphatasia. Am J Med 22:730–746

Friede RL (1966) A quantitative mapping of alkaline phosphatase in the brain of the rhesus monkey. J Neurochem 13:197–203

Goldstein DJ, Harris H (1981) Mammalian brain alkaline phosphatase: expression of liver/bone/kidney locus. Comparison of fetal and adult activities. J Neurochem 36:53–57

Gomori G (1939) Microtechnical demonstration of phosphatase in tissue sections. Proc Soc Exp Biol Med 42:23–26

Hahnel AC, Rappolee DA, Millan JL, Manes T, Ziomek CA, Theodosiou NG, Werb Z, Pedersen RA, Schultz GA (1990) Two alkaline phosphatase genes are expressed during early development in the mouse embryo. Development (Cambridge, England) 110:555–564

Hanics J, Barna J, Xiao J, Millan JL, Fonta C, Négyessy L (2012) Ablation of TNAP function compromises myelination and synaptogenesis in the mouse brain. Cell Tissue Res 349:459–471

Hofmann C, Liese J, Schwarz T, Kunzmann S, Wirbelauer J, Nowak J, Hamann J, Girschick H, Graser S, Dietz K, Zeck S, Jakob F, Mentrup B (2013) Compound heterozygosity of two functional null mutations in the ALPL gene associated with deleterious neurological outcome in an infant with hypophosphatasia. Bone 55:150–157

Iida K, Fukushi J, Fujiwara T, Oda Y, Iwamoto Y (2012) Adult hypophosphatasia with painful periarticular calcification treated with surgical resection. J Bone Miner Metab 30:722–725

Kennedy H, Bullier J, Dehay C (1985) Cytochrome oxidase activity in the striate cortex and lateral geniculate nucleus of the new born and adult macaque monkey. Exp Brain Res 61:204–209

Kermer V, Ritter M, Albuquerque B, Leib C, Stanke M, Zimmermann H (2010) Knockdown of tissue nonspecific alkaline phosphatase impairs neural stem cell proliferation and differentiation. Neurosci Lett 485:208–211

Kozlenkov A, Le Du MH, Cuniasse P, Ny T, Hoylaerts MF, Millan JL (2004) Residues determining the binding specificity of uncompetitive inhibitors to tissue-nonspecific alkaline phosphatase. J Bone Miner Res 19:1862–1872

Langer D, Ikehara Y, Takebayashi H, Hawkes R, Zimmermann H (2007) The ectonucleotidases alkaline phosphatase and nucleoside triphosphate diphosphohydrolase 2 are associated with subsets of progenitor cell populations in the mouse embryonic, postnatal and adult neurogenic zones. Neuroscience 150:863–879

Langer D, Hammer K, Koszalka P, Schrader J, Robson S, Zimmermann H (2008) Distribution of ectonucleotidases in the rodent brain revisited. Cell Tissue Res 334:199–217

Larsson H, Wallin M, Edstrom A (1979) Acid and alkaline phosphatases in a brain tubulin preparation. J Neurochem 32:155–161

Latker CH, Shinowara NL, Miller JC, Rapoport SI (1987) Differential localization of alkaline phosphatase in barrier tissues of the frog and rat nervous systems: a cytochemical and biochemical study. J Comp Neurol 264:291–302

Lebrand C, Cases O, Adelbrecht C, Doye A, Alvarez C, El Mestikawy S, Seif I, Gaspar P (1996) Transient uptake and storage of serotonin in developing thalamic neurons. Neuron 17:823–835

Leung EC, Mhanni AA, Reed M, Whyte MP, Landy H, Greenberg CR (2013) Outcome of perinatal hypophosphatasia in manitoba mennonites: a retrospective cohort analysis. JIMD reports 11:73–78

Levitt P (2003) Structural and functional maturation of the developing primate brain. J Pediatr 143: S35–S45

Litmanovitz Reish O, Dolfin T, Arnon S, Regev R, Grinshpan G, Yamazaki M, Ozono K (2002) Glu274Lys/Gly309Arg mutation of the tissue-nonspecific alkaline phosphatase gene in neonatal hypophosphatasia associated with convulsions. J Inherit Metab Dis 25:35–40

Matsuura S, Kishi F, Kajii T (1990) Characterization of a 5’-flanking region of the human liver/bone/kidney alkaline phosphatase gene: two kinds of mRNA from a single gene. Biochem Biophys Res Commun 168:993–1000

Mayahara H, Hirano H, Saito T, Ogawa K (1967) The new lead citrate method for the ultracytochemical demonstration of activity of non-specific alkaline phosphatase (orthophosphoric monoester phosphohydrolase). Histochemie 11:88–96

Miki A, Tanaka Y, Ogata S, Ikehara Y (1986) Selective preparation and characterization of membranous and soluble forms of alkaline phosphatase from rat tissues. A comparison with the serum enzyme. Eur J Biochem 160:41–48

Missler M, Wolff JR, Heger W, Merker HJ, Treiber AM, Scheib R, Crook GA (1992) Developmental biology of the common marmoset : proposal for a “postnatal staging”. J Med Primatol 21:285–298

Missler M, Eins S, Merker HJ, Rothe H, Wolff JR (1993a) Pre-and postnatal development of the primary visual cortex of the common marmoset. I. A changing space for synaptogenesis. J Comp Neurol 333:41–52

Missler M, Wolff A, Merker HJ, Wolff JR (1993b) Pre- and postnatal development of the primary visual cortex of the common marmoset. II. Formation, remodelling, and elimination of synapses as overlapping processes. J Comp Neurol 333:53–67

Mori S, Nagano M (1985a) Electron-microscopic cytochemistry of alkaline-phosphatase activity in endothelium, pericytes and oligodendrocytes in the rat brain. Histochemistry 82:225–231

Mori S, Nagano M (1985b) Electron microscopic cytochemistry of alkaline phosphatase in neurons of rats. Arch Histol Jpn 48:389–397

Narisawa S, Hasegawa H, Watanabe K, Millan JL (1994) Stage-specific expression of alkaline phosphatase during neural development in the mouse. Dev Dyn 201:227–235

Narisawa S, Frohlander N, Millan JL (1997) Inactivation of two mouse alkaline phosphatase genes and establishment of a model of infantile hypophosphatasia. Dev Dyn 208:432–446

Négyessy L, Xiao J, Kantor O, Kovacs GG, Palkovits M, Doczi TP, Renaud L, Baksa G, Glasz T, Ashaber M, Barone P, Fonta C (2011) Layer-specific activity of tissue non-specific alkaline phosphatase in the human neocortex. Neuroscience 172:406–418

Ng TB, Tam PP (1986) Changes of acid and alkaline phosphatase activities in the developing mouse brain. Biol Neonate 50:107–113

Norman MG, O'Kusky JR (1986) The growth and development of microvasculature in human cerebral cortex. J Neuropathol Exp Neurol 45:222–232

Nunes ML, Mugnol F, Bica I, Fiori RM (2002) Pyridoxine-dependent seizures associated with hypophosphatasia in a newborn. J Child Neurol 17:222–224

Oda K, Amaya Y, Fukushi-Irie M, Kinameri Y, Ohsuye K, Kubota I, Fujimura S, Kobayashi J (1999) A general method for rapid purification of soluble versions of glycosylphosphatidylinositol-anchored proteins expressed in insect cells: an application for human tissue-nonspecific alkaline phosphatase. J Biochem 126:694–699

Ovtscharoff W (1973) Ultracytochemical localisation of the alkaline phosphatase in the cerebral cortex of newborn rats. Histochemie 37:93–95

Paus T, Collins DL, Evans AC, Leonard G, Pike B, Zijdenbos A (2001) Maturation of white matter in the human brain: a review of magnetic resonance studies. Brain Res Bull 54:255–266

Pinner B, Campbell JB (1965) Alkaline phosphatase activity in incisures and nodes during degeneration and regenaration of peripheral nerve fibers. Exp Neurol 12:159–172

Przegalinski E, Bigajska K, Siwanowicz J (1980) Psychopharmacological profile of dexamisole. Pol J Pharmacol Pharm 32:21–29

Rebsam A, Seif I, Gaspar P (2002) Refinement of thalamocortical arbors and emergence of barrel domains in the primary somatosensory cortex: a study of normal and monoamine oxidase a knock-out mice. J Neurosci 22:8541–8552

Rogers RD (2011) The roles of dopamine and serotonin in decision making: evidence from pharmacological experiments in humans. Neuropsychopharmacology 36:114–132

Robins E, Smith DE, Eydt KM, McCaman RE (1956) The quantitative histochemistry of the cerebral cortex. II. Architectonic distribution of nine enzymes in the motor and visual cortices. J Neurochem 1:68–76

Scherer SS (1996) Molecular specializations at nodes and paranodes in peripheral nerve. Microsc Res Tech 34:452–461

Sergienko E, Su Y, Chan X, Brown B, Hurder A, Narisawa S, Millan JL (2009) Identification and characterization of novel tissue-nonspecific alkaline phosphatase inhibitors with diverse modes of action. J Biomol Screen 14:824–837

Shimizu N (1950) Histochemical studies on the phosphatase of the nervous system. J Comp Neurol 93:201–217

Spatz WB, Vogt DM, Illing RB (1993) Distribution of cytochrome oxidase in layers IV and V of the striate cortex in neonate monkeys. Exp Brain Res 95:183–186

Street SE, Kramer NJ, Walsh PL, Taylor-Blake B, Yadav MC, King IF, Vihko P, Wightman RM, Millan JL, Zylka MJ (2013) Tissue-nonspecific alkaline phosphatase acts redundantly with PAP and NT5E to generate adenosine in the dorsal spinal cord. J Neurosci 33:11314–11322

Studer M, Terao M, Gianni M, Garattini E (1991) Characterization of a second promoter for the mouse liver/bone/kidney-type alkaline phosphatase gene: cell and tissue specific expression. Biochem Biophys Res Commun 179:1352–1360

Sugimura K, Mizutani A (1979) Histochemical and cytochemical studies of alkaline phosphatase activity in the synapses of rat brain. Histochemistry 61:123–129

Takamatsu H (1939) Histologische und biochemische Studien über die Phosphatase. Histochemische Untersuchungsmethodik der Phosphatase und deren Verteilung in verschiedenen Organen und Geweben. Trans Soc Path Jap 29:492–498

Taketani T, Onigata K, Kobayashi H, Mushimoto Y, Fukuda S, Yamaguchi S (2014) Clinical and genetic aspects of hypophosphatasia in Japanese patients. Arch Dis Child 99:211–215

Toh Y, Yamamoto M, Endo H, Misumi Y, Ikehara Y (1989) Isolation and characterization of a rat liver alkaline phosphatase gene. A single gene with two promoters. Eur J Biochem 182:231–237

Van Belle H (1976) Alkaline phosphatase. I. Kinetics and inhibition by levamisole of purified isoenzymes from humans. Clin Chem 22:972–976

Vanhoutte PM, Van Nueten JM, Verbeuren TJ, Laduron PM (1977) Differential effects of the isomers of tetramisole on adrenergic neurotransmission in cutaneous veins of dog. J Pharmacol Exp Ther 200:127–140

Vardy ER, Kellett KA, Cocklin SL, Hooper NM (2012) Alkaline phosphatase is increased in both brain and plasma in Alzheimer's disease. Neuro-degenerative Dis 9:31–37

Vorbrodt AW, Lossinsky AS, Wisniewski HM (1986) Localization of alkaline phosphatase activity in endothelia of developing and mature mouse blood-brain barrier. Dev Neurosci 8:1–13

Waymire KG, Mahuren JD, Jaje JM, Guilarte TR, Coburn SP, MacGregor GR (1995) Mice lacking tissue non-specific alkaline phosphatase die from seizures due to defective metabolism of vitamin B-6. Nat Genet 11:45–51

Weiss MJ, Ray K, Henthorn PS, Lamb B, Kadesch T, Harris H (1988) Structure of the human liver/bone/kidney alkaline phosphatase gene. J Biol Chem 263:12002–12010

Whyte MP, Mahuren JD, Fedde KN, Cole FS, McCabe ER, Coburn SP (1988) Perinatal hypophosphatasia: tissue levels of vitamin B6 are unremarkable despite markedly increased circulating concentrations of pyridoxal-5'-phosphate. Evidence for an ectoenzyme role for tissue-nonspecific alkaline phosphatase. J Clin Investig 81:1234–1239

Whyte MP, Wenkert D, McAlister WH, Mughal MZ, Freemont AJ, Whitehouse R, Baildam EM, Coburn SP, Ryan LM, Mumm S (2009) Chronic recurrent multifocal osteomyelitis mimicked in childhood hypophosphatasia. J Bone Miner Res 24:1493–1505

Yamamoto H, Sasamoto Y, Miyamoto Y, Murakami H, Kamiyama N (2004) A successful treatment with pyridoxal phosphate for West syndrome in hypophosphatasia. Pediatr Neurol 30:216–218

Zhang D, Xiong W, Chu S, Sun C, Albensi BC, Parkinson FE (2012a) Inhibition of hippocampal synaptic activity by ATP, hypoxia or oxygen-glucose deprivation does not require CD73. PLoS ONE 7:e39772

Zhang H, Ke YH, Wang C, Yue H, Hu WW, Gu JM, Zhang ZL (2012b) Identification of the mutations in the tissue-nonspecific alkaline phosphatase gene in two Chinese families with hypophosphatasia. Arch Med Res 43:21–30

Zisapel N, Haklai R (1980) Localization of an alkaline phosphatase and other synaptic vesicle proteins. Neuroscience 5:2297–2303

Zisapel N, Levi M (1980) A calcium-stimulated alkaline phosphatase associated with synaptic vesicles. Brain Res 201:385–398

Chapter 6
The Retinal TNAP

Orsolya Kántor, Dorottya Cserpán, Béla Völgyi, Ákos Lukáts and Zoltán Somogyvári

Abstract Accumulating evidence from recent literature underline the important roles of tissue non specific alkaline phosphatase (TNAP) in diverse functions as well as diseases of the nervous system. Exploration of TNAP in well characterized neural circuits such as the retina, might significantly advance our understanding regarding neural TNAP's roles. This chapter reviews the scarce literature as well as our findings on retinal TNAP. We found that retinal TNAP activity was preserved and followed diverse patterns throughout vertebrate evolution. We have consistently observed TNAP activity (1) in retinal vessels, (2) in photoreceptors and (3) in the majority of the studied species in the outer (OPL) and inner plexiform layers (IPL), where synaptic transmission occurs. Importantly, in some species the IPL exhibits several TNAP positive strata. These strata exactly corresponded those seen after quadruple immunohistochemistry with four canonical IPL markers (tyrosine hydroxylase, choline acetyltransferase, calretinin, protein kinase C α). Diabetes results in diminishing retinal TNAP activity before changes in canonical markers

O. Kántor (✉)
Department of Anatomy, Histology and Embryology, Semmelweis University, Tűzoltó str. 58, Budapest 1094, Hungary
e-mail: kantor.orsolya@med.semmelweis-univ.hu

D. Cserpán · Z. Somogyvári
Theoretical Neuroscience and Complex Systems Research Group, Wigner Research Center for Physics, Hungarian Academy of Sciences, Budapest, Hungary

B. Völgyi
Department of Experimental Zoology and Neurobiology, University of Pécs, Pécs, Hungary

B. Völgyi
János Szentágothai Research Center, Pécs, Hungary

Á. Lukáts
Department of Human Morphology and Developmental Biology, Semmelweis University, Budapest, Hungary

C. Fonta and L. Négyessy (eds.), *Neuronal Tissue-Nonspecific Alkaline Phosphatase (TNAP)*, Subcellular Biochemistry 76,
DOI 10.1007/978-94-017-7197-9_6

could be observed in a rat model. The presence of TNAP activity at critical sites of neurotransmission suggests its important and evolutionary conserved role in vision. In diabetes, the decreased TNAP activity indicates neurological alterations adding further evidence for the role of TNAP in brain diseases.

Keywords TNAP · Retina · Inner plexiform layer · Visual processing · Diabetes

6.1 Brief Introduction to the Retina

The retina, which is part of the diencephalon, lies outside the brain and is connected through the optic nerve to it. In the mammalian retina approximately 50–60 different cell types serve the detection and initial processing of light impulses. With the exception of primate fovea, these cell types are distributed in a fairly even way, consisting of many parallel, anatomically equipotent microcircuits rather than of a few major pathways. Elaborate chemical as well as electrical synapses enable communication between the neurons. Schematized, information flows through vertical channels between photoreceptors, bipolars and ganglion cells (the sole retinal output neurons). This information flow is modified in the lateral direction through horizontal and amacrine cells. The diverse cell types are arranged in a crystallised anatomical structure. Photosensitive elements, cell bodies and interconnected fibers of the retinal neurons are strictly organized in nine neuronal layers: (1) photoreceptor layer (photosensitive parts of rods and cones), (2) outer limiting membrane (consisting of the outer processes of Müller glia), (3) inner nuclear layer (INL, cell bodies of rods and cones), (4) outer plexiform layer (OPL, highly specialized synapses between rods, cones and bipolar and horizontal cells), (5) inner nuclear layer (INL, cell bodies of horizontal, bipolar and amacrine cells as well as Müller glia), (6) inner plexiform layer (IPL, an ordered stack of synaptic planes between bipolars, amacrines and ganglion cells), (7) ganglion cell layer (GCL, cell bodies of ganglion cells), (8) nerve fiber layer (composed of the axons of ganglion cells running toward the optic disk), (9) inner limiting membrane (inner fibers of the Müller glia isolating the neural retina from the vitreous body). As compared to other parts of the central nervous system, retinal cell types and their connections are fairly well characterized. Well known connections, easy accessibility and stimulability of the neurons render the retina a widely used model system in neuroscience research.

Moreover, common systemic human diseases (most frequently hypertension and diabetes mellitus) can result in severe ocular complications leading (if untreated) often to blindness. The retina is the only part of the organism where arterioles (affected by the above diseases) can be observed directly as part of ophthalmological examination. Other common neurological diseases such as Alzheimer's and Parkinson's disease (see recent reviews Dehabadi et al. 2014; Bodis-Wollner et al. 2014) also have ocular manifestations, therefore, retinal research has a major impact also on human pathophysiology.

Although research tackling TNAP's involvement in functioning and malfunctioning of the nervous system is emerging, little attention was paid on the retina. In this chapter we summarize the sparse literature and our findings in this field.

6.2 Localisation of Retinal TNAP Activity in Different Mammalian and Non-Mammalian Species

TNAP activity was reported in the homogenized retina of chicken embryos as early as 1949 (Lindeman 1949) and on histological sections in 1963 (Rogers 1963). In mammals, a few controversial papers exist, which were debated even at their time because of unreliability of the used histochemical methods (Yoshida 1957; De Vincentiis and Testa 1959; Reis 1954). Although the reliable diazo-dye method was used in the work of Reis (1954) in ox, the paper was focused on retinal vessels instead of neural-glial elements. More recently, Iandiev et al. (2007) found no detectable TNAP activity in mouse retina. However, in their work the histochemical reaction was carried out at pH 7.4 which lies well under the pH optimum of alkaline phosphatase (pH 9.3; Fedde and Whyte 1990).

In our very recent work (Kántor et al. 2014), we investigated the presence and pattern of TNAP activity in the retinas of several species throughout the vertebral evolution: in zebrafish, convict cichlid, frog, chicken, rat, mice, rabbit, golden hamster, guinea pig, ferret, cat, dog, squirrel monkey. Additionally, post-mortem samples from human patients without reported eye disease were also investigated. Western blot (WB) analysis and TNAP enzyme histochemistry were carried out on the samples.

In the studied species, polyclonal rabbit TNAP antibody (1:2,000; Abcam, ab65834, Cambridge, UK; Díaz-Hernandez et al. 2010) detected a single band of approximately 90 kDa in the samples (Fig. 6.1), which corresponds to the molecular weight of TNAP dimer (NCBI Protein Database, Le Du and Millán 2002). Therefore, TNAP is expressed in the retina of a wide range of vertebrate species.

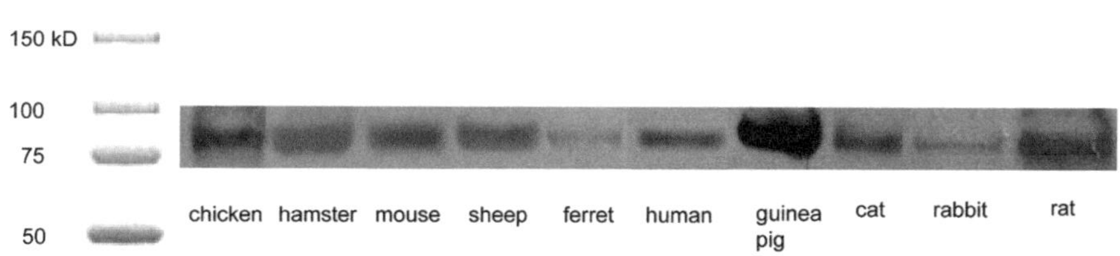

Fig. 6.1 Western blot analysis of TNAP in retinas of different species. A single band at 90 kDa was detected in each investigated species. *Left* molecular marker. Originally published in Cell and Tissue Research, vol. 358, p. 90, Fig. 6.1, TNAP activity is localized at critical sites of retinal neurotransmission across various vertebrate species, Kántor O, Varga A, Kovács-Öller T, Énzsöly A, Balogh L, Baksa G, Szepessy Z, Fonta C, Roe AW, Nitschke R, Szél Á, Négyessy L, Völgyi B, Lukáts Á. With kind permission from Springer Science and Business media

The strength of the TNAP bands varied across the different species, suggesting a difference in the expression levels.

After TNAP histochemistry, a tremendous variation of TNAP activity patterns were found in retinal sections of the examined species. The species were grouped according to their activity pattern and representative species are shown in Fig. 6.2.

6.2.1 All or None: Dog, Zebrafish

In the dog retina, no neural elements, only retinal vessels showed alkaline phosphatase activity (asterisks on Fig. 6.2a). Lutty and McLeod (2005) reported similar findings. In contrast, TNAP activity was present in all layers of the zebrafish retina (Fig. 6.2b). Somata in the nuclear layers as well as both plexiform layers were extensively outlined by the reaction product. Notably, four darkly and three intermingled relatively lightly stained strata of the inner plexiform layer (IPL) became visible. In the photoreceptors' layer, the triangular shaped outer segments of short single cones (arrows) were heavily, whereas long single cone, double cone and rod outer segments only weakly stained.

Retinal TNAP activity pattern in sheep and cichlid fish (not shown) was similar to that of the zebrafish, except that no stratification pattern of the IPL was found after the histochemical reaction.

6.2.2 TNAP Activity in the Outer Retina: Squirrel Monkey and Ferret

We found a pronounced difference between the staining intensity of the outer and inner retina of the squirrel monkey (Fig. 6.2c) and ferret (not shown). In both species, inner retinas remained unstained, whereas considerable TNAP activity was detected in the outer retinas. In the photoreceptor layer, rod outer segments displayed moderate TNAP activity, whereas their inner segments were relatively heavily labeled. In contrast, cone inner segments (arrows on Fig. 6.2c) displayed lower activity than the surrounding rod inner segments, thus seeming to be TNAP negative. Somata of photoreceptors in the outer nuclear layer were homogenously delineated by moderate enzymatic activity that rapidly declined from the OPL inward and resulted in very weak or no staining in the inner retina. Beside neuronal elements, blood vessels also displayed TNAP activity in the squirrel monkey and ferret.

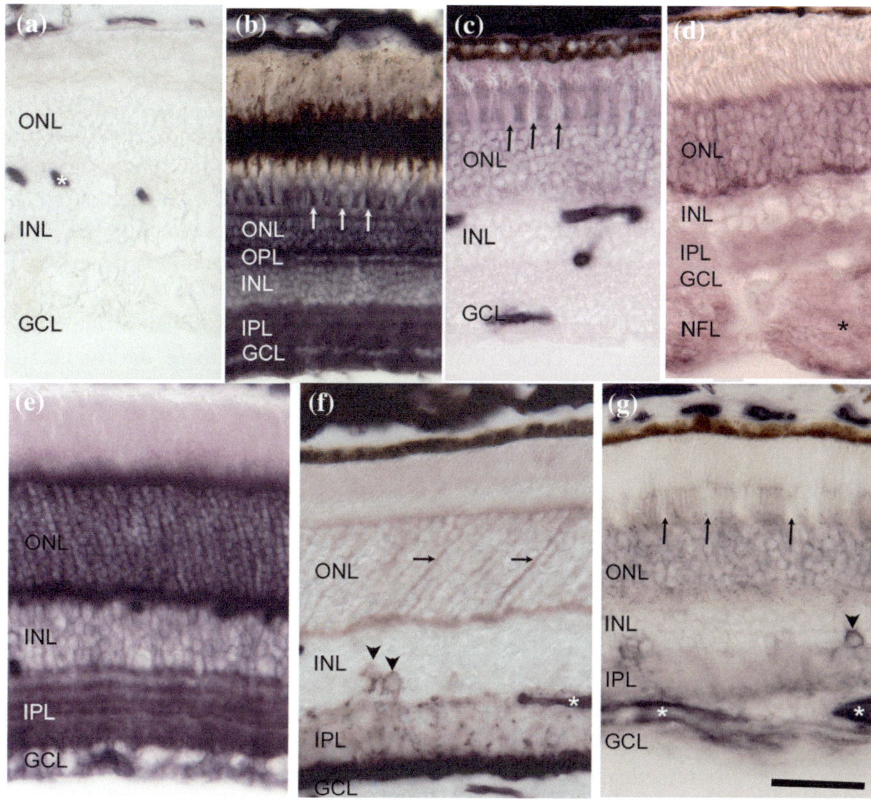

Fig. 6.2 Representative species showing different patterns of retinal TNAP activity. In dog (**a**), only retinal vessels (*asterisk*) displayed TNAP activity. In contrast, in zebrafish (**b**), all retinal layers were strongly stained. *Arrows* single short cones. In squirrel monkey (**c**), only the outer retina showed TNAP staining, cones (*arrows*) were much more weakly stained than the surrounding rod inner segments. OPL showed moderate TNAP activity, whereas the intensity of the staining dropped abruptly towards the inner retina. In rabbit (**d**), photoreceptor inner segments, somata and both plexiform layers were heavily stained. Moreover, at central locations, bundles of optic fibers (*asterisk*) also displayed moderate TNAP activity. In rat (**e**), TNAP activity was most prominent at the level of the outer segments, in the OPL and ONL. Moreover, in the IPL, five darkly labeled strata and a fainter stained stripe with intervened lighter strata became visible. In mice (**f**), photoreceptors were moderately, somata strongly stained. Occasionaly, some photoreceptors (*cones*, *arrows*) could be traced until their end feet embedded in the heavily stained OPL. Some amacrine cells (*arrowheads*) in the INL were also outstandingly labeled, with processes ending in the innermost stratum of the IPL. In contrast, in the human retina (**g**), inner segments of cones (*arrows*) were more weakly stained than surrounding rods. In the INL, some amacrine cells (*arrowheads*) were also heavily stained with processes ending also in the innermost layer of the IPL. *ONL* outer nuclear layer, *INL* inner nuclear layer, *GCL* ganglion cell layer, *OPL* outer plexiform layer, *IPL* inner plexiform layer, *NFL* nerve fiber layer. *Bar* 50 μm. Parts of the figure are originally published in Cell and Tissue Research, vol. 358, Fig. 3b, 4a, 5b, 8e, pp. 91, 92 and 96, respectively. TNAP activity is localized at critical sites of retinal neurotransmission across various vertebrate species, Kántor O, Varga A, Kovács-Öller T, Énzsöly A, Balogh L, Baksa G, Szepessy Z, Fonta C, Roe AW, Nitschke R, Szél Á, Négyessy L, Völgyi B, Lukáts Á. With kind permission from Springer Science and Business media

6.2.3 TNAP Activity in the Photoreceptor and Synaptic Layers: Cat, Rabbit and Guinea Pig

In rabbit, blood vessels are only present in the optic disc area and, as in previous species, they showed TNAP activity (picture not shown). Rod inner segments, photoreceptor somata and terminals became visible after TNAP histochemistry. The strongest TNAP activity was found at the level of rod outer segments and in the OPL. The IPL and (at central locations) bundles of optic fibers (asteriks on Fig. 6.2d) were moderately stained.

In cat and guinea pig the TNAP staining pattern was similar to that of the rabbit. Interestingly, in guinea pig, TNAP histochemistry revealed three darkly labeled strata in the IPL. Moreover, some amacrine cells were also outlined by the purple reaction product.

6.2.4 TNAP Activity in Multiple Strata of the IPL: Rat and Chicken

In the retina of all investigated rat strains (Sprague-Dawley, Wistar and Long Evans), blood vessels were darkly stained by TNAP enzyme histochemistry (Fig. 6.2e). Weak, homogenous TNAP staining was seen at the level of the outer segments, whereas the activity was more intense in the layer of the inner segments. Somata in the outer nuclear layer as well as photoreceptor terminals in the OPL were densely stained. Soma outlines in both the INL and GCL and the entire IPL displayed TNAP activity. TNAP histochemistry revealed five dark and one somewhat fainter TNAP positive layers in the IPL that were separated by thin TNAP negative stripes. The detailed description of the exact position of these strata is given in the following section.

In the adult chicken retina (picture not shown), similarly to rats, photoreceptors (especially inner segments) and both plexiform layers displayed strong TNAP activity. As seen in rats, several strata in the IPL became visible after TNAP histochemistry. Moreover, calretinin containing horizontal cells were also heavily stained. In central retinal locations, where the optic fiber layer is prominent, TNAP-labeled optic axons became visible.

6.2.5 TNAP Positivity in Retinal Amacrine Cells: Frog, Mouse, Golden Hamster and Human

Blood vessels were strongly TNAP stained in the mouse retina (asterisk on Fig. 6.2f). A moderate enzymatic activity was found in the photoreceptor layer. Here, outer segments displayed stronger staining than the inner segments.

Occasionally, cell bodies and photoreceptor inner fibers exhibited strong TNAP labeling (arrows on Fig. 6.2f). In these latter cases the TNAP stained photoreceptors were traceable to their pedicles embedded in the less prominently stained OPL. In the inner retina, a subpopulation of amacrine cells displayed marked TNAP activity (arrowheads on Fig. 6.2f). Their fine processes were apparent as they richly arborized in the IPL. Most of these TNAP labeled amacrine cell processes ended in the innermost strata of the IPL (Fig. 6.2f), where the strongest TNAP labeling was seen. In these subset of amacrine cells, TNAP activity colocalized with glutamic acid decarboxylase (GAD67) immunoreactivity showing that they corresponded to to GABAergic amacrine cells.

TNAP labeling in golden hamster (picture not shown) was similar to the pattern seen in mice. Interestingly, this was the only species where vessels displayed no detectable TNAP activity.

In the human retina, blood vessels exhibited a strong TNAP labeling (asterisks on Fig. 6.2g). Rod photoreceptor outer segments were only lightly stained, whereas a somewhat stronger TNAP labeling was seen at the level of the inner segments. Interestingly, both cone outer and inner segments (arrows on Fig. 6.2g) were much weaker (if at all) TNAP stained. Photoreceptor somata were clearly outlined by the TNAP reaction product while the staining intensity was only moderate in the OPL. In the inner retina, some amacrine cells became visible by the TNAP staining (arrowhead on Fig. 6.2g). Occasionally, dendritic processes of these amacrine cells were apparent as they entered the inner division of the IPL. As seen in mouse and hamster, the dendrites of these amacrine cells reached the innermost stratum of the IPL. Here, TNAP activity appeared to be somewhat stronger than in the outer parts of the layer.

TNAP staining pattern in the frog retina (data not shown) was similar to the other species in this subgroup. The IPL was divided into seven strata by three strongly and four weakly TNAP stained stripes.

In summary, retinal TNAP activity was retained throughout vertebrate evolution. The activity pattern in the examined species differed tremendously. However, some trends were found regarding the TNAP staining pattern. (1) Retinal blood vessels displayed strong TNAP activity in all examined species but the golden hamster. (2) Photoreceptors were consistenty stained, however, the staining was not constricted to either rods or cones. (3) In the majority of the examined species, OPL and IPL, layers where synaptic connections occur were also TNAP labeled. We do not have direct electron microscopical evidence but this latter finding suggests that, similarly to the vertebrate neocortex (Fonta et al. 2004; Négyessy et al. 2011), retinal TNAP activity also plays a role in synaptic functions. 4 In some studied species, the inner retina also displayed TNAP activity. In certain species (frog, guinea pig, mice, golden hamster, human) some amacrine cells were outlined by the TNAP staining, in other species (zebrafish, chicken, rat, mice, golden hamster, human) strata of the IPL were revealed by TNAP histochemistry. The variations of the staining patterns did not follow evolutionary relationships, as evolutionary distant species (e.g. frog and humans) displayed similar patterns, whereas closely related species (e.g. squirrel monkey and humans) were often different. The staining

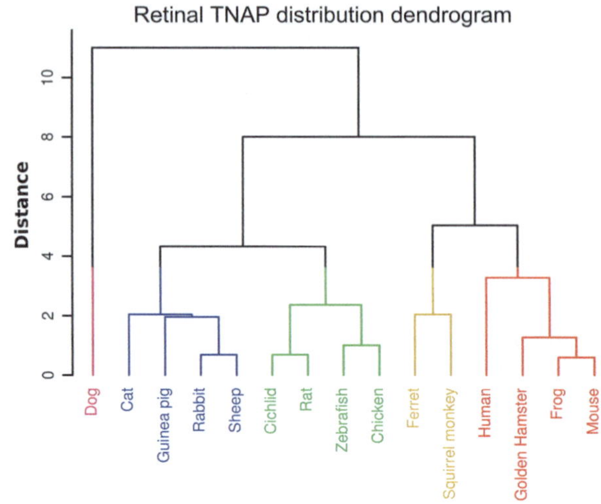

Fig. 6.3 Hierarchical clustering of the retinal TNAP distribution of the examined species resulted in five main clusters of species depicted by different colors

pattern did not vary according to the diurnal life style of the species, since crepuscular dog, cat and rodents showed great differences. The real cause of the described variations remains obscure.

In order to examine the structural similarities between retinal TNAP distribution among the examined species in an observer independent manner, we performed hierarchical cluster analysis on the retinal layer staining intensity profiles. The I_j layer staining intensities were described semiquantitatively on a 0–3 scale. The distance between the staining intensity profiles were measured as:

$$D_{jk} = \mathrm{sum}\left(\mathrm{abs}\left({I_j}^{0.75} - {I_k}^{0.75}\right)\right)$$

Then, the complete linkage hierarchical clustering algorithm were performed in R (http://cran.r-project.org), in which we merge in each step the two clusters with the smallest maximum pairwise distance. The resulted similarity tree is presented as a dendrogram on Fig. 6.3, where the 5 main branches are colored to emphasize the structural families. The resulting clusters of species largely corresponded those established by visual observations (Fig. 6.3).

6.3 The Stratified Organization of the Inner Plexiform Layer Is Revealed by TNAP Activity in the Rat Retina

The IPL was historically divided into 5 substrata (Cajal 1893; see in Kolb 2007) or in two sublaminae (ON and OFF). Recent electrophysiological studies expanded our knowledge about the functional organization of IPL. Accordingly, it was suggested that the visual world is differentially represented in at least ten parallel strata, thereby creating visual functions such as local edge detection, direction selectivity and looming detection (Roska and Werblin 2001). Therefore, information about lamina specific pattern of connectivity between retinal cell types in the IPL is pivotal for studies tackling retinal or downward visual processing in the brain.

Roska and Werblin (2001) defined ten IPL strata based on the dendritic arborization and physiological properties of different ganglion cell types in rabbits. Recently, Siegert et al. (2009) divided the mouse IPL into 10 strata by combining starburst amacrine cell marker choline acetyltransferase (ChAT), Ca^{2+} binding protein calretinin (CR), tyrosine hydroxylase (TH; marker of the dopaminergic amacrine cells) and rod bipolar marker protein kinase C alpha subunit (PKCα) immunohistochemical reactions. Notably, the innermost part of the IPL with strong PKCα staining was divided arbitrarily into two strata by the authors.

In rats, we have shown that TNAP histochemistry alone visualized five dark and one fainter TNAP positive strata in the IPL that were separated by thin TNAP negative stripes. These alternating TNAP-positive and negative layers divided the IPL into eleven sequential strata (Fig. 6.2g). The first four TNAP positive and the three TNAP negative strata in the distal IPL were easily discernible. The four additional TNAP positive and TNAP negative strata in the innermost IPL were not always obvious and could only be clearly separated by detailed mathematical analysis (Kántor et al. 2015). The strata revealed by TNAP histochemistry were numerated 1–11 in the disto-proximal direction of the IPL. In our scheme, strata 1, 3, 5, 7, 9 and 11 were TNAP positive and the intermediate TNAP negative layers were strata 2, 4, 6, 8 and 10. Moreover, we performed a further mathematical model based analysis to quantify the exact position of the above strata. Detailed description of the method can be found in Kántor et al. (2015). Briefly, individual TNAP staining intensity profiles of selected IPL areas were generated using Image J. The individual TNAP profiles were combined to generate a mean TNAP staining profile (Fig. 6.4a). Next, the BIC (Bayesian information criteria) was calculated from $N = 1$ to 12. The BIC measures the optimality of a model by balancing the fitting error and the model complexity. The minimum of the BIC was obtained at $N = 6$ indicating that the observed TNAP staining can be modeled best by the sum of 6 individual Gaussian curves. Adjusting the amplitude, center position and width of the Gaussians toward smaller fitting error, the best fitting parameter set of the 6 Gaussians emerged. Notably, the fitted model and the mean TNAP profile almost exactly matched (Fig. 6.4b). As a consequence, the fitted model consisted of 6 darker and 5 lighter TNAP stripes in the IPL. The 6 Gaussian curves had similar amplitudes. TNAP activity was highest in the middle part of the IPL, especially in stratum 7. Thereafter, IPL strata borders were determined as the positions of the inflection points of the fitted Gaussians. The most variable staining was observed at IPL depth encompassing strata 7–10. Therefore, as seen in Fig. 6.3a, b on the mean TNAP intensity curve, peaks corresponding to stratum 5 and 7 as well as 9 and 11 lie close together and the minimum value between them is not pronounced.

In rats we performed TH, ChAT, CR, PKCα immunoreaction combined with TNAP histochemistry in the same retinal specimen to determine whether IPL strata defined by TNAP staining correspond those seen with canonical IPL markers. The

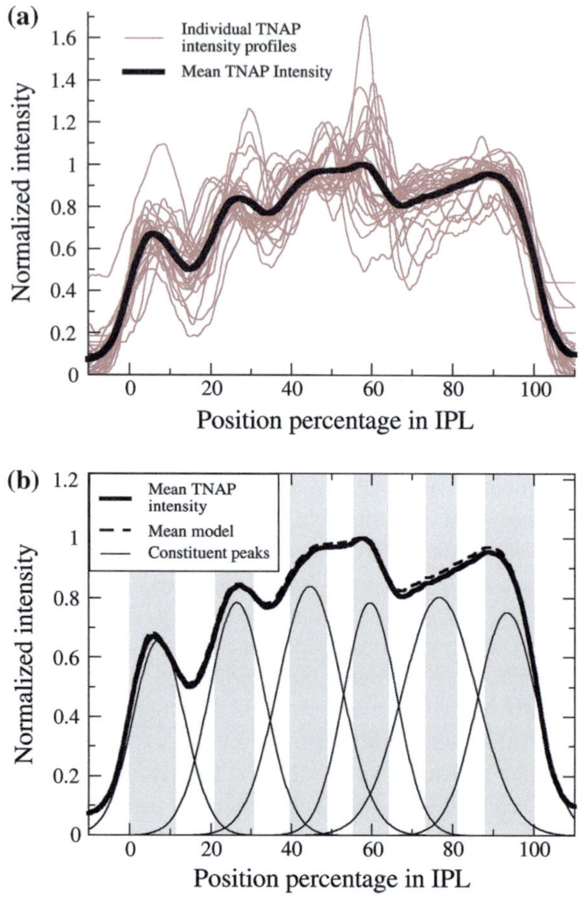

Fig. 6.4 Characterization of TNAP labeled IPL strata in the rat retina using model based analysis. **a** Mean profile of TNAP staining intensity (*dark line* in **a**, **b**) was created from plots of individual TNAP stainings (*light lines*). **b** Using Bayesian information criterion analysis, the best model of the mean TNAP intensity results from the sum of six Gaussian curves (*dotted line* in **b**). Indeed, the best fitting model (*dotted line*) almost exactly matched the mean TNAP intensity profile (*thick line*). Borders of the IPL strata were defined as the inflection points of the Gaussians. According to the model, 11 strata were delineated. Originally published in Cell and Tissue Research, vol. 359, p.415 Fig. 2c,e, Stratified organization and disorganization of inner plexiform layer revelaed by TNAP activity in healthy and diabetic rat retina, Kántor O, Varga A, Tóth R, Énzsöly A, Pálfi E, Kovács-Öller T, Nitschke R, Szél Á, Székely A, Völgyi B, Négyessy L, Somogyvári Z, Lukáts Á. With kind permission from Springer Science and Business media

colocalization of the TNAP strata with that of widely used markers of the IPL is shown in Fig. 6.4a, b. After the images of the above-mentioned different stainings were merged, an exact correspondence of the immunolabeled IPL strata and those labeled by TNAP histochemistry was found (Fig. 6.5a, b). The dark TNAP stratum 1 corresponded to the outermost IPL stratum, which was marked by dendritic

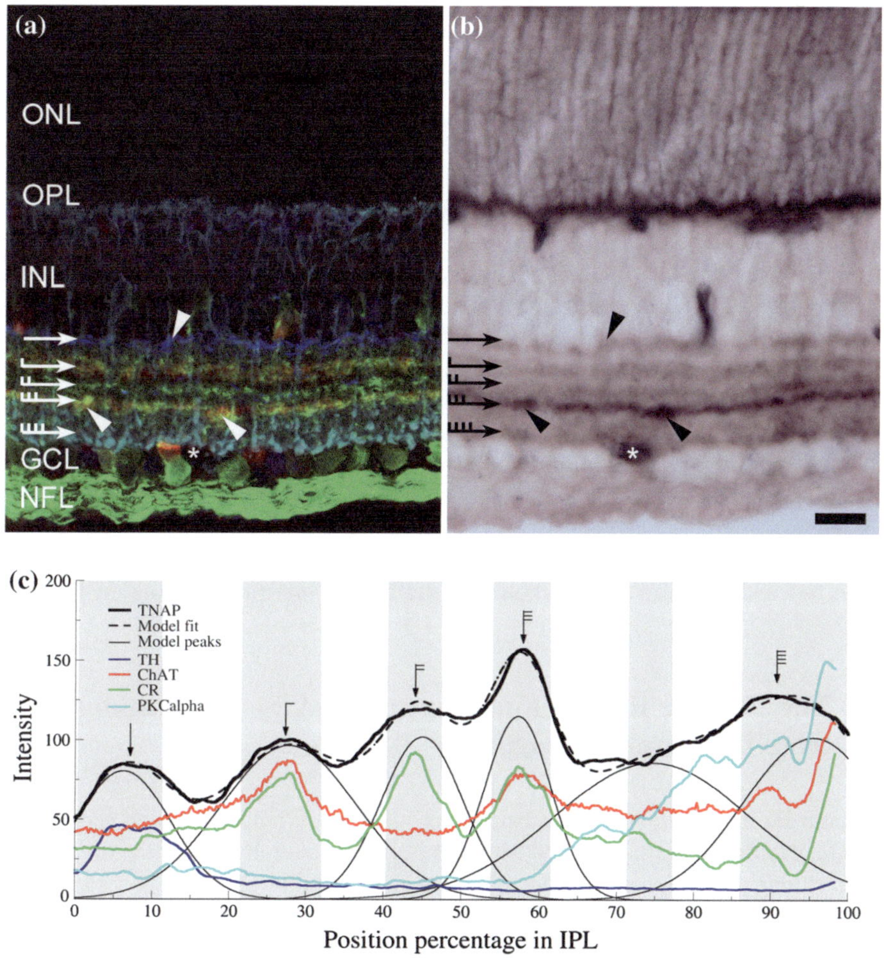

Fig. 6.5 Colocalization of TNAP stained strata of the IPL with canonical markers. Merged confocal image of the quadruple tyrosine hydroxylase (TH, *blue*), choline acetyltransferase (ChAT, *red*), calretinin (CR, *green*) and protein kinase C alpha (PKCα, turquoise) immunohistochemical reactions **a** with TNAP histochemistry **b** on the same specimen revealed the same IPL strata as indicated by the *arrows* (fletched in the same manner). Arrowheads in the same positions indicate corresponding structures in both pictures. Fitting the intensity profiles of the different immunohistochemical stainings (**c**, color coding as in **a**), TNAP histochemistry (*black line*) as well as mean TNAP staining (*dashed line*) showed that the strata revealed by TNAP indeed corresponded to those seen after the combined immunohistochemistry. Strata are depicted as *shaded* and *light* areas. *ONL*: outer nuclear layer, *OPL* outer plexiform layer, *INL* inner nuclear layer, *GCL* ganglion cell layer, *NFL* nerve fiber layer. *Asterisk*: same vessel in **a**, **b**. *Bar* 20 μm. Originally published in Cell and Tissue Research, vol. 359, p. 416 Fig. 3c Stratified organization and disorganization of inner plexiform layer revelaed by TNAP activity in healthy and diabetic rat retina, Kántor O, Varga A, Tóth R, Énzsöly A, Pálfi E, Kovács-Öller T, Nitschke R, Szél Á, Székely A, Völgyi B, Négyessy L, Somogyvári Z, Lukáts Á. With kind permission from Springer Science and Business media

processes of the TH positive amacrine cells. Similarly, dark TNAP strata 3, 5 and 7 exactly matched layers containing CR positive fibers. Strata 3 and 7 were also identified as ChAT positive layers that are generally known to contain dendrites of ON and OFF subpopulations of starburst amacrine cells. In contrast, strata 8–11 lacked all these previous stainings, but all showed strong PKCα labeling. However, the PKCα staining was homogenous and could not clearly delineate strata the way the TNAP staining did. The strongest PKCα labeling, corresponding to the terminals of rod bipolar cells, was observed in the dark TNAP stratum 11. The nine strata of the IPL delineated by the quadruple immunohistochemical reaction exactly corresponded to those seen after TNAP histochemistry alone (Fig. 6.5b). This finding was proven by model fitting analyses of the staining intensity profiles of the different immunolabelings (Fig. 6.5c). In contrast to TH, ChAT, CR, PKCα combined stainings, TNAP histochemistry alone divided the IPL into eleven strata thereby offering a useful alternative to label distinct IPL strata of the rat retina. In addition, with the quadruple immunohistochemical reaction, the innermost part of the IPL was labeled rather homogeneously with PKCα. On the contrary, using TNAP histochemistry alone, it was possible to divide this part of the IPL into three further strata (strata 9–11). Our work demonstrates that TNAP histochemistry is an easy, fast and cost-effective method to visualize the lamination of the rat IPL.

6.4 The Possible Retinal Functions of TNAP—Implications from the Cerebral Cortex

Vitamin B_6 is one of TNAP's substrates (Fedde and Whyte 1990) and serves as a GAD cofactor (Soghomonian and Martin 1998). However, by regulating the level of intracellular pyridoxal phosphate, TNAP can potentially affect other vitamin B_6-dependent enzymes, including those synthesizing glutamate and monoamines (Amadasi et al. 2007). Moreover, functional deficiency in aromatic-L-amino acid decarboxylase enzyme required for the synthesis of dopamine and serotonin has been associated with hypophosphatasia presented by two neonates (Balasubramaniam et al. 2010). Thus, TNAP can affect major neurotransmitter systems in the retina.

Regarding the possible functions, it is important to add the role of TNAP in purinergic signaling via its ectophosphatase activity on all forms (ATP, ADP, AMP) of adenosine phosphates in vitro (Abbracchio et al. 2009; Zhang et al. 2012). Since purinergic receptors are abundant in the retina and purinergic signaling is an important regulator of retinal transmission (Puthussery and Fletcher 2006; Puthussery et al. 2006; Housley et al. 2009) TNAP can possibly influence retinal purinergic signaling.

In vitro, knockdown of TNAP using RNA interference reduced proliferation and differentiation of neural stem cells into either neurons or oligodendrocytes (Kermer et al. 2010). In hippocampal culture blocking of TNAP inhibited the growth and branching of neurons, whereas addition of TNAP promoted axonal growth. TNAP colocalized with $P2X_7$ receptors in the growth cone (Díez-Zaera et al. 2011).

6.5 Possible Involvement of TNAP in Diabetic Retinopathy

In type 1 and type 2 diabetic patients without or with only minimal diabetic retinopathy, significant mitigation of the GCL/IPL was found using optical coherence tomography, whereas no changes in other neural retinal layers were detected, pointing to the early involvement of the inner retina in diabetic retinopathy (van Dijk et al. 2009, 2012). Moreover, visual functions in patients with minimal diabetic retinopathy, such as contrast and color sensitivity, are also disturbed (Ismail and Whitaker 1998; Adams and Bearse 2012).

As part of posttranslational modifications, TNAP isoenzymes became glycosylated. Removal of the N-linked sugar chains resulted in a full loss of enzymatic activity (Nosjean et al. 1997) indicating that altered states of protein glycosylation, as seen in diabetes, might affect TNAP activity. Furthermore, in a large, multiethnic population study, in non-diabetic patients with hyperinsulinemia and insulin resistance, elevated level of bone-type TNAP activity was found (Cheung et al. 2013). Similarly, in vitro, TNAP expression was elevated in an osteoblast cell line in the presence of high glucose and insulin concentrations (Cunha et al. 2014). Conversely, in a streptozotocin-induced rat model of type I diabetes, reduced TNAP activity was found in bone marrow stromal cells (Zhao et al. 2013).

In addition to research on retinal connectomics, TNAP histochemistry can also be useful to investigate the effects of diabetic retinopathy in rat models. In streptozotocin treated rats, 12 weeks after the induction of diabetes, we found an overall reduction of TNAP activity in all retinal layers (Fig. 6.6c). Moreover, in the IPL of diabetic animals, we observed diminished TNAP activity in a patchy pattern, which is in contrast to the continuous activity of the enzyme alongside this layer seen in the control rats (Kántor et al. 2015). This observation indicates that in diabetes, retinal TNAP activity shows regional impairments whereby affected areas are surrounded by regions with apparently preserved TNAP function. The stratified pattern of TNAP staining in the IPL was severely disrupted (46 % overall decrease) in diabetic animals, whereas no change in the stratification pattern of ChAT and CR labeled processes was detected.

Using our model fitting analysis, in diabetic animals, a significant decrease of TNAP activity was found at depths of 5–65 % (corresponding to TNAP strata 1–7) and at 88–100 % (corresponding to stratum 11) of the IPL (Fig. 6.6e, f). No significant difference was found between 65–88 % depth, corresponding to strata 8–10. Moreover, this reduced TNAP activity was also accompanied by a significant drop in the global retinal TNAP mRNA level. The corresponding quantitative PCR reaction revealed a significant reduction of TNAP mRNA in the retina of diabetic animals to 31 % of the control level. The extent of loss is comparable to the results of Zhao et al. (2013), who found a drop of TNAP mRNA level to 39 % of the control in bone marrow stromal cells in streptozotocin induced diabetic rats. This indicates that the reduced TNAP activity is due to (at least partially) the reduced TNAP expression in the retina.

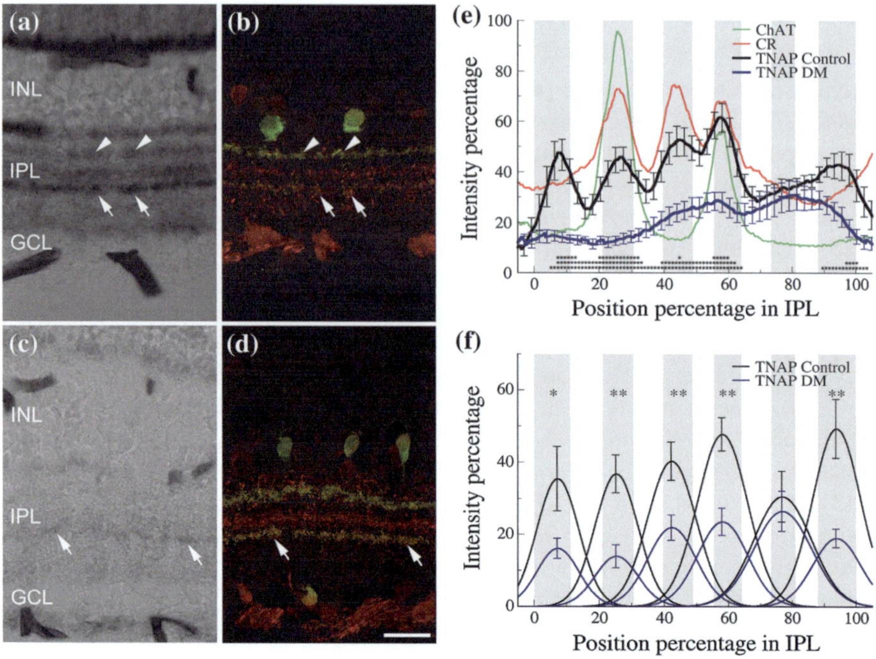

Fig. 6.6 Loss of TNAP activity in the IPL of diabetic rats. Representative confocal images from healthy control **a**, **b** and diabetic rats **c**, **d** depicting TNAP activity **a**, **c** as well as calretinin (CR, *red*) and choline acetyltransferase (ChAT, *green*) immunoreactivities **b**, **d** in the IPL. In diabetic animals, TNAP activity in the IPL was severely diminished, and IPL strata were not discernible. However, no changes in the CR and ChAT stratification patterns were detected at this stage of the disease. **e** Quantitative comparison of the percentage of mean TNAP staining intensity profiles and TNAP positive layers represented by Gaussians **f** between control (n = 7, *black line*/Gaussians) and diabetic retinas (n = 9, *blue line*/Gaussians). The intensity values are normalized to the staining intensity of the vessels in the same photograph. Note that significant decreases were found at the depth of 5–65 % (corresponding to TNAP layers 1–7, that is Gaussians 1–4) and 89–100 % (layer 11, 6th Gaussian) of the IPL. *Error bars* represent SEM values. *Stars* along the X axis represents the significance of the difference at any given position: 1, 2 and 3 *stars* (in vertical orientation) indicate p = 0.05; 0.01 and 0.001 significance levels, respectively *INL* inner nuclear layer, *IPL* inner plexiform layer, *GCL* ganglion cell layer. *DM* diabetes mellitus. *Arrows* and *arrowheads* indicate identical elements of the *IPL*. *Bar* 20 μm. Originally published in Cell and Tissue Research, vol. 359, p. 418 Fig. 5, Stratified organization and disorganization of inner plexiform layer revelaed by TNAP activity in healthy and diabetic rat retina, Kántor O, Varga A, Tóth R, Énzsöly A, Pálfi E, Kovács-Öller T, Nitschke R, Szél Á, Székely A, Völgyi B, Négyessy L, Somogyvári Z, Lukáts Á. With kind permission from Springer Science and Business media

6.6 Possible Role of TNAP in Retinal Development (Review of the Literature)

TNAP′s role in the retinal development is rather scarcely investigated. In chicken embryo, intraocular administration of levamisole (reversible, non-competitive inhibitor of TNAP) led to disturbances in photoreceptor development, reduction of thickness of the OPL and optic fiber layer, reduced number of photoreceptors and degenerating ganglion cells (Araki and Saito 1986; Araki et al. 1987). However, it is yet unknown whether TNAP plays a similar role in retinal development of mammals.

6.7 Retinal Pathology Related to TNAP Functions

In both hypo- and hyperphosphatasia, cases with ophthalmological symptoms can be found in the literature: atypical retinitis pigmentosa was associated to hypophosphatasia (Roxburgh 1983). Kerr et al. (2010) found progressive retinopathy in most of their studied patients (12–14) suffering from primary hyperphosphatasia, sometimes leading even to blindness. The role of phospho-tau protein was implicated in the pathophysiology of glaucoma (Gupta et al. 2008). Since TNAP dephosphorylates hyperphosphorylated tau, a key player in Alzheimer's and in other neurodegenerative diseases (Díaz-Hernandez et al. 2010), TNAP might play a role in the pathophysiology of glaucoma as well. Moreover, recent observations explored retinal neurodegeneration in Alzheimer disease (Guo et al. 2010).

In conclusion, although direct experimental evidence is missing, presence of retinal TNAP activity throughout the vertebrate evolution suggests a fundamental role of this ubiquitous enzyme in visual functions. The present works also summarized the first steps undertaken to reveal TNAP's role in diseases with retinal involvement.

Acknowledgment The work was supported by the Hungarian Scientific Research Fund (OTKA-K113147) to Zoltán Somogyvári.

Conflict of Interest The authors declare no conflict of interest.

References

Abbracchio MP, Burnstock G, Verkhratsky A, Zimmermann H (2009) Purinergic signalling in the nervous system: an overview. Trends Neurosci 32:19–29

Adams AJ, Bearse MA (2012) Retinal neuropathy precedes vasculopathy in diabetes: a function-based opportunity for early treatment intervention? Clin Exp Optom 95:256–265

Amadasi A, Bertoldi M, Contestabile R, Bettati S, Cellini B, di Salvo ML, Borri-Voltattorni C, Bossa F, Mozzarelli A (2007) Pyridoxal 5′-phosphate enzymes as targets for therapeutic agents. Curr Med Chem 14:1291–1324

Araki M, Saito T (1986) Alterations in the differentiation of chick retina caused by an intraocular injection of alkaline phosphatase inhibitor. Exp Eye Res 43:713–728

Araki M, Sato T, Saito T (1987) Effects of alkaline phosphatase inhibitors on chick neural retinal cell differentiation in vitro: ultracytochemical studies. Histochem J 19:579–588

Balasubramaniam S, Bowling F, Carpenter K, Earl J, Chaitow J, Pitt J, Mornet E, Sillence D, Ellaway C (2010) Perintal hypophosphatasia presenting as neonatal epileptic encephalopathy with abnormal neurotransmitter metabolism secondary to reduced co-factor pyridoxal-5′-phosphate availability. J Inherit Metab Dis 33:25–33

Bodis-Wollner I, Miri S, Glazman S (2014) Venturing into the no-man's land of the retina in Parkinsons's disease. Mov Disord 29:15–22

Cajal SR (1893) La retine des vertebres. Cellule 9:119–257

Cheung CL, Tan KC, Lam KS, Cheung BM (2013) The relationship between glucose metabolism, metabolic syndrome, and bone-specific alkaline phosphatase: a structural equation modeling approach. J Clin Endocrinol Metab 98:3856–3863

Cunha JS, Ferreira VM, Maquigussa E, Naves MA, Boim MA (2014) Effects of high glucose and high insulin concentrations on osteoblast function in vitro. Cell Tissue Res 358:249–256

De Vincentiis M, Testa M (1959) Distribution of alkaline and acid phosphatase in the particulate fractions of the retina. J Histochem Cytochem 7:393–394

Dehabadi MH, Davis BM, Wong TK, Cordeiro MF (2014) Retinal manifestations of Alzheimer's disease. Neurodegener Dis Manag 4:241–252

Díaz-Hernandez M, Gómez-Ramos A, Rubio A, Gómez-Villafuertes R, Naranjo JR, Miras-Portugal MT, Avila J (2010) Tissue-nonspecific alkaline phosphatase promotes the neurotoxicity effect of extracellular tau. J Biol Chem 285:32539–32548

Díez-Zaera M, Díaz-Hernández JI, Hernández-Álvarez E, Zimmermann H, Díez-Hernández M, Miras-Portugal MT (2011) Tissue-nonspecific alkaline phosphatase promotes axonal growth of hippocampal neurons. Mol Biol Cell 22:1014–1024

Fedde KN, Whyte MP (1990) Alkaline phosphatase (tissue-nonspecific isoenzyme) is a phosphoethanolamine and pyridoxal-5′-phosphate ectophosphatase: normal and hypophosphatasia fibroblast study. Am J Hum Genet 47:767–775

Fonta C, Négyessy L, Renaud L, Barone P (2004) Areal and subcellular localization of the ubiquitous alkaline phosphatase in the primate cerebral cortex: evidence for a role in neurotransmission. Cereb Cortex 14:595–604

Guo L, Duggan J, Cordeiro MF (2010) Alzheimer's disease and retinal neurodegeneration. Curr Alzheimer Res 7:3–14

Gupta N, Fong J, Ang LC, Yücel YH (2008) Retinal tau pathology in human glaucomas. Can J of Ophthamol 43:53–60

Housley GD, Bringmann A, Reichenbach A (2009) Purinergic signaling in special senses. TINS 32:128–141

Iandiev I, Wurm A, Pannicke T, Wiedemann P, Reichenbach A, Robson SC, Zimmermann H, Bringmann A (2007) Ectonucleotidases in Müller glial cells of the rodent retina: Involvement in inhibition of osmotic cell swelling. Purinergic Signal 3:423–433

Ismail GM, Whitaker D (1998) Early detection of changes in visual function in diabetes mellitus. Ophthalmic Physiol Opt 18:3–12

Kántor O, Varga A, Kovács-Öller T, Énzsöly A, Balogh L, Baksa G, Szepessy Z, Fonta C, Szél Á, Négyessy L, Völgyi B, Lukáts Á (2014) TNAP activity is localized at critical sites of retinal neurotransmission across various vertebrate species. Cell Tissue Res 358:85–98

Kántor O, Varga A, Tóth R, Énzsöly A, Pálfi E, Kovács-Öller T, Nitschke R, Szél Á, Székely A, Völgyi B, Négyessy L, Somogyvári Z, Lukáts Á (2015) The stratified organization and disorganization of the inner plexiform layer revealed by TNAP activity in healthy and diabetic rat retina. Cell Tissue Res 359:409–421

Kermer V, Ritter M, Albuquerque B, Leib C, Stanke M, Zimmermann H (2010) Knockdown of tissue non-specific alkaline phosphatase impairs neural stem cell proliferation and differentiation. Neurosci Lett 485:208–211

Kerr NM, Cassinelli HR, DiMeglio LA, Tau C, Tüysüz B, Cundy T, Vincent AL (2010) Ocular manifestations of juvenile Paget disease. Arch Ophthalmol 128:698–703

Kolb H (2007) Roles of amacrine cells. In: Kolb H, Fernandez E, Nelson R (eds) Webvision: the organization of the retina and visual system (Internet). University of Utah Health Sciences Center, Salt Lake City (UT)

Le Du MH, Millán JL (2002) Structural evidence of functional divergence in human alkaline phosphatases. J Biol Chem 277:49808–49814

Lindeman VF (1949) Alkaline and acide phosphatase activity of the embryonic chick retina. Proc Soc Exp Biol Med 71:435–437

Lutty GA, McLeod DS (2005) Phosphatase enzyme for studying vascular hierarchy, pathology, and endothelial dysfunction in retina and choroid. Vision Res 45:3504–3511

Négyessy L, Xiao J, Kántor O, Kovács GG, Palkovits M, Dóczi TP, Renaud L, Baksa G, Glasz T, Ashaber M, Barone P, Fonta C (2011) Layer-specific activity of tissue non-specific alkaline phosphatase in the human neocortex. Neuroscience 172:406–418

Nosjean O, Koyama I, Goseki M, Roux B, Komoda T (1997) Human tissue non-specific alkaline phosphatases: sugar-moiety-induced enzymic and antigenic modulations and genetic aspects. Biochem J 321:297–303

Puthussery T, Fletcher EL (2006) $P2X_2$ receptors on ganglion and amacrine cells in cone pathways of the rat retina. J Comp Neurol 496:595–609

Puthussery T, Yee P, Vingrys AJ, Fletcher EL (2006) Evidence for the involvement of purinergic $P2X_7$ receptors in outer retinal processing. Eur J Neurosci 24:7–19

Reis JL (1954) Histochemical localization of alkaline phosphatase in the retina. Br J Ophthalmol 38:35–38

Rogers KT (1963) Studies on chick brain differentiation V. Comparative histochemical alkaline phosphatase studies on chick retina, and blackbird, mouse, rabbit, cat and human brains. J Exp Zool 153:21–35

Roska B, Werblin F (2001) Vertical interactions across ten parallel, stacked representations in the mammalian retina. Nature 410:583–587

Roxburgh ST (1983) Atypical retinitis pigmentosa with hypophosphatasia. Trans Ophthalmol Soc UK 103:513–516

Siegert S, Gross Scherf B, Del Punta K, Didlovsky N, Heintz N, Roska B (2009) Genetic address book for retinal cell types. Nat Neurosci 12:1197–1206

Soghomonian JJ, Martin DL (1998) Two isoforms of glutamate decarboxylase: why? Trends Pharmacol Sci 19:500–505

van Dijk HW, Kok PHB, Garvin M, Sonka M, DeVries JH, Michels RPJ, van Velthoven MEJ, Schlingemann RO, Verbraak FD, Abramoff MD (2009) Selective loss of inner retinal layer thickness in type 1 diabetic patients with minimal diabetic retinopathy. IOVS 50:3404–3409. doi:10.1167/iovs.08-3143

van Dijk HW, Verbraak FD, Kok PH, Stehouver M, Garvin MK, Sonka M, DeVries JH, Schlingemann RO, Abramoff MD (2012) Early neurodegeneration in the retina of type 2 diabetic patients. IOVS 53:2715–2719. doi:10.1167/iovs.11-8997

Yoshida M (1957) The alkaline phosphatase activities in the isolated nuclei of ox retina. Jpn J Physiol 7:190–198

Zhang D, Xiong W, Chu S, Sun C, Albensi BC, Parkinson FE (2012) Inhibition of hippocampal synaptic activity by hypoxia or oxygen-glucose deprivation does not require CD73. PLoS ONE 7:e39772

Zhao YF, Zeng DL, Xia LG, Zhang SM, Jiang XQ, Zhang FQ (2013) Osteogenic potential of bone stromal cells derived from streptozotocin-induced diabetic rats. Int J Mol Med 31:614–620

Chapter 7
Tissue Non-specific Alkaline Phosphatase (TNAP) in Vessels of the Brain

Barbara Deracinois, Anne-Marie Lenfant, Marie-Pierre Dehouck and Christophe Flahaut

Abstract The microvessels of the brain represent around 3–4 % of the brain compartment but constitute the most important length (400 miles) and surface of exchange (20 m^2) between the blood and the parenchyma of brain. Under influence of surrounding tissues, the brain microvessel endothelium expresses a specific phenotype that regulates and restricts the entry of compounds and cells from blood to brain, and defined the so-called blood–brain barrier (BBB). Evidences that alkaline phosphatase (AP) is a characteristic feature of the BBB phenotype that allows differentiating capillary endothelial cells from brain to those of the periphery have rapidly emerge. Thenceforth, AP has been rapidly used as a biomarker of the blood-brain barrier phenotype. In fact, brain capillary endothelial cells (BCECs) express exclusively tissue non-specific alkaline phosphatase (TNAP). There are several lines of evidence in favour of an important role for TNAP in brain function. TNAP is thought to be responsible for the control of transport of some compounds across the plasma membrane of the BCECs. Here, we report that levamisole-mediated inhibition of TNAP provokes an increase of the permeability to Lucifer Yellow of the endothelial monolayer. Moreover, we illustrate the disruption of the cytoskeleton organization. Interestingly, all observed effects were reversible 24 h after levamisole removal and correlated with the return of a full activity of the TNAP. This reversible effect remains to be studied in details to evaluate the potentiality of a levamisole treatment to enhance the entry of drugs in the brain parenchyma.

Keywords Biomarker · Blood-brain barrier · Endothelial permeability

B. Deracinois · A.-M. Lenfant · M.-P. Dehouck · C. Flahaut (✉)
University of Lille Nord de France, F-59000, Lille, France
e-mail: christophe.flahaut@univ-artois.fr

B. Deracinois · A.-M. Lenfant · M.-P. Dehouck · C. Flahaut
University of Artois, LBHE, F-62307, Lens, France

B. Deracinois · A.-M. Lenfant · M.-P. Dehouck · C. Flahaut
IMPRT-IFR114, F-59000, Lille, France

C. Fonta and L. Négyessy (eds.), *Neuronal Tissue-Nonspecific Alkaline Phosphatase (TNAP)*, Subcellular Biochemistry 76,
DOI 10.1007/978-94-017-7197-9_7

7.1 Introduction

The microvasculature within the central nervous system (CNS) shows the most extreme form of phenotypic differentiation of any vasculature in the body. This phenotypic differentiation of endothelial cells comes from the up-regulation of an important number of features either poorly expressed or absent from peripheral vascular endothelium and inversely down regulation of some basic endothelial features (Abbott 2005). The concept of the blood-brain barrier (BBB) emerged in the beginning of the previous century with the combined works of Paul Ehrlich and his student Edwin Goldman (Ehrlich 1885; Goldmann 1913) who experimentally demonstrated using injections of tissular dyes in mice that cerebral compartment is isolated from the rest of the body. This barrier protects the brain from huge variations of compound concentrations that occur in blood and impedes the accessibility to brain of many substances and cells. In fact, the CNS, like all body tissues, is supplied by a blood vasculature network which in the CNS, however, is rather uniquely built. In human brain for instance, this capillary network is more than 400 miles long and spans a surface area of about 20 m^2 (Pardridge 2002). In the brain the capillaries have evolved to closely control the transfer between blood and brain of ions, small and large molecules as well as cells. Therefore, this interface has been named for the first time blood-brain barrier (BBB) by Lewandowsky in 1900 (Lewandowsky 1900) and the anatomical structure supporting the BBB is now clearly established as the brain capillary endothelial cells (BCECs).

The year 2014 will mark the 90th anniversary of the discovery of alkaline phosphatase (AP) (Whyte 1994). Since the 1930s, quantification of AP activity is a routine test in clinics and is likely one of the most frequently performed enzyme assay. Although numerous methods utilizing different substrates and reaction conditions have been applied to measurement of AP activity, the most widely used reference method utilizes 4-nitro-phenyl phosphate, a readily hydrolyzed, self-indicating AP substrate introduced in 1946 (Bessey et al. 1946). It consists of a colorimetric assay. Most tissues contain AP and AP activity is important in kidney, liver, bone, intestine, placenta and reticuloendothelial tissue, especially in the endothelial cells of brain (Djuricić et al. 1978; Vorbrodt et al. 1981). Moreover, specificity of AP towards BCECs and more specifically BBB phenotype has emerged only in the 1950s (Leduc and Wislocki 1952). In fact, numerous histochemical and biochemical studies from 1970s have shown cerebral microvessels to be rich in some phosphatases (Leduc and Wislocki 1952; Farkas-Bargeton and Arsenio-Nunes 1970; Joó and Karnushina 1973; Ciani et al. 1973; Renkawek et al. 1976; Pardridge and Oldendorf 1977; Meunier and Bouchaud 1978; Williams et al. 1980) and these enzymes are partly considered responsible for the control of transport across the plasma membrane of the BCECs (Torack and Barrnett 1964; Ginsbourg et al. 1973; Ginsbourg and Le Beau 1975; Jancsó et al. 1975; Vorbrodt et al. 1981). However, the majority of sustained elevated AP levels in the blood are associated with disorders of the liver or bone, or both (see Chap. 1).

7.2 Blood-Brain Barrier (BBB) and Tissue Non-specific Alkaline Phosphatase (TNAP)

7.2.1 The Blood-Brain Barrier

The CNS is protected by a structural, functional and metabolical barrier: the BBB. This later, consisting in the deep-brain microvascular endothelium, is the largest specialized, anatomical structure controlling cellular and molecular exchanges between the blood and the cerebrospinal fluid and thus into the brain and spinal cord. The BBB lets essential metabolites, such as oxygen and glucose, pass from the blood to the brain and CNS but blocks almost all molecules with a molecular mass >500 daltons. In other words, BBB is a biological low mass filtering device that refuses brain access to small biological molecules such as a large majority of hormones, neurotransmitters but also certain cells, virus, bacteria and controls the access of host immune cells [see (Wong et al. 2013) for recent review]. The human brain is comprised of around 15–30 % of extracellular space, 65–70 % of neurons, astrocytes, microglia, and pericytes and less than 5 % of blood vessels. Capillary in the brain may be as small as 7–10 μm in diameter and the intercapillary distance is about 40 μm.

In the brain microvasculature, cell-cell junctions are keys to maintaining the integrity of BBB and regulating transport between the BCECs. A major aspect of the BBB is the thin, flat endothelial cells which form the walls of brain capillaries (Fig. 7.1a). In the body, the endothelial cells are joined together by cellular junctions (adherens junctions and tight junctions) to form a tube. In the brain, BCECs are more tightly joined by the reinforcement of their tight junctions that therefore induce the polarization of the endothelial cells by restricting the plasma-membrane protein exchanges from the luminal to the abluminal faces and vice versa. The term luminal face indicates the face in contact with the bloodstream while abluminal face is used to specify the side of endothelial cells that repose on the basal membrane. The BCECs together with surrounding tissues and the basal membrane maintain

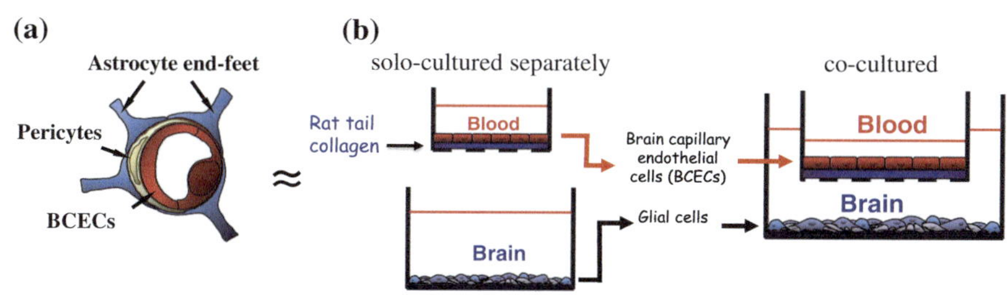

Fig. 7.1 Schematic representation of **a** the ultrastructure of the BBB and **b** our in vitro co-culture BBB model

this unique selective barrier. Thenceforth, the transendothelial transport is responsible for most molecular and cellular trafficking between the blood and the brain compartment.

The BBB is present in all brain regions, except in those regulating autonomic nervous system and endocrine glands of the body, where blood vessels permit diffusion of blood-borne molecules across the wall of the endothelium (Ballabh et al. 2004). The turnover of the BCECs is thought to be very low (Wong et al. 2013) and the increase in BCEC turnover due to angiogenesis and vascular remodelling as well as the BCEC response to apoptosis are currently unknown.

7.2.2 Architecture of the BCECs

The ultrastructure of the BBB is comprised, from the capillary lumen to the brain, of highly specialized BCECs that are surrounded by a dense basement membrane that engulfs pericytes (Fig. 7.1a). This cellular assembly is covered by astrocytes end-feet that form a sheath surrounding the basal membrane [see (Abbott et al. 2010) for details]. Together with neurons, the BBB comprises a construct termed neurovascular unit by Hawkins and Davis in 2005 (Hawkins and Davis 2005).

The BBB phenotype arises from the induction of a particular phenotype in the BCECs via molecular communication with the surrounding tissues. The morphology and functional properties of BCECs that form (with the other cells) the BBB are now well documented. These endothelial cells (**i**) lack fenestrations and are (**ii**) linked by reinforced tight-junctions forming continuous belt-like structure. They differs also from the peripheral capillaries of muscle and mesentery by showing (**iii**) a paucity of pinocytose vesicles, caveolae and transendothelial channels concomitantly to (**iv**) an increase in the number (and volume) of mitochondria (Oldendorf et al. 1977; Abbott et al. 2010; Nag 2011). Taken together, these characteristic features lead to a decrease in endothelial permeability both transcellular and paracellular. The transendothelial electrical resistance (TEER) is then greatly increased in comparison to those of peripheral endothelial cells. The paracellular barrier formed by interendothelial junctions restricts the free movement of water-soluble substances between adjacent cells while the transendothelial barrier comes from the low levels of endocytosis and transcytosis and inhibits transport of substance through the cytoplasm.

7.2.3 Role of the BBB

The BBB plays four key functions: (**i**) maintaining a constant environment (homeostasis) for the brain, (**ii**) constant supplying of nutrients by specific transport systems, (**iii**) shielding the brain from hormones and neurotransmitters and (**iv**) protecting the brain from "foreign substances" travelling in the blood that could injure the brain. *In fine*, the brain compartment is a sanctuary where substances need to be invited to enter.

The barrier functions result from a combination of physical, transport-related and metabolic barriers. BBB acts as a physical barrier because complex tight junctions between adjacent endothelial cells force most molecular traffic to take a transcellular way to across the BBB, rather than moving paracellularly through the junctions. BBB acts as a selective transport barrier, permitting or facilitating the entry of required nutrients, and excluding or effluxing potentially harmful compounds thanks to specific transport systems on the luminal and abluminal plasma membranes and vesicle membranes that regulate the transcellular traffic of small hydrophilic molecules. BBB acts also as a metabolic barrier thanks to a combination of intracellular and extracellular enzymes. The ecto-enzymes (nucleotidases, peptidases, hydrolase (such as AP) and intracellular enzymes (monoamine oxidase and cytochrome P450) afford a metabolic protection to the brain. Ecto-enzymes are involved in mechanisms that influence the extracellular composition surrounding cells whereas endo-enzymes are involved in the cellular detoxication mechanisms by inactivating many neuroactive and toxic compounds. However, the barrier functions are not perpetual and can be modulated under both physiological and pathological conditions.

7.2.4 AP Activity: A Very Useful Marker of the BBB Phenotype

The general presence of AP in the brain has emerged in the 1950s (Leduc and Wislocki 1952; Schiffer et al. 1962; Torack and Barrnett 1964). However, the specificity of AP towards BCECs arises from numerous cytochemical and biochemical studies showing that cerebral microvessels are rich in some phosphatases (Ciani et al. 1973; Farkas-Bargeton and Arsenio-Nunes 1970; Joó and Karnushina 1973; Meunier and Bouchaud 1978; Pardridge and Oldendorf 1977; Renkawek et al. 1976; Williams et al. 1980). Since 1970, Farkas-Bargeton and Arsenio-Nunes reported the change of enzymatic equipment in the walls of brain vessels by histochemical approach (Farkas-Bargeton and Arsenio-Nunes 1970). Some years later, several other groups published similar results. Ciani et al. (1973) and Jancsó et al. (1975) reported the ultrastructural location of AP in in vitro culture of nervous tissue and in brain smears, respectively. Nevertheless, the definitive demonstration that AP activity is linked to the brain vessel walls and more specifically to the plasma membrane of BCECs displaying BBB phenotype comes from studies published in 1980s. Betz et al., Williams et al. and Vorbrodt et al., were among the first to report biochemical evidences of the presence of AP at the luminal and abluminal side of BCECs (Betz et al. 1980; Williams et al. 1980; Vorbrodt et al. 1981; Inomata et al. 1984; Nishio et al. 1986). Clearly, the AP appears at this time as one of the enzymes which is highly active in the plasmalemma of endothelial cells of BBB-type microvessels (Meunier and Bouchaud 1978; Ehrlich et al. 1986; Plesner 1995). Inversely, the choroid plexuses were characterized by the absence of AP activity (Ginsbourg and Le Beau 1975). Therefore, AP activity has been used for many decades as a marker of brain microvessels displaying the BBB phenotype in mammalians (Bell and Ball 1985; Anstrom et al. 2002; Fonta and Imbert 2002).

Zoellner and Hunter were the first to report that AP activity of microvessels can be due to the TNAP isoform (Zoellner and Hunter 1989). More precise reports concerning the location of AP in the BCECs and in the cerebral parenchyma in general are more recent (Fonta et al. 2004, 2005). As demonstrated by Brun-Heath et al. (2011) the brain tissue does not express brain-specific transcript of AP. On the other hand, neuronal cells and BCECs exclusively express the bone TNAP transcript in human, marmoset, rat and mouse. *In fine*, TNAP is the only AP isoform found in the brain (Ermonval et al. 2009). This characteristic was also noted in several in vitro models of mammalian BBBs (Meyer et al. 1990; Nakazato et al. 1997; Sobue et al. 1999).

The AP family (EC. 3.1.3.1) catalyzes the hydrolysis of phosphomonoesters to produce inorganic phosphate and an alcohol from a variety of organic compounds: nucleosides (5′ -tri-, -di- and -mono- phosphates), inorganic pyrophosphate, pyridoxal-5′-phosphate and phosphorylated proteins (Chan and Stinson 1986; Millán 2006). Hence, the APs can control not only cellular ectonucleosidase but also protein dephosphorylation activities. However, despite its known phosphomonoesterase activity, TNAP's putative ability to dephosphorylate protein targets remains subject to debate. Direct proof of TNAP's action on the phosphorylation state of laminin has been reported (Ermonval et al. 2009) but contrasts with evidence to show that TNAP does not modulate the phosphorylation of plasma membrane proteins (Fedde et al. 1993). In Alzheimer's disease (AD), hyperphosphorylation of the tau protein plays a key role in progression of AD. Recently, Vardy et al. demonstrated that the TNAP activity increase in the hippocampus of patients with sporadic and familial AD is likely a consequence of AD-associated changes in the brain (Vardy et al. 2012). TNAP dephosphorylates the hyperphosphorylated extracellular tau protein once it is released upon neuronal death. Only dephosphorylated tau triggered neuronal death through an increase in intracellular calcium levels mediated by the activation of muscarinic M1 and M3 receptors (Díaz-Hernández et al. 2010). This activation enhances the expression of TNAP in return (for details, see Chap. 18). Since TNAP promotes the expression of the BBB phenotype in BCECs, one can legitimately suppose that reducing the BBB permeability can enhance tau-mediated brain lesions by reducing cerebral clearance of the hyperphosphorylated and dephosphoryleted tau protein.

Protein phosphorylation regulates function of transmembrane and accessory proteins of tight junctions (Sakakibara et al. 1997). Occludin phosphorylation controls the intracellular distribution of occludin and subsequent barrier properties where hyperphosphorylation of occludin was found at intercellular junctions and appeared necessary for maintaining the restrictive barrier function whereas decreased phosphorylation of occludin correlates with an increase of the endothelial permeability (Wong 1997; Andreeva et al. 2001; Sundstrom et al. 2009).

Concomitantly, phosphorylation of claudins (–4 and –5) are also involved in the regulation of the paracellular permeability of BCECs (Soma et al. 2004; Persidsky et al. 2006). However, no distinct clear-cut pattern has been established between the site of phosphorylation and the tight junction function. Hawkins and Davis (2005) suggested that most probably phosphorylation schemes of serine, tyrosine and threonine residues have distinct structural and functional effects.

7.3 Specificity of the BBB

7.3.1 Species Specificity

The architecture of the brain microvasculature is very similar across species, indicating that the human brain microvasculature is not physically unique. Indeed, many authors have reported the existence of BBB in many vertebrates (Abbott 2005). Several studies demonstrated the presence of AP activity in capillary endothelium of rat brain (Meunier and Bouchaud 1978; Liu et al. 1979; Inomata et al. 1984), mouse brain (Vorbrodt et al. 1981, 1986), goat brain (Vorbrodt and Trowbridge 1991), avian species (Stewart and Wiley 1981), frog brain (Latker et al. 1987; Lazzari et al. 1989), cat brain (Vorbrodt et al. 1985), primate brain (Shi and Audus 1994), bovine brain (Brunel et al. 1969). Recently, the impermeability of cerebral microvessels has been first reported in zebrafish following in vivo experimentation of permeability assays (sulfo-NHS-biotin and horseradish peroxydase) and immunoreactivity towards integral proteins (Claudin-5 and Zonula Occludens-1) of the tight junctions (Jeong et al. 2008). The cerebral microvessels share both structural and functional similarities with that of mammals (Fleming et al. 2013).

Many invertebrates do not have BBB. However, in those that do have (insect, crustacean, cephalopod molluscs, squid, octopus) the BBB has a glial origin as in the elasmobranch subclass. Among vertebrates, elasmobranch fish (sharks, rays, skates-cartilaginous fishes) have BBB not formed by the endothelium but formed by perivascular glial end-feet (Abbott et al. 1992). This suggests that glial barrier is the ancestral BBB, and during the evolution the barrier has shifted to the endothelium (Abbott 2005). However, an important question is whether the human BBB is functionally different from other species, for instance in the expression levels of plasma membrane proteins assuming the molecular transports and controlling the molecular exchanges.

7.3.2 Development and Maturation

More and more data are now known about the early development of the BBB. What we do know has been extrapolated from rodent and chicken (Sessa and Perez 1975; Stewart and Wiley 1981; Risau et al. 1986). AP appeared during the third week of chick embryo development and during the last week of quail embryonic development (Stewart and Wiley 1981). In fact, the maturation of the chick BBB is completed after the 15th day of incubation as revealed by exclusion of intravasally injected horseradish peroxydase (Wakai and Hirokawa 1978). Some years later, AP activity was found to be present in BCECs of mouse, quail and chick and appeared around embryonic day 17, 15 and 12, respectively (Risau et al. 1986). In the BCECs of mouse the cytochemically detectable AP activity is correlated with the setting-up of BBB phenotype (Vorbrodt et al. 1986). This observation first indicated that the maturation of the BBB was accompanied by the development of an enzymatic barrier where AP constituted one of the components of this barrier. In the rodent the development of the BBB occurs at embryonic day 9 (Engelhardt 2003). The BBB is formed during embryogenesis when the endothelial cells invade the CNS and pericytes are recruited to the nascent microvessels (Daneman et al. 2010b). The perineural vascular plexus covers the neuroectoderm and begins to invade it via vascular sprouts at embryonic day 11 and 2 in rodents and chicken, respectively. The vascular sprouting takes place around the 8th week of gestation in humans (Bär 1980; Ballabh et al. 2005). From 15 weeks gestation to ten years postnatal, AP activity is detected in intracortical vessels (Norman and O'Kusky 1986). Once again, during the maturation of the CNS the barrier shifts from the glial to endothelial cells (Abbott 2005). The maturation of BBB continues during late stages of embryogenesis and after the birth. *In fine*, various cell types modulate the phenotype of vessels during embryogenesis and might use similar mechanisms to induce BBB properties (Alvarez et al. 2013). Nevertheless, the role of glial cells and neurons still remains to be established. The signaling pathways are known to influence patterning and branching during embryogenesis. The Hedgehog pathway and the vascular endothelial growth factor pathway for example are involved in the maturation of the brain microvasculature (Alvarez et al. 2011).

Although zebrafish is a model organism (for multiple reasons) largely used in embryonic studies and the understanding of the gene function, its use in developmental studies dedicated to BBB is recent. Indeed, the expression of Claudin-5 and ZO-1 was detected in cerebral microvessels from 3 days post-fertilization, concomitant to the maturation of the BBB (Jeong et al. 2008). Fleming et al. have combined anatomical analysis, the use of fluorescent markers and compound uptake assays using mass spectrometry to demonstrate that maturation of the BBB in zebrafish occurs between 3 and 10 days post-fertilization and that this barrier shares both structural and functional similarities with that of mammals (Fleming et al. 2013). However, to our knowledge no AP activity study has been investigated in adult or embryo zebrafish.

The possible influence of TNAP activity on early stages of morphogenesis, histogenesis, cellular differentiation, or axonal guidance has been proposed but has not been directly established. TNAP involvement in each of these developmental steps may be related to the neurological disorders observed in TNAP-deficient individuals. Although impaired TNAP activity is associated with infantile, childhood, and adult forms of hypophosphatasia, TNAP activity and its regulation have not been investigated in the CNS during postnatal development (Fonta et al. 2005).

7.4 Cell Localisation, Nature and Activity of TNAP

7.4.1 Cell Localisation

Due to its central position in the cellular and molecular exchanges between the blood and the brain compartment, the plasma membrane is largely studied all around the world. Therefore, the best-characterized molecular changes issued of the BBB phenotype mainly concern plasma membrane proteins—especially those involved in tight-junction formation, receptor-mediated transcellular transport, efflux pumps and transporters (Daneman 2012). In this field, the proteomics efforts have been focused on the accurate quantification by mass spectrometry of plasma membrane proteins of special interest. Thus, the Terasaki's research team reported the quantification of around 30 mouse plasma membrane proteins in 2008 (Kamiie et al. 2008). This study was followed by the quantitative identification of 114 plasma membrane proteins (transporters and receptors) from human brain micro-vessels (Uchida et al. 2011). However, the presence of TNAP has never been investigated by such mass spectrometry-based technologies.

TNAP (gene name: ALPL) is a glycosylphosphatidyl inositol (GPI)-anchored membrane protein (see Chap. 16). To date, many different GPIs have been characterized; all of which share the conserved construct: NH_2–CH_2–CH_2–O–P(=O)(–OH)–6–O–Manα1–2Manα1–6Manα1–4GlcN$H_2\alpha$1–6–myo–inositol–1–O–phosphoglycerolipid. The addition of GPI on the TNAP occurs in the lumen of endoplasmic reticulum (ER) where the GPI biosynthesis takes place also [see details in (Yu et al. 2013)]. This attachement process of GPI on protein involves an enzymatic complex named GPI transaminidase. This enzymatic complex recognises a C-terminus signal of nascent proteins and replaces it by the GPI-anchor which is therefore always linked to the C-terminus extremity of proteins. Interestingly, mutated TNAP has been largely used as target for deciphering the C-terminus signal sequence of proteins post-translationally modified by a GPI-anchor.

TNAP is targeted to the outer leaflet of the cytoplasmic membrane, where the GPI serves as a membrane anchor. In the 80s researchers did not know that AP activity of BCECs was due only to TNAP. Accordingly Betz et al. demonstrated by electron microscopy examination that in vivo TNAP activity was located in both the luminal and the abluminal cytoplasmic membrane of the BCECs (Betz et al. 1980). Conversely, the TNAP activity was mainly detected in the abluminal plasma membrane of rat BCECs (Inomata et al. 1984), in luminal plasma membrane of mice brain capillaries and in both luminal and abluminal plasma membranes in arterioles (Vorbrodt et al. 1986). Again, the TNAP activity is detected both on the luminal and abluminal walls of BCECs in frog (Lazzari et al. 1989) and goat (Vorbrodt and Trowbridge 1991). This discrepancy of localisation comes likely from methods used to fix tissue that inactivate the TNAP. *In fine*, from bovin BCECs fractionnated in a discontinuous Ficoll gradient supplemental evidences were brought in favor of a distribution both in the luminal and abluminal plasma membrane (Sánchez del Pino et al. 1995).

However,Vorbrodt et al. reported also that a change in polar distribution occurs after BBB damages, shifting from the luminal side to the abluminal side (Vorbrodt et al. 1983). The same authors described also that the phosphatase activity (then TNAP) was mainly detected at the luminal plasma membrane of the endothelial cells of mice but changes in activity and distribution in scrapie-infected mice (Vorbrodt et al. 1981) and this is associated with the increased of endothelial permeability (Wisniewski et al. 1983). In the same manner, TNAP activity shifts from the luminal side to the abluminal one after single inorganic mercury injection in cerebral microvascular endothelial cells of rat (Albrecht et al. 1994).

7.4.2 Biochemical Nature of TNAP

TNAP possesses five potential N-glycosylation sites that are conserved between mammalian species (uniprot: P05186, PPBT_human; P09242, PPBT_mouse; P08289, PPBT_rat; P09487, PPBT_bovin; Q29486, PPBT_felca). However, very little is known about the glycosylation of TNAP, principally because of the great diversity of post-translational modifications of the various isoenzymes. The first attempt to characterise N-glycosylation of human TNAP has been carried out some years ago and demonstrated that all TNAP isoforms bear some N-glycans and, except the liver isoform, the bone and kidney isoforms are also O-glycosylated (Nosjean et al. 1997). Interestingly, N-glycosylation is absolutely essential for TNAP activity, and the oligosaccharide chain of the Asn-213 plays a key role in enzyme regulation. Recently, evidences that surface-linked TNAP from mouse cells is glycosylated on the two first and two last sites of potential N-glycosylation (Asn-Xaa-Ser/Thr) have been brought by strategies employing mass spectrometry through a global work dedicated to study N-linked cell surface glycoproteins (Wollscheid et al. 2009). To date, to our knowledge no recent study reports the N-/O-glycosylation patterns of TNAP prepared or issued from endothelial cells and especially BCECs (See also Chap. 9).

Mammalian alkaline phosphatases (AP; EC 3.1.3.1) make up a ubiquitous family of non-specific phosphomonoesterases (Millán 2006). Human APs are encoded by at least four different genes which express the kidney/bone/liver-type [tissue non-specific (TNAP)], placental-type (PLAP), intestinal-type (IAP) and germ-cell-type (or placental variant) enzymes, respectively. The gene locus for the TNAP has been mapped to chromosome 1p34-p36.1 (Smith et al. 1988). Genomic studies regarding the expression of gene in BCECs are not legion. Genomic methods used in BBB research field include gene microarray technologies, serial analysis of gene expression (SAGE) and suppression subtractive hybridization (SSH) (Calabria and Shusta 2006; Pottiez et al. 2009). Initial applications of the SSH methodology to the rat BBB led to the identification of transcripts selectively expressed at the BBB; these included novel gene sequences not found in existing databases, ESTs, and known genes that were not known to be selectively expressed at the BBB (Li et al. 2001, 2002; Boado 2003; Chu et al. 2008). The differential gene expression within the human brain microvasculature was evaluated using suppression subtractive hybridization with RNA isolated from human brain microvessels but the study was restricted to the first 71 differentially expressed clones (Shusta et al. 2002). A full and quantitative SAGE catalogue from rat BCECs has been constructed. Its comparison with SAGE catalogue not specific for BBB highlighted 864 genes more particularly related to the BBB (Enerson and Drewes 2006). More rencently Calabria and Shusta have reported a comparative study of the genomic expression from in vitro and in vivo rat brain microvascular endothelial cells (Calabria and Shusta 2008, 2009). Concomitantly, a comparable study has been done in mouse (Daneman et al. 2010a). Anyway, none of these studies reports the genomic expression of TNAP in BCECs.

AP activity detected in endothelial cells of brain from human, primate, rat, mouse results from the expression of the *alpl* gene. In fact, BCECs exclusively express the bone TNAP transcript in those species and this expression does not depend on a brain-specific TNAP promoter (Brun-Heath et al. 2011). This study led to the determination of the complete cDNA sequence of the marmoset TNAP (GQ465850: 2547 bp).

7.4.3 Activity of AP

AP (E.C. 3.1.3.1) is a phosphate-monoester phosphohydrolase that catalyses the removal of a phosphate group by cleavage of a phosphate-monoester binding. Therefore, the AP activity measurement is based on the liberation of an inorganic phosphate from p-nitrophenyl phosphate (pNPP) as a phosphatase substrate which turns yellow (λmax = 405 nm) when dephosphorylated by AP. This assay has been used in numerous studies dedicated to evaluate the AP activity of the BCECs. Rapid and convenient assays for the activity determination of AP in cultured endoethelial cells and other cells have furthermore been reported (Fukushima et al. 1990). The pNPP was chosen because it does not penetrate the cellular membrane (DePierre and Karnovsky 1974). Numerical values of AP activity was reported exclusively for isolated BCECs. Relative quantification of AP activity was reported from nine clones of mouse BCECs. The AP activity for brain capillaries was 3 times higher than in other brain microvessel cell types with a value of 160 ± 4 milliunits/mg of proteins but no correlation analyses with the mRNA and protein expression was performed (Tatsuta et al. 1992). In the same manner, a comparative study of cell cultures of BCECs and aortic endothelial cells established that the endothelium of brain origin had higher AP activity than the aortic one (Mizuguchi et al. 1994). Moreover, in a porcine model of in vitro BCEC culture, the total activity of AP can reach (up to sixfold increasing) the same level as detected in freshly isolated BCECs when medium culture was supplemented by cAMP (Beuckmann et al. 1995) suggesting that cAMP is a messenger involved in the induction of AP. *In fine*, in a rat immortalized cell line (RBE4) (Couraud et al. 2003), Km and Vmax were 1200 μM and 7.9 ± 0.3 nmol.min^{-1}/mg of proteins, respectively (Calhau et al. 2002).

Since the early 1960s, several specific inhibitors of AP isozymes have been uncovered. They include L-amino acids, such as L-phenylalanine, L-tryptophan, L-leucine, and L-homoarginine as well as some nonrelated compounds, such as levamisole (Van Belle 1976) and theophylline (Farley et al. 1980). The inhibition is of an uncompetitive type for levamisol but no information is available about the binding sites of the three commonly used inhibitors, that are L-homoarginine, levamisole, and theophylline. Until now, levamisole and theophilline are the most used inhibitors of TNAP with Ki values of 16 and 82 μM, respectively (Kozlenkov et al. 2004). Levamisole is considered as the specific-inhibitor of TNAP (see Chap. 12). AP activity of microvascular endothelial cells has been noted to be specifically inhibited by the levamisole (Lev and Griffiths 1982; Latker et al. 1987). Because TNAP is a potential therapeutic target for the treatment of soft tissue ossification abnormalities (including ankylosis, osteoarthritis, and arterial calcification) (Mebarek et al. 2011), the discovery of novel selective inhibitors of TNAP derived from pyrazole has been reported (Sidique et al. 2009). However, such pyrazole-derived inhibitors have not yet been evaluated in regards of the AP activity of BCECs.

7.5 In Vitro Models, Induction of BBB and TNAP

The works related to the isolation of capillaries from rat brain were the starting point of the crusade dedicated to the setting-up of in vitro cell culture models mimicking the in vivo situation (Joó and Karnushina 1973). Freshly isolated brain capillary endothelial cells display all the BBB-specific features. They contain furthermore high activity levels of the BBB marker enzymes AP and γ-glutamyl-transpeptidase. However, in primary culture the activities of these specific enzymes decrease with increasing cell proliferation and passage of cells. In these efforts to establish an in vitro model of the BBB, many research groups have reported that isolated, cultured brain capillary endothelial cells from diverse species lose most of their AP activity (Cardelli-Cangiano et al. 1987; Meyer et al. 1990; Rubin et al. 1991; Rauh et al. 1992) and this dedifferentiation is due to an arrest of de novo synthesis (Beuckmann et al. 1995).

In fine, when isolated, BCECs lose most of their BBB-specific properties. This loss of in vitro BBB characteristics is often assessed by monitoring properties such as transendothelial electrical resistance, small molecule permeability, transporter expression, tight-junction proteins and enzymes such as the AP. Sometimes enzymatic activities specific of the BBB such as AP, gamma glutamyl-transpeptidase are evaluated but the effects of culture manipulation and refinement on in vitro BBB model quality are often evaluated using the first above mentioned criteria.

7.5.1 BBB-Phenotype Induction by Surrounding Cells, Conditioned Media and Other Compounds

Beck et al. (1986) first mentioned that TNAP activity was markedly elevated when endothelial cells (ECs) were co-cultured with glial cells. One year later, the in vitro evidence that astrocytes, cultured in close contact to bovine BCECs, play a role in the formation and extend of tight-junctions in bovine BCECs was reported (Janzer and Raff 1987; Tao-Cheng et al. 1987). Astrocytes are therefore reported to induce capillary endothelial cells to differentiate into the specialized BBB phenotype (Janzer and Raff 1987; Tontsch and Bauer 1991; Meyer et al. 1991; Blasig et al. 2001). Nevertheless Meyer et al. reported that a direct contact between astrocytes and BCECs is necessary to induce the BBB features in BCECs (Meyer et al. 1991). The authors observed a spatial induction within a BCEC colony. Astrocyte-induced AP activity was first observed at an outer belt of BCEC colonies and progressed to the cell culture center. In fact, controversial studies demontrated that no direct contact between the cells is required [review in (Abbott et al. 2006)]. Astrocytes secret (a) product(s) that provoke(s) the cellular differentiation of BCECs. Astrocytes are also capable to induce some BBB-specific features such as a AP activity increase and the formation of tight junctions in non-BBB endothelial cells as human umbilical vein endothelial cells (Tio et al. 1990).

This highly differentiated BCEC phenotype is induced and maintained by the cross-talk with the surrounding cells such as astrocytes and pericytes. This fact was several times confirmed by many distinct research groups [see (Abbott et al. 2006; Paolinelli et al. 2011; Alvarez et al. 2013) for details]. Recently, Brun-Heath and co-authors confirmed by a study showing that bone-type TNAP is present and down-regulated in neurons and brain ECs in several species (Brun-Heath et al. 2011). The study revealed strongly down-regulated expression of the bone TNAP transcript when mouse BCECs were maintained in solo-culture for a few days (Brun-Heath et al. 2011). Interestingly, a variety of molecules are known to modulate TNAP expression [retinoic acid (Zhou et al. 1994; Lechardeur et al. 1995); cAMP (Beuckmann et al. 1995); glucocorticoids (Adams and Melnykovych 1985; Mulkins and Allison 1987); transforming growth factor-beta (Nakazato et al. 1997); IL-6 (Nakazato et al. 1997) and basic fibroblast growth factor (Roux et al. 1994; Sobue et al. 1999)] (see also Chap. 4) and brain capillary endothelial permeability. This non-exhaustive list clearly underlines the importance of the extracellular messengers secreted by both cells. Unfortunately, no study reports a differential analysis of secreted proteins in the medium that constitute the so-called secretome.

Pericyte-endothelial interactions are particularly prominent for BBB maintenance, with profound effects on basement membrane and endothelial tight-junction structure and function (Liu et al. 2012). The higher pericyte abundance in BCECs as compared to the periphery highlights the functional importance of pericytes in CNS vascular biology in both embryogenesis and adulthood (Daneman et al. 2010b; Bell et al. 2010). Pericytes act by regulating BBB-specific gene expression pattern in endothelial cells, and by inducing polarization of astrocyte end-feet surrounding CNS blood vessels (Balabanov and Dore-Duffy 1998; Armulik et al. 2010; Daneman et al. 2010b; Bonkowski et al. 2011). However, no correlation of pericyte influence on the TNAP gene expression pattern has been reported in these studies.

Transendothelial electrical resistance and permeability to large and small molecular tracers can be re-induced in BCECs when grown in the presence of astrocytes-conditioned media (Vinters et al. 1987; Neuhaus et al. 1991; Prat et al. 2001; Siddharthan et al. 2007). Re-induction of the AP activity using astrocyte-conditioned medium was first reported as ineffective in cultured BCECs (Meyer et al. 1991). Conversely, Rubin et al. reported in the same time the setting-up of a cell culture model of BBB using astrocytes-conditioned medium in combination with others agents (Rubin et al. 1991). It is now largely accepted that astrocytes-conditioned medium induces most of the BBB-specific phenotype in cultured BCECs, aortic endothelial cells and immortalized cell lines (Takemoto et al. 1994; Roux et al. 1994; Mizuguchi et al. 1995).

Astrocytes secrete several factors that promote the differentiation of BCECs (Haseloff et al. 2005; Abbott et al. 2006). Evidences that several signalling pathways (Wnt/β-catenin, glia-derived neurotrophic factor, transforming growth factor-beta) control and regulate the BBB-phenotype in BCECs are now legion [see (Alvarez et al. 2013) for details].

7.5.2 In Vitro BBB Model and AP Activity

Until the 1970s, studies on the BBB were conducted almost entirely on in vivo and in situ preparations. The introduction by Joó and Karnushina (1973) of methods to isolate cerebral capillaries marked the beginning of a new generation of studies in vitro and has led to the development of in vitro cell culture model to mimic the physiological BBB phenotype. However, when brain microvessels are isolated and their endothelial cells dissociated, isolated, seeded and grown in culture, the BCECs tend to revert towards the "default" phenotype of peripheral endothelial cells and dedifferentiate (Reichel et al. 2003).

In the frame of the deep understanding of the BBB and the development of drugs able to cross the BBB, large efforts have been done to obtain in vitro BBB model displaying the same features as in the in vivo situation. The idea that BBB can be reproduce in vitro (Goldstein 1988) comes from progress done to isolate and separately culture populations of brain-derived endothelial cells and astrocytes (Joó and Karnushina 1973; Frangakis and Kimelberg 1984). The close anatomic relationship between brain endothelial cells and astrocytes and influence of astrocytes on BCECs culture suggest a role for astrocytes as a source of the differentiation signals. Therefore, the evidences that the presence of astrocytes in co-culture with BCECs or astrocytes-conditioned medium induce almost all BBB-specific features in isolated BCECs provide the basis for a model of endothelial cell-astrocyte interaction that may reproduce several aspects of characteristic features and behavior of the BBB (Beck et al. 1986; Tao-Cheng et al. 1987; Vinters et al. 1987). The first co-culture BBB models are born in the 1990s (Isobe et al. 1996; Dehouck et al. 1990; Rubin et al. 1991; Rauh et al. 1992; Kondo et al. 1994; Hurst and Fritz 1996; Sobue et al. 1999). The results obtained by such in vitro approach have been reviewed (Rubin 1991; Joo 1993). Many in vitro BBB culture models have been developed and characterized in the last three decades [see (Deli et al. 2005) for details] and these in vitro BBB models continue to grow, to progress and to specialise (Stanness et al. 1999; Ramsauer et al. 2002; Lippmann et al. 2011; Vandenhaute et al. 2012). However, with the progress of the biology tools, the AP activity is becoming, over time, a differentiation criterion obsolete as BBB-specific marker in the BCECs, and to date, only few people report AP activity in the studies dedicated to BBB.

7.6 BBB, Diseases and TNAP

7.6.1 BBB-Diseases Relationship

CNS diseases include mental disorders, migraine, epilepsy, neurodegenerative diseases (e.g. Huntington's disease, Alzheimer's disease, Parkinson's disease, amyotrophic lateral sclerosis), cancer, trauma, infection (e.g. meningitis), inflammatory and cerebrovascular diseases (e.g. stroke, multiple sclerosis) (Hawkins and Davis 2005; Abbott et al. 2010; Daneman 2012). In these situations, due to its central position in the exchanges between the brain and the blood stream, the BBB plays two opposite role. On one hand, many diseases of the CNS can lead to the disruption of the BBB and on another hand the barrier function of the BBB represents a hurdle in delivering drugs to the brain compartment.

Therefore, over the last decade, it has become clear that the BBB plays a major role in the brain health, in a large number of diseases and in their therapeutic treatment. The BBB is now considered to be an active partner or prime participant (rather than a passive target) in diseases such as obesity, Alzheimer's disease, multiple sclerosis, stroke, brain cancer and diabetes mellitus. Paradoxically the BBB is the first obstacle to the therapeutic treatments of the brain compartment due to its specific phenotype of BBB that restricts the cellular and molecular exchanges between the blood and the brain compartment. Therefore, the value of many promising CNS drug candidates is diminished by the presence of the BBB because 95 % of drugs developed to treat diseases of the human brain (and showing in vitro efficacy) are unable to reach the brain compartment (Pardridge 2007). Fortunately the barrier function is not perpetual and can be modulated under either physiological and pathological conditions or therapeutic treatments, and an improved understanding of the mechanisms controlling this largest regulatory interface may provide novel and more effective strategies to treat neurological disorders.

As shown by this book arguments are accumulating for the involvement of TNAP in brain functions, in particular in view of the neurological symptoms accompanying human TNAP deficiencies. Indeed, there are several lines of evidence in favour of an important role for TNAP in brain function. Hypophosphatasia is one of several brain diseases that may be related (at least in part) to the BBB (Palmer 2010; Grammas et al. 2011). It results from various mutations of the TNAP gene and is associated with neurological disorders such as mental retardation, seizures and epilepsy (Mabry et al. 1970; Whyte 1994; Oby and Janigro 2006).

7.6.2 AP: An Old Enzyme Newly Discovered

Recently, from an in vitro co-culture BBB model, we reported proteomic evidences that TNAP is over-expressed in bovine BCECs following the glial cell-mediated re-induction of the BBB phenotype (Deracinois et al. 2012). Briefly, primary cultures of mixed glial cells were initiated from new-born rat cerebral cortex. Three weeks after seeding, the confluent glial cell cultures can be used for co-cultures. Concomitantly, bovine BCECs were isolated and characterized according to Méresse et al. (1989) and bovine BCECs were cultured for 12 days in absence of glial cells (i.e. in solo-culture, yielding limited BBB (Lim. BBB) functions to BCECs) or in presence of glial cells (i.e. in co-culture, yielding re-induced BBB (Re-ind. BBB) functions) on a Transwell™ insert coated with rat tail collagen as extracellular matrix protein in supplemented DMEM (Fig. 7.1b).

For the first time, we brought proteomic evidence that TNAP is present in bovine BCECs. Immunocytochemistry of solo-cultivated and co-cultivated BCEC monolayers revealed the presence with punctiform distribution of the TNAP inside the whole cell. No marked difference was observed between the two cell conditions (Fig. 7.2a). However, as demonstrated previously, its mRNA and protein expression were notably dissociable (Fig. 7.2b). The statistical analyses of RT-PCR experiments showed a transcript expression increase of 2.13 times in BCECs displaying Re-ind. BBB functions compared to those showing Lim. BBB functions. As regards the protein expression, this one follows the same trend (with a greater factor) as cells with Re-ind. BBB functions that had an expression 18.7 times higher than their counterpart with Lim. BBB functions (Deracinois et al. 2012).

However, the over-expression of TNAP (demonstrated by transcripts and protein expression analyses), does not reveal its activity level. Indeed, many factors influence the enzymatic activity (inhibitors, regulatory proteins, co-factors, allostery, etc.). Here, we show that the AP activity is also markedly increased (3.5 times) in BCECs displaying Re-ind. BBB functions (Fig. 7.3a) compared to ones with Lim. BBB functions. Subsequently, the cells were incubated 4 h with different concentrations (0–5 mM) of levamisole, an uncompetitive inhibitor specific for TNAP. The AP activity was quantified with an Alkaline Phosphatase Colorimetric Assay Kit from Abcam (Paris, France). As shown in Fig. 7.3a, the inhibition (measured by AP enzymatic activity) was proportional to the levamisole

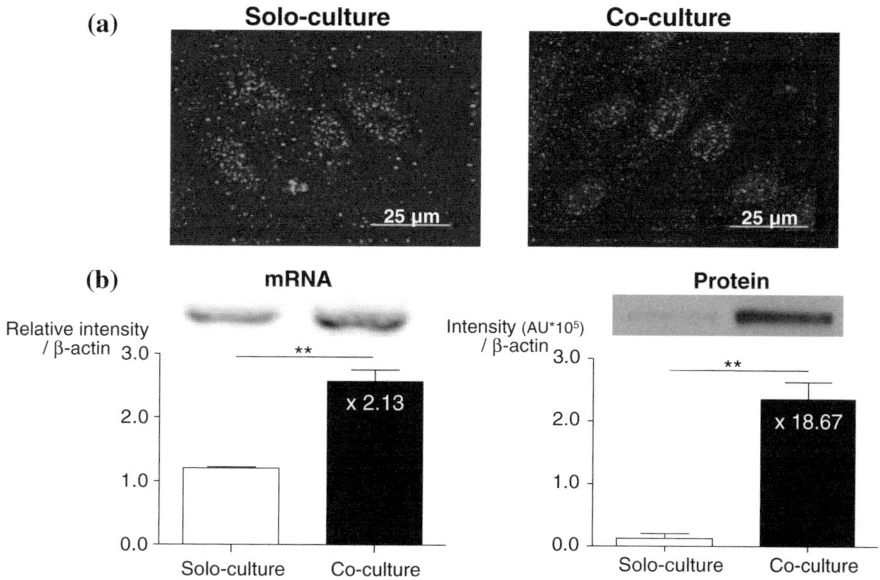

Fig. 7.2 mRNA transcript expression level, relative protein rate and immunolocalization of TNAP in solo-culture and co-culture of BCECs. **a** TNAP immunolocalization by confocal microscopy in solo-cultured BCECs displaying limited BBB properties (Lim. BBB) and in co-cultured BCECs showing re-induced BBB functions (Re-ind. BBB). **b** RT-PCR and western blot analyses of mRNA transcript (**b**, *left*) and protein (**b**, *right*) expressions from bovine BCECs solo-cultured (Lim. BBB) or co-cultivated (Re-ind. BBB). Quantitative and statistical analyses were performed with Quantity One (RT-PCR) or TotalLab 100 (western blots) and PRISM 5 software packages, respectively. The results correspond to the mean ± SEM from three distinct assays. $^{**}p < 0.002$ (in an unpaired t-test for RT-PCR and a paired t-test for western blots). The expression of β-actin was monitored as a sample quality control (Deracinois et al. 2012)

concentration added and reached 95 % in solo-cultivated BCECs and 86 % in co-cultivated BCECs incubated with 5 mM of levamisole. The influence of TNAP inhibition on the monolayer endothelial permeability for the Lucifer Yellow (LY) was measured (Fig. 7.3b). This one was not affected except for the concentration of 5 mM in solo-culture condition (condition where enzymatic activity was close to zero).

The permeability significantly rised from 0.772×10^{-3} cm min^{-1} (±0.032) for control to 3.106×10^{-3} cm min^{-1} (±0.227), suggesting a loss of monolayer integrity. The non-cytotoxicity of levamisole was evaluated by assaying of the lactate dehydrogenase (LDH) activity in the culture media (Fig. 7.3c). Endothelial cell death was assessed using the CytoTox-ONE™ Membrane Integrity Assay (Promega Corporation, USA). The results show no relevant cellular death even for high levamisole concentration, and therefore levamisole was not toxic for concentrations and times studied. It was also possible to visualize the TNAP inhibition influence with phase contrast microscopy and immunocytochemistry of vimentin,

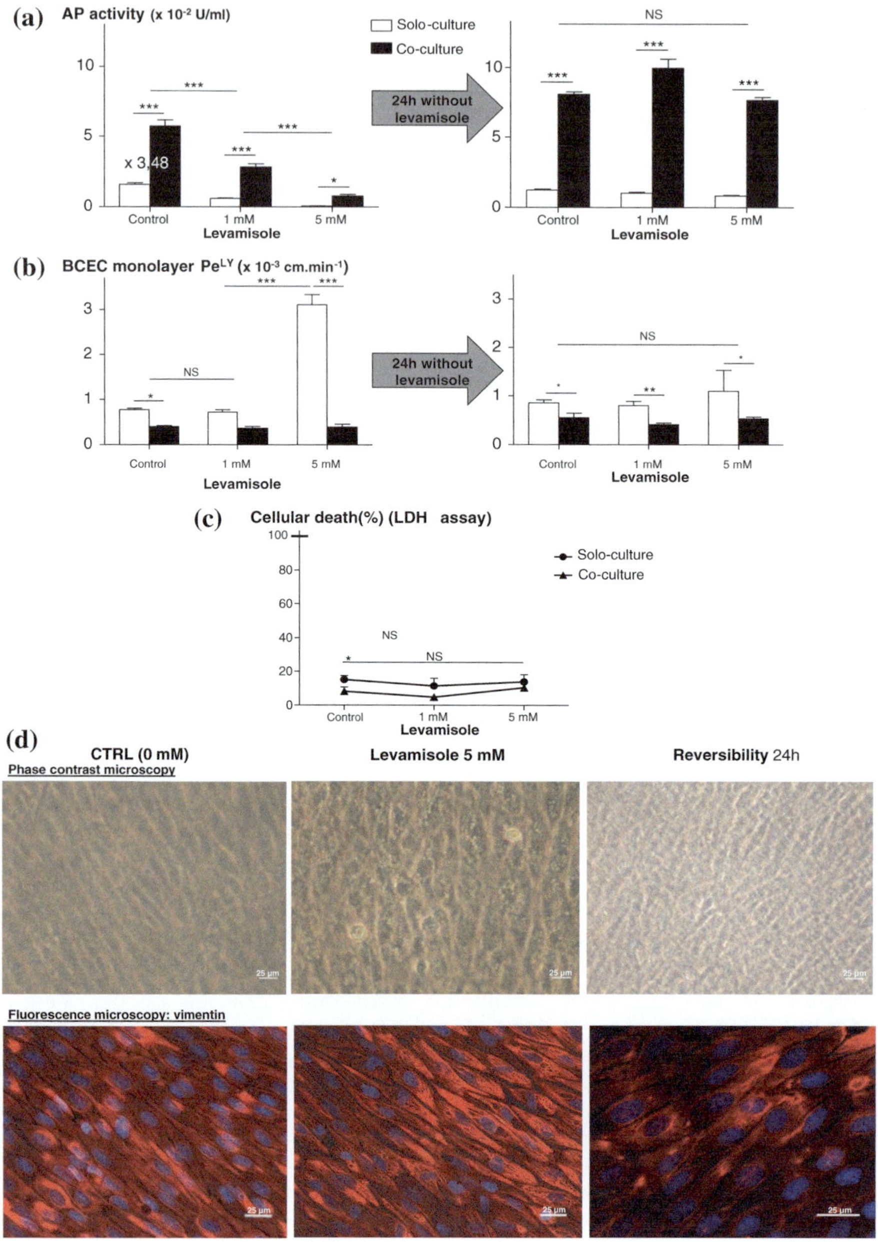
(a)
AP activity (x 10-2 U/ml)
Solo-culture
Co-culture
x 3,48
24h without levamisole
NS
Control
1 mM
5 mM
Levamisole
(b)
BCEC monolayer PeLY (x 10-3 cm.min-1)
NS
24h without levamisole
Control
1 mM
5 mM
Levamisole
(c)
Cellular death(%) (LDH assay)
Solo-culture
Co-culture
NS
Control
1 mM
5 mM
Levamisole
(d)
CTRL (0 mM)
Levamisole 5 mM
Reversibility 24h
Phase contrast microscopy
Fluorescence microscopy: vimentin
25 µm

◀ **Fig. 7.3** Levamisole-mediated inhibition of alkaline phosphatase. **a** Alkaline phosphatase (AP) activity, **b** BCEC monolayer permeability to Lucifer Yellow (Pe^{LY}) and **c** percentage of endothelial cell death (lactate dehydrogenase (LDH) assay from the culture medium): determined after 4 h of incubation of solo-cultivated and co-cultivated BCECs (Lim. BBB and Re-ind. BBB, respectively) with different concentrations of levamisole (0–5 mM) and 24 h after removal of levamisole. The extent of cell death in each experimental condition was expressed as percentage of full kill (where the latter corresponded to cells lysed with 9 % Triton X–100 (w/v) buffer). The statistical analysis was carried out using PRISM 5 software. The results correspond to the mean ± SEM (N = 3, n = 9 BCEC monolayers per treatment). The significances of between-experiment differences were tested in a two-way analysis of variance, followed by a Bonferroni post hoc test. *$p < 0.05$; ***$p < 0.001$; NS: non-significant. **d** Co-cultivated (Re-ind. BBB) phase contrast and fluorescence microscopy of vimentin after 4 h of incubation with levamisole (0 and 5 mM) and 24 h after levamisole removal (Deracinois et al. 2012)

an intermediate filament (Fig. 7.3d). The pictures show an intracellular perturbation of BCECs when cells were incubated with levamisole. The cytoskeletal organization is disrupted. This disruption is particularly marked in solo-culture in contrast to co-culture. The cells retract and dissociate at the level of adjacent cells (data not shown) and cause the monolayer integrity loss. Interestingly, all observed effects (permeability increase, disruption of the cytoskeletal organization) were reversible 24 h after levamisole removal and the return of a full activity of the AP (Fig. 7.3). Indeed, the AP activity re-increases to level identical to level observed in control conditions (Fig. 7.3a, right panel). The BCEC monolayer permeability to Lucifer Yellow (Pe^{LY}) reaches the threshold value of 1×10^{-3} cm.min^{-1} (Fig. 7.3b, right panel), value characteristic of the re-induction of the BBB phenotype in BCECs. Moreover, the disruption of the cytoskeleton organisation appears also reversed 24 h after levamisole removal (Fig. 7.3d).

In fine, these results demonstrate that the expression level and activity of TNAP is clearly enhanced in bovine BCECs displaying re-ind. BBB functions compared to bovine BCECs without such features. However, immunolocalization of TNAP remains subject to debate since TNAP is a GPI-anchored protein. Moreover, the glial cells display a protection role on BCECs in regard to levamisole-induced cell effects. Indeed, incubation of BCECs in presence of levamisole leads to (i) a decrease of the AP activity (as expected), (ii) an increase of the BCEC monolayer permeability and (iii) a disruption of the cytoskeleton organization. Furthermore, the cellular effects of levamisole seem to be proportional to the levamisole concentration. Remarkably, all levamisole-induced effects were reversible 24 h after levamisole removal. Anyway, a differential proteomics approach dedicated to phosphoproteins and phosphopeptides could bring crucial information in regard to the role of TNAP in the microvessels of the brain and its implication to the BBB phenotype.

Acknowledgments This research was funded by the Ministère de la Recherche et de l'Enseignement Supérieur. We are also grateful to Johan Hachani for his technical expertise and continuous help of every day.

Conflict of Interest The authors declare no conflict of interest.

References

Abbott NJ (2005) Dynamics of CNS barriers: evolution, differentiation, and modulation. Cell Mol Neurobiol 25:5–23. doi:10.1007/s10571-004-1374-y

Abbott NJ, Hughes CC, Revest PA, Greenwood J (1992) Development and characterisation of a rat brain capillary endothelial culture: towards an in vitro blood-brain barrier. J Cell Sci 103(Pt 1): 23–37

Abbott NJ, Patabendige AA, Dolman DE et al (2010) Structure and function of the blood-brain barrier. Neurobiol Dis 37:13–25

Abbott NJ, Ronnback L, Hansson E, Rönnbäck L (2006) Astrocyte-endothelial interactions at the blood-brain barrier. Nat Rev Neurosci 7:41–53. doi:10.1038/nrn1824

Adams SE, Melnykovych G (1985) Synergistic stimulation of alkaline phosphatase activity in bovine aortic endothelial cells grown in the presence of retinoids and glucocorticoids. J Cell Physiol 124:120–124. doi:10.1002/jcp.1041240119

Albrecht J, Szumanska G, Gadamski R, Gajkowska B (1994) Changes of activity and ultrastructural localization of alkaline phosphatase in cerebral cortical microvessels of rat after single intraperitoneal administration of mercuric chloride. Neurotoxicology 15:897–902

Alvarez JI, Dodelet-Devillers A, Kebir H et al (2011) The Hedgehog pathway promotes blood-brain barrier integrity and CNS immune quiescence. Science 334:1727–1731. doi:10.1126/science.1206936

Alvarez JI, Katayama T, Prat A (2013) Glial influence on the blood brain barrier. Glia 61:1939–1958. doi:10.1002/glia.22575

Andreeva aY, Krause E, Müller EC et al (2001) Protein kinase C regulates the phosphorylation and cellular localization of occludin. J Biol Chem 276:38480–38486. doi:10.1074/jbc.M104923200

Anstrom JA, Brown WR, Moody DM et al (2002) Temporal expression pattern of cerebrovascular endothelial cell alkaline phosphatase during human gestation. J Neuropathol Exp Neurol 61:76–84

Armulik A, Genove G, Mae M et al (2010) Pericytes regulate the blood-brain barrier. Nature 468:557–561

Balabanov R, Dore-Duffy P (1998) Role of the CNS microvascular pericyte in the blood-brain barrier. J Neurosci Res 53:637–644

Ballabh P, Braun A, Nedergaard M (2004) The blood-brain barrier: an overview: structure, regulation, and clinical implications. Neurobiol Dis 16:1–13

Ballabh P, Hu F, Kumarasiri M et al (2005) Development of tight junction molecules in blood vessels of germinal matrix, cerebral cortex, and white matter. Pediatr Res 58:791–798. doi:10.1203/01.PDR.0000180535.14093.FB

Bär T (1980) The vascular system of the cerebral cortex. Adv Anat Embryol Cell Biol 59:I–VI, 1–62

Beck DW, Roberts RL, Olson JJ (1986) Glial cells influence membrane-associated enzyme activity at the blood-brain barrier. Brain Res 381:131–137

Bell MA, Ball MJ (1985) Laminar variation in the microvascular architecture of normal human visual cortex (area 17). Brain Res 335:139–143

Bell RD, Winkler EA, Sagare AP et al (2010) Pericytes control key neurovascular functions and neuronal phenotype in the adult brain and during brain aging. Neuron 68:409–427. doi:10.1016/j.neuron.2010.09.043

Bessey O, Lowry OH, Brock MJ (1946) A method for the rapid determination of alkaline phosphates with five cubic millimeters of serum. J Biol Chem 164:321–329

Betz AL, Firth JA, Goldstein GW (1980) Polarity of the blood-brain barrier: distribution of enzymes between the luminal and antiluminal membranes of brain capillary endothelial cells. Brain Res 192:17–28

Beuckmann C, Hellwig S, Galla HJ (1995) Induction of the blood/brain-barrier-associated enzyme alkaline phosphatase in endothelial cells from cerebral capillaries is mediated via cAMP. Eur J Biochem 229:641–644

Blasig IE, Giese H, Schroeter ML et al (2001) *NO and oxyradical metabolism in new cell lines of rat brain capillary endothelial cells forming the blood-brain barrier. Microvasc Res 62:114–127

Boado RJ (2003) Blood-brain barrier genomics. Methods Mol Med 89:401–418. doi:10.1385/1-59259-419-0:401

Bonkowski D, Katyshev V, Balabanov RD et al (2011) The CNS microvascular pericyte: pericyte-astrocyte crosstalk in the regulation of tissue survival. Fluids Barriers CNS 8:8. doi:10.1186/2045-8118-8-8

Brunel C, Cathala G, Saintot M (1969) Purification and properties of alkaline phosphatase from beef brain. Biochim Biophys Acta 191:621–635

Brun-Heath I, Ermonval M, Chabrol E et al (2011) Differential expression of the bone and the liver tissue non-specific alkaline phosphatase isoforms in brain tissues. Cell Tissue Res 343:521–536. doi:10.1007/s00441-010-1111-4

Calabria AR, Shusta EV (2006) Blood-brain barrier genomics and proteomics: elucidating phenotype, identifying disease targets and enabling brain drug delivery. Drug Discov Today 11:792–799. doi:10.1016/j.drudis.2006.07.006

Calabria AR, Shusta EV (2008) A genomic comparison of in vivo and in vitro brain microvascular endothelial cells. J Cereb Blood Flow Metab 28:135–148. doi:10.1038/sj.jcbfm.9600518

Calabria AR, Shusta EV (2009) A genomic comparison of in vivo and in vitro brain microvascular endothelial cells. Blood 28:135–148. doi:10.1038/sj.jcbfm.9600518.A

Calhau C, Martel F, Soares-da-Silva P et al (2002) Regulation of [(3)H]MPP(+) transport by phosphorylation/dephosphorylation pathways in RBE4 cells: role of ecto-alkaline phosphatase. Naunyn Schmiedebergs Arch Pharmacol 365:349–356. doi:10.1007/s00210-002-0542-1

Cardelli-Cangiano P, Fiori A, Cangiano C et al (1987) Isolated brain microvessels as in vitro equivalents of the blood-brain barrier: selective removal by collagenase of the A-system of neutral amino acid transport. J Neurochem 49:1667–1675

Chan JR, Stinson Ra (1986) Dephosphorylation of phosphoproteins of human liver plasma membranes by endogenous and purified liver alkaline phosphatases. J Biol Chem 261:7635–7639

Chu C, Li JY, Boado RJ et al (2008) Blood-brain barrier genomics and cloning of a novel organic anion transporter. J Cereb Blood Flow Metab 28:291–301. doi:10.1038/sj.jcbfm.9600538

Ciani F, Contestabile A, Minelli G, Quaglia A (1973) Ultrastructural localization of alkaline phosphatase in cultures of nervous tissue in vitro. J Neurocytol 2:105–116

Couraud P, Greenwood J, Roux F, Adamson P (2003) Development and characterization of immortalized cerebral endothelial cell lines. Methods Mol Med 89:349–364

Daneman R (2012) The blood-brain barrier in health and disease. Ann Neurol 72:648–672. doi:10.1002/ana.23648

Daneman R, Zhou L, Agalliu D, et al (2010a) The mouse blood-brain barrier transcriptome: a new resource for understanding the development and function of brain endothelial cells. 5:1–16. doi:10.1371/journal.pone.0013741

Daneman R, Zhou L, Kebede A, Barres BA (2010b) Pericytes are required for blood–brain barrier integrity during embryogenesis. 468:562–566. doi:10.1038/nature09513.Pericytes

Dehouck MP, Meresse S, Delorme P et al (1990) An easier, reproducible, and mass-production method to study the blood-brain barrier in vitro. J Neurochem 54:1798–1801

Deli MA, Abrahám CS, Kataoka Y et al (2005) Permeability studies on in vitro blood-brain barrier models: physiology, pathology, and pharmacology. Cell Mol Neurobiol 25:59–127

DePierre JW, Karnovsky ML (1974) Ecto-enzymes of the guinea pig polymorphonuclear leukocyte. I. Evidence for an ecto-adenosine monophosphatase, adenosine triphosphatase, and -p-nitrophenyl phosphates. J Biol Chem 249:7111–7120

Deracinois B, Duban-Deweer S, Karamanos Y et al (2012) TNAP and EHD1 are over-expressed in bovine brain capillary endothelial cells after the re-induction of blood-brain barrier properties. PLoS ONE 7:e48428. doi:10.1371/journal.pone.0048428

Díaz-Hernández M, Gómez-Ramos A, Rubio A et al (2010) Tissue-nonspecific alkaline phosphatase promotes the neurotoxicity effect of extracellular tau. J Biol Chem 285:32539–32548. doi:10.1074/jbc.M110.145003

Djuricić BM, Rogac L, Spatz M et al (1978) Brain microvessels. I. Enzymic activities. Adv Neurol 20:197–205

Ehrlich P (1885) Das Sauerstoff-bedürfniss des Organismus. Ein farbenanalytische Studie. Hirschwald, Berlin

Ehrlich YH, Davis TB, Bock E et al (1986) Ecto-protein kinase activity on the external surface of neural cells. Nature 320:67–70. doi:10.1038/320067a0

Enerson BE, Drewes LR (2006) The rat blood-brain barrier transcriptome. J Cereb Blood Flow Metab 26:959–973. doi:10.1038/sj.jcbfm.9600249

Engelhardt B (2003) Development of the blood-brain barrier. Cell Tissue Res 314:119–129

Ermonval M, Baudry A, Baychelier F et al (2009) The cellular prion protein interacts with the tissue non-specific alkaline phosphatase in membrane microdomains of bioaminergic neuronal cells. PLoS ONE 4:e6497. doi:10.1371/journal.pone.0006497

Farkas-Bargeton E, Arsenio-Nunes ML (1970) Maturation of enzymatic equipment in the vessel walls of the nervous system. Histochemical study. Acta Neuropathol 15:251–271

Farley JR, Joel L, Baylink DJ (1980) Human Skeletal Alkaline Phosphatase. J Biol Chem 10:4680–4686

Fedde KN, Michel MP, Whyte MP (1993) Evidence against a role for alkaline phosphatase in the dephosphorylation of plasma membrane proteins: hypophosphatasia fibroblast study. J Cell Biochem 53:43–50. doi:10.1002/jcb.240530106

Fleming A, Diekmann H, Goldsmith P (2013) Functional characterisation of the maturation of the blood-brain barrier in larval zebrafish. PLoS ONE 8:e77548. doi:10.1371/journal.pone.0077548

Fonta C, Imbert M (2002) Vascularization in the primate visual cortex during development. Cereb Cortex 12:199–211

Fonta C, Negyessy L, Renaud L, Barone P (2005) Postnatal development of alkaline phosphatase activity correlates with the maturation of neurotransmission in the cerebral cortex. J Comp Neurol 486:179–196. doi:10.1002/cne.20524

Fonta C, Négyessy L, Renaud L, Barone P (2004) Areal and subcellular localization of the ubiquitous alkaline phosphatase in the primate cerebral cortex: evidence for a role in neurotransmission. Cereb Cortex 14:595–609. doi:10.1093/cercor/bhh021

Frangakis MV, Kimelberg HK (1984) Dissociation of neonatal rat brain by dispase for preparation of primary astrocyte cultures. Neurochem Res 9:1689–1698

Fukushima H, Fujimoto M, Ide M (1990) Quantitative detection of blood-brain barrier-associated enzymes in cultured endothelial cells of porcine brain microvessels. Vitro Cell Dev Biol 26:612–620

Ginsbourg M, Foncin JF, Le Beau J (1973) Vascular abnormalities in cerebral tumours. A comparative study of enzyme activity and permeability to fluorescent tracers. Rev Neurol (Paris) 129:275–288

Ginsbourg M, Le Beau J (1975) Histochemical correlations of vascular permeability and enzyme activity in the choroid plexus in man. C R Seances Soc Biol Fil 169:518–521

Goldmann E (1913) Vitalfärbung am zentral nerven System. Bietrag zur Physio-Pathologie des Plexus Choroidus und der Hirnhuate. Abhandlungen der koniglich preubischen Akademie der Wissenschaften. Physikalisch-Mathematische Classe 1:1–64

Goldstein GW (1988) Endothelial cell-astrocyte interactions. A cellular model of the blood-brain barrier. Ann N Y Acad Sci 529:31–39

Grammas P, Martinez J, Miller B (2011) Cerebral microvascular endothelium and the pathogenesis of neurodegenerative diseases. Expert Rev Mol Med 13:e19. doi:10.1017/S146239941 1001918

Haseloff RF, Blasig IE, Bauer HC, Bauer H (2005) In search of the astrocytic factor(s) modulating blood-brain barrier functions in brain capillary endothelial cells in vitro. Cell Mol Neurobiol 25:25–39

Hawkins BT, Davis TP (2005) The blood-brain barrier/neurovascular unit in health and disease. Pharmacol Rev 57:173–185. doi:10.1124/pr.57.2.4.173
Hurst RD, Fritz IB (1996) Properties of an immortalised vascular endothelial/glioma cell co-culture model of the blood-brain barrier. J Cell Physiol 167:81–88. doi:10.1002/(SICI) 1097-4652(199604)167:1<81:AID-JCP9>3.0.CO;2-8
Inomata K, Yoshioka T, Nasu F, Mayahara H (1984) Ultracytochemical studies of capillary endothelial cells in the rat central nervous system. Acta Anat (Basel) 118:243–248
Isobe I, Watanabe T, Yotsuyanagi T, et al (1996) Astrocytic contributions to blood-brain barrier (BBB) formation by endothelial cells: a possible use of aortic endothelial cell for in vitro BBB model. Neurochem Int 28:523–33
Jancsó G, Tóth L, Karcsú S (1975) Histochemical localization of capillary enzyme activities in brain smears. Z Mikrosk Anat Forsch 89:983–990
Janzer RC, Raff MC (1987) Astrocytes induce blood-brain barrier properties in endothelial cells. Nature 325:253–257. doi:10.1038/325253a0
Jeong J-Y, Kwon H-B, Ahn J-C et al (2008) Functional and developmental analysis of the blood-brain barrier in zebrafish. Brain Res Bull 75:619–628. doi:10.1016/j.brainresbull.2007.10.043
Joo F (1993) The blood-brain barrier in vitro: the second decade. Neurochem Int 23:499–521
Joó F, Karnushina I (1973) A procedure for the isolation of capillaries from rat brain. Cytobios 8:41–48
Kamiie J, Ohtsuki S, Iwase R et al (2008) Quantitative atlas of membrane transporter proteins: development and application of a highly sensitive simultaneous LC/MS/MS method combined with novel in-silico peptide selection criteria. Pharm Res 25:1469–1483
Kondo T, Imaizumi S, Kato I, Yoshimoto T (1994) Isolation and culture of brain endothelial cells and establishment of in vitro blood-brain barrier model. Cell Transplant 3(Suppl 1):S35–S37
Kozlenkov A, Le Du MH, Cuniasse P et al (2004) Residues determining the binding specificity of uncompetitive inhibitors to tissue-nonspecific alkaline phosphatase. J Bone Miner Res 19:1862–1872. doi:10.1359/JBMR.040608
Latker CH, Shinowara NL, Miller JC, Rapoport SI (1987) Differential localization of alkaline phosphatase in barrier tissues of the frog and rat nervous systems: a cytochemical and biochemical study. J Comp Neurol 264:291–302. doi:10.1002/cne.902640302
Lazzari M, Franceschini V, Ciani F, Minelli G (1989) Cytochemical localization of alkaline phosphatase and Na+, K+-ATPase activities in the blood-brain barrier of Rana esculenta. Basic Appl Histochem 33:113–120
Lechardeur D, Schwartz B, Paulin D, Scherman D (1995) Induction of blood-brain barrier differentiation in a rat brain-derived endothelial cell line. Exp Cell Res 220:161–170. doi:10.1006/excr.1995.1302
Leduc EH, Wislocki GB (1952) The histochemical localization of acid and alkaline phosphatases, non-specific esterase and succinic dehydrogenase in the structures comprising the hemato-encephalic barrier of the rat. J Comp Neurol 97:241–279
Lev R, Griffiths WC (1982) Colonic and small intestinal alkaline phosphatase. A histochemical and biochemical study. Gastroenterology 82:1427–1435
Lewandowsky M (1900) Zur lehre von der Cerebrospinalflüssigkeit. Z Klin Med 480–494
Li JY, Boado RJ, Pardridge WM (2001) Blood-brain barrier genomics. J Cereb Blood Flow Metab 21:61–68
Li JY, Boado RJ, Pardridge WM (2002) Rat blood-brain barrier genomics. II. J Cereb Blood Flow Metab 22:1319–1326. doi:10.1097/00004647-200211000-00005
Lippmann ES, Weidenfeller C, Svendsen CN, Shusta EV (2011) Blood-brain barrier modeling with co-cultured neural progenitor cell-derived astrocytes and neurons. J Neurochem 119:507–520. doi:10.1111/j.1471-4159.2011.07434.x
Liu HC, Lee JC, Bakay L (1979) Experimental cerebral concussion. A histochemical study. Acta Neurochir (Wien) 47:105–122
Liu S, Agalliu D, Yu C, Fisher M (2012) The role of pericytes in blood-brain barrier function and stroke. Curr Pharm Des 18:3653–3662

Mabry CC, Bautista A, Kirk RF et al (1970) Familial hyperphosphatase with mental retardation, seizures, and neurologic deficits. J Pediatr 77:74–85

Mebarek S, Hamade E, Thouverey C et al (2011) Ankylosing spondylitis, late osteoarthritis, vascular calcification, chondrocalcinosis and pseudo gout: toward a possible drug therapy. Curr Med Chem 18:2196–2203

Meresse S, Dehouck MP, Delorme P et al (1989) Bovine brain endothelial cells express tight junctions and monoamine oxidase activity in long-term culture. J Neurochem 53:1363–1371

Meunier MT, Bouchaud C (1978) Histoenzymological contribution to the study of the rat blood-brain barrier. Arch Anat Microsc Morphol Exp 67:81–98

Meyer J, Mischeck U, Veyhl M et al (1990) Blood-brain barrier characteristic enzymatic properties in cultured brain capillary endothelial cells. Brain Res 514:305–309

Meyer J, Rauh J, Galla HJ (1991) The susceptibility of cerebral endothelial cells to astroglial induction of blood-brain barrier enzymes depends on their proliferative state. J Neurochem 57:1971–1977

Millán JL (2006) Alkaline Phosphatases: structure, substrate specificity and functional relatedness to other members of a large superfamily of enzymes. Purinergic Signal 2:335–341. doi:10.1007/s11302-005-5435-6

Mizuguchi H, Fujii A, Utoguchi N et al (1995) Glial-conditioned medium elevates alkaline phosphatase activity in cultured bovine aortic endothelial cells. Microvasc Res 50:129–132

Mizuguchi H, Hashioka Y, Utoguchi N et al (1994) A comparison of drug transport through cultured monolayers of bovine brain capillary and bovine aortic endothelial cells. Biol Pharm Bull 17:1385–1390

Mulkins MA, Allison AC (1987) Recombinant human interleukin-1 inhibits the induction by dexamethasone of alkaline phosphatase activity in murine capillary endothelial cells. J Cell Physiol 133:539–545. doi:10.1002/jcp.1041330315

Nag S (2011) Morphology and properties of brain endothelial cells. Methods Mol Biol 686:3–47

Nakazato H, Deguchi M, Fujimoto M, Fukushima H (1997) Alkaline phosphatase expression in cultured endothelial cells of aorta and brain microvessels: induction by interleukin-6-type cytokines and suppression by transforming growth factor betas. Science 80(61):2065–2072

Neuhaus J, Risau W, Wolburg H (1991) Induction of blood-brain barrier characteristics in bovine brain endothelial cells by rat astroglial cells in transfilter coculture. Ann N Y Acad Sci 633:578–580

Nishio S, Egami H, Fukui M et al (1986) Ultrastructural and cytochemical study of microvascular enzyme activity in experimental brain tumors of rat. Neurol Med Chir (Tokyo) 26:527–533

Norman MG, O'Kusky JR (1986) The growth and development of microvasculature in human cerebral cortex. J Neuropathol Exp Neurol 45:222–232

Nosjean O, Koyama I, Goseki M, et al (1997) Human tissue non-specific alkaline phosphatases: sugar-moiety-induced enzymic and antigenic modulations and genetic aspects. Biochem J 321 (Pt 2):297–303

Oby E, Janigro D (2006) The blood-brain barrier and epilepsy. Epilepsia 47:1761–1774

Oldendorf WH, Cornford ME, Brown WJ (1977) The large apparent work capability of the blood-brain barrier: a study of the mitochondrial content of capillary endothelial cells in brain and other tissues of the rat. Ann Neurol 1:409–417. doi:10.1002/ana.410010502

Palmer AM (2010) The role of the blood-CNS barrier in CNS disorders and their treatment. Neurobiol Dis 37:3–12. doi:10.1016/j.nbd.2009.07.029

Paolinelli R, Corada M, Orsenigo F, Dejana E (2011) The molecular basis of the blood brain barrier differentiation and maintenance. Is it still a mystery? Pharmacol Res 63:165–171. doi:10.1016/j.phrs.2010.11.012

Pardridge WM (2002) Drug and gene targeting to the brain with molecular Trojan horses. Nat Rev Drug Discov 1:131–139

Pardridge WM (2007) Blood-brain barrier delivery. Drug Discov Today 12:54–61

Pardridge WM, Oldendorf WH (1977) Transport of metabolic substrates through the blood-brain barrier. J Neurochem 28:5–12

Persidsky Y, Heilman D, Haorah J et al (2006) Rho-mediated regulation of tight junctions during monocyte migration across the blood-brain barrier in HIV-1 encephalitis (HIVE). Blood 107:4770–4780. doi:10.1182/blood-2005-11-4721

Plesner L (1995) Ecto-ATPases: identities and functions. Int Rev Cytol 158:141–214

Pottiez G, Flahaut C, Cecchelli R, Karamanos Y (2009) Understanding the blood-brain barrier using gene and protein expression profiling technologies. Brain Res Rev 62:83–98. doi:10.1016/j.brainresrev.2009.09.004

Prat A, Biernacki K, Wosik K, Antel JP (2001) Glial cell influence on the human blood-brain barrier. Glia 36:145–155

Ramsauer M, Krause D, Dermietzel R (2002) Angiogenesis of the blood-brain barrier in vitro and the function of cerebral pericytes. FASEB J 16:1274–1276. doi:10.1096/fj.01-0814fje

Rauh J, Meyer J, Beuckmann C, Galla HJ (1992) Development of an in vitro cell culture system to mimic the blood-brain barrier. Prog Brain Res 91:117–121

Reichel A, Begley DJ, Abbott NJ (2003) An overview of in vitro techniques for blood-brain barrier studies. blood brain barrier. Biol Res Protoc 89:307–324

Renkawek K, Murray MR, Spatz M, Klatzo I (1976) Distinctive histochemical characteristics of brain capillaries in organotypic culture. Exp Neurol 50:194–206

Risau W, Hallmann R, Albrecht U (1986) Differentiation-dependent expression of proteins in brain endothelium during development of the blood-brain barrier. Dev Biol 117:537–545

Roux F, Durieu-Trautmann O, Chaverot N et al (1994) Regulation of gamma-glutamyl transpeptidase and alkaline phosphatase activities in immortalized rat brain microvessel endothelial cells. J Cell Physiol 159:101–113. doi:10.1002/jcp.1041590114

Rubin LL (1991) The blood-brain barrier in and out of cell culture. Curr Opin Neurobiol 1:360–363

Rubin LL, Hall DE, Porter S et al (1991) A cell culture model of the blood-brain barrier. J Cell Biol 115:1725–1735

Sakakibara A, Furuse M, Saitou M et al (1997) Possible involvement of phosphorylation of occludin in tight junction formation. J Cell Biol 137:1393–1401

Sánchez del Pino MM, Hawkins RA, Peterson DR (1995) Biochemical discrimination between luminal and abluminal enzyme and transport activities of the blood-brain barrier. J Biol Chem 270:14907–14912

Schiffer D, Vesco C, Piazza L (1962) Contribution to the histochemical demonstration and distribution of phosphatases and non-specific esterase in human nervous tissue. Psychiatr Neurol (Basel) 144:34–47

Sessa G, Perez MM (1975) Biochemical changes in rat brain associated with the development of the blood-brain barrier. J Neurochem 25:779–782

Shi F, Audus KL (1994) Biochemical characteristics of primary and passaged cultures of primate brain microvessel endothelial cells. Neurochem Res 19:427–433

Shusta EV, Boado RJ, Mathern GW, Pardridge WM (2002) Vascular genomics of the human brain. J Cereb Blood Flow Metab 22:245–252. doi:10.1097/00004647-200203000-00001

Siddharthan V, Kim YV, Liu S, Kim KS (2007) Human astrocytes/astrocyte-conditioned medium and shear stress enhance the barrier properties of human brain microvascular endothelial cells. Brain Res 1147:39–50. doi:10.1016/j.brainres.2007.02.029

Sidique S, Ardecky R, Su Y et al (2009) Design and synthesis of pyrazole derivatives as potent and selective inhibitors of tissue-nonspecific alkaline phosphatase (TNAP). Bioorg Med Chem Lett 19:222–225. doi:10.1016/j.bmcl.2008.10.107

Smith M, Weiss MJ, Griffin CA et al (1988) Regional assignment of the gene for human liver/bone/kidney alkaline phosphatase to chromosome 1p36.1–p34. Genomics 2:139–143

Sobue K, Yamamoto N, Yoneda K et al (1999) Induction of blood-brain barrier properties in immortalized bovine brain endothelial cells by astrocytic factors. Neurosci Res 35:155–164

Soma T, Chiba H, Kato-Mori Y et al (2004) Thr(207) of claudin-5 is involved in size-selective loosening of the endothelial barrier by cyclic AMP. Exp Cell Res 300:202–212. doi:10.1016/j.yexcr.2004.07.012

Stanness KA, Neumaier JF, Sexton TJ et al (1999) A new model of the blood–brain barrier: co-culture of neuronal, endothelial and glial cells under dynamic conditions. Neuroreport 10:3725–3731

Stewart PA, Wiley MJ (1981) Structural and histochemical features of the avian blood-brain barrier. J Comp Neurol 202:157–167. doi:10.1002/cne.902020203

Sundstrom JM, Tash BR, Murakami T et al (2009) Identification and analysis of occludin phosphosites: a combined mass spectrometry and bioinformatics approach. J Proteome Res 8:808–817. doi:10.1021/pr7007913

Takemoto H, Kaneda K, Hosokawa M, et al (1994) Conditioned media of glial cell lines induce alkaline phosphatase in cultured artery endothelial cells of interleukin-6 as an induction factor activity Identification. 350:99–103

Tao-Cheng J, Nagy Z, Brightman M (1987) Tight junctions astroglia of brain endothelium in vitro are enhanced by astroglia. J Neurosci 7:3293–3299

Tatsuta T, Naito M, Oh-hara T et al (1992) Functional involvement of P-glycoprotein in blood-brain barrier. J Biol Chem 267:20383–20391

Tio S, Deenen M, Marani E (1990) Astrocyte-mediated induction of alkaline phosphatase activity in human umbilical cord vein endothelium: an in vitro model. Eur J Morphol 28:289–300

Tontsch U, Bauer HC (1991) Glial cells and neurons induce blood-brain barrier related enzymes in cultured cerebral endothelial cells. Brain Res 539:247–253

Torack R, Barrnett R (1964) The fine structural localization of nucleoside phosphatase activity in the blood-brain barrier. J Neuropathol Exp Neurol 23:46–59

Uchida Y, Ohtsuki S, Katsukura Y et al (2011) Quantitative targeted absolute proteomics of human blood-brain barrier transporters and receptors. J Neurochem 117:333–345. doi:10.1111/j.1471-4159.2011.07208.x

Van Belle H (1976) Alkaline phosphatase. I. Kinetics and inhibition by levamisole of purified isoenzymes from humans. Clin Chem 22:972–976

Vandenhaute E, Culot M, Gosselet F et al (2012) Brain pericytes from stress-susceptible pigs increase blood-brain barrier permeability in vitro. Fluids Barriers CNS 9:11. doi:10.1186/2045-8118-9-11

Vardy ERLC, Kellett KAB, Cocklin SL, Hooper NM (2012) Alkaline phosphatase is increased in both brain and plasma in Alzheimer's disease. Neurodegener Dis 9:31–37. doi:10.1159/000329722

Vinters HV, Reave S, Costello P et al (1987) Isolation and culture of cells derived from human cerebral microvessels. Cell Tissue Res 249:657–667

Vorbrodt aW, Trowbridge RS (1991) Ultracytochemical characteristics of cultured goat brain microvascular endothelial cells [corrected]. J Histochem Cytochem 39:1555–1563. doi:10.1177/39.11.1655877

Vorbrodt AW, Lossinsky AS, Wisniewski HM et al (1981) Ultrastructural cytochemical studies of cerebral microvasculature in scrapie infected mice. Acta Neuropathol 53:203–211

Vorbrodt AW, Lossinsky AS, Wisniewski HM (1986) Localization of alkaline phosphatase activity in endothelia of developing and mature mouse blood-brain barrier. Dev Neurosci 8:1–13

Vorbrodt AW, Lossinsky AS, Wisniewski HM et al (1985) Ultrastructural observations on the transvascular route of protein removal in vasogenic brain edema. Acta Neuropathol 66:265–273

Vorbrodt AW, Lossinsky AS, Wisniewski HM (1983) Enzyme cytochemistry of blood-brain barrier (BBB) disturbances. Acta Neuropathol Suppl 8:43–57

Wakai S, Hirokawa N (1978) Development of the blood-brain barrier to horseradish peroxidase in the chick embryo. Cell Tissue Res 195:195–203

Whyte MP (1994) Hypophosphatasia and the role of alkaline phosphatase in skeletal mineralization. Endocr Rev 15:439–461

Williams SK, Gillis JF, Matthews MA et al (1980) Isolation and characterization of brain endothelial cells: morphology and enzyme activity. J Neurochem 35:374–381

Wisniewski HM, Lossinsky AS, Moretz RC et al (1983) Increased blood-brain barrier permeability in scrapie-infected mice. J Neuropathol Exp Neurol 42:615–626

Wollscheid B, Bausch-Fluck D, Henderson C et al (2009) Mass-spectrometric identification and relative quantification of N-linked cell surface glycoproteins. Nat Biotechnol 27:378–386. doi:10.1038/nbt.1532

Wong AD, Ye M, Levy AF et al (2013) The blood-brain barrier: an engineering perspective. Front Neuroeng 6:7. doi:10.3389/fneng.2013.00007

Wong V (1997) Phosphorylation of occludin correlates with occludin localization and function at the tight junction. Am J Physiol 273:C1859–C1867

Yu S, Guo Z, Johnson C, et al (2013) Recent progress in synthetic and biological studies of GPI anchors and GPI-anchored proteins. Curr Opin Chem Biol 1–8. doi:10.1016/j.cbpa.2013.09.016

Zhou H, Manji SS, Findlay DM et al (1994) Novel action of retinoic acid. Stabilization of newly synthesized alkaline phosphatase transcripts. J Biol Chem 269:22433–22439

Zoellner HF, Hunter N (1989) Histochemical identification of the vascular endothelial isoenzyme of alkaline phosphatase. J Histochem Cytochem 37:1893–1898. doi:10.1177/37.12.2584695

Part III
TNAP Functions in the Nervous System

Chapter 8
What Can We Learn About the Neural Functions of TNAP from Studies on Other Organs and Tissues?

José Luis Millán

Abstract To-date, the function of tissue-nonspecific alkaline phosphatase (TNAP) has largely been defined through studies in patients and mice affected by hypophosphatasia (HPP), a rare inborn-error-of-metabolism caused by mutation(s) in the TNAP gene (*ALPL*). The skeletal disease in HPP can be explained by alterations in the P_i/PP_i ratio, with accumulation in the concentration of the mineralization inhibitor PP_i as the culprit in preventing propagation of mineralization onto the collagenous extracellular matrix in bones and teeth. Accumulation of phosphorylated osteopontin increases the severity of HPP, at least in mice. Disruption in the metabolism of vitamin B6 leads to intracellular deficiency of pyridoxal, and this causes vitamin B6-responsive seizures in patients with the severe forms of the disease. Recent findings also implicate TNAP in the metabolism of ATP, in the production of adenosine and in the dephosphorylation of the bacterial toxin lipopolysaccharide, all molecules known to be involved in inflammation. The role of TNAP in establishing the ATP/adenosine ratio is important for purinergic signaling, and these mechanisms could be significant in determining axonal growth in the brain. Finally, the potential involvement of TNAP in dephosphorylating tau protein and its role in the pathogenesis of Alzheimer's disease is intriguing.

Keywords Rickets · Osteomalacia · Tooth mineralization · Seizures · Physiological substrates

J.L. Millán (✉)
Sanford-Burnham Medical Research Institute, La Jolla, CA, USA
e-mail: millan@sanfordburnham.org

C. Fonta and L. Négyessy (eds.), *Neuronal Tissue-Nonspecific Alkaline Phosphatase (TNAP)*, Subcellular Biochemistry 76,
DOI 10.1007/978-94-017-7197-9_8

8.1 What We Know About TNAP During Skeletal and Dental Mineralization

The study of "hypophosphatasia" (HPP)—the inborn error of metabolism caused by missense mutations in the *ALPL* gene heralding rickets in children or osteomalacia in adults—has provided an unparalleled opportunity for understanding the physiological roles of TNAP (Whyte et al. 2012). The study of HPP patients, and subsequently studies in *Alpl* knockout mice (Waymire et al. 1995; Narisawa et al. 1997; Fedde et al. 1999), have conclusively demonstrated a role for TNAP in modulating inorganic pyrophosphate (PP_i) concentrations and thus, regulating skeletal and dental mineralization as well as in vitamin B6 metabolism where insufficient hydrolysis of pyridoxal-5′-phosphate (PLP) leads to pyridoxal (PL)-responsive seizures. While elevated levels of urinary phosphoethanolamine are also found in patients and mice with HPP, there is little information regarding the pathway that is affected by lack of TNAP leading to PEA accumulation and there is evidence that PEA may be a poor substrate for TNAP (Halling-Linder et al. 2009), so this compound will not be discussed further in this chapter.

In agreement with the clinical observation of elevated PP_i in the blood and urine of HPP patients (Whyte 2010), studies using single and double knockout mice have shown that PP_i, produced enzymatically by the action of ectonucleotide pyrophosphatase/phosphodiesterase-1 (NPP1) (Johnson et al. 2000; Hessle et al. 2002) and also transported from the cytosol to the extracellular milieu by the ankylosis protein (ANK) transporter (Harmey et al. 2004), is hydrolyzed by TNAP. This hydrolysis restricts the concentrations of the mineralization inhibitor PP_i (Meyer 1984), while also contributing to the P_i pool available for deposition in the formation of bone mineral (Millán 2013). Thus, it is appropriate to refer to TNAP function as controlling a P_i/PP_i ratio conducive to normal skeletal mineralization. Changes in this ratio can lead either to soft bones disease or to soft tissue calcification. Pioneering work by H Clarke Anderson and collaborators in the late 60 s (reviewed in Anderson et al. 2005) had pointed to chondrocyte-derived matrix vesicles (MVs) as the site of initiation of skeletal mineralization. The Buchet laboratory reported that formation of apatite by chondrocyte-derived MVs is optimal when the P_i/PP_i molar ratio is above 140, but is totally inhibited when the ratio decreases below 70 (Thouverey et al. 2009; Buchet et al. 2013). The retardation of any mineral formation appears maximal at a P_i/PP_i ratio of around 30, and crystals of calcium PP_i dihydrate (CPPD) are exclusively produced by MVs when the ratio is below 6. This likely explains why some patients with adult HPP suffer from CPPD deposition (aka chondrocalcinosis), occasionally with attacks of pseudogout (Whyte 2010).

The fact that mineralization is inhibited by PP_i and that TNAP promotes apatite crystal formation by hydrolyzing PP_i questions why mineralization does not also occur in other soft tissues, which also express TNAP. Using different loss- and gain-of-function approaches in vitro and in vivo, Murshed and collaborators demonstrated that mineralization requires the co-localization of TNAP and a fibrilar collagen, such as type I (in bone and dentin) or type II (in growth plate cartilage) (Murshed et al. 2005). Collagen fibrils are necessary to provide a scaffold for propagation of apatite crystals. It is now well established that mineralizing cells (chondrocytes, osteoblasts, odontoblasts) calcify the extracellular matrix via the production of MVs and regulation of the P_i/PP_i ratio, as demonstrated through the analysis of *Alpl*$^{-/-}$ mice and the success of enzyme replacement using mineral-targeting TNAP to correct the skeletal and dental phenotype in this mouse model of infantile HPP (Millán et al. 2008; McKee et al. 2011; Foster et al. 2013). Similarly to dentin, the acellular cementum is under strict regulation by the P_i/PP_i ratio (Foster et al. 2012), and this ratio is normalized, and acellular cementum corrected, in HPP mice treated with enzyme replacement (McKee et al. 2011). Enamel defects are also present in HPP mice and these defects are preventable by enzyme replacement with mineral-targeting TNAP (Yadav et al. 2012), arguing that the P_i/PP_i ratio is also involved in the regulation of enamel matrix mineralization. The accumulation of PP_i in the extracellular matrix leads to arrest in the propagation of bone mineral onto the fibrilar collagenous matrix, leading to the soft bones phenotype of HPP. Thus, TNAP is involved in the propagation of mineralization but not necessarily in the initiation of mineralization, as studies in MVs derived from HPP patients (Anderson et al. 2002) and from *Alpl*$^{-/-}$ mice (Anderson et al. 2004) have demonstrated the presence of hydroxyapatite crystal inside TNAP-deficient MVs. Work in our laboratory has shown that another phosphatase, PHOSPHO1, is involved in the initiation of MV-mediated calcification (Roberts et al. 2007; Yadav et al. 2011), at least in mice. PHOSPHO1 appears to scavenge P_i by hydrolyzing phosphocholine and phosphoethanolamine derived from the MV membranes, thus affecting the intravesicular P_i/PP_i ratio in favor of initiation of calcification (Yadav et al. 2011; Millan 2013). Abolishing PHOSPHO1 function leads to osteomalacia and severe scoliosis, while the simultaneous ablation of both TNAP and PHOSPHO1 function completely prevent skeletal mineralization (Yadav et al. 2011), demonstrating the non-redundant roles of these two phosphatases in the mechanisms controlling skeletal mineralization. Interestingly, it was through the study of the interrelated roles of TNAP and PHOSPHO1 that the role of TNAP, and also of NPP1, as potent ATPase came into focus. That MVs have ATPase activity has been known for some time (Majeska and Wuthier 1975; Pizauro et al. 1998), but it was Ciancaglini et al. that showed conclusively that TNAP is the major ATPase in MVs, followed by NPP1, and enzyme that beside producing PP_i from ATP, as already described, can also act as a backup ATPase as well as PP_iase, in the absence of TNAP (Ciancaglini et al. 2010; Simão et al. 2010). The roles of TNAP as an ATPase, generating P_i needed for mineralization, and as a PP_iase inactivating the mineralization inhibitor PP_i have been entertained for many years as contradicting hypotheses. But these recent data, obtained through the analysis of the [*Phospho1*$^{-/-}$; *Alpl*$^{-/-}$] double knockout phenotype support an inclusive model of the initiation of mineralization that takes into account both roles for TNAP. This unified model starts with the MVs as the site of initiation of mineralization (Fig. 8.1). Hydroxyapatite crystals appear inside the MVs favored by P_i accumulation resulting from a dual mechanism, i.e., PHOSPHO1-mediated intravesicular

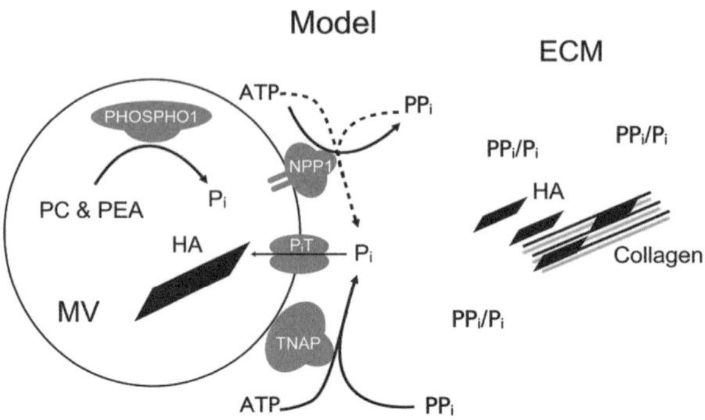

Fig. 8.1 Model of initiation of skeletal mineralization including the function of PHOSPHO1, TNAP, NPP1 and phosphate transporters. The first step of MV-mediated mineralization involves the convergence of two independent biochemical pathways. **a** Intravesicular P_i generation by the enzymatic action of PHOSPHO1 and **b** influx of P_i, generated in the peri-vesicular space by the activities of TNAP and NPP1, via P_i-transporters. *PC* phosphocholine, *PEA* phosphoethanolamine, *MV* matrix vesicles, *P_iT* phosphate transporter 1, *HA* hydroxyapatite, *ECM* extracellular matrix, *P_i* inorganic phosphate, *PP_i* inorganic pyrophosphate

production and transporter-mediated influx of P_i produced extravesicularly primarily by TNAP's ATPase activity or, secondarily in the absence of TNAP, by NPP1's ATPase activity. Organophosphate compounds (ATP) and perhaps PP_i are the source of P_i for this initial step of calcification. Then, extravesicular calcification is supported primarily by TNAP's pyrophosphatase activity and secondarily by NPP1's pyrophosphatase activity and is driven by the availability of P_i and the presence of a collagenous fibrilar scaffold and guided by other extracellular matrix mineral-binding proteins.

Osteopontin (OPN, encoded by the *Spp1* gene) is another potent inhibitor of mineralization highly expressed in osteoblasts, chondrocytes and also osteoclasts (Boskey et al. 1993; Hunter et al. 1994; Sodek et al. 2000; Hunter 2013). OPN is an Arg-Gly-Asp motif-containing matricellular protein (Oldberg et al. 1986) that harbors a multitude of putative phosphorylation sites (Sorensen et al. 1995) and that can inhibit hydroxyapatite deposition in vitro when phosphorylated (Boskey et al. 1993). Furthermore, phosphorylation-dependent inhibition of mineralization by OPN appears to be mediated by ASARM peptides released from native OPN by phosphate regulating endopeptidase homolog, X-linked (PHEX) cleavage (Addison et al. 2010). We (Johnson et al. 2003; Harmey et al. 2004, 2006), and others (Addison et al. 2007) have shown that OPN expression is regulated by extracellular concentrations of PP_i. Indeed, there is a linear correlation between PP_i and OPN concentrations in mutant mouse strains displaying dysregulation of PP_i metabolism (Harmey et al. 2006). For example we showed increased levels of plasma OPN in *Alpl*$^{-/-}$ mice, correlating with their increased levels of plasma PP_i (Harmey et al.

2006). Ablating *Spp1* in an *Alpl* null background however only led to a partial rescue of the HPP phenotype (Harmey et al. 2006), confirming that the elevated PP_i concentrations were the culprit in the development of HPP and that OPN only contributed partially to this phenotype. Of interest, however, the double ablation of *Spp1* and *Phospho1* improved the skeletal phenotype, including the severe scoliosis, characteristic of the *Phospho1*$^{-/-}$ model of osteomalacia (Yadav et al. 2014). Thus, while *Alpl* and *Phospho1* deficiencies lead to similar skeletal phenotypes and comparable changes in the expression levels of PP_i and OPN, there is a clear dissociation in the hierarchical roles of these potent inhibitors of mineralization, with elevated PP_i and elevated p-OPN levels causing the respective skeletal phenotypes in *Alpl*$^{-/-}$ and *Phospho1*$^{-/-}$ mice. It is clear however that the levels of plasma OPN correlate with the severity of the HPP phenotype in mice (Fig. 8.2a). We recently demonstrated that OPN is a physiological substrate of TNAP, having identified at least two preferred site of TNAP in the OPN molecule, and we showed accumulation of phosphorylated forms of OPN in *Alpl*$^{-/-}$ HPP mice (Narisawa et al. 2013). In recent experiments examining the dentin phenotype in *Alpl*$^{-/-}$ mice we were able to visualize the binding of OPN to HA mineral produced by MV as it is exposed to the ECM (Fig. 8.2b).

Thus, from the above presentation it is clear that, at least in mice, PHOSPHO1, NPP1 and OPN are molecules able to modify the severity of the HPP phenotype. It remains to be demonstrated whether these genes act as modifiers of disease severity in HPP patients.

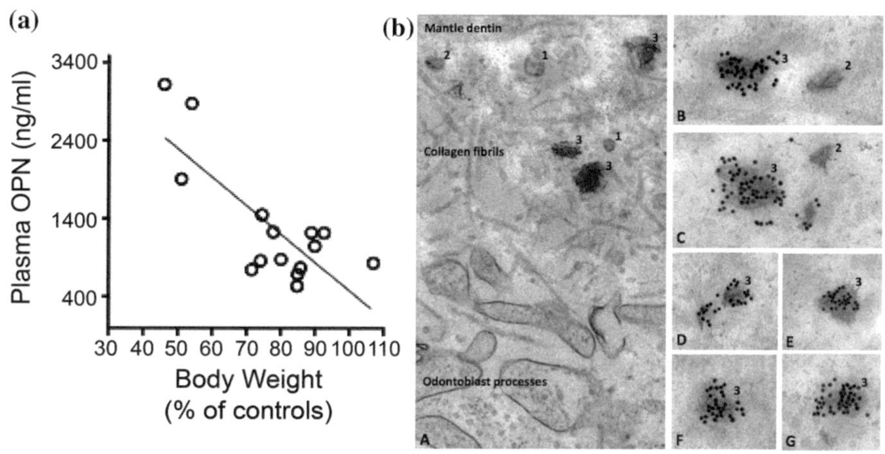

Fig. 8.2 **a** Correlation between body weight and plasma OPN in 11-day-old *Alpl*$^{-/-}$ mice (figure taken from Narisawa et al. 2013). **b** Immunogold labeling for osteopontin in mantle dentin of *Alpl*$^{-/-}$ mice. Immunogold labeling demonstrates high levels of OPN localization at sites of mineralization-ruptured MVs (stage 3 of mineralization in range 1 of 3, where 3 is most advanced), but not at non-ruptured MVs (stages 1 or 2), in mantle dentin of *Alpl*$^{-/-}$ mouse molars. Abbreviations as the same as in Fig. 8.1

8.2 What We Know About TNAP Function and Pyridoxal-Responsive Seizures

Neuronal abnormalities in HPP are likely due to deregulated metabolism of vitamin B6, and in particular in the pyridoxal-5′-phosphate (PLP) vitamer (Thompson et al. 2006), which appears to be a physiological substrate for TNAP in serum and/or in leukocytes (Wilson et al. 1983). After ingestion, phosphorylated vitamin B6 vitamers are dephosphorylated by intestinal alkaline phosphaatse (IAP) in the gut, and the resulting hydrophobic vitamers pyridoxine, pyridoxal (PL) and pyridoxamine are then rapidly absorbed by a carrier-mediated diffusion process (Gospe 2006). In the liver, these vitamers are converted via pyridoxal kinase and pyridoxamine 5′-phosphate oxidase back to their phosphorylated forms (see Chap. 11). In TNAP-deficient mice, absence of TNAP activity results in intracellulardeficiency of B6 vitamers and decreased ability to sustain vitamin B6-dependent cellular homeostasis, including glutamate decarboxylase activity and the synthesis of neurotransmitters such as serotonin or gamma-aminobutyric acid (GABA) in the brain (See Chap. 14). Thus, *Alpl*$^{-/-}$ mice die from seizures before weaning (Waymire et al. 1995; Narisawa et al. 1997, 2001). The most likely cause of death in *Alpl*$^{-/-}$ animals appears to be apnea, which occurs in conjunction with epileptic seizures (Millán 2006; Narisawa et al. 1997). Seizures are temporarily suppressed in *Alpl*$^{-/-}$ mice after PL administration (Narisawa et al. 2001; Waymire et al. 1995). Elevated plasma PLP levels may also be the cause of epileptic seizures experienced by patients with HPP (Baumgartner-Sigl et al. 2007). When seizures are present, they may respond to treatment with vitamin B6 (Mornet 2007).

8.3 Hypophosphatasia—Beyond the Seizures, Skeletal and Dental Disease

The development of a successful, life-saving, enzyme replacement therapy for HPP (Millán et al. 2008; Whyte et al. 2012) now offers the prospects of increased life-expectancy and greatly improved quality of life. Still, there are some complications of HPP that we are yet to understand and treat successfully (see Chap. 15). For example the development of craniosynostosis and nephrocalcinosis, almost always present in patients with the most severe forms of the disease (Whyte 2010; Whyte et al. 2012). Then, there is the question of whether TNAP has important functions in other tissues, where it is known to be expressed, but where we have a paucity of information about function, such as the liver, kidneys, white blood cells and the central and peripheral nervous system, the focus of this book.

TNAP is expressed in human hepatocytes, and bile acids increase its activity (Hatoff and Hardison 1981) and secretion in the bile (Hatoff and Hardison 1982). TNAP in rat hepatocytes is predominantly localized in the bile canalicular domain of the plasma membrane (Chida and Taguchi 2005, 2011), but it can be found in the basolateral membrane in the presence of high levels of bile acid (Ogawa et al. 1990). In contrast, mouse hepatocytes do not express TNAP (Halling Linder et al. 2013). In humans, liver TNAP may be expressed both at the sinusoidal and biliary pole of the hepatocyte. This explains why a significant proportion of TNAP activity in the circulation of healthy individuals originates from the liver. Elevated TNAP serum levels can be an indication of cholestasis. TNAP levels increase due to retrograde reflux of biliary alkaline phosphatase, enhance hepatic synthesis and enzyme release into the serum, and induction of the IAP isozyme (Hatoff and Hardison 1982; Kaplan and Righetti 1970; Poelstra et al. 1997a). It has been suggested that TNAP inhibits bile secretion (Alvaro et al. 2000). Intraluminal injection of alkaline phosphatase inhibits the basal activity of the Cl^-/HCO_3^- exchanger (Alvaro et al. 2000). It has also been proposed that TNAP acts in the extracellular compartment by hydrolyzing ATP and modulating purinergic receptor signaling (Alvaro et al. 2000). Evidence for a role of TNAP in purinergic signaling is mounting (Fig. 8.3). Picher and collaborators published on the role of TNAP in dephosphorylating AMP to produce adenosine, needed for proper mucociliary clearance and inflammatory responses in the airways that help prevent lung infections (Picher and Boucher 2001; Picher et al. 2003). Recently, Street et al. demonstrated that TNAP, together with prostatic acid phosphatase and 5′ nucleotidase is responsible for adenosine production in the dorsal root ganglia where it helps establish the ATP/adenosine ratio required for nociception (Street et al. 2013) (See also Chap. 13). In addition, Pettengill et al. demonstrated that in neonates, blood TNAP is involved in the production of anti-inflammatory adenosine (Pettengill et al. 2013). As will be discussed by Dr. Narisawa in her chapter, the common practice of using placental alkaline phosphatase (PLAP) as a tag in transgenic experiments for brain studies may need to be revisited, as it is likely that the overexpression of PLAP is not without physiological consequences.

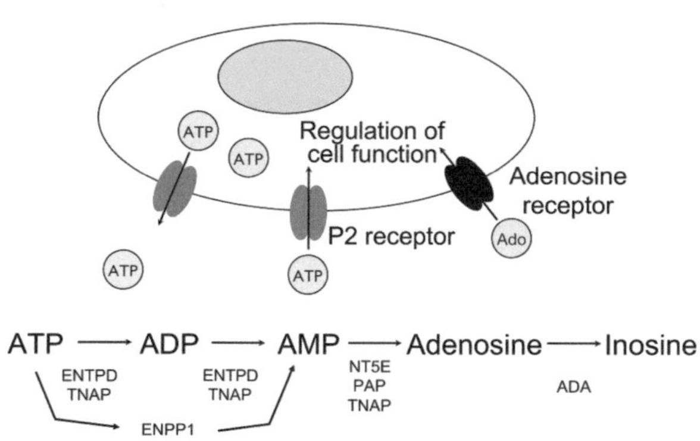

Fig. 8.3 Cartoon depicting the enzymes involved in the catabolism of ATP to form adenosine, and thus regulate the ATP/adenosine ratio important for purinergic signaling. ENPP1 Ectonucleotide pyrophosphatase/phosphodiesterase, ENTPD/CD39 Ectonucleotide triphosphate diphosphohydrolase, NT5E/CD73 Ecto-5′-nucleotidase, PAP prostatic acid phosphatase, ADA Adenosine deaminase, TNAP tissue-nonspecific alkaline phosphatase

The liver is the major organ that removes lipopolysaccharide (LPS), a toxin produced by gram-negative bacteria. The majority of systemic LPS is removed from the bloodstream by Kupffer cells and excreted in the bile by hepatocytes (Hori et al. 1998). One important function of TNAP in the liver may be to dephosphorylate endotoxins. The toxic part of lipopolysaccharide (LPS) is located in the lipid A part of the molecule, which is the most well conserved moiety among LPS serotypes from a variety of gram-negative bacteria. Two phosphate groups coupled to two glucosamine in the lipid A part largely determine the toxicity of LPS (Kanistanon et al. 2012). Removal of a single phosphate group results in the formation of a non-toxic monophosphoryl lipid A moiety (Bentala et al. 2002; Rietschel et al. 1994). Administration of LPS in rats stimulates the expression of TNAP by hepatocytes (Tuin et al. 2006) and TNAP is able to dephosphorylate LPS (Poelstra et al. 1997b). Poelstra et al. (1997b) proposed that TNAP function is to dephosphorylate endotoxin, whose levels are elevated in cholestasis. The phagocytic activity of Kupffer cells may be suppressed during cholestasis and the normal enterohepatic route for endotoxin removal via the biliary system blocked in this disease. The enhanced TNAP activity during cholestasis may reflect a physiological response of the liver upon bile duct obstruction (Poelstra et al. 1997b).

In human kidney, TNAP is expressed along the proximal tubule in segments S1, S2 and S3 (Nouwen and De Broe 1994). IAP is also expressed in the kidney but is restricted to the pars recta (S3) of the proximal tubule (Nouwen and De Broe 1994). The expression of TNAP in the kidney is counterintuitive since TNAP potently hydrolyses PP_i, and PP_i is a well-known inhibitor of mineralization in the urine (March et al. 2001). Indeed, individuals prone to form calcium stones appear to show reduced urinary PP_i excretion (Baumann et al. 1977), and in a group of 107 patients with recurrent calcium stones, PP_i:creatinine ratios were reduced compared with control subjects (Laminski et al. 1990). Actually, it would seem that PP_i is generated in the segments of the nephron downstream of TNAP-expressing proximal segments. However, TNAP can be shed from cells on the surface of nanosized vesicles and thus can conceivably act downstream from its site of synthesis, as has been shown to be the case for IAP in the gut (Goldberg et al. 2008). Whereas PP_i levels in urine are around 10 μM (March et al. 2001), PP_i is present in plasma at a concentration of 1–6 μM (Russell et al. 1976). Intravenous $^{32}PP_i$ is rapidly hydrolyzed in plasma, with PP_i also being filtered at the glomerulus and subject to further hydrolysis within the kidney; only less than 5 % of intravenous $^{32}PP_i$ appears in urine. These data indicate that by far the largest source of PP_i in the kidney is local generation (Moochhala et al. 2008). In the kidney, PP_i may be generated from ATP by nucleoside triphosphate diphosphohydrolase (NTPD) (Moochhala et al. 2008). However, the largest source of PP_i generation in the kidney may be intracellular. Expression of ANK occurs in the segments of the nephron downstream from the alkaline phosphatase-expressing segments (Carr et al. 2007). Since ANK is located both at the apical and basolateral membranes, it may function to inhibit mineralization within both the renal interstitium and/or the tubule lumen. While further studies are needed, to-date the function of TNAP in the kidney appears to be linked to PP_i metabolism and LPS detoxification (Kapojos et al. 2003).

8.4 Potential Roles of TNAP in the Nervous System

Clearly, three of the pathways discussed so far could be relevant to understanding the function of TNAP in the brain and peripheral nervous system: the ability of TNAP to mediate neurotransmitter synthesis requiring PL as cofactor, as is the case for GABA, serotonin has been amply demonstrated (See Chaps. 10 and 14). The ability of TNAP to modulate the ATP/adenosine ratio and thus affect cellular processes via purineric signaling regulation is an area of extreme interest as a mechanism mediating axonal growth (see Chaps. 4 and 18). Finally, the ability of TNAP to dephosphorylate extracellular matrix proteins is also becoming relevant in the brain, as TNAP has recently been suspected to play a role in Alzheimer's disease (AD). TNAP activity in patients with AD is inversely correlated with cognitive functions (Vardy et al. 2012), suggesting that TNAP may be involved in the degenerative process (see Chap. 17). Diaz-Hernandez et al. recently reported that cell membrane-anchored TNAP is able to dephosphorylate extracellular tau released upon cell death in neuronal cells (Díaz-Hernández et al. 2010). They observed that once dephosphorylated, tau behaves as an agonist of muscarinic M1 and M3 receptors, provoking a robust and sustained intracellular calcium increase finally triggering neuronal death (Díaz-Hernández et al. 2010). Interestingly, TNAP expression and activity are up-regulated by extracellular dephosphorylated tau in human neuronal cells (Díaz-Hernández et al. 2010), suggesting a vicious circle during which TNAP is stimulated by the product it forms (see Chap. 18). Thus, the association between TNAP expression and Alzheimer's disease is intriguing and merits careful investigation.

Conflict of Interest The authors declare no conflict of interest.

References

Addison WN, Azari F, Sorensen ES et al (2007) Pyrophosphate inhibits mineralization of osteoblast cultures by binding to mineral, up-regulating osteopontin, and inhibiting alkaline phosphatase activity. J Biol Chem 282:15872–15883

Addison WN, Masica DL, Gray JJ et al (2010) Phosphorylation-dependent inhibition of mineralization by osteopontin ASARM peptides is regulated by PHEX cleavage. J Bone Miner Res 25:695–705

Alvaro D, Benedetti A, Marucci L et al (2000) The function of alkaline phosphatase in the liver: regulation of intrahepatic biliary epithelium secretory activities in the rat. Hepatology 32:174–184

Anderson HC, Garimella R, Tague SE (2005) The role of matrix vesicles in growth plate development and biomineralization. Front Biosci 10:822–837

Anderson HC, Reynolds PR, Hsu HH et al (2002) Selective synthesis of bone morphogenetic proteins-1, -3, -4 and bone sialoprotein may be important for osteoinduction by Saos-2 cells. J Bone Miner Metab 20:73–82

Anderson HC, Sipe JB, Hessle L et al (2004) Impaired calcification around matrix vesicles of growth plate and bone in alkaline phosphatase-deficient mice. Am J Pathol 164:841–847

Baumann JM, Bisaz S, Felix R et al (1977) The role of inhibitors and other factors in the pathogenesis of recurrent calcium-containing renal stones. Clin Sci Mol Med 53:141–148

Baumgartner-Sigl S, Haberlandt E, Mumm S et al (2007) Pyridoxine-responsive seizures as the first symptom of infantile hypophosphatasia caused by two novel missense mutations (c.677T > C, p. M226T; c.1112C > T, p.T371I) of the tissue-nonspecific alkaline phosphatase gene. Bone 40:1655–1661
Bentala H, Verweij WR, Huizinga-Van Der Vlag A et al (2002) Removal of phosphate from lipid A as a strategy to detoxify lipopolysaccharide. Shock 18:561–566
Boskey AL, Maresca M, Ullrich W et al (1993) Osteopontin-hydroxyapatite interactions in vitro: inhibition of hydroxyapatite formation and growth in a gelatin-gel. Bone Miner 22:147–159
Buchet R, Millan JL, Magne D (2013) Multisystemic functions of alkaline phosphatases. Methods Mol Biol 1053:27–51
Carr G, Sayer JA, Simmons NL (2007) Expression and localisation of the pyrophosphate transporter, ANK, in murine kidney cells. Cell Physiol Biochem 20:507–516
Chida K, Taguchi M (2005) Localization of alkaline phosphatase and proteins related to intercellular junctions in primary cultures of fetal rat hepatocytes. Anat Embryol (Berl) 210:75–80
Chida K, Taguchi M (2011) Localization of alkaline phosphatase and cathepsin D during cell restoration after colchicine treatment in primary cultures of fetal rat hepatocytes. Acta Histochem Cytochem 44:155–158
Ciancaglini P, Yadav MC, Simão AM et al (2010) Kinetic analysis of substrate utilization by native and TNAP-, NPP1-, or PHOSPHO1-deficient matrix vesicles. J Bone Miner Res 25:716–723
Díaz-Hernández M, Gómez-Ramos A, Rubio A et al (2010) Tissue-nonspecific alkaline phosphatase promotes the neurotoxicity effect of extracellular tau. J Biol Chem 285:32539–32548
Fedde KN, Blair L, Silverstein J et al (1999) Alkaline phosphatase knock-out mice recapitulate the metabolic and skeletal defects of infantile hypophosphatasia. J Bone Miner Res 14:2015–2026
Foster BL, Nagatomo KJ, Nociti FH Jr et al (2012) Central role of pyrophosphate in acellular cementum formation. PLoS One 7:e38393
Foster BL, Nagatomo KJ, Tso HW et al (2013) Tooth root dentin mineralization defects in a mouse model of hypophosphatasia. J Bone Miner Res 28:271–282
Goldberg RF, Austen WG Jr, Zhang X et al (2008) Intestinal alkaline phosphatase is a gut mucosal defense factor maintained by enteral nutrition. Proc Natl Acad Sci USA 105:3551–3556
Gospe SM (2006) Pyridoxine-dependent seizures: new genetic and biochemical clues to help with diagnosis and treatment. Curr Opin Neurol 19:148–153
Halling-Linder C, Narisawa S, Millán JL, Magnusson P (2009) Glycosylation differences contribute to distinct catalytic properties among bone alkaline phosphatase isoforms. Bone 45:987–993
Halling-Linder C, Englund UH, Narisawa S et al (2013) Isozyme profile and tissue-origin of alkaline phosphatases in mouse serum. Bone 53:399–408
Harmey D, Hessle L, Narisawa S et al (2004) Concerted regulation of inorganic pyrophosphate and osteopontin by *Akp2*, *Enpp1*, and *Ank*: an integrated model of the pathogenesis of mineralization disorders. Am J Pathol 164:1199–1209
Harmey D, Johnson KA, Zelken J et al (2006) Elevated skeletal osteopontin levels contribute to the hypophosphatasia phenotype in Akp2(-/-) mice. J Bone Miner Res 21:1377–1386
Hatoff DE, Hardison WG (1981) Bile acids modify alkaline phosphatase induction and bile secretion pressure after bile duct obstruction in the rat. Gastroenterology 80:666–672
Hatoff DE, Hardison WG (1982) Bile acid-dependent secretion of alkaline phosphatase in rat bile. Hepatology 2:433–439
Hessle L, Johnson KA, Anderson HC et al (2002) Tissue-nonspecific alkaline phosphatase and plasma cell membrane glycoprotein-1 are central antagonistic regulators of bone mineralization. Proc Natl Acad Sci USA 99:9445–9449
Hori Y, Takeyama Y, Ueda T et al (1998) Impaired transport of lipopolysaccharide across the hepatocytes in rats with cerulein-induced experimental pancreatitis. Pancreas 16:148–153
Hunter GK (2013) Role of osteopontin in modulation of hydroxyapatite formation. Calcif Tissue Int (epub before print)

Hunter GK, Kyle CL, Goldberg HA (1994) Modulation of crystal formation by bone phosphoproteins: structural specificity of the osteopontin-mediated inhibition of hydroxyapatite formation. Biochem J 300(Pt 3):723–728

Johnson K, Goding J, Van Etten D et al (2003) Linked deficiencies in extracellular PP(i) and osteopontin mediate pathologic calcification associated with defective PC-1 and ANK expression. J Bone Miner Res 18:994–1004

Johnson KA, Hessle L, Vaingankar S et al (2000) Osteoblast tissue-nonspecific alkaline phosphatase antagonizes and regulates PC-1. Am J Physiol Regul Integr Comp Physiol 279: R1365–R1377

Kanistanon D, Powell DA, Hajjar AM et al (2012) Role of Francisella lipid A phosphate modification in virulence and long-term protective immune responses. Infect Immun 80:943–951

Kaplan MM, Righetti A (1970) Induction of rat liver alkaline phosphatase: the mechanism of the serum elevation in bile duct obstruction. J Clin Invest 49:508–516

Kapojos JJ, Poelstra K, Borghuis T et al (2003) Induction of glomerular alkaline phosphatase after challenge with lipopolysaccharide. Int J Exp Pathol 84:135–144

Laminski NA, Meyers AM, Sonnekus MI et al (1990) Prevalence of hypocitraturia and hypopyrophosphaturia in recurrent calcium stone formers: as isolated defects or associated with other metabolic abnormalities. Nephron 56:379–386

Majeska RJ, Wuthier RE (1975) Studies on matrix vesicles isolated from chick epiphyseal cartilage. Association of pyrophosphatase and ATPase activities with alkaline phosphatase. Biochim Biophys Acta 391:51–60

March JG, Simonet BM, Grases F (2001) Determination of pyrophosphate in renal calculi and urine by means of an enzymatic method. Clin Chim Acta 314:187–194

Mckee MD, Nakano Y, Masica DL et al (2011) Enzyme replacement therapy prevents dental defects in a model of hypophosphatasia. J Dent Res 90:470–476

Meyer JL (1984) Can biological calcification occur in the presence of pyrophosphate? Arch Biochem Biophys 231:1–8

Millan JL (2006) Mammalian alkaline phosphatases: from biology to applications in medicine and biotechnology. Wiley VCH Verlag, Weinheim

Millan JL (2013) The role of phosphatases in the initiation of skeletal mineralization. Calcif Tissue Int 93(4):299–306

Millán JL, Narisawa S, Lemire I et al (2008) Enzyme replacement therapy for murine hypophosphatasia. J Bone Miner Res 23:777–787

Moochhala SH, Sayer JA, Carr G et al (2008) Renal calcium stones: insights from the control of bone mineralization. Exp Physiol 93:43–49

Mornet E (2007) Hypophosphatasia. Orphanet J Rare Dis 2:40

Murshed M, Harmey D, Millán JL et al (2005) Unique coexpression in osteoblasts of broadly expressed genes accounts for the spatial restriction of ECM mineralization to bone. Genes Dev 19:1093–1104

Narisawa S, Fröhlander N, Millán JL (1997) Inactivation of two mouse alkaline phosphatase genes and establishment of a model of infantile hypophosphatasia. Dev Dyn 208:432–446

Narisawa S, Wennberg C, Millan JL (2001) Abnormal vitamin B6 metabolism in alkaline phosphatase knock-out mice causes multiple abnormalities, but not the impaired bone mineralization. J Pathol 193:125–133

Narisawa S, Yadav MC, Millán JL (2013) In vivo over-expression of tissue-nonspecific alkaline phosphatase increases skeletal mineralization and affects the phosphorylation status of osteopontin. J Bone Miner Res 28:1587–1598

Nouwen EJ, De Broe ME (1994) Human intestinal versus tissue-nonspecific alkaline phosphatase as complementary urinary markers for the proximal tubule. Kidney Int Suppl 47:S43–S51

Ogawa H, Mink J, Hardison WG et al (1990) Alkaline phosphatase activity in hepatic tissue and serum correlates with amount and type of bile acid load. Lab Invest 62:87–95

Oldberg Å, Franzen A, Heinegård R (1986) Cloning and sequence analysis of rat bone sialoprotein (osteopontin) cDNA reveals an Arg-Gly-Asp cell-binding sequence. Proc Natl Acad Sci USA 83:8819–8823

Pettengill M, Robson S, Tresenriter M et al (2013) Soluble Ecto-5'-nucleotidase (5'-NT), alkaline phosphatase, and adenosine deaminase (ADA1) activities in neonatal blood favor elevated extracellular adenosine. J Biol Chem 288:27315–27326

Picher M, Boucher RC (2001) Metabolism of extracellular nucleotides in human airways by a multi-enzyme system. Drug Dev Res 52:66–75

Picher M, Burch LH, Hirsh AJ et al (2003) Ecto 5'-nucleotidase and nonspecific alkaline phosphatase. Two AMP-hydrolyzing ectoenzymes with distinct roles in human airways. J Biol Chem 278:13468–13479

Pizauro JM, Demenis MA, Ciancaglini P, Leone FA (1998) Kinetic characterization of a membrane specific ATPase from rat osseous plate and its possible significance on endochodral ossification. Biochim Biophys Acta 1368:108–114

Poelstra K, Bakker WW, Klok PA et al (1997a) A physiologic function for alkaline phosphatase: endotoxin detoxification. Lab Invest 76:319–327

Poelstra K, Bakker WW, Klok PA et al (1997b) Dephosphorylation of endotoxin by alkaline phosphatase in vivo. Am J Pathol 151:1163–1169

Rietschel ET, Kirikae T, Schade FU et al (1994) Bacterial endotoxin: molecular relationships of structure to activity and function. FASEB J 8:217–225

Roberts S, Narisawa S, Harmey D et al (2007) Functional involvement of PHOSPHO1 in matrix vesicle-mediated skeletal mineralization. J Bone Miner Res 22:617–627

Russell RG, Bisaz S, Fleisch H (1976) The influence of orthophosphate on the renal handling of inorganic pyrophosphate in man and dog. Clin Sci Mol Med 51:435–443

Simão AM, Yadav MC, Narisawa S et al (2010) Proteoliposomes harboring alkaline phosphatase and nucleotide pyrophosphatase as matrix vesicle biomimetics. J Biol Chem 285:7598–7609

Sodek J, Ganss B, Mckee MD (2000) Osteopontin. Crit Rev Oral Biol Med 11:279–303

Sorensen ES, Hojrup P, Petersen TE (1995) Posttranslational modifications of bovine osteopontin: identification of twenty-eight phosphorylation and three O-glycosylation sites. Protein Sci 4:2040–2049

Street SE, Kramer NJ, Walsh PL et al (2013) Tissue-nonspecific alkaline phosphatase acts redundantly with PAP and NT5E to generate adenosine in the dorsal spinal cord. J Neurosci 33:11314–11322

Thompson MD, Killoran A, Percy ME et al (2006) Hyperphosphatasia with neurologic deficit: a pyridoxine-responsive seizure disorder? Pediatr Neurol 34:303–307

Thouverey C, Bechkoff G, Pikula S et al (2009) Inorganic pyrophosphate as a regulator of hydroxyapatite or calcium pyrophosphate dihydrate mineral deposition by matrix vesicles. Osteoarthr Cartil 17:64–72

Tuin A, Huizinga-Van Der Vlag A, Van Loenen-Weemaes AM et al (2006) On the role and fate of LPS-dephosphorylating activity in the rat liver. Am J Physiol Gastrointest Liver Physiol 290: G377–G385

Vardy ER, Kellett KA, Cocklin SL et al (2012) Alkaline phosphatase is increased in both brain and plasma in Alzheimer's disease. Neurodegener Dis 9:31–37

Waymire KG, Mahuren JD, Jaje JM et al (1995) Mice lacking tissue non-specific alkaline phosphatase die from seizures due to defective metabolism of vitamin B-6. Nat Genet 11:45–51

Whyte MP (2010) Physiological role of alkaline phosphatase explored in hypophosphatasia. Ann N Y Acad Sci 1192:190–200

Whyte MP, Greenberg CR, Salman NJ et al (2012) Enzyme-replacement therapy in life-threatening hypophosphatasia. N Engl J Med 366:904–913

Wilson PD, Smith GP, Peters TJ (1983) Pyridoxal 5'-phosphate: a possible physiological substrate for alkaline phosphatase in human neutrophils. Histochem J 15:257–264

Yadav MC, De Oliveira RC, Foster BL et al (2012) Enzyme replacement prevents enamel defects in hypophosphatasia mice. J Bone Miner Res 27:1722–1734

Yadav MC, Simão AM, Narisawa S et al (2011) Loss of skeletal mineralization by the simultaneous ablation of PHOSPHO1 and alkaline phosphatase function: a unified model of the mechanisms of initiation of skeletal calcification. J Bone Miner Res 26:286–297

Yadav MC, Huesa C, Narisawa S et al (2014) Ablation of osteopontin improves the skeletal phenotype of *Phospho1*$^{-/-}$ mice. J Bone Miner Res 29(11):2369–2381

Chapter 9
TNAP, an Essential Player in Membrane Lipid Rafts of Neuronal Cells

Myriam Ermonval, Florence Baychelier and Caroline Fonta

Abstract The tissue non-specific alkaline phosphatase (TNAP) is a glycosyl-phosphatidylinositol (GPI) anchored glycoprotein which exists under different forms and is expressed in different tissues. As the other members of the ecto-phosphatase family, TNAP is targeted to membrane lipid rafts. Such micro domains enriched in particular lipids, are involved in cell sorting, are in close contact with the cellular cytoskeleton and play the role of signaling platform. In addition to its location in functional domains, the extracellular orientation of TNAP and the fact this glycoprotein can be shed from plasma membranes, contribute to its different phosphatase activities by acting as a phosphomonoesterase on various soluble substrates (inorganic pyrophosphate -PPi-, pyridoxal phosphate -PLP-, phosphoethanolamine -PEA-), as an ectonucleotidase on nucleotide-phosphate and presumably as a phosphatase able to dephosphorylate phosphoproteins and phospholipids associated to cells or to extra cellular matrix. More and more data accumulate on an involvement of the brain TNAP both in physiological and pathological situations. This review will summarize what is known and expected from the TNAP localization in lipid rafts with a particular emphasis on the role of a neuronal microenvironment on its potential function in the central nervous system.

Keywords TNAP activity · Neuronal function · Extra cellular matrix · Protein network · Microenvironment

---

M. Ermonval (✉) · F. Baychelier
Département de Virologie, Institut Pasteur, Unité Des Stratégies Antivirales, F-75015, Paris, France
e-mail: myriam.ermonval@pasteur.fr

C. Fonta
Centre de Recherche Cerveau et Cognition (CerCo), Université de Toulouse UPS; CNRS UMR 5549, Toulouse, France

C. Fonta and L. Négyessy (eds.), *Neuronal Tissue-Nonspecific Alkaline Phosphatase (TNAP)*, Subcellular Biochemistry 76,
DOI 10.1007/978-94-017-7197-9_9

9.1 The Tissue Non Specific Alkaline Phosphatase, a GPI-Anchored Glycoprotein Present Under Different Forms in Different Cellular Spaces

9.1.1 The Different Forms of TNAP

The tissue non specific alkaline phosphatase (TNAP) belongs to a large family of GPI-anchored ecto-phosphatases whose main function is that of a phospho-monoesterase with an optimal in vitro enzymatic activity at alkaline pH (Halling Linder et al. 2012). In contrast to the tissue specific intestinal and placental alkaline phosphatases, TNAP is found in different organs (Hoshi et al. 1997), and exhibits variable amount of expression. For instance, a higher amount of TNAP is expressed in the bone than in the liver and the kidney. Moreover, studies on the regulation of TNAP expression during embryonic life and development (Fonta et al. 2004; Narisawa et al. 1994), are in support of its role during differentiation of the nervous system (Delic and Zimmermann 2011). Depending on promoter regulation and the usage of the first non coding exon, the bone or the liver transcript will be found, leading to a first level of variability in TNAP expression depending on cell specific transcriptional factors (Brun-Heath et al. 2011).

TNAP is a membrane protein maturating in the secretory pathway. It is anchored in the outer leaflet of the plasma membrane through a glycosyl-phosphatidylinositol moiety attached at its C terminus and, as such, localized in raft micro domains. In addition, TNAP is found as a soluble protein in the serum and is also present in extracellular spaces where it exerts functions either associated to matrix vesicles during bone mineralization (Ciancaglini et al. 2006; Wuthier and Lipscomb 2011) or to mucosal surfactant as a defense to pathogens in the lung (Koyama et al. 2002). Another level of diversity is linked to post-translational modifications. TNAP contains 5 potential N-glycosylation sites and a C-terminal GPI moiety. Not only the number of N-glycans relating to the variable occupancy of asparagin sites in the ER (Jones et al. 2005), but also the N-glycan (Stanley 2011) and the GPI (Maeda and Kinoshita 2011) structures maturating during trafficking in the Golgi apparatus, can vary according to the cellular type. Therefore, different glycoforms of TNAP having the same amino-acid composition are expected and various forms of TNAP have indeed been described by HPLC and electrophoresis analyses (Halling Linder et al. 2013). In this respect, the TNAP N-glycans of the bone isoform contain sialic acid while the liver one does not, and the TNAP function could vary accordingly (Halling Linder et al. 2009).

9.1.2 Implication of a GPI Anchoring

The glycosyl-phosphatidylinositol (GPI) core present at the C terminus of different proteins can insert into plasma membrane leaflets. This lipid moiety not only allows the protein attachment to the cell membrane but is also directly involved in cell signaling and trafficking. In the absence of a cytoplasmic domain, intracellular signaling cascade downstream a GPI-protein must be initiated through interaction of the ectodomain of such anchored protein with extracellular domains of transmembrane proteins or of the GPI region with intramembranous scaffold proteins such as caveolin present in specialized lipid rafts (Mouillet-Richard et al. 2000). Indeed, the GPI, by its capacity to interact with cholesterol, is responsible for the localization of such anchored proteins into membrane lipid rafts and replacement of a GPI-anchor by a transmembrane sequence affects raft localization and regulates uptake of receptors. Conversely, the structure of the GPI may affect lipid rafts and membrane composition (Chatterjee and Mayor 2001). Moreover, the complete extracellular orientation of GPI anchored glycoproteins makes them able to interact with extracellular matrix (ECM) and soluble compounds present in the extracellular space.

GPI-anchored proteins can be shed from the membrane by phophatidylinositol-phospholipases (PIPLC, PIPLD) that specifically recognize GPI structures, then releasing the protein moiety from its anchor. Soluble TNAP is indeed detected in the serum. In contrast, proteins with their GPI (Kooyman et al. 1998) have been demonstrated to be able to integrate the plasma membrane of other cells, a mechanism known as the "painting" of membranes. This illustrates the possibility for such GPI anchored proteins to transfer between cells as well as their affinity for lipids. In vitro experiments indicate that such exogenously added proteins may integrate different micro domains (Legler et al. 2005) of new membranes without loss of their activity as shown for different GPI anchored proteins (Bate and Williams 2011; Premkumar et al. 2001). However, most of these observations have been obtained experimentally and with artificial membranes, and whether TNAP secreted in the circulation is released by phospholipases or may carry its GPI is not often documented.

As already noticed, the GPI moiety can be differentially modified according to tissues and cell types expressing the anchored protein. Importance of this structure in the case of TNAP has been highlighted by the fact that mutations affecting different steps in the synthesis of the glycolipid anchor are involved in human diseases of hyperphosphatasia. Details on GPI synthesis are exposed in the Chap. 16.

During biosynthesis,maturation of lipid rafts takes place in the secretory pathway where GPI anchor proteins are sorted from other proteins and packaged into ER-derived vesicles (Kinoshita et al. 2013). Besides raft association, the GPI lipid moiety confers specific membrane trafficking properties to the attached protein, such as apical targeting in polarized cells and axonal and dendritic localization in neuronal cells (Chatterjee and Mayor 2001). Mechanisms of GPI anchored protein endocytosis are less defined and could depend on the cell type and the protein. While different pathways have been described, endocytosis of GPI-proteins is suspected to be important in the regulation of cellular signaling (Lakhan et al. 2009). TNAP has also been found in secreted vesicles during mineralization. These vesicles are formed either intracellularly from multi vesicular body released as exosomes or directly at the plasma membrane as is the case with the matrix vesicles forming by interaction with components of the ECM, including binding of membrane anchored TNAP with matrix collagens II and X (Wuthier and Lipscomb 2011). Briefly, secreted exosomes are formed within the multi vesicular bodies (MVBs) arising from the conjunction of vesicular transport between the endocytic and the secretory pathway. The first step is the formation of endocytic vesicles and the sequestration of early endosomes targeted to MVBs where they are incorporated as intraluminal vesicles by invagination of the MVBs membrane. MVBs are either degraded by fusion with the lysosome or released in the extracellular space as exosomes upon fusion with the plasma membrane. Extracellular vesicles (EVs) are also secreted from neuronal cells and have recently been shown to be important for cell to cell communication operating both during physiological or pathological process in the central nervous system. For instance, in normal situation EVs are involved in the maintenance of axonal integrity during myelination by allowing a bidirectional communication between neurons and oligodendrocytes. They are also suggested to have a role in interneural exchange of information within functional brain networks. EVs which are present in high amount in some brain disorders such as neurodegenerative diseases have the capability to propagate the pathology by allowing cell transmission of toxic proteins as described for the prion protein or the amyloid β in Alzheimer disease (Rajendran et al. 2014). EVs harbor different composition according to their cell origin, and different signatures according to the neurodegenerative disorder. Therefore, it would be of interest to look whether TNAP could be present at the surface of such EVs.

9.1.3 TNAP in Membrane Lipid Rafts

Since TNAP, as a GPI anchored protein, is enriched in membrane lipid rafts (MLR), its function must also relate to some features of these specialized micro domains. Although the definition of lipid rafts is still into debate, their existence as functional signaling platforms is fully recognized (Lindner and Naim 2009). These micro domains, which are enriched in cholesterol and sphingolipids, form lipid ordered patches at the plasma membrane surrounded by disordered lipid phase containing phospholipids and unsaturated fatty acids. This confers differences in the lateral diffusion of membrane proteins in raft and non raft domains and in their biochemical properties, such as the solubility in non ionic detergents, one of the characteristics used to define rafts (Silvius 2005). The specific lipid and protein composition of rafts is in support of the involvement of membrane micro domains in different cellular processes including cell sorting, endocytosis and signaling (Head et al. 2014).

A contact with the actin and tubulin components of the cytoskeleton is required to maintain, beside cell integrity, the organization and distribution of micro domains at the plasma membrane. Interaction of actin filament with MLR allows their correct repartition during polarization of epithelial, endothelial and neuronal cells, in particular during growth cone formation important for axonal guidance. Such stabilization and coalescence of membrane subdomains lead to the formation of regions of high protein density and specific composition involved in adherence and cell signaling (Head et al. 2014).

Composition and dynamics of MLR will also depend on the cell type and ECM microenvironment. The numerous cellular functions associated to MLR depend on a fine interplay with the cytoskeleton but also with the ECM, both being important to cellular communication and regulatory processes. However, much has still to be learnt about the way subcellular assembly in particular in neurons occurs and regulates functional properties. GPI anchored proteins, such as TNAP, may contribute by providing scaffold properties to this organization.

9.2 TNAP, Different Forms, Different Functions

9.2.1 The Tissue Specificity

Little is known about the regulation and the role of the different TNAP isoforms. However, since the enzymes responsible for post-translational modifications in the glycosylation pathway may vary between organs and even between tissues of an organ, such modifications could contribute to variation in the potential of the extracellular domain of TNAP to interact with different substrates. The role of the glycosylation on TNAP activity has been suggested (Halling Linder et al. 2009). In addition, the trafficking, degradation and function of TNAP have been shown to be affected by different mutations (Chap. 2) interfering with either the GPI or N-glycosylation pathways and responsible for genetic disorders of hyperphosphatasia (Chap. 16) or hypophosphatasia (Ishida et al. 2003; Ito et al. 2002; Sultana et al. 2013), respectively.

The tissue in which TNAP is expressed may also influence its function by the fact that encountered substrates may differ whatever the enzyme being circulating or associated to raft micro domains. Therefore, TNAP exhibits specific or shared activities in different organs (see Chap. 8), which are impacted by the different ligands that can be dephosphorylated by TNAP. Beside circulating phosphomonoester substrates, such as PPi important in bone mineralization, and PLP (a derivative of vitamin B6) or nucleotide-phosphate involved in neurotransmitter metabolism, some phosphoproteins and phospholipids have also been described to be TNAP substrates therefore, enlarging the potential roles of this phosphatase. Dephosphorylation of such substrates may be involved in functions as different as (i) defense against pathogen by detoxification of LPS endotoxin or by counteracting inflammatory processes in the lung, intestine or kidney (Goldberg et al. 2008; Kapojos et al. 2003; Koyama et al. 2002), (ii) regulation of membrane ion channel such as CFTR in cystic fibrosis (Becq et al. 1994) or in secretory activity of the liver bilary epithelium (Alvaro et al. 2000), (iii) transport activities as is the case of insulin in brain micro vessels or drug uptake and detoxification (Calhau et al. 2002a, b; Martel et al. 1996), and (iv) yet unknown functions involving interaction with other plasma membrane proteins (Scheibe et al. 2000) or matrix protein, such as laminin important for neuronal differentiation (Ermonval et al. 2009a). As reviewed by Diaz-Hernandez (Chap. 18), TNAP also acts on extracellular phosphorylated tau with implication in Alzheimer' disease (Diaz-Hernandez et al. 2010).

Importantly, although TNAP substrates are circulating in blood vessels and TNAP is expressed in different organs, TNAP exhibits functions that are tissue specific. This is supported by the observation of an effect of vitamin B6 on neurological dysfunction (seizure) but not on mineralization or other aspects of TNAP deficiency (Narisawa et al. 2001). In line with this, human newborns with perinatal hypophosphatasia did not exhibit epileptic symptoms anymore when treated with pyridoxin (Yamamoto et al. 2004) but still died rapidly from further complications independent of PLP specificity (Nunes et al. 2002). Moreover, while TNAP allows bone mineralization by usage of its PPi substrate, it is only in abnormal situation that ectopic calcification can be observed. It is the case of vascular calcification (Lomashvili et al. 2014), kidneys' stones (Moochhala et al. 2008), breast cancer micro calcification (Cox et al. 2012) or after traumatic brain injury (Toffoli et al. 2008). A case of vessel mineralization has also been documented in the central nervous system of cervids suffering of the chronic wasting disease (CWD) caused by prion protein aggregation (Hamir et al. 2008). This raises the question of a possible implication of TNAP in the course of this pathological CWD disease and in link with the prion protein, the agent of the disease.

9.2.2 The Impact of Membrane on TNAP Activity

In view of the various forms of TNAP, and the fact it can be associated to cells or be secreted as a soluble or vesicle associated protein, it is important to know whether the secreted and the GPI forms of TNAP exhibit the same activity. While a local effect can be expected from the GPI anchored protein, the secreted TNAP may have a more global effect.

An influence of plasma membranes on enzymatic activity has indeed been demonstrated for GPI anchored enzymes. The 5′ectonucleotidase (Lehto and Sharom 1998), the dipeptidase (Brewis et al. 1994) and the TNAP (Kothekar et al. 2014), exhibit higher specific activities after release from the membrane by phospholipase C cleavage of the GPI anchor. However, it does not seem to be always the case. In particular, in the upper mentioned studies, the group of Brewis using similar pig kidney membrane model as the group of Kothekar, did not find any impact of membranes on TNAP activity (Brewis et al. 1994). Besides, artificial membrane reconstitutions have shown an impact of the lipid composition and membrane curvature on alkaline phosphatase activity (Sesana et al. 2008).

As expected from the role of matrix vesicles formed at the site of bone mineralization the orientation of TNAP present at their surfaces is the same as at the plasma membrane since both are released by PIPLC treatment. Two types of vesicles either secreted in the medium (medium vesicles, MeV), or forming at the membrane of chondrocytes (matrix vesicles, MVs), have been described and both expressed high level of TNAP. However, only the MVs type is able to induce mineralization (Wuthier and Lipscomb 2011). This emphasizes the importance of the membrane micro environment for the function of TNAP.

9.3 The Case of the Neuronal TNAP: A Modulator Role of the Neuronal TNAP Depending on Its Microenvironment?

9.3.1 Lipid Rafts and Neuronal Functions

Membrane lipid rafts (MLR) constitute signaling platforms linking extracellular compounds from milieu or matrix to intracellular compartments through membrane proteins acting as receptors (Head et al. 2014). Evidence of a role of these micro domains in different types of neurotransmitter signaling has been obtained (Allen et al. 2007), including: (i) ionotropic and G-protein coupled neurotransmitter receptors such as serotonin (Bjork et al. 2010), (ii) downstream effectors such as ions channels, phosphodiesterase, adenyl cyclase, phospholipases (Head et al. 2006), (iii) neurotransmitter transporters (Cuddy et al. 2014). It has also been established that MLR are important in synaptic transmission and plasticity (Newpher and Ehlers 2009; Sebastiao et al. 2013) and in axonal growth and guidance (Guirland and Zheng 2007). In addition, lipid micro domains are in close contact with the cytoskeleton, what allows to maintain and regulate precise localization of proteins at the synapse and in dendritic spine (Chen and Sabatini 2012). Many of these studies were concluded by combining different approaches that included (i) localization of actors of neurotransmission, (ii) use of agonists and antagonists and (iii) the effect of the perturbation of raft organization on neurotransmission by drugs interfering with their lipidic constituents, in particular, the cholesterol (main lipid in MLR). Another domain of interest is the node of Ranvier rich in GPI anchored proteins, revealing the role of lipid microenvironment in the regulation of trafficking and association of axo-glial complexes constituting axonal nodal subdomains (Labasque and Faivre-Sarrailh 2010). This is in support with the fact that TNAP activity has been demonstrated to localize in the white matter at myelin-free axonal region, namely at the node of Ranvier, and at the level of synaptic contact (Fonta et al. 2005).

Importance of the extracellular matrix (ECM) in such neuronal activities is now well recognized (Dityatev et al. 2010). Components of the ECM are secreted by different cell types and in the case of the central nervous system (CNS), they build perineural nets of specific compositions associated to glial cells, neuronal cells or oligodendrocytes. The ECM, which fills extracellular spaces in the CNS, has the particularity to contain little collagen and fibronectin as compared to other tissues. By contrast, it contains a high level of carbohydrates carried by glycosaminoglycans, either linked to proteins such as the chondroitin and heparan proteoglycans family that can be sulfated (CSPG and HSPG), or free, as is the hyaluronic acid. It also contains diverse glycoproteins such as laminin. These molecules interact with each other to form a net but also with cell surface receptors thereby providing some scaffolds that promote clustering of cell surface proteins as required for some signaling function in lipid rafts. Diverse molecules of the ECM have been shown to be involved in proliferation, migration and cell differentiation. The perineural nets in the CNS enwrap dendrites protruding from cell bodies and synaptic contacts as well as nodes of Ranvier, therefore participating in neuronal functions important for synaptic plasticity in adult brain, which include regulation of long term potentiation by interaction of ECM molecules with different types of receptors (Dityatev and Schachner 2003). Ion channels, neurotransmitter receptors and other compounds present in lipid rafts may be differentially regulated by ECM molecules produced by different neuronal cell types to further activate tyrosine kinase and downstream signaling cascades coupled to the raft through the cytoskeleton. In this context, a glycotranscriptome study revealed differential expression of genes involved in glycosylation pathways while comparing serotoninergic and noradrenergic derived neuronal cells. In particular, 9 among the 44 transcripts differentially expressed in these bioaminergic neurons (out of the 375 glycosylation relating transcripts tested) relate to glycosaminoglycan metabolism associated to some ECM molecules (Ermonval et al. 2009b). The ECM will contribute then to shape synaptic plasticity by imposing diffusion constraints for neuronal factors (neurotransmitters, trophic factors) and to regulate different neuronal functions in a spatio-temporal manner.

9.3.2 Role of the Neuronal TNAP

TNAP is expressed in different cell types (endothelial, neuronal) in the brain, and in different areas of the central nervous system (CNS) (see Chap. 5). The different known activities of TNAP, its various isoforms and locations in intracellular and extracellular spaces in the CNS as well as its spatio-temporal regulation during neuronal development, identified this ectophosphatase with an important role in the CNS. Studies on hypophosphatasia deficiencies and associated neuronal complications have added information supporting this view. However, understanding the contribution of TNAP in neuronal functioning is at an early stage, and further investigations are required to learn more about it.

Different studies have brought some clues on the way neuronal TNAP could exert activities in the brain. The difficulties are in integrating all the data obtained in relation with the different specificities and regulations of TNAP. Questions then arise of how these activities may be influenced by the local environment of the TNAP and, conversely, does TNAP influence its partners in close vicinity in lipid rafts. As already mentioned and without going into details given in other chapters, the phosphatase activities of TNAP, mainly linked to its phosphomonoesterase and nucleotidase functions, are contributing to neuronal homeostasis. Regulation of the

PLP substrate, a cofactor of many enzymes among which decarboxylases are key enzymes involved in the metabolism of different monoamine neurotransmitters (serotonin, norepinephrine, GABA), is essential for bioaminergic neuron activities. Another important role of TNAP, relates to the regulation of the adenosine signaling through purinergic receptors that can be obtained locally by sequential dephosphorylation of extracellular ATP, ADP, AMP, to adenosine (Matsuoka and Ohkubo 2004). However, TNAP can be found in the blood circulation and adenosine could be generated in the brain microvasculature (Pettengill et al. 2013) as well as the pyridoxal since the PLP can not cross the blood brain barrier and/or enter cells and must be dephosphorylated first (detailed in Chap. 11). It is thus not an easy task to establish whether these regulatory cofactors that act locally are produced in the near surrounding of their receptors. Interestingly, TNAP is a metalloenzyme dependent on zinc for its stability and acquisition of its functionality raising the question of the requirement of zinc transporter for activation of TNAP (Fukunaka et al. 2011). It is noticeable that in the CNS, metals and in particular Zn have to be tightly regulated. Considering only the enzymatic activities of TNAP and its differential expression already raised important questions such as: (i) what are the substrates found around the TNAP in extracellular space, in particular at the synaptic cleft, (ii) are the receptors of the products of TNAP in close vicinity with the enzyme to ensure local effects, (iii) can TNAP act on membrane compounds of surrounding cells or on matrix phosphoproteins, or at distance through the generation of circulating dephosphorylated products.

9.3.3 Role of Microenvironment on TNAP Function, an Hypothesis on Its Role in the Central Nervous System

Other neuronal functions could relate to the ability of TNAP to dephosphorylate membrane or matrix phosphoproteins and to its participation to protein complexes of high density in lipid rafts known to be important in neuronal signaling. Therefore, it can be expected to see an impact of micro domain composition in differentiated cells on the functional outcomes of TNAP in relation with soluble substrates, membrane receptors or matrix proteins. In this context, TNAP has been shown to be expressed both in serotonergic and noradrenergic neurons upon neuronal differentiation of the neuroectodermic 1C11 cell line. TNAP expression in lipid rafts of bioaminergic neurons correlates with the detection of its specific activity whose inhibition impacts the metabolism of the corresponding neurotransmitter (Ermonval et al. 2009a). Also, TNAP colocalizes with the cellular prion protein, PrP^C, in micro domains of bioaminergic 1C11 derivatives. PrP^C was previously shown to participate in a signaling plateform with the caveolin and fyn (Mouillet-Richard et al. 2000), modulating the agonist response of serotonin (5-HT) receptors (Mouillet-Richard et al. 2007). The complex network of interactions between actors of neurotransmission in lipid rafts is also illustrated by the fact that the G-protein coupled serotonin receptors and the serotonin transporter are both localizing in lipid rafts (Sebastiao et al. 2013). Moreover, 5-HT receptor activation allows neurite outgrowth and neuronal survival by triggering downstream signaling cascades (Fricker et al. 2005). It has to be added that the 5-HT neurotransmitter exerts its function through 14 different subtypes of 5-HT receptors and dysregulated metabolism is implicated in anxiety and depression disorders (Bjork et al. 2010). Other studies have pointed to the role of TNAP enzymatic activity in neuronal differentiation (Delic and Zimmermann 2011) and neuritogenesis by promoting axonal growth of hippocampal neurons (Diez-Zaera et al. 2011). In this latter study, a coordinated action of TNAP and purinergic receptors colocalizing at the level of the axonal growth cone was required for axonal growth to occur.

While interaction of TNAP with matrix collagen during mineralization has been largely investigated, it is only recently that the neuronal TNAP has been shown to be able to dephosphorylate the phospho-laminin. In addition, the inactivation of TNAP by levamisol stabilizes the phosphorylated form of the laminin and influences its binding to PrP^C (Ermonval et al. 2009a). The laminin by interacting with different membrane proteins such as integrins, PrP^C, the laminin receptor, have been ascribed roles in differentiation, survival and migration of neurons but also in neuronal plasticity (Nishimune et al. 2004; Sperling et al. 2012).

Altogether these studies revealed the complexity of the interplay between different partners required for neuronal functioning and positioned TNAP as an essential player in the multi-molecular complexes constituting lipid rafts, whose composition may depend on the cellular context and environment and lead to diverse functions. All these compounds and their ligands are tightly regulated and imbalanced conditions may lead to dysfunction of the whole system, as in the case of seizures associated to severe hypophosphatasia. A role of TNAP on extracellular missorted phosphorylated Tau (Diaz-Hernandez et al. 2010; Zempel and Mandelkow 2014) a protein involved in a neurodegenerative disorder (the Alzheimer disease, AD) has also recently been reported (Kellett et al. 2011; Vardy et al. 2012). A link has also been made between AD and lipid rafts by the group of Hooper, adding to the complexity of interactions between partners in micro domains. Indeed, their studies show that amyloid ß oligomers bind to PrP^C and activate fyn kinase only when PrP^C is localized in lipid rafts. They also documented a role of PrP^C in zinc uptake besides its role in homeostasis of other metals such as Cu^{2+}. Interestingly, both a diminution of Zn transport and an alteration of raft composition were occurring with aging, one of the identified risk factor in AD (Watt et al. 2014). These different studies illustrate how clusters in MLR of protein complexes of different compositions may have different outcomes.

It is clear that TNAP may have different functional impacts on different neuronal cells according to its micro environment in the CNS. An hypothesis could be that this enzyme exerts a modulator role on its partners within lipid raft complexes, whose composition may vary as well as those of the extracellular matrix and the secreted compounds able to serve as substrate or receptor ligands. The determination of the composition of protein complexes associated with TNAP in the rafts, in different situations, should provide a better understanding of its neuronal functions.

9.4 Conclusion

Although increasing evidence has been obtained for a direct role of TNAP during neuronal development, in blood brain barrier function and for neurotransmission, the physio-pathological involvement of TNAP in the central nervous system is far from being fully understood. The differential role that this ubiquitous ecto-phosphatase harbors in various organs, points to the importance of TNAP regulation occurring at different levels of expression (transcriptional, post translational, compartment targeting) and depending on its environment in the rafts and available ligands, altogether impacting its activity and function. In addition, TNAP being part of dynamic complexes in the rafts, it may also exert a modulator role on its partners in micro domains contributing to neuronal activity. This is also true for other molecules, as shown for the PrP^C, present in multi-molecular complexes among micro domains whose composition varies in different neuronal cell types. Interplays between different compounds, as shown in Fig.1, may then converge towards signaling cascades important in neuronal functions such as cell adhesion and survival, neurite outgrowth and neuronal homeostasis.

It can be envisioned that TNAP activities may vary according to raft dynamics and extracellular composition at the level of the synapse or the perineural net constituting a particular matrix. Neuronal homeostasis also relates to neurotransmitter concentration, whose equilibrium may change according to the neuronal state, the brain area and whether pathological situation occurs such as the one met in some neurodegenerative diseases when extracellular matrix is affected. The abundance of extracellular phosphorylated tau protein in Alzheimer disease is an example. In this context, the new opening field of neuronal vesicles released in theextracellular space and their role in neuronal communication as well as in neurodegenerative diseases may have an impact on the outcome of TNAP activity.

Importance of the microenvironment leads to think it will be of great interest to look at the TNAP as part of networks, whose function may depend on its different forms, cellular localization and therefore, accessibility to its different substrates. It will be determinant to know what are the different partners of TNAP, specially in lipid rafts, to understand the potential role this ecto enzyme may have in the modulation of neuronal functions. The establishment of the link of TNAP with different key players of signaling neuronal networks (Chap. 10) but also other networks of interaction with the ECM, the cytoskeleton and intracellular trafficking processes, as well as with other systems such as the immune system where for instance adenosine also play a key role as regulator of immune functions (Pettengill et al. 2013), will allow to model the impact of specific TNAP substrates either in physiological or pathological situations. At this point, it is crucial to understand

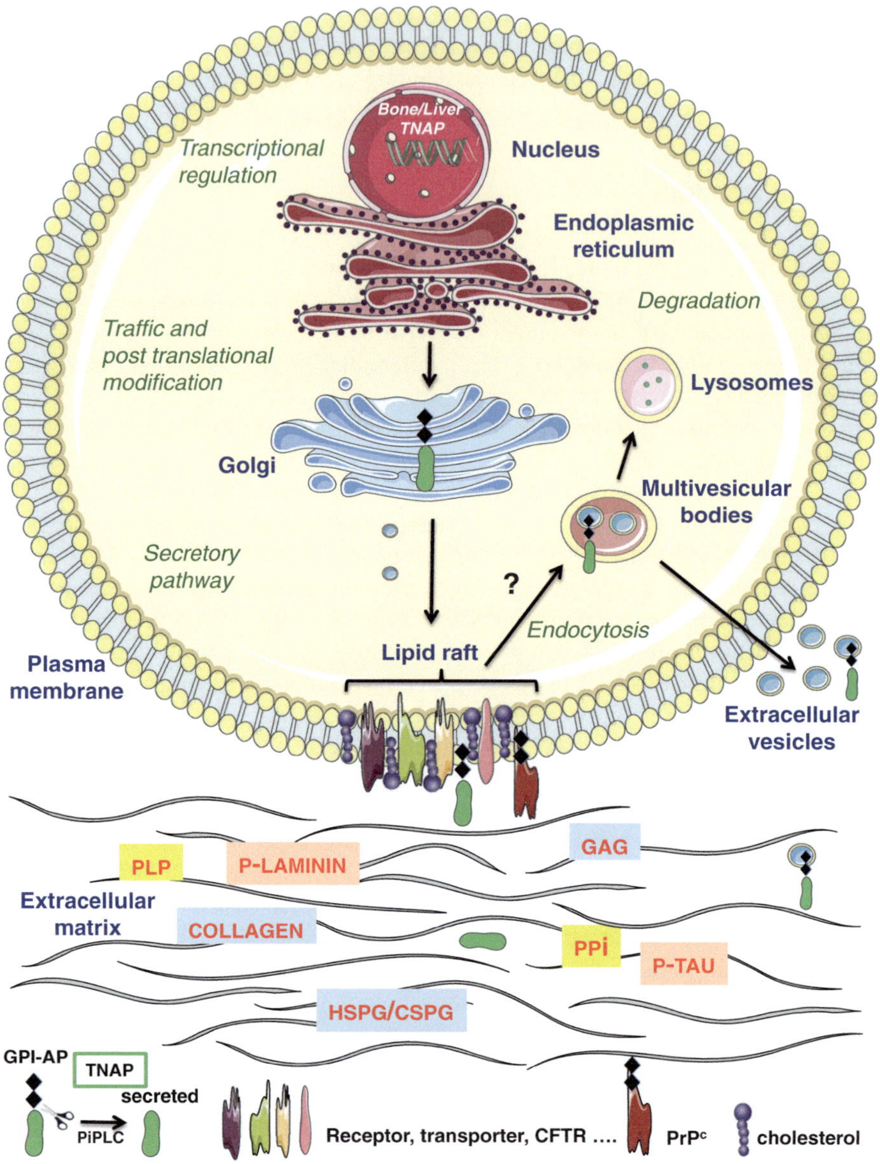
Bone/Liver TNAP
Nucleus
Transcriptional regulation
Endoplasmic reticulum
Degradation
Traffic and post translational modification
Lysosomes
Golgi
Multivesicular bodies
Secretory pathway
?
Endocytosis
Lipid raft
Plasma membrane
Extracellular vesicles
PLP
P-LAMININ
GAG
Extracellular matrix
COLLAGEN
PPi
P-TAU
HSPG/CSPG
GPI-AP
TNAP
secreted
PiPLC
Receptor, transporter, CFTR ....
PrPc
cholesterol

◀ **Fig. 1** Schematic representation of the different environments of TNAP found during its trafficking in intra- and extra-cellular compartments. The different levels of regulation that affect TNAP expression in a cell type specific manner appear in green letters. This includes the transcriptional level involving factors that interact with the bone and liver promoters, the different post-translational modifications impacting TNAP glycoform composition all along the secretory pathway and the expression level and stability of TNAP in its final destination, i.e. the membrane lipid rafts. TNAP in lipid rafts can indeed be submitted to regulation following PiPLC (phophatidylinositol-phospholipase C) release or endocytosis. Endosomes can then reach multivesicular bodies (MVBs the sequestration of early endosomes targeted) from which TNAP could be either sent to degradation or be released on extracellular vesicles (EVs). The different cell compartments are named in blue letters. The extracellular orientation of the GPI anchored protein (GPI-AP) at the plasma membrane or at the surface of micro vesicles, as well as of the secreted form of TNAP, allows this enzyme to be in contact with its various substrates or ligands (shown in red letters) circulating in the extracellular space such as the phosphomonoester substrates (inorganic pyrophosphate -PPi-, pyridoxal phosphate -PLP-) or compounds of the extracellular matrix or perineural net (collagen, laminin, GAG and proteoglycans…) involved in different cellular functions. In addition, its clustering in specialized microdomains supports its participation in different signalling functions depending on the microenvironment constituted by its close partners. In the case of neuronal cells, the proximity of TNAP with receptors, transporters or other membrane channels present in lipid rafts and themselves depending on regulatory processes could contribute to neuronal homeostasis as part of neuronal networks. The abbreviations in the figure are as follows: GAG for glycosaminoglycan, HSPG/CSPG, respectively for heparan sulfate proteoglycans and chondroitin sulfate proteoglycans, CFTR for cystic fibrosis transmembrane receptor and PrPC for the cellular prion protein

what is the resilience of the system and therefore, to identify which of the network links can be modified without perturbation and which ones are essential and will affect the whole system and lead to disorders in case of homeostatic imbalance. In this respect, the case of epilepsy that may involve TNAP corresponds to a loss of homeostasis triggering synchronization of neuronal network leading the system to become out of control. This supports a modulator role of TNAP associated to the fine tuning of neuronal homeostasis in the context of connected network.

Conflict of Interest The author declares no conflict of interest.

References

Allen JA, Halverson-Tamboli RA, Rasenick MM (2007) Lipid raft microdomains and neurotransmitter signalling. Nat Rev Neurosci 8(2):128–140. doi:nrn2059

Alvaro D, Benedetti A, Marucci L, Delle Monache M, Monterubbianesi R, Di Cosimo E, Perego L, Macarri G, Glaser S, Le Sage G, Alpini G (2000) The function of alkaline phosphatase in the liver: regulation of intrahepatic biliary epithelium secretory activities in the rat. Hepatology 32(2):174–184. doi:10.1053/jhep.2000.9078

Bate C, Williams A (2011) Monoacylated cellular prion protein modifies cell membranes, inhibits cell signaling, and reduces prion formation. J Biol Chem 286(11):8752–8758. doi: M110.186833

Becq F, Jensen TJ, Chang XB, Savoia A, Rommens JM, Tsui LC, Buchwald M, Riordan JR, Hanrahan JW (1994) Phosphatase inhibitors activate normal and defective CFTR chloride channels. Proc Natl Acad Sci USA 91(19):9160–9164

Bjork K, Sjogren B, Svenningsson P (2010) Regulation of serotonin receptor function in the nervous system by lipid rafts and adaptor proteins. Exp Cell Res 316(8):1351–1356. doi:10.1016/j.yexcr.2010.02.034

Brewis IA, Turner AJ, Hooper NM (1994) Activation of the glycosyl-phosphatidylinositol-anchored membrane dipeptidase upon release from pig kidney membranes by phospholipase C. Biochem J 303(Pt 2):633–638

Brun-Heath I, Ermonval M, Chabrol E, Xiao J, Palkovits M, Lyck R, Miller F, Couraud PO, Mornet E, Fonta C (2011) Differential expression of the bone and the liver tissue non-specific alkaline phosphatase isoforms in brain tissues. Cell Tissue Res 343(3):521–536. doi:10.1007/s00441-010-1111-4

Calhau C, Martel F, Hipolito-Reis C, Azevedo I (2002a) Modulation of uptake of organic cationic drugs in cultured human colon adenocarcinoma Caco-2 cells by an ecto-alkaline phosphatase activity. J Cell Biochem 87(4):408–416. doi:10.1002/jcb.10306

Calhau C, Martel F, Pinheiro-Silva S, Pinheiro H, Soares-da-Silva P, Hipolito-Reis C, Azevedo I (2002b) Modulation of insulin transport in rat brain microvessel endothelial cells by an ecto-phosphatase activity. J Cell Biochem 84(2):389–400. doi:10.1002/jcb.10027

Chatterjee S, Mayor S (2001) The GPI-anchor and protein sorting. Cell Mol Life Sci 58(14):1969–1987

Chen Y, Sabatini BL (2012) Signaling in dendritic spines and spine microdomains. Curr Opin Neurobiol 22(3):389–396. doi:10.1016/j.conb.2012.03.003

Ciancaglini P, Simao AM, Camolezi FL, Millan JL, Pizauro JM (2006) Contribution of matrix vesicles and alkaline phosphatase to ectopic bone formation. Braz J Med Biol Res 39(5):603–610. doi:S0100-879X2006000500006

Cox RF, Hernandez-Santana A, Ramdass S, McMahon G, Harmey JH, Morgan MP (2012) Microcalcifications in breast cancer: novel insights into the molecular mechanism and functional consequence of mammary mineralisation. Br J Cancer 106(3):525–537. doi:bjc2011583

Cuddy LK, Winick-Ng W, Rylett RJ (2014) Regulation of the high-affinity choline transporter activity and trafficking by its association with cholesterol-rich lipid rafts. J Neurochem 128(5):725–740. doi:10.1111/jnc.12490

Delic J, Zimmermann H (2011) Nucleotides affect neurogenesis and dopaminergic differentiation of mouse fetal midbrain-derived neural precursor cells. Purinergic Signal 6(4):417–428. doi:10.1007/s11302-010-9206-7

Diaz-Hernandez M, Gomez-Ramos A, Rubio A, Gomez-Villafuertes R, Naranjo JR, Miras-Portugal MT, Avila J (2010) Tissue-nonspecific alkaline phosphatase promotes the neurotoxicity effect of extracellular tau. J Biol Chem 285(42):32539–32548. doi:10.1074/jbc.M110.145003

Diez-Zaera M, Diaz-Hernandez JI, Hernandez-Alvarez E, Zimmermann H, Diaz-Hernandez M, Miras-Portugal MT (2011) Tissue-nonspecific alkaline phosphatase promotes axonal growth of hippocampal neurons. Mol Biol Cell 22(7):1014–1024. doi:10.1091/mbc.E10-09-0740

Dityatev A, Schachner M (2003) Extracellular matrix molecules and synaptic plasticity. Nat Rev Neurosci 4(6):456–468. doi:10.1038/nrn1115

Dityatev A, Schachner M, Sonderegger P (2010) The dual role of the extracellular matrix in synaptic plasticity and homeostasis. Nat Rev Neurosci 11(11):735–746. doi:10.1038/nrn2898

Ermonval M, Baudry A, Baychelier F, Pradines E, Pietri M, Oda K, Schneider B, Mouillet-Richard S, Launay JM, Kellermann O (2009a) The cellular prion protein interacts with the tissue non-specific alkaline phosphatase in membrane microdomains of bioaminergic neuronal cells. PLoS ONE 4(8):e6497. doi:10.1371/journal.pone.0006497

Ermonval M, Petit D, Le Duc A, Kellermann O, Gallet PF (2009b) Glycosylation-related genes are variably expressed depending on the differentiation state of a bioaminergic neuronal cell line: implication for the cellular prion protein. Glycoconj J 26(4):477–493. doi:10.1007/s10719-008-9198-5

Fonta C, Negyessy L, Renaud L, Barone P (2004) Areal and subcellular localization of the ubiquitous alkaline phosphatase in the primate cerebral cortex: evidence for a role in neurotransmission. Cereb Cortex 14(6):595–609. doi:10.1093/cercor/bhh021

Fonta C, Negyessy L, Renaud L, Barone P (2005) Postnatal development of alkaline phosphatase activity correlates with the maturation of neurotransmission in the cerebral cortex. J Comp Neurol 486(2):179–196. doi:10.1002/cne.20524

Fricker AD, Rios C, Devi LA, Gomes I (2005) Serotonin receptor activation leads to neurite outgrowth and neuronal survival. Brain Res Mol Brain Res 138(2):228–235. doi:10.1016/j.molbrainres.2005.04.016

Fukunaka A, Kurokawa Y, Teranishi F, Sekler I, Oda K, Ackland ML, Faundez V, Hiromura M, Masuda S, Nagao M, Enomoto S, Kambe T (2011) Tissue nonspecific alkaline phosphatase is activated via a two-step mechanism by zinc transport complexes in the early secretory pathway. J Biol Chem 286(18):16363–16373. doi:M111.227173

Goldberg RF, Austen WG Jr, Zhang X, Munene G, Mostafa G, Biswas S, McCormack M, Eberlin KR, Nguyen JT, Tatlidede HS, Warren HS, Narisawa S, Millan JL, Hodin RA (2008) Intestinal alkaline phosphatase is a gut mucosal defense factor maintained by enteral nutrition. Proc Natl Acad Sci USA 105(9):3551–3556. doi:0712140105 [pii]

Guirland C, Zheng JQ (2007) Membrane lipid rafts and their role in axon guidance. Adv Exp Med Biol 621:144–155. doi:10.1007/978-0-387-76715-4_11

Halling Linder C, Englund UH, Narisawa S, Millan JL, Magnusson P (2013) Isozyme profile and tissue-origin of alkaline phosphatases in mouse serum. Bone 53(2):399–408. doi:10.1016/j.bone.2012.12.048

Halling Linder C, Narisawa S, Millan JL, Magnusson P (2009) Glycosylation differences contribute to distinct catalytic properties among bone alkaline phosphatase isoforms. Bone 45 (5):987–993. doi:10.1016/j.bone.2009.07.009

Halling Linder C, Narisawa S, Millan JL, Magnusson P (2012) Characterization of alkaline phosphatase in mice. Bull Group Int Rech Sci Stomatol Odontol 51(1):e32

Hamir AN, Palmer MV, Kunkle RA (2008) Wasting and neurologic signs in a white-tailed deer (Odocoileus virginianus) not associated with abnormal prion protein. J Wildl Dis 44(4):1045–1050. doi:44/4/1045

Head BP, Patel HH Insel PA (2014) Interaction of membrane/lipid rafts with the cytoskeleton: impact on signaling and function: membrane/lipid rafts, mediators of cytoskeletal arrangement and cell signaling. Biochim Biophys Acta 1838(2):532–545. doi:10.1016/j.bbamem.2013.07.018

Head BP, Patel HH, Roth DM, Murray F, Swaney JS, Niesman IR, Farquhar MG, Insel PA (2006) Microtubules and actin microfilaments regulate lipid raft/caveolae localization of adenylyl cyclase signaling components. J Biol Chem 281(36):26391–26399. doi:M602577200

Hoshi K, Amizuka N, Oda K, Ikehara Y, Ozawa H (1997) Immunolocalization of tissue non-specific alkaline phosphatase in mice. Histochem Cell Biol 107(3):183–191

Ishida Y, Komaru K, Ito M, Amaya Y, Kohno S, Oda K (2003) Tissue-nonspecific alkaline phosphatase with an Asp(289) –> Val mutation fails to reach the cell surface and undergoes proteasome-mediated degradation. J Biochem 134(1):63–70

Ito M, Amizuka N, Ozawa H, Oda K (2002) Retention at the cis-Golgi and delayed degradation of tissue-non-specific alkaline phosphatase with an Asn153 –> Asp substitution, a cause of perinatal hypophosphatasia. Biochem J 361(Pt 3):473–480

Jones J, Krag SS, Betenbaugh MJ (2005) Controlling N-linked glycan site occupancy. Biochim Biophys Acta 1726(2):121–137. doi:10.1016/j.bbagen.2005.07.003

Kapojos JJ, Poelstra K, Borghuis T, Van Den Berg A, Baelde HJ, Klok PA, Bakker WW (2003) Induction of glomerular alkaline phosphatase after challenge with lipopolysaccharide. Int J Exp Pathol 84(3):135–144. doi:345

Kellett KA, Williams J, Vardy ER, Smith AD, Hooper NM (2011) Plasma alkaline phosphatase is elevated in Alzheimer's disease and inversely correlates with cognitive function. Int J Mol Epidemiol Genet 2(2):114–121

Kinoshita T, Maeda Y, M Fujita (2013) Transport of glycosylphosphatidylinositol-anchored proteins from the endoplasmic reticulum. Biochim Biophys Acta 1833(11):2473–2478. doi:10.1016/j.bbamcr.2013.01.027
Kooyman DL, Byrne GW, Logan JS (1998) Glycosyl phosphatidylinositol anchor. Exp Nephrol 6 (2):148–151. doi:exn06148 [pii]
Kothekar D, Bandivdekar A, Dasgupta D (2014) Increased activity of goat liver plasma membrane alkaline phosphatase upon release by phosphatidylinositol-specific phospholipase C. Indian J Biochem Biophys 51(4):263–270
Koyama I, Matsunaga T, Harada T, Hokari S, Komoda T (2002) Alkaline phosphatases reduce toxicity of lipopolysaccharides in vivo and in vitro through dephosphorylation. Clin Biochem 35(6):455–461. doi:S0009912002003302 [pii]
Labasque M, Faivre-Sarrailh C (2010) GPI-anchored proteins at the node of Ranvier. FEBS Lett 584(9):1787–1792. doi:S0014-5793(09)00652-8
Lakhan SE, Sabharanjak S, De A (2009) Endocytosis of glycosylphosphatidylinositol-anchored proteins. J Biomed Sci 16:93. doi:1423-0127-16-93 [pii]
Legler DF, Doucey MA, Schneider P, Chapatte L, Bender FC, Bron C (2005) Differential insertion of GPI-anchored GFPs into lipid rafts of live cells. FASEB J 19(1):73–75. doi:03-1338fje [pii]
Lehto MT, Sharom FJ (1998) Release of the glycosylphosphatidylinositol-anchored enzyme ecto-5'-nucleotidase by phospholipase C: catalytic activation and modulation by the lipid bilayer. Biochem J 332(Pt 1):101–109
Lindner R, Naim HY (2009) Domains in biological membranes. Exp Cell Res 315(17):2871–2878. doi:S0014-4827(09)00332-2 [pii]
Lomashvili KA, Narisawa S, Millan JL, O'Neill WC (2014) Vascular calcification is dependent on plasma levels of pyrophosphate. Kidney Int 85(6):1351–1356. doi:ki2013521 [pii]
Maeda Y, Kinoshita T (2011) Structural remodeling, trafficking and functions of glycosylphosphatidylinositol-anchored proteins. Prog Lipid Res 50(4):411–424. doi:10.1016/j.plipres.2011.05.002
Martel F, Martins MJ, Hipolito-Reis C, Azevedo I (1996) Inward transport of [3H]-1-methyl-4-phenylpyridinium in rat isolated hepatocytes: putative involvement of a P-glycoprotein transporter. Br J Pharmacol 119(8):1519–1524
Matsuoka I, Ohkubo S (2004) ATP- and adenosine-mediated signaling in the central nervous system: adenosine receptor activation by ATP through rapid and localized generation of adenosine by ecto-nucleotidases. J Pharmacol Sci 94(2):95–99
Moochhala SH, Sayer JA, Carr G, Simmons NL (2008) Renal calcium stones: insights from the control of bone mineralization. Exp Physiol 93(1):43–49. doi:expphysiol.2007.040790 [pii]
Mouillet-Richard S, Ermonval M, Chebassier C, Laplanche JL, Lehmann S, Launay JM, Kellermann O (2000) Signal transduction through prion protein. Science 289(5486):1925–1928
Mouillet-Richard S, Schneider B, Pradines E, Pietri M, Ermonval M, Grassi J, Richards JG, Mutel V, Launay JM, Kellermann O (2007) Cellular prion protein signaling in serotonergic neuronal cells. Ann NY Acad Sci 1096:106–119. doi:10.1196/annals.1397.076
Narisawa S, Hasegawa H, Watanabe K, Millan JL (1994) Stage-specific expression of alkaline phosphatase during neural development in the mouse. Dev Dyn 201(3):227–235. doi:10.1002/aja.1002010306
Narisawa S, Wennberg C, Millan JL (2001) Abnormal vitamin B6 metabolism in alkaline phosphatase knock-out mice causes multiple abnormalities, but not the impaired bone mineralization. J Pathol 193(1):125–133. doi:10.1002/1096-9896(2000)9999:9999
Newpher TM, Ehlers MD (2009) Spine microdomains for postsynaptic signaling and plasticity. Trends Cell Biol 19(5):218–227. doi:10.1016/j.tcb.2009.02.004 [pii]
Nishimune H, Sanes JR, Carlson SS (2004) A synaptic laminin-calcium channel interaction organizes active zones in motor nerve terminals. Nature 432(7017):580–587. doi:nature03112 [pii]
Nunes ML, Mugnol F, Bica I, Fiori RM (2002) Pyridoxine-dependent seizures associated with hypophosphatasia in a newborn. J Child Neurol 17(3):222–224

Pettengill M, Robson S, Tresenriter M, Millan JL, Usheva A, Bingham T, Belderbos M, Bergelson I, Burl S, Kampmann B, Gelinas L, Kollmann T, Bont L, Levy O (2013) Soluble ecto-5'-nucleotidase (5'-NT), alkaline phosphatase, and adenosine deaminase (ADA1) activities in neonatal blood favor elevated extracellular adenosine. J Biol Chem 288 (38):27315–27326. doi:10.1074/jbc.M113.484212

Premkumar DR, Fukuoka Y, Sevlever D, Brunschwig E, Rosenberry TL, Tykocinski ML, Medof ME (2001) Properties of exogenously added GPI-anchored proteins following their incorporation into cells. J Cell Biochem 82 (2):234–245. doi:10.1002/jcb.1154

Rajendran L, Bali J, Barr MM, Court FA, Kramer-Albers EM, Picou F, Raposo G, van der Vos KE, van Niel G, Wang J, Breakefield XO (2014) Emerging roles of extracellular vesicles in the nervous system. J Neurosci 34(46):15482–15489. doi:10.1523/JNEUROSCI.3258-14.2014

Scheibe RJ, Kuehl H, Krautwald S, Meissner JD, Mueller WH (2000) Ecto-alkaline phosphatase activity identified at physiological pH range on intact P19 and HL-60 cells is induced by retinoic acid. J Cell Biochem 76(3):420–436. doi:10.1002/(SICI)1097-4644(20000301)76:3<420::AID-JCB10>3.0.CO;2-F

Sebastiao AM, Colino-Oliveira M, Assaife-Lopes N, Dias RB, Ribeiro JA (2013) Lipid rafts, synaptic transmission and plasticity: impact in age-related neurodegenerative diseases. Neuropharmacol 64:97–107. doi:10.1016/j.neuropharm.2012.06.053

Sesana S, Re F, Bulbarelli A, Salerno D, Cazzaniga E, Masserini M (2008) Membrane features and activity of GPI-anchored enzymes: alkaline phosphatase reconstituted in model membranes. Biochem 47(19):5433–5440. doi:10.1021/bi800005s

Silvius JR (2005) Partitioning of membrane molecules between raft and non-raft domains: insights from model-membrane studies. Biochim Biophys Acta 1746(3):193–202. doi:10.1016/j.bbamcr.2005.09.003

Sperling LE, Klaczinski J, Schutz C, Rudolph L, Layer PG (2012) Mouse acetylcholinesterase enhances neurite outgrowth of rat R28 cells through interaction with laminin-1. PLoS ONE 7 (5):e36683. doi:10.1371/journal.pone.0036683

Stanley P (2011) Golgi glycosylation. Cold Spring Harb Perspect Biol 3(4). doi:cshperspect.a005199 [pii]

Sultana S, Al-Shawafi HA, Makita S, Sohda M, Amizuka N, Takagi R, Oda K (2013) An asparagine at position 417 of tissue-nonspecific alkaline phosphatase is essential for its structure and function as revealed by analysis of the N417S mutation associated with severe hypophosphatasia. Mol Genet Metab 109(3):282–288. doi:10.1016/j.ymgme.2013.04.016

Toffoli AM, Gautschi OP, Frey SP, Filgueira L, Zellweger R (2008) From brain to bone: evidence for the release of osteogenic humoral factors after traumatic brain injury. Brain Inj 22(7–8):511–518. doi:10.1080/02699050802158235

Vardy ER, Kellett KA, Cocklin SL, Hooper NM (2012) Alkaline phosphatase is increased in both brain and plasma in Alzheimer's disease. Neurodegener Dis 9(1):31–37. doi:10.1159/000329722

Watt NT, Griffiths HH, Hooper NM (2014) Lipid rafts: linking prion protein to zinc transport and amyloid-beta toxicity in Alzheimer's disease. Front Cell Dev Biol 2:41. doi:10.3389/fcell.2014.00041

Wuthier RE, Lipscomb GF (2011) Matrix vesicles: structure, composition, formation and function in calcification. Front Biosci (Landmark edn) 16:2812–2902. doi:3887 [pii]

Yamamoto H, Sasamoto Y, Miyamoto Y, Murakami H, Kamiyama N (2004) A successful treatment with pyridoxal phosphate for West syndrome in hypophosphatasia. Pediatr Neurol 30 (3):216–218. doi:10.1016/j.pediatrneurol.2003.08.003

Zempel H, Mandelkow E (2014) Lost after translation: missorting of Tau protein and consequences for Alzheimer disease. Trends Neurosci. doi:10.1016/j.tins.2014.08.004

Chapter 10
Signal Transduction Pathways of TNAP: Molecular Network Analyses

László Négyessy, Balázs Győrffy, János Hanics, Mihály Bányai, Caroline Fonta and Fülöp Bazsó

Abstract Despite the growing body of evidence pointing on the involvement of tissue non-specific alkaline phosphatase (TNAP) in brain function and diseases like epilepsy and Alzheimer's disease, our understanding about the role of TNAP in the regulation of neurotransmission is severely limited. The aim of our study was to integrate the fragmented knowledge into a comprehensive view regarding neuronal functions of TNAP using objective tools. As a model we used the signal transduction molecular network of a pyramidal neuron after complementing with TNAP related data and performed the analysis using graph theoretic tools. The analyses show that TNAP is in the crossroad of numerous pathways and therefore is one of the key players of the neuronal signal transduction network. Through many of its connections, most notably with molecules of the purinergic system, TNAP serves as

L. Négyessy (✉)
Theoretical Neuroscience and Complex Systems Research Group, Wigner Research Center for Physics, Budapest, Hungary
e-mail: negyessy.laszlo@wigner.mta.hu

L. Négyessy · J. Hanics
Department of Anatomy, Histology and Embryology, Semmelweis University Medical School, 1094 Budapest, Hungary

B. Győrffy
Laboratory of Proteomics and MTA-ELTE NAP B Neuroimmunology Research Group, Institute of Biology, Eötvös Loránd University, Budapest, Hungary

M. Bányai
Computational Systems Neuroscience Lab, Wigner Research Center for Physics, Hungarian Academy of Sciences, 1121 Budapest, Hungary

C. Fonta
Centre de Recherche Cerveau et Cognition (CerCo), Université de Toulouse UPS; CNRS UMR 5549, Toulouse, France

F. Bazsó
Theoretical Neuroscience and Complex Systems Research Group, Wigner Research Center for Physics, Hungarian Academy of Sciences, Budapest, Hungary

F. Bazsó
SU-Tech College of Applied Sciences, Subotica, Serbia

C. Fonta and L. Négyessy (eds.), *Neuronal Tissue-Nonspecific Alkaline Phosphatase (TNAP)*, Subcellular Biochemistry 76,
DOI 10.1007/978-94-017-7197-9_10

a controller by funnelling signal flow towards a subset of molecules. TNAP also appears as the source of signal to be spread via interactions with molecules involved among others in neurodegeneration. Cluster analyses identified TNAP as part of the second messenger signalling cascade. However, TNAP also forms connections with other functional groups involved in neuronal signal transduction. The results indicate the distinct ways of involvement of TNAP in multiple neuronal functions and diseases.

Keywords Second messenger · Centrality · Cluster analyses · Purinergic · Alzheimer's disease

10.1 Introduction

Understanding neuronal signal transduction is essential to explore the biological bases of brain functions and diseases. This recognition motivated molecular and cellular neurobiology to generate an extraordinary amount of data about the structure, interactions and functions of different molecules. Meanwhile it became evident that many molecules partake in multiple interactions and that molecular interactions form a complex machinery network which can be studied by graph theoretic approaches. Ma'ayan et al. (2005) published the first comprehensive neuronal signal transduction network by collating literature data derived from the hippocampal CA1 pyramidal neuron. Using network analytical approach, this study (Ma'ayan et al. 2005) presented interesting data about the basic regulatory mechanisms of neurotransmission.

Recent discovery of the highly specific localization of tissue non-specific alkaline phosphatase (TNAP) in the cerebral cortex and retina and its neuronal activity dependent function evoked a renewable interest towards the neuronal functions of this enzyme (see Chaps. 5 and 6). It has long been known that hypo- and hyper-phosphatasia, associated respectively with a too low or too high blood level of TNAP, causes serious neurological symptoms, most notably epilepsy (see Chaps. 14 and 16). There is growing evidence showing decreased levels of GABA and some monoamines in hypophosphatasia suggesting that this might cause the seizures (see Chaps. 1 and 14). Accordingly, these seizures were often found to be pyridoxal dependent as it supplements B6 vitamin requirement for the proper functioning of GABA and monoamine synthesising enzymes (Surtees et al. 2006; see also Chaps. 11, 14 and 16). However, recent studies also provided evidence for the role of TNAP in neuronal development and purinergic transmission (see Chaps. 4, 13 and 18). There are also observations indicating that TNAP can interact with the cellular prion protein PrP^C and TAU (see Chaps. 9, 17 and 18), which play central role in degenerative brain diseases in the aggregated form. TNAP exerts its manifold functions via its phosphohydrolase activity on very diverse types of molecules including organic (B6 vitamin, proteins, nucleotides) and inorganic phosphates (see Chap. 8). Since phosphorylation/dephosphorylation is a fundamental way of regulating the activity of proteins, and that TNAP is required for normal B6 metabolism and therefore for the sufficient functioning of the large family of pyridoxine dependent enzymes, which include amino acid (GABA, glutamate) neurotransmitter synthases and monoamine synthases, TNAP can be an important player of regulating neurotransmission. This assumption is supported by the localization of the enzyme within the cleft of excitatory and inhibitory synapses of the cerebral cortex (see Chap. 5). However, these assumptions are based on fragmentary experimental observations and need to be supported by further evidence. Considering the complexity of interactions and cellular processes a comprehensive and objective view of the molecular machinery including TNAP can significantly support additional future experimental studies to a better understanding of the role of TNAP in neuronal functions.

Network analysis became one of the most important approaches of complex systems studies (see e.g. Strogatz 2001; Watts 2004). Therefore, not surprisingly, network analysis is used extensively to study cellular functions considering that the underlying molecular machinery is composed of several hundreds or thousands of molecules and typically several times more links connecting molecules with known interactions (for comprehensive reviews see e.g. Christensen et al. 2007; Ma'ayan 2009; Csermely et al. 2013). Usually, the links between interacting molecules have known directions and this allows the analysis not only of particular relationship between (multiple) pairs of molecules but also offers the opportunity to investigate chains of reactions, so called pathways. This can be made in an objective way via identifying the shortest paths between pairs of elements. Considering the shortest path is an important criterion for two main reasons: the shortest paths are assumed to be particularly relevant in terms of information transfer and that network topology determines which elements are included within particular signalling paths. Graph theory also provides tools to characterize the importance of the network elements, i.e. nodes (molecules) and links/edges (relationship between molecules). Centrality indices define how important a particular node or link is within the network. Degree centrality is merely the number of links, either outgoing or incoming, usually of a node: the higher the degree centrality, the more important the node is because of the large number of interactions within the network. However, there are other indices such as e.g. closeness, Eigenvector, page rank, which measure other characteristics, i.e. centrality of the network element from another perspective. Among the indices betweenness centralities are particularly interesting, giving the number of shortest paths passing through the network element being either a node (node betweenness, NB) or an edge (edge betweenness, EB). A large value suggests that the network element is involved in numerous chains of interactions. Please note, that in contrast to the degree centrality, betweenness centrality is a global measure computed by considering all interactions within the network. Similarly, the shortest path analyses helps quantifying the degree of convergence or divergence via the edges of the network only if the network is directed (the links have a direction). In short, convergence degree (CD) of an edge computes the relative size (number of nodes) of the input to the target set of nodes (Négyessy et al. 2008; Bányai et al. 2011). A centrality measure of the nodes can also be derived from the convergence degree of the directed edges of a node (Négyessy et al. 2008; Bányai et al. 2011). Simply saying, network elements with high centrality indices can be considered as hubs. Cluster analysis is another important technique, which is used to identify closely related groups of elements. A cluster is usually defined by the high density of connections meaning more intensive interactions within the cluster as compared to that within the full network. Clustering is very important to identify functional modules subserving different cell functions including proliferation, apoptosis, signal transduction, transport and many more (see e.g. Ulitsky and Shamir 2007). By identifying clusters, one can locate the elements involved in interactions between the clusters of interest. There is a lot of more ways of analysing the networks; here we described those which are the most relevant in the present study. Comprehensive overviews are provided by the reviews cited above (for a more general overview see Newman 2001).

The aim of our study was to provide an integrated and objective understanding of the neural functions of TNAP via its multiple molecular interactions based on our current state of knowledge. To this end we applied network analytic tools to study the neuronal signal transduction machinery published by Ma'ayan et al. (2005) complemented by the interactions of TNAP irrespective of the cell types where the interactions were described. However, all new molecules added to the original network can be found in neuronal cells. We used regular centrality measures and clustering techniques complemented by database driven functional annotations to identify the pathways and functional groups of interacting molecules where TNAP plays significant roles.

10.2 Methods

10.2.1 Collection of Data

Interactions of TNAP were retrieved via search engine from the Pathway Studio 9.0 software environment with ResNet 9 Mammalian Database (Elsevier Life Science Solutions) and then validated by published data by searching NCBI PubMed. The list of TNAP and connected molecules embedded into the network analysed (published online as Supplementary Material by Ma'ayan et al. 2005) is provided in Table 10.1. Several indirect interactions were omitted in the analyses, except the one with mitogen-activated protein kinase (MAPK) and the muscarinic acetylcholine receptor M1R. MAPK regulates TNAP through bone morphogenetic protein (BMP) and SMAD as shown in osteoblasts (Osyczka and Leboy 2005). Unfortunately, in contrast to SMADs, which were included into the analyses, BMP could not be sufficiently integrated into the pyramidal neuronal signal transduction molecular network due to the lack of relevant data in the databases used. However, MAPK is an important regulator of neuronal functions. Therefore, considering also the tight relationship between MAPK and TNAP (Osyczka and Leboy 2005) adding this interaction could provide results, inspiring future studies. Regarding the interaction with M1R, available literature data suggest that TNAP regulates M1R actions via dephosphorylating hyperphosphorylated TAU (see Chaps. 17 and 18). However, studies also point out the importance of ectophosphorylation residues for modulation of receptor and ion channel functions (Redegeld et al. 1999) and we were interested in the importance of such a pathway. M1R seemed to be a good example, which possesses extracellular phosphorylation site and is indirectly influenced by TNAP via TAU. Therefore with limited modification of the network (adding one link) we could obtain new insight about a predicted function. A list of other possible targets for ectophoshorylation based neuromodulation is provided in Table 10.2, which was constructed by data obtained from the UniProt (http://www.uniprot.org/) and PhosphoSite databases (http://www.phosphosite.org).

10.2.2 Data Analysis

The neuronal signal transduction network was treated as a directed graph; the molecules and the interactions between them were represented by vertices (nodes) and edges (links), respectively. Although, some interactions without known directionality as e.g. binding with structure proteins, were represented as bidirectional links. Computations were performed with self-developed programs and scripts along with Microsoft Excel, the NodeXL template for MS Excel (http://nodexl.codeplex.com/) and the igraph network analysis package (http://igraph.org/).

The **basic characteristics** of the network were determined by calculating its edge density, transitivity and diameter. The edge density or simply density of the network is the total number of connections divided by the number of all possible connections. The transitivity of a node is the proportion of links between its neighbouring nodes divided by the number of links that could possibly exist between them; transitivity of the whole graph is the mean of the transitivities for each individual node (Watts and Strogatz 1998). The distance between any two nodes is the length of the shortest path connecting them (i.e. the number of edges forming the shortest path); the largest distance is called the diameter of the network.

Betweenness centrality is computed using all the shortest paths in the network. First, we find all the shortest paths between all possible pairs of nodes s and t. For each s–t pair, we count the number of shortest paths between them that pass through a given node (node betweenness, NB) or a given edge (edge betweenness, EB) and divide them by the number of all the shortest paths between s and t. Finally, we sum these ratios over all s–t pairs. Less formally, the betweenness centrality is the number of shortest paths in the network that pass through a given node or edge, but if there are k distinct shortest paths between nodes s and t, each of them weigh only 1/k in the final betweenness score.

Table 10.1 List of molecules and connections involving TNAP, which were embedded into the pyramidal cell signal transduction network published by Ma'ayan et al. (2005)

Source	Id	Target	Id	PubMed Id
TNAP	P05186	EGFR	P00533	10585492
TNAP	P05186	GABAAR	P14867	1335848
TNAP	P05186	M1R	P11229	20634292
TNAP	P05186	D1R	P21728	21803776
TNAP	P05186	AMPAR	P42262	18665261
TNAP	P05186	TAU	P10636	20634292
TNAP	P05186	NIK	Q99558	15208311, 21173796
TNAP	P05186	P2X7	Q99572	21289095
TNAP	P05186	PLP	NA	7550313
TNAP	P05186	ADENOSIN	NA	22761898
TNAP	P05186	AMP	NA	22761898
TNAP	P05186	ADP	NA	19008000
TNAP	P05186	ATP	NA	19008000
SLC30A7	Q8NEW0	TNAP	P05186	15525635
MAPK	P28482	TNAP	P05186	15905316
P2X7	Q99572	TNAP	P05186	21289095
FIVEHT2BR	P41595	TNAP	P05186	20573958

Id UniProt accession number

The edge related indices **convergence- and overlap degrees** were computed as described in Bányai et al. (2011). Convergence degree (CD) computes the difference between the size of input and target areas of an edge identified by computing the shortest paths through the given edge.

$$CD_{(i,j)} = \frac{|In_{i,j}| - |Out_{i,j}|}{|In_{i,j} \cup Out_{i,j}|}$$

CD provides values between minus one and plus one. Positive (negative) values mean convergence (divergence) as the number of input (target) nodes exceeds that of the target (input) nodes. Summing or averaging the CD of the incoming and outgoing connections of a node results in the node-reduced CD that characterizes the role of the given node in the information flow in the network (Négyessy et al. 2008; Bányai et al. 2011).

Overlap between the input- and target sets of nodes of a given edge was computed as described in Bányai et al. (2011).

$$Ovl_{(i,j)} = \frac{|In_{i,j} \cap Out_{i,j}|}{|In_{i,j} \cup Out_{i,j}|}$$

Table 10.2 Proteins responsible for control of neurotransmission with extracellular phosphorylation sites

Receptors	Id	Channels	Id
P2X1	P51575	AMPAR	P42261
P2X4	Q99571	GIRK	P48549
P2X6	O15547	IP3R	Q14643
P2Y1	P47900	KAR	P39086
A1R	P30542	KIR21	P48049
A2AR	P29274	KIR23	P48050
ALPHA1AR	P35348	KIR41	P78508
ALPHA2AR	P08913	KV11	Q09470
ALPHA7NACHR	P36544	KV12	P16389
BETA2AR	P07550	KV14	P22459
BR1R	P46663	KV41	Q9NSA2
CB1R	P21554	KV42	Q9NZV8
CB2R	P34972	LTYPECA	Q13936
D1R	P21728	NTYPECA	Q00975
D2R	P14416	PQCaCh	O00555
D3R	P35462	RTYPECA	Q15878
DOPR	P41143	RYR	P21817
EGFR	P00533	SEC61P	P61620
EPHB2	P52799	VDAC2	P45880
ERBB	P04626		
FAS	P25445		
FIVEHT1AR	P08908		
FIVEHT2AR	P28223		
FIVEHT4R	Q13639		
GABAAR	P14867		
GABABR	Q9UBS5		
IGF1R	P08069		
IR	P06213		
KOPR	P41145		
LRP	Q9BQ69		
M1R	P11229		
M2R	P08172		
M4R	P08173		
MGLUR5	Q13255		
MGLUR7	Q14831		
MOPR	P35372		
NMDAR	Q12879		
NOPR	P41146		
PAFR	P25105		
PAR2	P55085		
PDGFR	P09619		

(continued)

Table 10.2 (continued)

Receptors	Id	Channels	Id
RET	P09455		
SRPR	P08240		
SSTR1	P30872		
SSTR2	P30874		
SYNDECAN	P18827		
TRKA	P04629		
TRKB	Q16620		

Id UniProt accession number

Note that neither CD nor overlap considers the distance, i.e. the lengths of the shortest paths connecting the nodes of interest.

Three different algorithms were applied to identify **clusters**. Markov clustering (MCL) is based on simulated stochastic flows in a directed weighted graph (Van Dongen 2008). The key idea of the community walktrap (CW) clustering algorithm is that it is more probable for a random walk to stay in a densely connected component than to flow into other components (Enright et al. 2002). Edge betweenness values have been utilized as the weights of edges. The clusters defined by an inflation (granularity) value of 3.3 have been selected. The Clauset-Newman-Moore (CNM) method is a hierarchical agglomeration algorithm, which was developed to work on very large graphs (Clauset et al. 2004).

10.3 Results

10.3.1 Neuronal Signal Transduction Network: General Properties

The addition of the 11 new molecules and 44 connections (referred in the text as TNAP related data) did not considerably change the diameter, average path lengths of the hippocampal pyramidal neuron signal transduction network (Table 10.3). Similarly, neither average degrees nor betweenness centralities changed noticeably after complementing the original network with the TNAP related data (Table 10.3). Also, degree and betweenness centralities exhibited very similar distributions in the original and updated networks (Fig. 10.1a–c). Some decrease of the clustering coefficient was apparent after complementing the network with the TNAP related data (see transitivity values in Table 10.3).

Table 10.3 Comparison of general network characteristics of the original and updated networks

	Original	With TNAP*
Number of nodes	551	563
Number of edges	1806	1839
Density	0.006	0.0058
Diameter	20	20
Average path length	5.9	6.1
Transitivity (local, average)**	0.147	0.143
Transitivity (global)**	0.0576	0.0567
In-degree (average)	2.30	2.32
Out-degree (average)	2.30	2.32
Node betweenness (average)	1773.6	1780.01
Edge betweenness (average)	602.06	485.14

*Including all new molecules and their connections. **Undirected graph; computed following symmetrisation of the network

10.3.2 Centrality of TNAP in the Signal Transduction Network

When compared to network average and median, TNAP differed in all indices studied (Table 10.4). Regarding degree centrality, a local network property, both in-degree and out-degree of TNAP exceeded the network average with out-degree exhibiting a several fold larger value compared either to in-degree of TNAP or the average of the network. However, both degree centralities were well below the maximum values observed in the network (Fig. 10.1a; Table 10.4). Similar observations were made by considering global network measures such as betweenness centralities (Fig. 10.1b, c). Both NB and EB of TNAP were higher than the corresponding average and median values of the total network (Table 10.4). However, similarly to the degree centralities, both betweenness centralities were below the maximum values of the network: TNAP had the 12th largest NB and its maximum EB was the 2nd in the network (Tables 10.4 and 10.5). Even the smallest EB value of TNAP was many folds larger than the minimum value in the network (Table 10.4). The NB of TNAP and the EB of roughly half of its links fell within the upper quartiles of the corresponding values in the full network (larger than NB = 1063.93 and EB = 469.2) (Fig. 10.1b, c; Table 10.4). Only two edges exhibited EB within the lower quartile of the network values (less than 76.45) (Fig. 10.1c; Table 10.4). The highest values of the indices were exhibited mostly by molecules and interactions involved in intracellular second messenger mechanisms such as Ca^{2+}, G-protein, cAMP and kinases (Table 10.5). Among the synaptic proteins only NMDAR appeared among the key molecules (Table 10.5). TNAP also exhibited high convergence degree (CD) through the majority of its connections (10 out of 17) with values close to the maximum of $|1|$ in most of the cases (Fig. 10.1d). An important observation was that the majority of

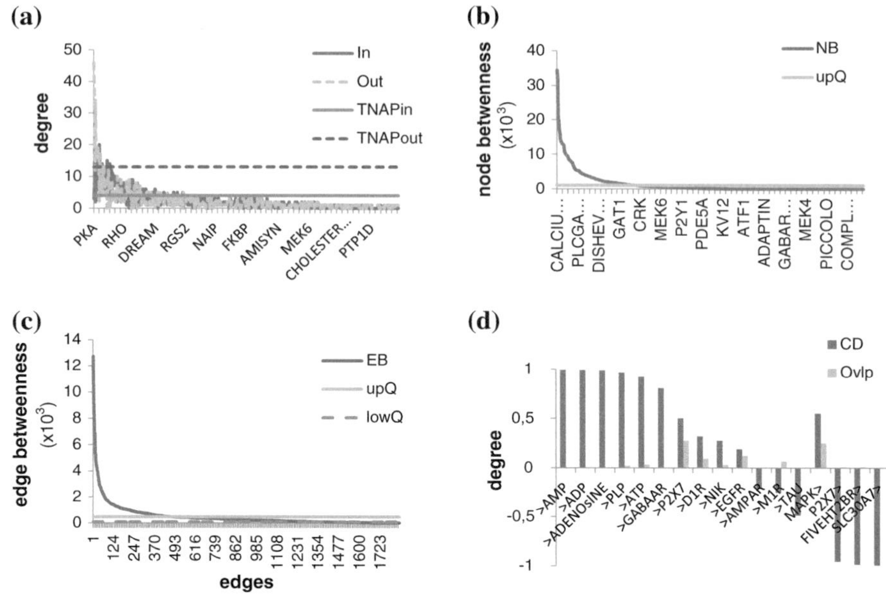

Fig. 10.1 Positioning TNAP on the basis of centrality measures in the signal transduction network. **a** Degree centrality. Distribution of in-degree (*In*) and out-degree (*Out*) centralities. For comparison in- and out-degrees of TNAP (*TNAPin* and *TNAPout*, respectively) are indicated by the *straight lines*. **b** Distribution of node betweenness centrality (*NB*). For reference the upper quartile (*upQ*) is shown by the *straight line*. **c** Distribution of edge betweenness centrality (*EB*) values in the network. Upper and lower quartiles (*upQ* and *lowQ*, respectively) are shown by the *straight lines* for reference. Note that the *lowQ* line overlies the abscissa. **d** Convergence degree (*CD*) and overlap degree (*Ovlp*) of the interactions of TNAP in the network. Inward and outward connections of TNAP are shown by > after and before the interacting molecules, respectively

outgoing edges of TNAP exhibited positive CD values (i.e. convergent in nature) and, with one exception, incoming edges of TNAP had negative CD (i.e. divergent type) values (Fig. 10.1d). Therefore, TNAP is one the controllers of signal flow in the network (Négyessy et al. 2008; Bányai et al. 2011).

10.3.3 Regulatory Pathways of TNAP in the Neuronal Signal Transduction Network

10.3.3.1 Weighting the Connections of TNAP

Analyses of EB yielded several interactions of TNAP with high values (Table 10.6). Having the 2nd highest EB value the MAPK → TNAP interaction indicates the central role of this link in the cascades of molecular interactions within the neuronal signal transduction network. In other interactions of TNAP with high EB values

Table 10.4 Centrality measures of TNAP compared to those of the total network

	Total network		TNAP
In-degree	Average	2.32	
	Median	1	4
	Maximum	30	
Out-degree	Average	2.32	
	Median	1	13
	Maximum	46	
Node betweenness	Average	1780.08	
	Median	277.89	14055.6
	Maximum	49230.22	
Edge betweenness	Average	485.14	1537.63
	Median	222.88	402
	Maximum	12744.65	12078.69
	Minimum	0.5	30.88
Clustering coefficient	Average	0.054	0.013
	Median	0	
	Maximum	0.5	

(within the upper quartile of the network values) TNAP was the origin of the links (Table 10.6). Among these interactions the majority (ATP, ADP, adenosine and P2X7) is involved in purinergic signalling. Interaction with pyridoxal phosphate (PLP), the active form of vitamin B6, also shows the high importance of this interaction in the signal transduction machinery. One more notable interaction of TNAP is with the epidermal growth factor receptor (EGFR) (Kim et al. 1999), which together with the interaction with MAPK shows the role of TNAP in fundamental cellular functions (Wong and Guillaud 2004). Finally, although the interaction of TNAP with M1R exhibited higher EB than that with TAU, neither interaction appeared significant in the signal transduction network (Table 10.6). On the contrary, the TAU → M1R interaction had relatively high EB (3259.08) falling within the upper quartile of the network.

Analyses of CD showed that interactions of TNAP, which implicate it in purinergic transmission (i.e. with AMP, ADP, adenosine and ATP), were highly convergent (Fig. 10.1d). Similar observation was made in case of the TNAP → PLP interaction (Fig. 10.1d). Although the interaction of TNAP with GABAA receptor (GABAAR) also exhibited high convergence, this link showed very low EB in the network (Fig. 10.1d). Among the highly divergent interactions of TNAP, two, one with P2X7 (purinergic) and the other via FIVEHT2BR (serotoninergic) were related to neurotransmission and the third one with the zinc transporter SLC30A7, which is an important regulator of enzymatic functions (McCall et al. 2000; Fukunaka et al. 2011) (Fig. 10.1d). The interaction of MAPK with TNAP exhibited unusual CD characteristics: this was the only incoming edge of TNAP, which exhibits convergent characteristics (Fig. 10.1d). This interaction was also different from most of the others of TNAP by exhibiting relatively high degree of overlap (0.25 in contrast

Table 10.5 Network elements exhibiting the highest values of indices

Node betweenness	Edge betweenness	In degree	Out degree
CALCIUM	CAMP → PKA	GBETAGAMMA	PKA
PKC	MAPK → TNAP	GALPHAI	PKC
GBETAGAMMA	CALCIUM → CALMODULIN	NMDAR	CALMODULIN
MAPK	CALCIUM → PKC	GALPHAO	CAMKII
NMDAR	RAF1 → MEK1	DYNAMIN	CALCIUM
CASPASE3			
CAMP			
PIP2			
MEK1			
PKA			
CALMODULIN			

Elements are listed in decreasing order

to the average of 0.05 and the median of 0.009 considering only the connections of TNAP), which provides large range of possibilities to modulate interactions. The highest overlap degree of the interactions of TNAP was found in case of TNAP → P2X7 (0.28). The connection between TNAP and P2X7 was one of the two reciprocal connections in the full network (the other was between CAMPGEFII and RIM). The TNAP → P2X7 showed high EB and divergent CD characteristics,

Table 10.6 Edge betweenness centralities (EB) of TNAP in decreasing order

Edge		EB
From	To	
MAPK	*TNAP*	*12078.68*
TNAP	*PLP*	*4013.00*
TNAP	*ATP*	*3238.31*
TNAP	*P2X7*	*1267.88*
TNAP	*ADENOSINE*	*1201.00*
TNAP	*ADP*	*803.00*
TNAP	*EGFR*	*770.06*
FIVEHT2BR	TNAP	465.41
TNAP	AMP	402.00
TNAP	NIK	377.38
SLC30A7	TNAP	351.19
TNAP	D1R	349.04
TNAP	AMPAR	347.56
P2X7	TNAP	194.52
TNAP	M1R	182.17
TNAP	*GABAAR*	*67.52*
TNAP	*TAU*	*30.88*

Values within the upper and lower quartiles are printed in italic while the reciprocal pair of this interaction had smaller EB, little overlap degree and high degree of convergence (Table 10.6, Fig. 10.1d). Considering the outgoing edges of TNAP, three exhibited divergent characteristics with negative CD contrary to the majority of the outputs (Fig. 10.1d). The divergent outgoing links targeted AMPAR, M1R and TAU. The incoming edges of M1R from TNAP and TAU exhibited marked differences as that from TNAP had a low level divergence while that from TAU was moderately convergent (0.5, see also Fig. 10.1d). Overlap indices also differed: the TNAP → M1R interaction exhibited low value compared to the TAU → M1R interaction, which had a relatively high degree of overlap (0.29, see also Fig. 10.1d).

10.3.3.2 The Role of TNAP in Interactions of the Functional Groups

The different clustering techniques exhibited different sensitivities and detected 23 (Community Walktrap, abbr. CW), 19 (Markov Clustering, MCL) and 13 (Clauset–Newman–Moore, CNM) groups in the neuronal signal transduction network. Cluster size exhibited steeply decreasing distribution in all three cases (Fig. 10.2). However, group size distribution was more homogenous in case of CNM as compared to CW or MCL (Fig. 10.2). Accordingly, CNM detected only one cluster with less than 10 members. In cases of MCL and CW more than half of the clusters included less than 10 participants (11 and 13 clusters, respectively). Also, several clusters included less than four members after clustering with CW and MCL.

Cluster functions identified by functional annotation in DAVID (Database for Annotation, Visualization and Integrated Discovery, v6.7, http://david.abcc.ncifcrf.gov/; Huang da et al. 2009a, b) were used to identify groups of clusters in the three cases (Table 10.7). A more cellular pathway orientated analysis was simultaneously performed using Reactome Pathway Database (http://www.reactome.org/; Croft et al. 2014). Analyses in both databases provided similar results, which indicate the database-independent attributes of the clusters in a functional manner. The

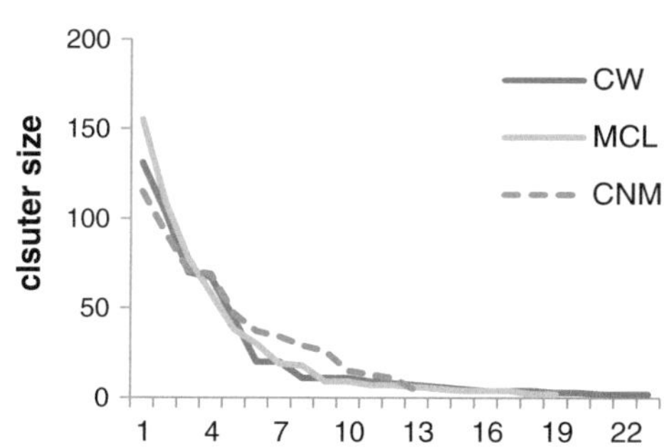

Fig. 10.2 Distribution of cluster sizes (number of molecules within the individual clusters) obtained by the three different clustering methods: community walktrap (*CW*), Markov clustering (*MCL*) and the Clauset–Newman–Moore (*CNM*)

Table 10.7 Results of the functional annotation of the clusters obtained by the Community Walktrap (CW), Markov Clustering (MCL) and the Clauset, Newman, Moore (CNM) techniques

Functional group	CW*	MCL	CNM
1. Second messenger signalling cascade	***3***, 7	***3***	***3***, 10
2. Synaptic transmission	**2**	**1**	**1**, ***7***, ***9***
3. Intracellular protein kinase cascade matrix	**1**, ***4***, ***5***	**2**, ***4***	**2**, ***4***
4. Cell proliferation/growth/survival	8, *11*, *12*, *17*, *18*	7, *9*, *13*, *14*, *15*, *17*	12, *13*
5. Regulation of transcription and translation	10, *13*, *14*, 22	8, *11*, 18	***8***
6. Apoptosis	6, 9, *15*, 19, 21	***5***, *16*, 19	***5***
7. Cytoskeleton/transport/extracellular matrix	*16*, 23	***6***, *10*, *12*	***6***, 11

Clusters are numbered according to size (i.e. the number of members) in decreasing order
Cluster size marked by the type of characters: **Bold**: ≥100, ***bold italic***: ≥50, normal underlined: ≥10, *italic*: ≥4, small normal: <4
*no functional terms were annotated to cluster 20 including 3 members

functional grouping indicated that many of the clusters had overlapping cellular functions. We identified seven major functional groups of clusters such as 1. second messenger signalling cascade, 2. synaptic transmission, 3. intracellular protein kinase cascade, 4. cell proliferation/growth/survival, 5. regulation of transcription and translation, 6. apoptosis and 7. cytoskeleton/transport/extracellular matrix (Table 10.7). Each group included at least one cluster containing more than 50 molecules in addition to several smaller ones (Table 10.7). Only two functional groups, the synaptic transmission and intracellular protein kinase cascade were formed exclusively by large clusters i.e. with more than 50 members (Table 10.7). Importantly, using each of the three clustering algorithms TNAP was grouped with molecules of the second messenger cascade functional group, which include G-protein and cAMP related signal transduction within the neuron (Table 10.7).

While TNAP formed several connections within its own cluster (8 in CW3, 7 in MCL3 and 8 in CNM3) it established connections with several other clusters (4, 6 and 4 in CW, MCL and CNM, respectively). Connections of TNAP within and between clusters are shown on Fig. 10.3 using CNM clustering as this method resulted in the least number of small fragmented clusters. Within cluster each technique found ADENOSINE, ADP, AMP, D1R, FIVEHT2BR, M1R and PLP connected to TNAP. Most of the between-cluster connections of TNAP were established with clusters belonging to synaptic transmission and intracellular protein kinase cascade (Table 10.7). CW and MCL clustering detected connections of TNAP with the functional group of cell proliferation/growth/survival. No direct connections of TNAP were observed with members of the functional groups of apoptosis and cytoskeleton/transport/extracellular matrix with the techniques used (Table 10.7). Connection with the functional group of regulation of transcription and translation was found only in one occasion via TAU by using clustering CW (Table 10.7). Using CW and CNM grouped TAU with molecules involved in intracellular protein kinase cascade. All the three clustering methods identified

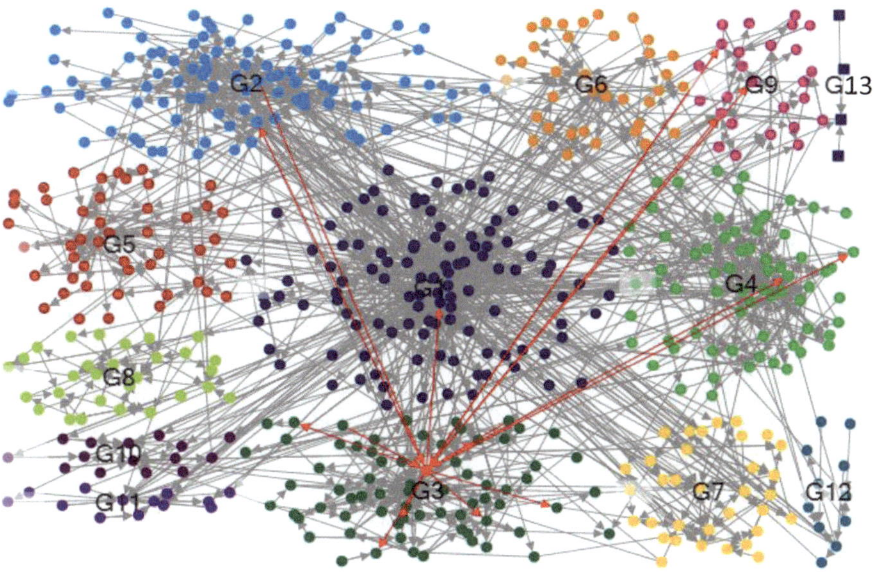

Fig. 10.3 The clustered organization of the pyramidal neuron signal transduction network as revealed by the Clauset–Newman–Moore method. Colors indicate cluster assignments of the molecules. Clusters are labelled by *G1–G13*. TNAP included in cluster *G3* and its connections are shown in *red*. *Arrows* indicate the direction of the interactions

MAPK as a member of the intracellular protein kinase cascade functional group (clusters CW4, MCL4 and CNM2) (Table 10.7). Finally, it should also be noted that while M1R was clustered together with TNAP by all three techniques, TAU was grouped together with MAPK into the intracellular protein kinase cascade group by CW and CNM while MCL grouped TAU with molecules included into the regulation of transcription and translation functional group.

10.4 Discussion

10.4.1 General Remarks

This study is a first step in the exploration of the role of TNAP in neuronal signal transduction on the basis of our current knowledge about the underlying molecular machinery as a whole. The high centrality measures, especially the node and edge betweenness suggest that TNAP is one of the key molecules, including those with centrality measures in the upper quartile of the population, of regulating neuronal functions. Although, TNAP exhibited mixed characteristics in regard to the sign of CD of the incoming and outgoing links, the combination of the dominantly convergent outputs with divergent inputs is characteristic of the control-like properties as opposed to source- and router-like functions when compared to other real world networks (Négyessy et al. 2008; Bányai et al. 2011). Similarly, TNAP appeared as an important node establishing connections not only within the cluster it participated, but with molecules of several other functional groups of the network studied. These observations are in line with the increasing number of experimental findings showing the interaction of TNAP with molecules of diverse neuronal functions such as e.g. prion proteins, TAU and ion channels (see Chaps. 9 and 12). The most important interaction of TNAP was that with MAPK. MAPK has the fourth highest NB and its link with TNAP exhibited the second highest EB values in the network. However, as far as we know, TNAP and MAPK have no direct link as MAPK regulates TNAP via BMP and SMAD (Osyczka and Leboy 2005). The finding in regard to the MAPK-TNAP interaction therefore sheds light on the importance of studying indirect interactions especially between molecules with high centrality measures in the future. All with this in mind, our assumption about the importance of studying regulatory pathways including MAPK and TNAP is supported by the present analyses and points to the high biological relevance of this question.

10.4.2 Possible Pitfalls and Weaknesses

It is important to emphasize that the computed measures were derived from a sub-graph of the whole cellular molecular network, and further inclusion of interactions and molecules would result in the rise of a different set of molecules and functions from those identified here. It is also highly probable that nodes with zero in- or out degrees and the small fragmental clusters found with all the three clustering techniques is the sign of the incompleteness of the data studied. However, studies on functionally meaningful sub-graphs can be more efficient than using big complex networks with diverse functions in understanding real world phenomena (Ma'ayan et al. 2005; Christensen et al. 2007; Ma'ayan 2008). In fact, we had to omit several interactions of TNAP, which are involved in cellular functions outside of signal transmission. However, the lack of yet known connections of TNAP, most notably incoming edges, which were relatively few in the network analysed, could be a more serious issue. Regarding the present observations, indices of network elements studied, such as betweenness centralities, could change by including more data in the future. However, considering the long tailed distribution of the different measures used here, partial drop of high values, which indicate importance in the network, would probably result in redistribution of the rank order only among the key elements. In contrast, significant increase of low values would provide substantial new information. The software based data retrieval technique could result in further inconsistencies as exemplified by MAPK and also M1R, which seems to have an indirect link from TNAP via TAU (Díaz-Hernández et al. 2010). Future studies on the network updated with novel data will be necessary to get a more complete view about the role of TNAP in neuronal signal transduction.

Finally, it is important to note that the network studied is hypothetical at least according to our present knowledge, as it may be highly specific for the CA1 hippocampal pyramidal neuron and it is unclear if TNAP plays substantial role in the functions of this particular type of neuron. Also, analyses of the hippocampal pyramidal signal transduction network does not provide information about the functioning of the GABAergic neurons since important molecules including GAD, which interact with TNAP via PLP (Erlander and Tobin 1991; Battaglioli et al. 2003), are missing from the network analysed. However, we should note that the excitatory pyramidal neurons, which are responsible for the long range connections, form roughly 80 % of the neurons as opposed to the local circuit, inhibitory GABAergic neurons consisting the remaining 20 % in the cerebral cortex. Also, several observations suggest the involvement of TNAP in excitatory neurotransmission such that TNAP activity was shown within excitatory-like synapses (Fonta et al. 2004) and also synthesis of glutamate, the major neurotransmitter of the pyramidal neurons, also depends on PLP hence TNAP (Percudani and Peracchi 2003; see also Chaps. 11 and 14).

10.4.3 The Role of TNAP in Neurotransmission

The molecular network analysed forms three major functional groups of clusters: the intracellular protein kinase cascade, the synaptic transmission group and the group with molecules involved in second messenger signalling. This forms the core of the neuronal signal transduction machinery. The other, smaller functional groups include molecules, which serve other cellular functions like proliferation, apoptosis, housekeeping functions etc.

All three clustering techniques identified TNAP as part of the group of second messenger signalling cascade, which was the third biggest cluster after that of intracellular protein kinase cascade and synaptic transmission. These findings clearly show that TNAP is an integral part of the machinery responsible for neuronal signal transduction. This observation is also supported by the connectivity of TNAP, which was not restricted within cluster but contributed to form connections between clusters mostly with that of the synaptic transmission and intracellular kinase cascade groups. Within its cluster TNAP connected molecules belonging to the major neuro-modulatory systems: the purinergic, the monoaminergic and the cholinergic. The role of TNAP in purinergic and monoaminergic transmission is now well documented. TNAP hydrolyses all forms of adenosine phosphates and therefore provides ligands for purinergic receptors. Accordingly, several studies found that TNAP modulates purinergic transmission (see Chaps. 4, 8, 13 and 18). The effect of TNAP on monoaminergic transmission is presumably exerted via PLP as aromatic L-amino acid decarboxylase (AADC), which is a key enzyme in monoamine synthesis, belongs to the family of B6-enzymes (Percudani and Peracchi 2003; Ermonval et al. 2009; see also Chap. 14). The role of TNAP in cholinergic transmission is of high interest because it is implicated in Alzheimer's disease (see Chap. 18). Interestingly, a possible interaction of TNAP → M1R has much lower EB than the TAU → M1R interaction, which later was within the upper quartile of the network. This result suggests that pathways to M1R through TAU could play a more significant role in Alzheimer's disease than those directly via TNAP. In agreement with this observation, the divergent and convergent nature of inputs of M1R from TNAP and TAU, respectively, indicate very different types of relationships (Négyessy et al. 2008; Bányai et al. 2011): in one case TNAP is the source of signal to spread via M1R in the part of the network affected by this interaction via the shortest paths. In contrast, TAU is a sink in its connections to M1R, which provides a way of funnelling signal flow to the given set of molecules in the network. These relationships suggest that TNAP is subordinate to M1R (similarly to TAU as shown in Fig. 10.1d), while in case of the relationship with TAU M1R plays the subordinate role (Négyessy et al. 2008; Bányai et al. 2011). In summary, both betweenness and convergence degrees suggest that a direct TNAP → M1R link, if it exists, would not be the primary cause of neurodegeneration in Alzheimer's disease. Interestingly, CD data also show, that in contrast to all the other connections where TNAP appears as a controller node focusing information flow within the network (via receiving divergent inputs and providing convergent outputs), in a subset of connections TNAP serves as a source where chains of interactions are originating via the convergent incoming edge from MAPK and the divergent outgoing links through AMPAR, M1R and TAU. Together with experimental evidences showing the role of AMPAR (Chang et al. 2006; Chang et al. 2012) and MAPK (Leugers et al. 2013) in Alzheimer's disease in addition to TAU and M1R, this observation suggests that TNAP could be a possible target molecule for novel drugs against neurodegenerative diseases. What is further notable in regard to the TAU → M1R relationship is the relatively large overlap between the input and target molecules of the shortest paths through this connection. It raises questions about the probable role of self-strengthening loops in the disease process. In light of the possible role of TNAP in neurodegenerative diseases it seems important that TNAP via purinergic transmission also promotes neuronal proliferation and differentiation (see Chaps. 4 and 18).

Among the inter-cluster connections of TNAP those with AMPA and GABAA receptors (R) are especially interesting when considering that these receptors form the major excitatory and inhibitory receptors in the brain, respectively (Show and Lanius 1992). It is known that TNAP can regulate the level of the neurotransmitters glutamate and GABA via the synthetizing enzymes, which depend on PLP metabolism where TNAP is a key player (Percudani and Peracchi 2003; see also Chaps. 11 and 14). However, Show and Lanius (1992) reported that TNAP treatment negatively affected the level of binding of isotope labelled antagonists to AMPAR and GABAAR proteins apparently by decreasing the number of the receptors rather than their affinity. Further studies are needed to clarify the precise role of TNAP in GABAergic and glutamatergic transmission. In addition to the interactions of TNAP with molecules directly involved in synaptic transmission, TNAP may also be involved in intracellular protein kinase cascades via EGFR (and indirectly via MAPK). While EGFR is a cell surface receptor, which in theory could be accessible to the ectoenzyme TNAP, it is unclear if TNAP and EGFR interact in vivo especially in the brain (Kim et al. 1999; Zhu et al. 2011). All these effects of TNAP on the modulatory systems are thought to be performed as an ectoenzyme.

It is important to mention that TNAP can modulate neurotransmission in other ways than discussed here. Many ion channels and transmitter receptors have residues for extracellular phosphorylation/dephosphorylation, which can be targeted by the ectoenzyme TNAP (for a review on the topic see Redegeld et al. 1999). A list of such molecules is summarized in Table 10.2. The comparison of interactions of TNAP with M1R to that of TAU resulted in higher EB, which, in accordance to the expectations, suggest that the direct effect on the receptor has a more significant result in neurotransmission. It is also notable, that in contrast to TAU, TNAP and M1R belonged to the same cluster, which, considering the high density of connections within cluster, indicates a closer functional relationship i.e. more intense interactions during signal transduction.

Taken together with the results of the present analyses the wide functional variety represented by the molecules of the list in Table 10.2 (which is most probably incomplete) suggests that TNAP is a regulatory protein affecting diverse cellular functions including neurotransmission. However, the highly specific histological and subcellular localization of TNAP as shown in the central nervous system (see Chaps. 5 and 6) can make this enzyme a highly specific regulator of neural functions. An interesting question is whether brain and cellular localizations play any role in the involvement of TNAP in different brain diseases such as epilepsy and Alzheimer's disease. Notably, epilepsy and Alzheimer's disease are linked to opposite functional states of TNAP: too low enzymatic activity or the lack of it could results in epilepsy, while a relatively high TNAP activity is assumed in Alzheimer's disease. These are two sides of the coin, which have to be considered when using drugs targeting TNAP. TNAP therefore seems to be a good candidate to study the complexity of brain functions and diseases.

Acknowledgments We are grateful to Dr M. Ermonval and Dr O. Kántor for suggestions on the manuscript.

Conflict of Interest The authors state no conflict of interest.

References

Bányai M, Négyessy L, Bazsó F (2011) Organisation of signal flow in directed networks. J Stat Mech Theory Exp P06001

Battaglioli G, Liu H, Martin DL (2003) Kinetic differences between the isoforms of glutamate decarboxylase: implications for the regulation of GABA synthesis. J Neurochem 86:879–887

Chang EH, Savage MJ, Flood DG, Thomas JM, Levy RB, Mahadomrongkul V, Shirao T, Aoki C, Huerta PT (2006) AMPA receptor downscaling at the onset of Alzheimer's disease pathology in double knockin mice. Proc Natl Acad Sci USA 103:3410–3415

Chang PK, Verbich D, McKinney RA (2012) AMPA receptors as drug targets in neurological disease–advantages, caveats, and future outlook. Eur J Neurosci 35:1908–1916

Christensen C, Thakar J, Albert R (2007) Systems-level insights into cellular regulation: inferring, analysing, and modelling intracellular networks. IET Syst Biol 1:61–77

Clauset A, Newman MEJ, Moore C (2004) Finding community structure in very large networks. Phys Rev E 70:066111

Croft D, Mundo AF, Haw R, Milacic M, Weiser J, Wu G, Caudy M, Garapati P, Gillespie M, Kamdar MR, Jassal B, Jupe S, Matthews L, May B, Palatnik S, Rothfels K, Shamovsky V, Song H, Williams M, Birney E, Hermjakob H, Stein L, D'Eustachio P (2014) The Reactome pathway knowledgebase. Nucleic Acids Res 42(Database issue):D472-D477

Csermely P, Korcsmáros T, Kiss HJ, London G, Nussinov R (2013) Structure and dynamics of molecular networks: a novel paradigm of drug discovery: a comprehensive review. Pharmacol Ther 138:333–408

Díaz-Hernández M, Gómez-Ramos A, Rubio A, Gómez-Villafuertes R, Naranjo JR, Miras-Portugal MT, Avila J (2010) Tissue-nonspecific alkaline phosphatase promotes the neurotoxicity effect of extracellular tau. J Biol Chem 285:32539–32548

Erlander MG, Tobin AJ (1991) The structural and functional heterogeneity of glutamic acid decarboxylase: a review. Neurochem Res 16:215–226

Ermonval M, Baudry A, Baychelier F, Pradines E, Pietri M, Oda K, Schneider B, Mouillet-Richard S, Launay JM, Kellermann O (2009) The cellular prion protein interacts with the tissue non-specific alkaline phosphatase in membrane microdomains of bioaminergic neuronal cells. PLoS ONE 4:e6497

Fonta C, Négyessy L, Renaud L, Barone P (2004) Areal and subcellular localization of the ubiquitous alkaline phosphatase in the primate cerebral cortex: evidence for a role in neurotransmission. Cereb Cortex 14:595–609

Fukunaka A, Kurokawa Y, Teranishi F, Sekler I, Oda K, Ackland ML, Faundez V, Hiromura M, Masuda S, Nagao M, Enomoto S, Kambe T (2011) Tissue nonspecific alkaline phosphatase is activated via a two-step mechanism by zinc transport complexes in the early secretory pathway. J Biol Chem 286:16363–16373

Huang da W, Sherman BT, Lempicki RA (2009a) Systematic and integrative analysis of large gene lists using DAVID bioinformatics resources. Nat Protoc 4:44–57

Huang da W, Sherman BT, Lempicki RA (2009b) Bioinformatics enrichment tools: paths toward the comprehensive functional analysis of large gene lists. Nucleic Acids Res 37:1–13

Kim SO, Houtman JC, Jiang J, Ruppert JM, Bertics PJ, Frank SJ (1999) Growth hormone-induced alteration in ErbB-2 phosphorylation status in 3T3-F442A fibroblasts. J Biol Chem 274:36015–36024

Leugers CJ, Koh JY, Hong W, Lee G (2013) Tau in MAPK activation. Front Neurol 4:161

Ma'ayan A (2008) Network integration and graph analysis in mammalian molecular systems biology. IET Syst Biol 2:206–221

Ma'ayan A (2009) Insights into the organization of biochemical regulatory networks using graph theory analyses. J Biol Chem 284:5451–5455

Ma'ayan A, Jenkins SL, Neves S, Hasseldine A, Grace E, Dubin-Thaler B, Eungdamrong NJ, Weng G, Ram PT, Rice JJ, Kershenbaum A, Stolovitzky GA, Blitzer RD, Iyengar R (2005) Formation of regulatory patterns during signal propagation in a Mammalian cellular network. Science 309:1078–1083

McCall KA, Huang C, Fierke CA (2000) Function and mechanism of zinc metalloenzymes. J Nutr 130:1437S–1446S

Négyessy L, Nepusz T, Zalányi L, Bazsó F (2008) Convergence and divergence are mostly reciprocated properties of the connections in the network of cortical areas. Proc Biol Sci 275:2403–2410

Newman MEJ (2001) The structure and function of complex networks. SIAM Review 45:167–256

Osyczka AM, Leboy PS (2005) Bone morphogenetic protein regulation of early osteoblast genes in human marrow stromal cells is mediated by extracellular signal-regulated kinase and phosphatidylinositol 3-kinase signaling. Endocrinology 146:3428–3437

Percudani R, Peracchi A (2003) A genomic overview of pyridoxal-phosphate-dependent enzymes. EMBO Rep 4:850–854

Redegeld FA, Caldwell CC, Sitkovsky MV (1999) Ecto-protein kinases: ecto-domain phosphorylation as a novel target for pharmacological manipulation? Trends Pharmacol Sci 20:453–459
Shaw CA, Lanius RA (1992) Reversible kinase and phosphatase regulation of brain amino acid receptors in postnatal development. Brain Res Dev Brain Res 70:153–161
Strogatz SH (2001) Exploring complex networks. Nature 410:268–276
Surtees R, Mills P, Clayton P (2006) Inborn errors affecting B6 metabolism. Future Neurol 1:615–620
Ulitsky I, Shamir R (2007) Identification of functional modules using network topology and high-throughput data. BMC Syst Biol 1:8
Van Dongen S (2008) Graph clustering via a discrete uncoupling process. SIAM J Matrix Anal Appl 30:121–141
Watts DJ (2004) The "new" science of networks. Annu Rev Sociol 30:243–270
Watts DJ, Strogatz SH (1998) Collective dynamics of 'small-world' networks. Nature 393:440–442
Wong RW, Guillaud L (2004) The role of epidermal growth factor and its receptors in mammalian CNS. Cytokine Growth Factor Rev 15:147–156
Zhu J, Shimizu E, Zhang X, Partridge NC, Qin L (2011) EGFR signaling suppresses osteoblast differentiation and inhibits expression of master osteoblastic transcription factors Runx2 and Osterix. J Cell Biochem 112:1749–1760

Chapter 11
Vitamin B-6 Metabolism and Interactions with TNAP

Stephen P. Coburn

Abstract Two observations stimulated the interest in vitamin B-6 and alkaline phosphatase in brain: the marked increase in plasma pyridoxal phosphate and the occurrence of pyridoxine responsive seizures in hypophosphatasia. The increase in plasma pyridoxal phosphate indicates the importance of tissue non-specific alkaline phosphatase (TNAP) in transferring vitamin B-6 into the tissues. Vitamin B-6 is involved in the biosynthesis of most of the neurotransmitters. Decreased gamma-aminobutyrate (GABA) appears to be most directly related to the development of seizures in vitamin B-6 deficiency. Cytosolic pyridoxal phosphatase/chronophin may interact with vitamin B-6 metabolism and neuronal development and function. Ethanolaminephosphate phospholyase interacts with phosphoethanolamine metabolism. Extracellular pyridoxal phosphate may interact with purinoceptors and calcium channels. In conclusion, TNAP clearly influences extracellular and intracellular metabolism of vitamin B-6 in brain, particularly during developmental stages. While effects on GABA metabolism appear to be the major contributor to seizures, multiple other intra- and extra-cellular metabolic systems may be affected directly and/or indirectly by altered vitamin B-6 hydrolysis and uptake resulting from variations in alkaline phosphatase activity.

Keywords Vitamin B-6 · Seizures · Hypophosphatasia · Gamma-aminobutyrate · Phosphoethanolamine

S.P. Coburn (✉)
Department of Chemistry, Indiana University–Purdue University,
2101 E. Coliseum Blvd, Fort Wayne, IN 46805, USA
e-mail: Coburn@ipfw.edu

C. Fonta and L. Négyessy (eds.), *Neuronal Tissue-Nonspecific Alkaline Phosphatase (TNAP)*, Subcellular Biochemistry 76,
DOI 10.1007/978-94-017-7197-9_11

11.1 Introduction to Vitamin B-6 Compounds and Metabolism

Studies of the interaction between vitamin B-6 and alkaline phosphatase have been reported for many years. One of the reasons for exploring vitamin B-6 metabolism in the current volume was the occurrence of pyridoxine responsive seizures in hypophosphatasia, a rare genetic disorder which inactivates tissue non-specific alkaline phosphatase (TNAP) (See Chaps. 1, 2, 3 and 14). Several reports suggest that Bethenod et al. (1967) reported pyridoxine responsive seizures in hypophosphatasia in 1967. This appears to be an error. Bethenod et al. did administer pyridoxine but it was not effective in relieving the seizures. Since then there have been multiple reports in which vitamin B-6 supplementation was beneficial (Sia et al. 1975; Litmanovitz et al. 2002; Nunes et al. 2002; Yamamoto et al. 2004; Arun et al. 2005; Baumgartner-Sigl et al. 2007; Balasubramaniam et al. 2010; Demirbilek et al. 2012; Belachew et al. 2013). Vitamin B-6 supplementation also corrected seizures in mice lacking alkaline phosphatase (Waymire et al. 1995; Narisawa et al. 2001). Finally, pyridoxal phosphate concentrations in plasma are markedly elevated in hypophosphatasia (Whyte et al. 1985) demonstrating an important physiological role for alkaline phosphatase in vitamin B-6 metabolism.

Vitamin B-6 is the name for a group of compounds with similar biological activity. The major forms in humans are pyridoxine, pyridoxal, pyridoxamine, their corresponding 5′-phosphate derivatives and the metabolic end product, 4-pyridoxic acid (Table 11.1; Fig. 11.1). Additional metabolites have been detected in humans receiving megadoses of vitamin B-6 (Mahuren et al. 1991) as well as in cats (Coburn and Mahuren 1987) and dogs (Ericson et al. 2008). The interconversion of the various forms combined with the wide variety of reactions in which vitamin B-6 is utilized makes detailed study challenging. Pyridoxal phosphate is a cofactor for over 150 enzymes representing every major class of enzymes except ligases (http://enzyme.expasy.org). This accounts for about 4 % of known enzyme activity (Percudani and Peracchi 2003). These enzymes catalyze reactions including transamination, decarboxylation, racemization, aldol cleavage, and beta and gamma eliminations and replacement reactions leading Jansonius (1998) to conclude that pyridoxal phosphate is the most versatile organic catalyst known. It has been suggested that the most ancient reaction of aerobic metabolism was the synthesis of pyridoxal phosphate or pyridoxal (Kim et al. 2012). Much information about the structure, sequence and function of vitamin B-6 enzymes is contained in the vitamin B-6 database (http://bioinformatics.unipr.it/cgi-bin/bioinformatics/B6db/home.pl) (Percudani and Peracchi 2009). Although reactions involving amino acid metabolism seem to attract the most attention, 70–80 % of the vitamin B-6 in mammals is located in skeletal muscle (Coburn et al. 1988a) associated with glycogen phosphorylase where the phosphate group rather than the amine or aldehyde group is the catalytic center (Palm et al. 1990). Also, while many studies of vitamin B-6 focus on pyridoxal phosphate, in most tissues other than muscle the amount of pyridoxal phosphate is equaled or exceeded by the amount of pyridoxamine phosphate

Table 11.1 Vitamin B-6 compounds and major enzymes associated with the interconversion of B-6 vitamers in animals

Vitamin B-6 metabolites	Abbreviation	CAS Registry No.
Pyridoxine	PN	65-23-6
Pyridoxal	PL	66-72-8
Pyridoxamine	PM	85-87-0
Pyridoxine 5′-phosphate	PNP	447-05-2
Pyridoxal 5′-phosphate	PLP	54-47-7
Pyridoxamine 5′-phosphate	PMP	529-96-4
4-pyridoxic acid	PA	82-82-6
Enzymes	Enzyme commision No.	CAS registry No.
Pyridoxal kinase (1)[a]	2.7.1.35	9025-42-0
Pyridoxal phosphatase (2)	3.1.3.74	9076-92-0
Alkaline phosphatase (2)	3.1.3.1	9001-78-9
Pyridoxal 5′-phosphate synthase (3)	1.4.3.5	9055-72-5
Multiple aminotransferases (4)	2.6.1	–
Aldehyde oxidase (5)	1.2.3.1	9029-07-6
Aldehyde dehydrogenase (NAD^+) (6)	1.2.1.3	9028-86-8
Aldehyde dehydrogenase ($NADP^+$) (6)	1.2.1.4	9028-87-9
Aldehyde dehydrogenase ($NAD(P)^+$) (6)	1.2.1.5	9028-88-0

[a]Number in parentheses indicates the associated reaction in Fig. 11.1. In some cases the enzyme involved may vary depending on the species, tissue and cellular locations. Enzymes which have been identified only in microorganisms are not listed

(Coburn et al. 1988b). There is little information on the rate of interchange of these forms. Unless enzymatic activity is inhibited during homogenization, pyridoxamine phosphate in liver is quickly converted to pyridoxal phosphate (Coburn et al. 1988b). That suggests that the two compounds may be sequestered separately. It also raises questions about the accuracy of observed vitamer distribution in some studies. The commonly accepted pathway of vitamin B-6 metabolism in humans (Fig. 11.1) involves phosphorylation of the 5′-position by pyridoxal kinase (EC 3.7. 1.35). The kinase can phosphorylate all three vitamers. Pyridoxine phosphate and pyridoxamine phosphate may be oxidized to pyridoxal phosphate. The oxidizing enzyme is usually named pyridoxine (pyridoxamine) phosphate oxidase. However, the name recommended by the Enzyme Commission is pyridoxal 5′-phosphate synthase (EC 1.4.3.5). The synthase can oxidize both pyridoxine phosphate and pyridoxamine phosphate. It is inhibited by its substrate, pyridoxine phosphate (Ki = 60 μM) (Kwon et al. 1991) and product, pyridoxal phosphate (Ki = 23 μM) (Kwok and Churchich 1980). Pyridoxal 5′-phosphate synthase activity varies considerably between tissues and between species (Fonda 1988). Mouse and human erythrocytes have appreciably higher activity than rat, hamster and rabbit erythrocytes. This means that the role of erythrocytes in vitamin B-6 metabolism will vary significantly between species. Pyridoxamine phosphate and pyridoxal phosphate can also be interconverted by aminotransferases. The phosphate group can be

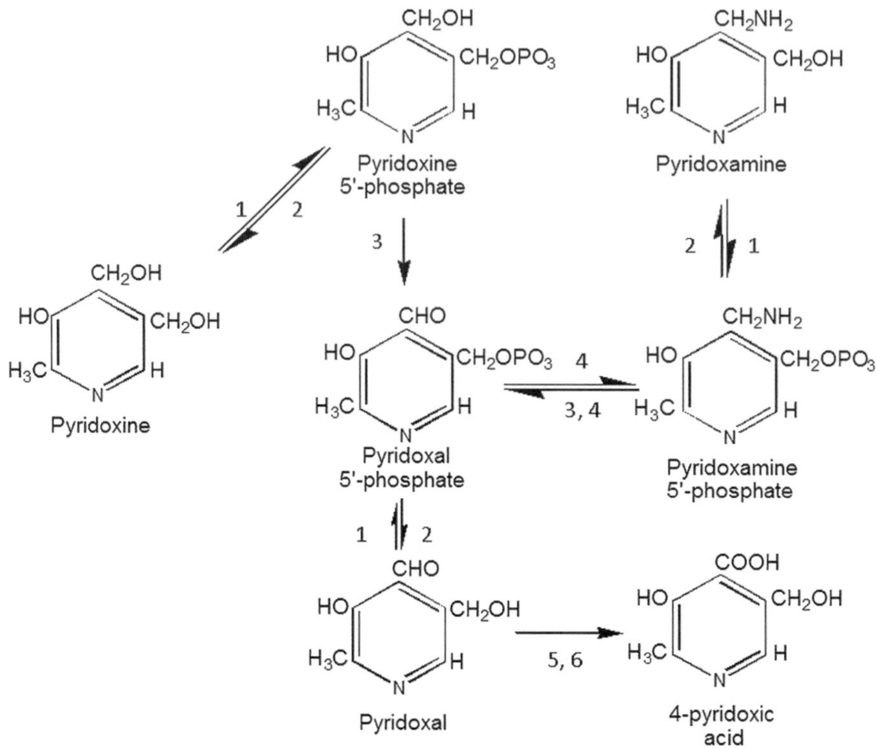

Fig. 11.1 Major pathways of vitamin B-6 metabolism. The enzymes associated with the reactions are listed in Table 11.1

removed by non-specific phosphatases or by pyridoxal phosphatase (EC 3.1.3.74). Pyridoxal can be oxidized to 4-pyridoxic acid by aldehyde oxidase (EC 1.2.3.1) which has been detected in a variety of tissues (Moriwaki et al. 2001). Stanulovic et al. (1976) found that rats lacking aldehyde oxidase were able to produce pyridoxic acid using a NAD dependent aldehyde dehydrogenase. The identity of this dehydrogenase remains unclear. Although Stanulovic et al. (1976) suggested that it is probably the nonspecific aldehyde dehydrogenase EC 1.2.1.3, the Enzyme Nomenclature database (http://www.chem.qmul.ac.uk/iubmb/enzyme/index.html) lists three nonspecific aldehyde dehydrogenases (EC 1.2.1.3 using NAD, EC 1.2.1. 4 using NADP and EC 1.2.1.5 using both) plus pyridoxal dehydrogenase (EC 1.1.1. 107) described in bacteria (Burg and Snell 1969). Although Stanulovic et al. (1976) used NAD, Bode et al. (1991) detected pyridoxal dehydrogenase activity in rats using NADP which they assigned to EC 1.2.1.4. Chemical Abstracts assigned the Stanulovic rat enzyme the same CAS number as the bacterial enzyme. However, there is no obvious justification for that. Pyridoxal dehydrogenase activity was not detected in human liver using NAD or NADP (Merrill et al. 1984). Therefore, it seems unlikely to occur in human brain.

11.2 Absorption and Transport of Vitamin B-6

Intestinal absorption of unphosphorylated vitamin B-6 appears to involve diffusion and/or facilitated diffusion (Mooney et al. 2009). While it is frequently assumed that the ionic nature of pyridoxal phosphate prevents it from crossing cell membranes without prior hydrolysis, perfusion studies in rats demonstrated that pyridoxal phosphate can be transferred intact from the lumen of the intestine into the circulation (Mehansho et al. 1979). Although liver is a major site of vitamin B-6 metabolism, small doses of pyridoxine and pyridoxamine can be converted to pyridoxal in the intestine prior to release into the portal circulation in the mouse (Sakurai et al. 1987, 1988). Similar results were reported with cultured Caco-2 cells (Albersen et al. 2013). The main transport forms of vitamin B-6 in blood are pyridoxal and pyridoxal phosphate, which is released from the liver associated primarily with albumin (Lumeng et al. 1974). Hydrolysis by TNAP is important in removal of pyridoxal phosphate from plasma as indicated by a marked increase in plasma concentrations of pyridoxal phosphate when TNAP activity is limited in hypophosphatasia (Whyte et al. 1985). Significant amounts of pyridoxine and pyridoxamine do not occur in peripheral plasma with normal vitamin B-6 intake (Albersen et al. 2014). Therefore, intravenous administration of pyridoxine produces atypical metabolism. For example, skeletal muscle in mice and rats contains pyridoxal kinase but little, if any, pyridoxal 5′-phosphate synthase (Fonda 1988). Therefore, pyridoxine can be taken up by muscle and phosphorylated but cannot be converted to pyridoxal phosphate. It must be hydrolyzed and released back into the circulation. This is not necessarily harmful but it does not reflect the normal kinetic processes followed by pyridoxal and pyridoxal phosphate.

Early attempts to isolate phosphatases produced variable results. Saraswathi and Bachhawat (1963) isolated from human brain an alkaline phosphatase for which pyridoxal phosphate was the best substrate with a Km of 0.430 mM at pH 9. Magnesium was required for optimal activity. Inorganic phosphate was a competitive inhibitor with Ki of 1.65 mM. Brunel et al. (1969) and Zisapel and Levi (1980) independently isolated from bovine brain alkaline phosphatases which had little activity toward pyridoxal phosphate. Brun-Heath et al. (2011) recently demonstrated that the alkaline phosphatase expressed in human neuronal and endothelial cells corresponds to the *ALPL* gene driven by the bone promoter. The interactions between alkaline phosphatase, pyridoxal phosphate, pH and inorganic phosphate are complex. Examination of the membrane bound alkaline phosphatase of rat liver demonstrated that at pH 7.4 the Km for pyridoxal phosphate was about 2 μM, which is approximately the concentration of free pyridoxal phosphate in liver (Lumeng and Li 1975). The enzyme was very sensitive to substrate inhibition by pyridoxal phosphate but not by p-nitrophenyl phosphate. Inorganic phosphate is a well-recognized inhibitor of alkaline phosphatase. Under physiological conditions alkaline phosphatase activity toward pyridoxine phosphate was reduced about 50 % by normal plasma phosphate concentrations and will increase or decrease significantly in response to changes in phosphate concentrations within the ranges observed clinically (Coburn et al. 1998). The measurements utilized pyridoxine phosphate rather than pyridoxal phosphate in order to minimize substrate inhibition and non-specific protein binding. The article includes a nomogram for predicting pyridoxine phosphate phosphatase activity in undiluted serum from routine clinical measurements of inorganic phosphate and alkaline phosphatase. Seargeant and Stinson (1979) estimated that inorganic phosphate concentrations in liver would be high enough to inhibit human liver alkaline phosphatase if the phosphate was accessible to the enzyme.

Spector and Johanson (2007) noted that there are five general processes that must be understood to explain mechanistically vitamin homeostasis in brain:

(i) Transport from plasma in and out of the extracellular space of brain and cerebrospinal fluid;
(ii) Uptake and release of transported vitamers by brain cells;
(iii) Vitamer activation by kinases or other enzymes, i.e. incorporation into cofactors and transfer into appropriate compartments in the brain cells;
(iv) When appropriate, reformation of the transported vitamer for release from the brain cell and then the central nervous system; and
(v) Finally, mechanisms in kidney and gut including saturable uptake, excretion, and secretion (kidney) mechanisms that tend to keep plasma vitamin levels constant, thus contributing to brain homeostasis.

Uptake by the choroid plexus primarily involves diffusion or facilitated diffusion of non-phosphorylated vitamers followed by trapping by pyridoxal kinase (Spector and Greenwald 1978). Dual label experiments with [^{3}H, ^{32}P]pyridoxine phosphate indicated that only 10–15 % of the absorbed vitamer was absorbed without hydrolysis. Since TNAP is expressed at high levels in the choroid plexus (Nishihara et al. 1994), reduced activity in hypophosphatasia could significantly impair vitamin B-6 uptake. During a 30 min period 56 % of accumulated [^{3}H] was released from isolated choroid plexus and 46 % from brain slices (Spector 1978b). However, 76 % of the label released by choroid plexus was phosphorylated compared to 1 % from brain slices. Therefore, choroid plexus accounts for essentially all of the phosphorylated B-6 vitamers in the cerebrospinal fluid (CSF). The rabbit choroid plexus can also absorb unphosphorylated vitamers from the CSF. Since the non-phosphorylated vitamers are removed from the CSF faster than pyridoxal phosphate, Spector and Johanson (2007) suggested that the phosphorylated form serves as a reservoir. [^{3}H]pyridoxic acid or [^{3}H]pyridoxic acid lactone accounted for less than 10 % of the label in plasma or CSF from in vivo experiments (Spector 1978a) and less than 5 % of the label released from brain and choroid plexus slices from in vitro experiments (Spector 1978b). Spector and Johanson (2007) gave the following data from rabbits: total vitamin B-6 in plasma is 0.30 μmol/L 55 % non-phosphorylated; 0.39 μmol/L in cisternal CSF 26 % nonphosphorylated; 8.9 μmol/kg in brain 23 % nonphosphorylated; brain turnover 17 %/d; CSF turnover 24 %/3 h; permeability-surface area product <10 compared to 0.2 for mannitol (absorbed by simple diffusion) and 158 for leucine (absorbed by facilitated diffusion); Kt ~ 1–2 μmol/L (half saturation concentration for transport through the blood brain barrier). Spector and Johanson (2007) concluded that since the Kt is similar to plasma concentration the carrier mediated process plays a crucial transport and regulatory role because simple diffusion is relatively unimportant.

In human cerebrospinal fluid pyridoxal is the dominant B-6 vitamer (Albersen et al. 2012; van der Ham et al. 2012; Albersen et al. 2014). Median values for pyridoxal and pyridoxal phosphate in the cerebrospinal fluid of infants declined over 50 % between a postmenstrual age of 30–37 and ≥42 weeks (Albersen et al. 2012). Two studies which measured only pyridoxal phosphate found that the decline continued for at least 12 months after birth (Ormazabal et al. 2008; Footitt et al. 2011). In adult subjects the median values for pyridoxal and pyridoxal phosphate were 10.5 and 55.9 nmol/L in plasma and 30.0 and 16.1 nmol/L in cerebrospinal fluid, respectively (Albersen et al. 2014). They reported a correlation of 0.629 between pyridoxal phosphate in cerebrospinal fluid and plasma. This might be interpreted to indicate direct transport of pyridoxal phosphate from plasma to cerebrospinal fluid. However, an alternative explanation could be that hydrolysis of pyridoxal phosphate in plasma is the primary source of the pyridoxal used to synthesize the pyridoxal phosphate released into the cerebrospinal fluid by the choroid plexus.

11.3 Vitamin B-6 Transporters

Evidence that vitamin B-6 uptake may involve facilitated diffusion implies the existence of transporters. A vitamin B-6 transporter has been identified in S. cervisiae (Stolz and Vielreicher 2003). The Tpn1p protein is a member of the purine-cytosine permease family within the major facilitator superfamily and functions as a proton symporter in the plasma membrane. In S. pombe the pyridoxine transporter was a different proton symporter encoded by the *car1*$^+$ (changed amyloride resistance) gene which Stolz et al. (2005) proposed renaming *bsu1*$^+$ for vitamin B-6 uptake. Carrier mediated pyridoxine transport sensitive to amiloride has been reported in Caco-2 cells (Said et al. 2003; Stolz et al. 2005). However, Stolz et al. (2005) stated that no mammalian homologue of Bsu1p has been identified in database searches and Said (2011) acknowledges that nothing is known about the molecular structure of the intestinal pyridoxine transporter of any mammalian species. It has recently been proposed that the transporter could be used to enhance transfection of cancer cells by attaching a ligand to pyridoxal phosphate (Pandey et al. 2013). The hypothesis is that the pyridoxal phosphate will bind to the transporter but the large size will prevent transport resulting in endocytosis of the entire complex. This protocol did express luciferase in mouse brain suggesting that a vitamin B-6 transporter does act in brain.

11.4 Intracellular Pyridoxal Phosphatase (EC 3.1.3.74)

TNAP is a membrane bound ectoenzyme which has a major role in controlling extracellular pyridoxal phosphate concentrations. In rat neurons pyridoxal phosphatase activity at pH 5.0 was cytochemically located in a specialized region of smooth endoplasmic reticulum that is located at the trans face of the stack of Golgi elements (GERL) (Spater et al. 1978). These locations differed from the usual sites of both acid and alkaline phosphatase activity. In 1992 Fonda (1992) observed that human erythrocytes contained alkaline phosphatase activity associated with the stroma and a soluble acid phosphatase which hydrolyzed vitamin B-6 phosphates. The optimal pH was 6.5. At pH 7.4 it hydrolyzed vitamin B-6 phosphates but not nucleotide phosphates, phosphoamino acids or other phosphorylated compounds. At pH 7.4 the Km was 1.5 μM. Activity was inhibited by inorganic phosphate (Ki = 0.8 mM), fluoride and weakly by pyridoxal. It was not inhibited by levamisole, L-phenylalanine or L-tartrate. Based on tryptic fragments obtained by Gao and Fonda (1994), Jang et al. (2003) were able to identify the pyridoxal phosphatase in a cDNA library from human brain and express it in *E. coli*. The mouse enzyme was also cloned. BLAST searches identified similar proteins in organisms ranging from bacteria to mammals. The enzyme is a dimer requiring magnesium for optimal activity. The Km values for the human enzyme were 2.5, 43.4 and 80.6 for pyridoxal phosphate, pyridoxine phosphate and pyridoxamine phosphate, respectively. There was little activity with p-nitrophenylphosphate. Phosphatase mRNA was detected in essentially all tissues tested but was highest in adult and fetal brain. Liver and testis were higher than tissues other than brain. Expression in spinal cord was much lower than in brain. The structure shows no evidence of integral membrane domains which is consistent with the erythrocyte enzyme being found in the soluble fraction. Based on these data in 2004 the Enzyme Commission created EC 3.1.3.74, pyridoxal phosphatase, which has the systematic name, pyridoxal 5′-phosphate phosphohydrolase.

Kang et al. (2004) examined the relative mRNA expression of pyridoxal 5′-phosphate synthase (PNPO), pyridoxal kinase (PDXK) and pyridoxal phosphatase (PDXP) in a wide variety of human tissues which are partially listed in Table 11.2. Maximal expression of the synthase, kinase and phosphatase was in the liver, testis and cerebral cortex, respectively. Expression of pyridoxal phosphatase mRNA in most brain regions was greater than 80 % of the maximum while expression of oxidase and kinase mRNA was below 35 % of the maximum. Boe et al. (2004) identified autoantigens to pyridoxal phosphatase in patients with paraneoplastic neurologic disorders. They confirmed that the pyridoxal phosphatase gene, PDXP, was most strongly expressed in various regions of the brain. Expression was also significant in liver, kidney and testis.

Gohla et al. (2005) found that pyridoxal phosphatase also activates cofilin, a key regulator of cytoskeletal dynamics. They named the pyridoxal phosphatase protein, chronophin. Lack of this enzyme causes major cell division defects. Northern blot analysis revealed high expression in brain, heart, skeletal muscle, liver and kidney.

Table 11.2 Relative mRNA expression of pyridoxine phosphate oxidase (PNPO), pyridoxal kinase (PDXK) and pyridoxal phosphatase (PDXP) in human tissues (Kang et al. 2004)

Tissue	PNPO	PDXK	PDXP
Whole brain	23.8	29.0	83.6
Cerebral cortex	25.9	27.0	100.0
Frontal lobe	12.4	14.3	81.8
Parietal lobe	13.6	32.2	82.4
Occipital lobe	14.7	25.3	89.1
Temporal lobe	12.6	24.4	84.6
Paracentral gyrus of cerebral cortex	9.5	18.5	73.2
Pons	6.2	10.3	47.9
Cerebellum, left	13.3	21.9	90.5
Cerebellum, right	25.0	31.1	91.1
Corpus callosum	18.4	19.8	37.7
Amygdala	13.0	22.6	94.2
Caudate nucleus	19.2	28.4	74.2
Hippocampus	12.3	24.6	81.8
Medulla oblongata	5.0	15.8	41.7
Putamen	2.0	17.9	54.2
Accumbens nucleus	5.2	19.4	70.8
Thalamus	10.8	23.1	85.2
Spinal cord	1.6	3.7	10.8
Heart	3.7	9.1	18.7
Kidney	85.6	58.0	31.4
Skeletal muscle	34.1	15.4	20.7
Testis	8.4	100.0	49.3
Liver	100.0	56.0	61.8

Highest expressing tissue for each enzyme was set to 100

Kim et al. (2008) found that pyridoxal phosphatase/chronophin was localized primarily in postsynaptic compartments and not in presynaptic compartments in rat hippocampus. They concluded that the interaction of pyridoxal phosphatase/chronophin with actin may have an important role in seizure activity. Hwang et al. (2007) examined the effect of ischemia produced by 5 min occlusion of the carotid arteries on vitamin B-6 metabolizing enzymes in the gerbil hippocampus. Ischemia had no significant effect on the amount of pyridoxal 5′-phosphate synthase in the CA2/3 regions. In the CA1 region relative optical density of the pyridoxal 5′-phosphate synthase spot vs sham was highest 12 h after ischemia, declined slightly until 2 days and returned to baseline by 4 days. Relative optical density of the pyridoxal phosphatase spot declined about 50 % at 3 h and peaked at about twice the sham spot after 2 days. Administration of pyridoxal phosphate 30 min prior to the ischemic event minimized the loss of pyramidal cells 4 days later. Hwang et al. suggested that this protective effect may have resulted from increased GABA. Koh (2010) examined the response of rats to middle cerebral artery occlusion for 24 h. The pyridoxal phosphatase spot from the cerebral cortex was significantly reduced at 18 % of the sham group. Koh did not cite the earlier results from Hwang et al. The difference in the response of pyridoxal phosphatase could reflect the difference in duration of ischemia and/or the portion of the brain examined. Furukawa et al. (2010) found that pyridoxal phosphatase activity was increased in the limbic forebrain of 3 month old senescence-accelerated mouse prone 10 (SAMP10) and suggested that morphological impairments in cytoplasmic processes of microglia may be related to the altered expression of pyridoxal phosphatase since this enzyme is known to be involved in brain cytoskeleton formation and associated with acute and chronic neurodegenerative conditions.

Pyridoxal phosphatase/chronophin dephosphorylates pyridoxal phosphate and actin-depolymerizing factor giving it a role in regulating both vitamin B-6 and F-actin (Kim et al. 2009). Kim et al. (2009) concluded that pyridoxal phosphatase/chronophin mediated actin dynamics may play an important role in the changes of morphological properties (dendritic spine reorganization) of the hippocampus in long-term potentiation. (Kim et al. 2008; Kwak et al. 2009).

11.5 Regulation of Vitamin B-6 Metabolism

Determining how vitamin B-6 metabolism is regulated is challenging. While it is frequently assumed that water soluble vitamins are readily lost from the body, vitamin B-6 seems to be efficiently conserved by most tissues. Pyridoxal kinase complexes with aspartate aminotransferase resulting in channeling newly formed pyridoxal phosphate to the aminotransferase (Kim et al. 1988). During vitamin B-6 deficiency in rats the large pool of pyridoxal phosphate sequestered in muscle glycogen phosphorylase was not released until anorexia and weight loss occurred (Black et al. 1978). When young adult men were placed on a controlled diet providing about 0.5 mg vitamin B-6/d, within 2 weeks their urinary excretion had dropped to approximate their intake with a loss of only 4 % of their body pool (Coburn et al. 1991). While plasma content dropped markedly, muscle content showed little change after 6 weeks. If the body conserves vitamin B-6, we hypothesized that the vitamin B-6 requirement for growth could be determined by multiplying the daily gain by the average body content of vitamin B-6 (15 nmol/g) (Coburn 1994). This approach provided reasonable agreement with experimentally determined requirements for species ranging from mammals to birds and fish. The total body pool in humans is about 1,000 μmol (Coburn et al. 1988a). We proposed that the turnover is determined primarily by vitamin B-6 intake. As noted below at very low intakes brain is adversely affected. If existing tissue efficiently conserves vitamin B-6, the fact that most studies of vitamin B-6 metabolism involve growing rats raises the possibility that the observed reductions in tissue vitamers are magnified by the addition of new tissue and do not necessarily represent changes that would occur in an animal maintaining a constant weight.

Rats receiving vitamin B-6 intakes ranging from 2 to 200 times the daily requirement for 20 days showed no significant difference in the pyridoxal phosphate content of liver or brain indicating that maximal pyridoxal phosphate content is well regulated (Li et al. 1974). Since pyridoxal 5′-phosphate synthase from liver (Wada and Snell 1961; Merrill et al. 1984) or brain (Kwok and Churchich 1980) is very sensitive to product inhibition by pyridoxal phosphate, the synthase was proposed as a regulatory factor (Snell and Haskell 1971; Merrill et al. 1978). Therefore, Li et al. (1974) tested the effect of pyridoxal phosphate on pyridoxal phosphate production by rat liver cytosol in vitro. They concluded there was no effect because the total produced was similar with and without added pyridoxal phosphate. However, Merrill et al. (1978) noted that the data of Li et al. did show a slightly decreased rate of production with added pyridoxal phosphate indicating that there was inhibition. Li et al. observed that adding 80 mM sodium phosphate to inhibit phosphatase activity resulted in a marked increase in net pyridoxal phosphate production in hepatocytes or plasma membranes but not cytosol. Therefore, Li et al. concluded that under normal circumstances pyridoxal phosphate which is not protected by protein binding is rapidly hydrolyzed thus maintaining a steady state. Since the phosphatase activity was in the membrane fraction but not the cytosol, it presumably represented alkaline phosphatase rather than pyridoxal phosphatase. We now know that alkaline phosphatase is an ectoenzyme. One explanation of the results of Li et al. might be that the pyridoxal phosphate produced by the hepatocytes was released into the medium before being hydrolyzed. The description of the sample processing does not indicate that the hepatocytes were separated from the medium before pyridoxal phosphate analysis. Liver routinely releases pyridoxal phosphate into the plasma (Lumeng et al. 1974) but brain may not (Spector 1978b). Perhaps in brain pyridoxal phosphatase controls the intracellular pyridoxal phosphate concentration.

Ohkawa et al. (1994) used immunochemistry to examine the distribution of pyridoxal kinase in rabbit brain with the following results: "The cytoplasm of neuronal cells and neuroglial cells in the cerebral cortex, hippocampal region, brain nuclei and cerebellar cortex showed positive staining with various degrees of intensity. The neuronal cells and surrounding fibers in some brain nuclei, such as the area tegmentalis ventralis or the substantia nigra, showed intense staining. The neuronal cells of the hippocampal region showed somewhat weak reactivity, but some with intense reactivity were found sparsely distributed and positive staining fiber networks of a very low density were also observed." Pyridoxal kinase activity in brain changes with dietary intake. Spector and Shikuma (1978) suggested that a report of a negative correlation between pyridoxal phosphate and pyridoxal kinase in rabbit brain (Ebadi et al. 1970) was due to errors caused by using pyridoxal rather than pyridoxine as the substrate for analyzing kinase activity. When various areas of normal rat brain were compared, there was a positive correlation between pyridoxal kinase activity and pyridoxal phosphate concentration (Ebadi and Bifano 1978). An early report suggested that dietary deficiency of vitamin B-6 in rats tended to increase kinase activity in brain slightly (McCormick et al. 1961). Later work by Meisler and Thanassi (1980) indicated that vitamin B-6 deficiency reduced both pyridoxal phosphate and pyridoxal kinase activity in rat brain. Pyridoxal 5′-phosphate synthase and phosphatase activities did not show any consistent changes. The animals weighed about 100 g at the start of the experiment. By the 4th week brain kinase activity in both the control and deficient groups increased and the group differences were no longer statistically significant. Restoring normal vitamin B-6 intake quickly restored kinase activity to control values in the liver. As noted above kinase activity in brain had already increased and supplementation had little additional effect. Rasmussen et al. (1979) also found that pyridoxal synthase activity was not affected by vitamin B-6 deficiency. Without citing any specific references Meisler and Thanassi (1980) state that studies reported by others have shown that pyridoxal 5′-phosphate synthase is involved in regulating pyridoxal phosphate in B-6 sufficient animals and that their current results indicated pyridoxal kinase is a regulatory factor in vitamin B-6 deficiency. Those suggestions have recently been cited by Kwak et al. (2009). It is important to note that product inhibition of pyridoxal 5′-phosphate synthase will occur only when the substrate is pyridoxine phosphate or pyridoxamine phosphate. Pyridoxal 5′-phosphate synthase is not required for the production of pyridoxal phosphate from pyridoxal. In unsupplemented humans there is little or no pyridoxine or pyridoxamine in the peripheral plasma or cerebrospinal fluid (Albersen et al. 2014). Therefore, most of the pyridoxal phosphate in brain arises from pyridoxal which would limit the role of pyridoxal 5′-phosphate synthase as a regulator. In addition we would suggest that since pyridoxal kinase activity responds rapidly to changes in vitamin B-6 intake, vitamin B-6 availability rather than pyridoxal kinase is the regulator.

The vitamin B-6 analog, 4′-deoxypyridoxine, can act as either a competitive inhibitor or a competing substrate depending on the reaction in question. Unfortunately, the distinction is frequently ignored in the literature. Ebadi et al. (1970) noted that pyridoxal kinase activity was increased in the presence of deoxypyridoxine and that the increase was prevented by actinomycin D or puromycin. While Ebadi et al. (1970) interpreted the results as an inverse relationship between pyridoxal phosphate and kinase activity, we suspect that since deoxypyridoxine is a substrate for the kinase, the presence of deoxypyridoxine may have stimulated the increased synthesis of pyridoxal kinase. Pyridoxal kinase from rat brain is also inhibited by dopa, dopamine, norepinephrine, tyramine, serotonin, histamine, and GABA (Ebadi and Govitrapong 1979). Kerry et al. (1986) attributed an increase in kinase activity during isolation from sheep brain to removal of interfering substances such as amines. Based on the above discussion we suggest that vitamin B-6 intake, protein binding, phosphatase activity and biogenic amines may be the prime regulators of intracellular pyridoxal phosphate concentration in brain. Post mortem examination of over 100 human brains detected a 1000-fold range of pyridoxal kinase activity (Snell 1964). Snell noted that pyridoxal kinase is quite stable in frozen samples but does not provide any documentation that all of the brains were subject to the same post mortem protocol prior to freezing the samples.

11.6 Vitamin B-6 and Seizures

The earliest symptoms of vitamin B-6 deficiency are skin disorders, seizures and anemia. Brain does not conserve vitamin B-6 as well as muscle during severe dietary vitamin B-6 deficiency. Early studies of vitamin B-6 deficiency reported seizures in dogs (Fouts et al. 1938), pigs (Chick et al. 1938) and rats (Chick et al. 1940). In humans seizures have been observed as a result of dietary vitamin B-6 deficiency (Coursin 1955) and genetic disorders of vitamin B-6 metabolism (Footitt et al. 2013; Dakshinamurti and Dakshinamurti 2014). Electroencephalographic changes were noted within 2 weeks in women fed a diet containing essentially no vitamin B-6 (Kretsch et al. 1991). The changes were corrected with an intake of 0.5 mg vitamin B-6/d. Gregory et al. (2013) recently fed human subjects 0.37 mg vitamin B-6/d for 28 days. with no reported clinical effects. They did not report electroencephalographic data. In a milder restriction in which rats were fed 60 % as much as controls for up to 40 weeks there were no significant differences in the pyridoxal phosphate or pyridoxamine phosphate content of brain (Wei 1999).

Sakurai et al. (1980b) examined the interactions between vitamin B-6 and penicillamine, thiosemicarbazide and semicarbazide in mouse brain. Doses of 2000 μmol penicillamine/kg, 165 μmol thiosemicarbazide or 1790 μmol semicarbazide/kg were lethal. Simultaneous subcutaneous administration of 250 μmol pyridoxine, pyridoxal or pyridoxamine/kg prevented both seizures and death from penicillamine or thiosemicarbazide but did not protect against semicarbazide. All three drugs reduced brain pyridoxal phosphate 20–50 % from the control value of about 10 nmol/g. This decrease following penicillamine or thiosemicarbazide was prevented by administration of 250 μmol pyridoxine, pyridoxal or pyridoxamine/kg. The authors do not speculate about why the B-6 vitamers did not prevent the effects of semicarbazide. Intramuscular administration of pyridoxine or pyridoxamine alone did not increase pyridoxal phosphate above the baseline value of about 10 nmol/g. Administration of pyridoxal caused a slight decrease in pyridoxal phosphate during the peak pyridoxal concentration, possibly indicating substrate inhibition. Pyridoxamine phosphate remained about 14 nmol/g in all treatments suggesting that pyridoxamine phosphate is not in rapid equilibrium with pyridoxal phosphate and cannot compensate for a rapid decrease in pyridoxal phosphate. Peak brain content of pyridoxal was about 27, 8 and 2 nmol/g following intramuscular administration of 250 μmol/kg of pyridoxal, pyridoxine and pyridoxamine, respectively. Peak content of pyridoxamine was less than 2 nmol/g with each vitamer. Pyridoxamine was the only vitamer which increased pyridoxamine phosphate. Even though pyridoxamine alone produced very little change from the baseline vitamin B-6 content of the brain, it was just as effective as pyridoxine and pyridoxal in preventing the decrease in pyridoxal phosphate produced by penicillamine and thiosemicarbazide. It seems likely although there was not a large increase in pyridoxal, there was enough to correct the deficit produced by the drugs. Mizuno et al. (1980) found that the vitamin B-6 analogs which were the most effective competitive substrates for pyridoxal kinase also were most effective in producing seizures in mice thus demonstrating the importance of pyridoxal phosphate in preventing seizures.

As noted earlier a role for TNAP was suggested by the occurrence of seizures in neonatal hypophosphatasia. This was confirmed in knockout mice lacking TNAP which died from seizures (Waymire et al. 1995). Pyridoxal phosphate in cerebrospinal fluid was above the normal range in three infants with hypophosphatasia (Baumgartner-Sigl et al. 2007; Hofmann et al. 2013; Belachew et al. 2013) supporting the hypothesis that low alkaline phosphatase activity restricted uptake of vitamin B-6 by the brain. Waymire et al. (1995) hypothesized that the seizures resulted from decreased concentration of the inhibitory neurotransmitter, gamma-aminobutyrate (GABA). GABA is produced from glutamic acid by glutamate decarboxylase (EC 4.1.1.15), which requires pyridoxal phosphate. After being released into the synaptic cleft GABA is taken up by adjoining astrocytes where it is converted to succinate semialdehyde and glutamate by 4-aminobutyrate-2-oxoglutarate transaminase (EC 2.6.1.19) (Baxter 2003), which also requires pyridoxal phosphate. The glutamate is then converted to glutamine by glutamate-ammonia ligase (EC 6.3.1.2) which is located primarily in the astrocytes (Suarez et al. 2002). The glutamine may return to the neurons for reuse.

Two forms of glutamic acid decarboxylase (EC 4.1.1.15) with complex regulation occur in GABA-ergic neurons (Martin and Rimvall 1993). GAD67 is cytosolic and distributed throughout the cell body. GAD65 is concentrated in the nerve terminal and is reversibly anchored to the membrane of synaptic vesicles. GAD65 in brain accounts for the large fraction of the enzyme which occurs in brain as the apoenzyme thus serving as a reserve supply. GAD67 is more highly saturated with cofactor. ATP stimulates the formation of apoenzyme while inorganic phosphate facilitates formation of the holoenzyme. Elimination of GAD67 results in lethal developmental defects including cleft palate and respiratory failure (Kash et al. 1997). Mice lacking GAD65 appear normal at birth with normal GABA and holo-GAD activity. However, the increase in activity associated with added pyridoxal phosphate is reduced indicating a decrease in total enzyme protein. The mice have increased susceptibility to seizures. Genetic studies in mice led Kash et al. (1997) to conclude that the relative contribution of GABA synthesized by GAD65 is genetically determined and that the resulting seizure activity involves the limbic region. GABA may also have a role in the synaptic integration of newly generated neurons in adult brain (Ge et al. 2006). Sharma and Dakshinamurti (1992) presented a thorough discussion of the electroencephalographic changes associated with vitamin B-6 deficiency. Seizures induced by picrotoxin, pentylenetetrazol or domoic acid in control or vitamin B-6 deficient rats could be prevented by local administration of pyridoxine or GABA in the thalamic VPL region or the hippocampal CA3 region indicating that GABA deficiency was a significant factor in seizure development.

At 10–14 days postnatal brain GABA concentrations were reduced 50 % in TNAP knockout mice compared to normal controls (Waymire et al. 1995). Similarly, in an infant with hypophosphatasia and West syndrome GABA was less than 0.005 nmol/mL in the cerebrospinal fluid (Yamamoto et al. 2004). GABA concentrations were also low in the cerebrospinal fluid of infants with pyridoxine dependent seizures (Kurlemann et al. 1992) and West syndrome (Kurlemann et al. 1997) prior to treatment with pyridoxine. Initial attempts to rescue the mice with pyridoxine had limited success presumably because the primary metabolic pathway involved phosphorylation followed by oxidation to pyridoxal phosphate. With oral administration this occurs primarily in the liver which releases pyridoxal phosphate into the plasma. Hydrolysis of the resulting plasma pyridoxal phosphate would be limited by the lack of alkaline phosphatase thus reducing the availability of pyridoxal for transfer into the brain. Plasma pyridoxal was reduced 10-fold in the knockout mice. Use of pyridoxal in place of pyridoxine was much more effective since it could be transported directly into the brain. In liver, kidney, heart and muscle of the knockout mice pyridoxal phosphate was 50–60 % of control values. In brain it was only 35 % of control indicating that brain was affected more than other tissues. Pyridoxal phosphate concentrations in the brains of two infants with hypophosphatasia were slightly lower than in a single normal infant (Whyte et al. 1988). Litmanovitz et al. (2002) suggested that the normal concentrations in the infants might reflect higher residual alkaline phosphatase activity. The ratio of pyridoxal phosphate in plasma/cerebrospinal fluid was about 2.5 in normal subjects ranging in age from 0 to 68 years (Shin et al. 1984). Patients with vitamin B-6 related seizures tended to have lower cerebrospinal fluid values and higher plasma values yielding a plasma/cerebrospinal fluid ratio of 21. It is interesting to note that an infant with hypophosphatasia and pyridoxine responsive seizures had a ratio of 36 (Baumgartner-Sigl et al. 2007) even though the concentration of pyridoxal phosphate in the spinal fluid was 57 nmol/L compared with their cited normal range of 7–39 nmol/L (Shin et al. 1984). This suggests that the ratio may be a more reliable marker of potential seizures than the cerebrospinal fluid value alone. Baumgartner-Sigl et al. (2007) noted that all reported cases of hypophosphatasia with neonatal seizures died by 18 months of age. Di Mauro et al. (2002) found that various TNAP mutations differed in their ability to hydrolyze pyridoxal phosphate. This could help to explain the variable occurrence of seizures in hypophosphatasia.

While vitamin B-6 deficiency can cause seizures, excesses are also neurotoxic. Pyridoxal phosphate causes seizures in immature mice by blocking GABA receptors (Ishioka et al. 1995). The authors suggested the immature blood brain barrier allowed passage of pyridoxal phosphate. Such leakage does not appear to occur in hypophosphatasia. High doses of pyridoxine cause neuropathy in rats (Callizot et al. 2001), dogs (Chung et al. 2008) and humans (Gdynia et al. 2008). There is one case report of pyridoxine responsive seizures in an infant whose mother consumed 80 mg pyridoxine/d during her pregnancy (South 1999b). However, the causal relationship was questioned (Baxter and Aicardi 1999; South 1999a). It is interesting that while pyridoxamine has been proposed as a treatment for the side effects of diabetes produced by advanced glycation end products (Chen and Francis 2012), pyridoxine intoxication has been recommended as a model for diabetic sensory neuropathy (Callizot et al. 2001).

11.7 Other Aspects of Vitamin B-6 Metabolism in Brain

While reduced synthesis of GABA resulting from decreased intracellular availability of pyridoxal phosphate appears to be a major factor in the seizures observed in neonatal hypophosphatasia, it is appropriate to be aware of several additional aspects of vitamin B-6 metabolism in brain. Two good summaries of earlier work are available (Ebadi and Costa 1972; Ebadi 1981). Ebadi (1981) noted that vitamin B-6 is involved in the synthesis or metabolism of multiple neurotransmitters in addition to GABA including dopamine, norepinephrine, serotonin, tyramine, tryptamine, taurine, histamine and even acetyl choline indirectly. He also noted that vitamin B-6 is involved in the synthesis of growth hormone, follicle-stimulating hormone, luteinizing hormone in the brain as well as aldosterone, glucagon, cortisol, estradiol, testosterone and epinephrine in other tissues. The following discussion will focus on information that is more recent or particularly relevant to the focus of this book.

11.7.1 Additional Regulatory Mechanisms

In the commentary following the 1981 review (Ebadi 1981) Metzler (1981) discusses the complexities of vitamin B-6 chemistry and potential interactions. Pyridoxamine phosphate is usually bound to enzymes less tightly than pyridoxal phosphate. Glutamate decarboxylase in brain is rapidly deactivated in the presence of glutamate (Miller et al. 1978; Seligmann et al. 1978). Metzler suggests that this could be the result of a side reaction involving decarboxylative transamination which would convert pyridoxal phosphate to pyridoxamine phosphate followed by dissociation of the cofactor from the enzyme. Reactivation of the enzyme would depend on the availability of pyridoxal phosphate which in turn would depend at least in part on oxidation of pyridoxamine phosphate back to pyridoxal phosphate. Metzler also notes that the apoenzymes are often degraded more rapidly than the holoenzyme. Therefore, the interactions between pyridoxal phosphate and pyridoxamine phosphate could have some complex regulatory functions. ATP and ADP at concentrations comparable to intracellular values inhibit the association of pyridoxal phosphate with glutamate dehydrogenase raising the possibility that decarboxylase activity might be influenced by energy metabolism (Seligmann et al. 1978). Glucose deprivation of cultured hippocampal cells causes reduced ATP, increased extracellular glutamate with NMDA receptor activation and release of lactic dehydrogenase indicating cell death (Geng et al. 1997). Pyridoxal phosphate prevented these changes by increasing pyruvate presumably through increased alanine aminotransferase activity. Pyridoxal phosphate also inhibits the citrate transporter thus keeping citrate in the mitochondria where it can be used to generate ATP. Maintaining adequate ATP prevented the other consequences of glucose deprivation. Based on these observations Geng et al. (1997) suggested that pyridoxal phosphate treatment might help to minimize neuronal death in some metabolic disorders.

11.7.2 Developmental Effects

There is a variety of evidence suggesting that limiting the vitamin B-6 supply in the brain can have effects in addition to reduced production of GABA. In studies in knockout mice lacking alkaline phosphatase Narisawa et al. (2001) found that pyridoxal treatment improved the appearance of lumbar nerve roots. Creation of a dietary vitamin B-6 deficiency in normal mice also caused seizures and thinning of nerve roots. Vitamin B-6 deficiency reduces synthesis of sphingolipids and interferes with formation of myelin (Stephens and Dakshinamurti 1975) raising the possibility that limited vitamin B-6 in neonatal hypophosphatasia could have developmental effects in addition to causing seizures. Narisawa et al. (2001) cited unpublished results showing that TNAP is produced in the neonatal mouse intestine until it is replaced by the intestinal isoenzyme around post-natal day 14–15 suggesting that TNAP knockout mice might have reduced absorption of pyridoxal phosphate from milk. We could not locate any data on the pyridoxal phosphate content of mouse milk. There is an inverse relationship between alkaline phosphatase activity and the pyridoxal phosphate content of milk in rats, goats, pigs, and cows (Coburn et al. 1992) indicating that these species release significant amounts of pyridoxal phosphate into the milk. However, human milk was low in both alkaline phosphatase activity and pyridoxal phosphate suggesting that very little pyridoxal phosphate is released into the milk.

Analysis of cerebrospinal fluid in infants with hypophosphatasia detected reduced 5-hydroxyindoleacetic acid and homovanillic acid in two out of four cases (Balasubramaniam et al. 2010; Hofmann et al. 2013; Belachew et al. 2013). 3-O-methyldopa and 5-hydroxytryptophan were elevated in the only cases where they were measured (Baumgartner-Sigl et al. 2007; Balasubramaniam et al. 2010). These data suggest that multiple vitamin B-6 pathways in the brain are probably altered in hypophosphatasia.

There has been some interesting work on the role of vitamin B-6 during development in rats. When female rats were fed a diet containing 0.7 mg pyridoxine hydrochloride/kg during gestation and lactation, the pups exhibited seizures from day 12 to day 18 after which the seizures stop (Guilarte 1989). The same diet failed to produce seizures in adult rats. At day 14 the total vitamin B-6 content of most brain areas in low intake animals was about 40 % of controls. However, by day 56 most areas had increased to 60 % of controls with substantia nigra and cerebellum exceeding 80 % of control values even though the rats remained on the low intake diet. GABA was significantly below control values in all brain regions at both 14 and 28 days even though there were no seizures at 28 days. By 56 days only the pons/medulla region was significantly below control GABA levels. These data suggest that decreased GABA concentrations may be a necessary but not sufficient factor in causing seizures. Glutamate was significantly reduced in the cortex, hippocampus, substantia nigra and pons/medulla at 14 days but not at 28 days. Glycine was significantly elevated in the hippocampus and caudate/putamen areas at 14 days but not at 28 days. Taurine was significantly reduced in the cortex, caudate/putamen, and substantia nigra at 14 days but not at 28 days. Interactions between these and other metabolites may interact with GABA to generate seizures.

When dams and pups were fed a diet containing 0.6 mg pyridoxine hydrochloride/kg, pyridoxal phosphate content of various brain regions at 7 days postnatal was 66–90 % of control values (Groziak et al. 1984). By day 15 the values declined to 15–33 % of controls after which they gradually increased. In control animals pyridoxal phosphate concentration approximately doubled during day 7–15. Brain weight increased by about 80 % in both groups. Therefore, total brain pyridoxal phosphate in the controls increased almost 4 times from day 7 to day 15. However, total pyridoxal phosphate in the 0.6 group remained essentially constant. This suggests that inability to supply adequate vitamin B-6 to new tissue resulted in a decline in vitamin B-6 content which may have contributed to the seizures Guilarte (1989) observed from 12 to 18 days. Pyridoxal kinase activity in rat brain increases about three times from birth to day 15 (Ebadi 1972). This increase may have allowed the tissues to accumulate sufficient vitamin B-6 to eliminate the seizures after 18 days even with limited vitamin B-6 intake. Four additional reports (Loo 1972; Bayoumi and Smith 1976; Bhagavan et al. 1977; Guilarte 1989) also found increasing pyridoxal phosphate in postnatal rat brain making it difficult to reconcile the data from Ebadi (1972) showing a marked decrease over the same period.

Pyridoxal kinase activity in human brain increases about 20-fold from 6 to 11 weeks of gestation (Bukin and Ivanova 1976). Activity in the medulla oblongata was about twice whole brain activity by 8 weeks and that difference was maintained into the second trimester. Activity in the medulla changed little from the second to third trimester while other brain areas increased to match or surpass it. By the second trimester activity in the hippocampus was also significantly higher than in the frontal lobe. By the third trimester activity in most brain areas had increased to eliminate these differences. However, activity in the cerebellar cortex was significantly below the frontal lobe. In bovine brain the cerebellar medulla rather than the cortex had low kinase activity (Table 11.3).

Although some reports (Bayoumi et al. 1972; Ruth and Morgan 1986) did not observe seizures in vitamin B-6 deficient adult rats, Sharma and Dakshinamurti

Table 11.3 Pyridoxal kinase content of bovine brain (McCormick et al. 1961)

Tissue	Activity (nmol/h·g)
Cerebrum	
Hemisphere	1620
Diencephalon	1184
Mesencephalon	1008
Cerebellum	
Cortex	1368
Medulla	235
Brain stem	
Pons	436
Medulla oblongata	364

(1992) detected electroencephalographic changes indicating that seizures would likely have developed if the deficiencies were extended. Based on these observations the current author suggests that the decreased susceptibility to seizures in adult rats is influenced by the slower rate of growth and the severity of the deficiency.

11.7.3 Pyridoxamine Phosphate

As noted earlier pyridoxamine phosphate is a significant component of most tissues other than muscle. Perhaps the most extreme example is in the neocortex of 30 days old rat pups in which pyridoxamine phosphate was 4–5 times higher than pyridoxal phosphate (Groziak and Kirksey 1987). In mouse brain most of the pyridoxamine phosphate was in the mitochondria while pyridoxal phosphate was found primarily in the cytosol as was most of the pyridoxal kinase activity (Rott et al. 1960). Pyridoxal phosphatase activity was mainly in the mitochondria. Lyon et al. (1962) found that 75 % of the pyridoxamine phosphate in mouse brain was associated with the mitochondria. When I-strain mice, which are susceptible to audiogenic seizures, received a vitamin B-6 deficient diet, pyridoxal phosphate in the brain declined but pyridoxamine phosphate remained constant. Lyon et al. concluded that pyridoxamine phosphate was not serving as a storage form since it was not released during vitamin B-6 deficiency. Similarly, after growing, pair-fed rats received a diet completely lacking vitamin B-6 for 2 weeks total brain B-6 declined about 26 % (Sampson and O'Connor 1989). However, pyridoxal phosphate declined about 46 % while pyridoxamine phosphate declined only 13 %. In the brains of TNAP deficient knockout mice pyridoxal phosphate was decreased to 35 % of control values while pyridoxamine phosphate was 65 % of controls (Waymire et al. 1995). While pyridoxamine phosphate may be more resistant to depletion, when rats were exposed to a low vitamin B-6 diet throughout gestation and lactation, all B-6 vitamers were significantly reduced in the neocortex at 30 days postnatal (Groziak and Kirksey 1987). Since tracer studies in mice indicated very limited production of pyridoxamine, Colombini and McCoy (1970) concluded that pyridoxamine phosphate was produced primarily from pyridoxal phosphate.

In Guinea pig cortex the major B-6 vitamers were pyridoxamine phosphate (7.5 nmol/g), pyridoxal phosphate (4.75 nmol/g) and pyridoxal (1.4 nmol/g) (Loo and Mack 1971). The subcellular distribution for pyridoxamine phosphate was nucleus (18.1 %), crude mitochondria (63.4 %) and supernatant (18.5 %) while pyridoxal phosphate was nucleus (12.4 %), mitochondria 39.1 % and supernatant 48.5 %. Fractionation of the crude mitochondria indicated that both pyridoxamine phosphate and pyridoxal phosphate were divided about equally between the synaptosomes and the extra terminal mitochondria. In the synaptosomal organelles pyridoxamine phosphate (2.83 nmol/g) was detected primarily in the cytoplasm (19.6 %), presynaptic membranes (29.3 %) and the mitochondria (47.3 %) while pyridoxal phosphate (2.40 nmol/g) was distributed in the cytoplasm (41.6 %), presynaptic membranes (12.0 %) and mitochondria (41.6 %). Rat tissue gave

Table 11.4 Subcellular distribution of vitamin B-6 in mouse brain (Sakurai et al. 1980a)

Fraction	Protein (mg/g brain)	Vitamer (pmol/mg protein)					
		PL	PLP	PM	PMP	PN	PNP
Whole brain	122	5.26	80.7	0.638	60.6	0.091	0.436
Nucleus	43.2	3.87	60.9	0.484	41.2	0.079	0.371
Crude mitochondria	49.4	5.67	104.7	0.619	62.9	0.096	0.413
Microsomes	7.5	6.73	30.7	2.39	27.0	0.194	0.634
Cytosol	22.6	23.68	128.3	3.17	42.2	0.251	0.866
Fractionation of crude mitochondria							
Myelin	3.77	2.96	37.2	5.77	48.3	0.199	1.93
Synaptosomes	24.12	4.82	51.4	1.53	51.5	0.097	0.325
Extraterminal mitochondria	11.36	6.06	129.6	3.82	129.8	0.173	0.774
Fractionation of synaptosomes							
Synaptoplasma	2.89	48.2	172	4.61	147	0.966	1.13
Synaptic vesicles	1.21	9.77	53	14.7	67.7	1.43	1.97
Intraterminal mitochondria	5.35	3.65	105	2.72	75.6	0.241	0.382

Abbreviations for vitamers are listed in Table 11.1

similar results. Loo et al. (1971) suggested that the occurrence of pyridoxamine phosphate and pyridoxal phosphate in the membranes might indicate a carrier function for amino and keto acids. Sakurai et al. (1980a) provided a detailed examination of the subcellular distribution of vitamin B-6 in mouse brain (Table 11.4). Using microbiological analysis pyridoxal phosphate was the dominant vitamer in whole brain. In the crude mitochondrial fraction pyridoxamine phosphate was highest in the extraterminal mitochondria while in the synaptosomal fraction it was higher in the synaptoplasma than in the intraterminal mitochondria. Although pyridoxine phosphate concentrations were no more than 5 % of pyridoxal phosphate or pyridoxamine phosphate, the fact that pyridoxine phosphate concentrations were highest in the myelin and synaptic vesicle fractions led Sakurai et al. to suggest that pyridoxine phosphate may have a role in nervous system conduction.

11.7.4 Phosphoethanolamine Metabolism

One of the consequences of hypophosphatasia is increased phosphoethanolamine in serum and/or urine. Phosphoethanolamine excretion can be influenced by vitamin B-6 as well as alkaline phosphatase. Ethanolamine-phosphate phospho-lyase (EC 4.2.3.2) utilizes pyridoxal phosphate to convert phosphoethanolamine to acetaldehyde, ammonia and inorganic phosphate (Fleshood and Pitot 1970). Enzyme isolated from rabbit liver had an optimal pH of 7.8, Km of 0.61 mM for phosphoethanolamine, Km of 0.27 μM for pyridoxal phosphate and Ki of 1.3 mM for inorganic phosphate. Rats fed a high protein, vitamin B-6 deficient diet developed osteoporotic changes and markedly increased excretion of phosphoethanolamine with no changes in serum alkaline phosphatase activity (Benke et al. 1972). Recently, AGXT2L1, a pyridoxal phosphate brain protein related to alanine glyoxylate aminotransferase, has been shown to be an ethanolamine-phosphate phospho-lyase (Veiga-da-Cunha et al. 2012; Schiroli et al. 2013). *AGXT2L1* was the most consistently upregulated gene in the brains of deceased patients with schizophrenia or bipolar disorder (Shao and Vawter 2008) and was the most upregulated gene in mouse brain after treatment with lithium (McQuillin et al. 2007). This is an example of an enzyme from a generalist family [class II aminotransferases (Mehta and Christen 2000)] which has become highly specific. Nam et al. (2012) noted that specialist enzymes are frequently essential, maintain higher metabolic flux, and require more regulation to control metabolic flux in dynamic environments. Schiroli et al. (2013) suggested that the need for strict regulation in the case of AGXT2L1 might result from the fact that the reaction is irreversible and generates potentially toxic products. They noted that phosphoethanolamine is a product of sphingolipid degradation and a substrate for glycerophospholipid biosynthesis. Ethanolamine and phosphoethanolamine can interfere with mitochondrial function (Modica-Napolitano and Renshaw 2004). The fact that AGXT2L1 requires pyidoxal phosphate raises the possibility that the increased phosphoethanolamine observed in hypophosphatasia could be exacerbated by low vitamin B-6 in addition to the lack of alkaline phosphatase particularly in infants where seizures suggest the existence of vitamin B-6 deficiency. We are unaware of any data on AGXT2L1 activity in hypophosphatasia. Although alterations in phosphoethanolamine metabolism have been associated with neuropsychiatric disorders (Miller et al. 2012) and most of the previous references to AGXT2L1 deal with such disorders, there is little evidence that neuropsychiatric problems are increased in hypophosphatasia. Since signs of vitamin B-6 deficiency in hypophosphatasia seem to be limited to infants, the duration of any reductions in AGXT2L1 might be too short to cause neuropsychiatric problems in surviving infants. The Uniprot data base (http://www.uniprot.org/uniprot/?query=ethanolamine-phosphate+phospho-lyase&sort=score) contains 5 reviewed entries for AGXT2L1 from human, mouse, bovine, zebrafish and African clawed frog all indicating the subcellular location as mitochondrion. Sia et al. (1975) reported that pyridoxine supplementation reduced phosphoethanolamine excretion in the parents of an infant with neonatal hypophosphatasia by 25–69 %. Administration of pyridoxine to the infant reduced the seizures but did not reduce phosphoethanolamine excretion. Prior to pyridoxine therapy neither the infant nor the parents had detectable degradation of phosphoethanolamine in serum. In vitro degradation of phosphoethanolamine was maximally stimulated by addition of 10^{-5} M pyridoxal phosphate. Control serum required 10^{-3} M to obtain a comparable effect. Addition of pyridoxal phosphate had no effect on the degradation of p-nitrophenol-phosphate. Based on these results Sia et al. (1975) recommended further study of the benefits of pyridoxine supplementation in hypophosphatasia.

11.7.5 Inflammation

It has been suggested that inflammation increases the demand for vitamin B-6 from plasma for increased metabolism of tryptophan to anthranilic acid and 3-hydroxyanthranilic acid (Paul et al. 2013; Reginaldo et al. 2013). A study of inflammation in hypophosphatasia showed that calcium pyrophosphate but not pyridoxal phosphate increased prostaglandin E2 synthesis in cultured skin and dura mater cells (Girschick et al. 2006). The authors concluded that high pyridoxal phosphate in hypophosphatasia did not cause inflammation. However, their data show that prostaglandin production by hypophosphatasia tissues in the presence of calcium pyrophosphate tended to be slightly lower when pyridoxal phosphate was administered with calcium pyrophosphate. This raises the possibility that the pyridoxal phosphate may have reduced rather than enhanced the response. Since the hypophosphatasia tissues presumably have limited ability to hydrolyze pyridoxal phosphate, it would be interesting to repeat the studies using pyridoxal in place of pyridoxal phosphate to determine whether increased availability of pyridoxal would reduce the inflammatory response.

11.7.6 Long Term Potentiation

Vitamin B-6 appears to be associated with long-term potentiation which in turn is associated with learning and memory. Immunoreactivity of pyridoxal 5′-phosphate synthase but not pyridoxal kinase was increased in the dentate gyrus of rats after induction of long-term potentiation (Kwak et al. 2009). Injection of pyridoxine, pyridoxal or pyridoxamine had no effect on development of long term potentiation in rat brain. Injecting pyridoxal phosphate produced multiple population spikes which were attributed to disruption of $GABA_A$ receptor mediated inhibition. Transduction of Tat-pyridoxal kinase did not affect induction of long-term potentiation. However, transduction of Tat-pyridoxal 5′-phosphate synthase increased the efficiency of long-term potentiation induction. The authors concluded that because Tat-pyridoxal 5′-phosphate synthase injection did not affect fast or late paired-pulse inhibitions, the upregulation of pyridoxal 5′-phosphate synthase may not be related to GABAergic inhibition. They did not report any measurements of pyridoxal phosphate concentrations. Serine racemase (EC 5.1.1.18) provides D-serine for the N-methyl-D-aspartate receptor which mediates neurotransmission, synaptic plasticity, cell migration and long term potentiation. Altered D-serine concentrations are associated with neuropathologies, aging deficits, and disorders including schizophrenia, amyotrophic lateral sclerosis, Alzheimer's disease and stroke (Campanini et al. 2013). The involvement of pyridoxal phosphatase/chronophin was mentioned earlier. It is interesting that alkaline phosphatase activity has also been associated with activity dependent functions in the cortex (Fonta et al. 2004).

11.7.7 Homocysteine

Vitamin B-6 is also related to homocysteine metabolism. A genome-wide assessment of the relationship between vitamin B-6, folate, vitamin B-12 and homocysteine detected a significant relationship between the C allele of the *ALPL* gene and reduced vitamin B-6 (Tanaka et al. 2009). Recent results showed that supplementation with folate, vitamin B-6 and vitamin B-12 reduced cerebral atrophy in gray matter in elderly subjects with increased dementia risk and elevated homocysteine (Douaud et al. 2013).

11.7.8 Lysosomal Enzymes

Vitamin B-6 deficiency increased lysosomal enzymes in rat brain (DiPaolo et al. 1975). Faber et al. (1991) suggested that vitamin B-6 might be involved in the synthesis of mannose phosphate marker or receptor proteins on lysosomes which dictate the disposition of lysosomal enzymes.

11.8 Pyridoxal Phosphate and Purinoceptors

The interaction between alkaline phosphatase and purinoceptors is discussed elsewhere in this book (See Chaps. 4 and 13). Pyridoxalphosphate-6-azophenyl-2′,4′-disulphonic acid is widely used as a purinoceptor antagonist raising the possibility that there might be some physiological interactions between purinoceptors and pyridoxal phosphate. In discussing metabolic errors in vitamin B-6 metabolism Surtees (2006) asked whether pyridoxal phosphate might have a noncofactor role, possibly interacting with ion channels or pumps. Trezise et al. (1994) concluded that pyridoxal phosphate acts as a specific inhibitor of P2 purinoceptors in the vagus nerve and vas deferens of rats. Pyridoxal phosphate had an effect at concentrations equal to those of α, β-methylATP. Wang et al. (1999) found that pyridoxal phosphate was an ATP receptor antagonist in rat cardiomyocytes. Binding of 10 μM ATP-S to low affinity binding sites was significantly reduced by 1 μM pyridoxal phosphate. Wang et al. (1999) concluded that since normal plasma concentrations of pyridoxal phosphate range from 24 to 81 nM, pyridoxal phosphate probably would not function as an inhibitor under physiological conditions. However, they also noted that the higher ATP concentrations that were needed in the intact heart and cardiomyocytes than in the sarcolemmal membranes might result from ATP hydrolysis thus suggesting that the effective concentration of ATP might be lower than the administered amount. $P2X_2$ and $P2X_{2/3}$ receptors expressed in HEK293 cells were inhibited by pyridoxal phosphate with an IC_{50} of 7 and 13 μM, respectively (Theriault et al. 2014). In a phase 2 clinical trial pyridoxal phosphate (250 mg or 750 mg/d) was administered orally beginning 3 to 10 h prior to coronary bypass graft surgery and continuing for 30 days (Tardif et al. 2007). The 250 mg treatment did not significantly affect the prespecified primary end point but was associated with a significant reduction in perioperative myocardial infarction with creatine kinase-myocardial band ≥100 ng/mL. The 250 mg dose was more effective than 750 mg. An editorial discussing those results suggested that improved efficacy may require targeting multiple receptors or perhaps a central control point in the relevant metabolic pathway (Verrier 2007). Pyridoxal phosphate may also interact with calcium channels (Dakshinamurti et al. 1998). Microinjection of pyridoxal phosphate causes seizures which were attributed to extracellular pyridoxal phosphate interacting with $GABA_A$ receptor (Salazar and Tapia 2001).

11.9 Hyperphosphatasia

Considering the marked elevation of plasma pyridoxal phosphate in hypophosphatasia, one might expect markedly reduced concentrations in hyperphosphatasia resulting from a defect in the biosynthesis of the phosphatidylinositol glycan anchor which attaches the enzyme to the membrane (Thompson et al. 2012) (see also Chap. 16). However, that does not appear to be the case. Thompson et al. (2010) reported only one out five patients with low pyridoxal phosphate and pyridoxine responsive seizures. That patient had the lowest relative phosphatase activity of any of the patients. Hydrolysis of pyridoxal phosphate was increased in plasma of subjects with increased alkaline phosphatase activity due to liver or bone disorders (Anderson et al. 1980). There was a negative correlation between endogenous plasma pyridoxal phosphate and alkaline phosphatase activity ($r = -0.893$). Increased hydrolysis of plasma pyridoxal phosphate is not necessarily a problem unless it increases the net loss of vitamin B-6 from the body. Additional data will be needed to clarify vitamin B-6 metabolism in hyperphosphatasia.

Acknowledgements The author appreciates the comments and suggestions provided by Caroline Fonta, Joshua Miller and Michael Whyte.

Conflict of Interest The author has no conflict of interest.

References

Albersen M, Bosma M, Knoers NVVA et al. (2013) The intestine plays a substantial role in human vitamin B-6 metabolism: a Caco-2 cell model. Plos One 8:e54113

Albersen M, Bosma M, Luykx JJ et al (2014) Vitamin B-6 vitamers in human plasma and cerebrospinal fluid. Am J Clin Nutr 100:587–592

Albersen M, Groenendaal F, van der Ham M et al (2012) Vitamin B-6 vitamer concentrations in cerebrospinal fluid differ between preterm and term newborn infants. Pediatrics 130:E191–E198

Anderson BB, O'Brien H, Griffin GE et al (1980) Hydrolysis of pyridoxal-5′-phosphate in plasma in conditions with raised alkaline phosphatase. Gut 21:192–194
Arun R, Khazim R, Webb JK et al (2005) Scoliosis in association with infantile hypophosphatasia: a case study in two siblings. Spine 30:E471–E476
Balasubramaniam S, Bowling F, Carpenter K et al (2010) Perinatal hypophosphatasia presenting as neonatal epileptic encephalopathy with abnormal neurotransmitter metabolism secondary to reduced co-factor pyridoxal-5′-phosphate availability. J Inherit Metab Dis 33(Supp 3):25–33
Baumgartner-Sigl S, Haberlandt E, Mumm S et al (2007) Pyridoxine-responsive seizures as the first symptom of infantile hypophosphatasia caused by two novel missense mutations (c.677T > C, p. M226T; c.1112C > T, p.T371I) of the tissue-nonspecific alkaline phosphatase gene. Bone 40:1655–1661
Baxter P (2003) Pyridoxine-dependent seizures: a clinical and biochemical conundrum. Biochim Biophys Acta 1647:36–41
Baxter P, Aicardi J (1999) Neonatal seizures after pyridoxine use. Lancet 354:2082–2083
Bayoumi RA, Kirwan JR, Smith WR (1972) Some effects of dietary vitamin B 6 deficiency and 4-deoxypyridoxine on γ-aminobutyric acid metabolism in rat brain. J Neurochem 19:569–576
Bayoumi RA, Smith WRD (1976) Pyridoxal 5′-phosphate levels in rat brain assayed by a modified method using enzymic decarboxylation of L-[^{14}C]tyrosine. J Neurochem 26:405–407
Belachew D, Kazmerski T, Libman I et al (2013) Infantile hypophosphatasia secondary to a novel compound heterozygous mutation presenting with pyridoxine-responsive seizures. JIMD Rep 11:17–24
Benke PJ, Fleshood HL, Pitot HC (1972) Osteoporotic bone disease in the pyridoxine-deficient rat. Biochem Med 6:526–535
Bethenod M, Cotte MF, Collombel C et al (1967) Neonatal discovery of hypophosphatasia. Bone improvement Fatal convulsant encephalopathy. Ann Pediatr (Paris) 14:835–841
Bhagavan HN, Coursin DB, Stewart CN (1977) Regional distribution of pyridoxal 5′-phosphate in developing and mature brains and its depletion in pyridoxine deficiency. Neurochem Res 2:681–687
Black AL, Guirard BM, Snell EE (1978) The behavior of muscle phosphorylase as a reservoir for vitamin B-6 in the rat. J Nutr 108:670–677
Bode W, Mocking JAJ, van den Berg H (1991) Influence of age and sex on vitamin-B-6 vitamer distribution and on vitamin-B-6 metabolizing enzymes in Wistar rats. J Nutr 121:318–329
Boe AS, Bredholt G, Knappskog PM et al (2004) Pyridoxal phosphatase is a novel cancer autoantigen in the central nervous system. Br J Cancer 91:1508–1514
Brun-Heath I, Ermonval M, Chabrol E et al (2011) Differential expression of the bone and the liver tissue non-specific alkaline phosphatase isoforms in brain tissues. Cell Tissue Res 343:521–536
Brunel C, Cathala G, Saintot M (1969) Purification and properties of alkaline phosphatase from beef brain. Biochim Biophys Acta Enzymol 191:621–635
Bukin Y, Ivanova AL (1976) Distribution and activity of pyridoxal kinase in human brain during ontogenesis. Biokhimiya 41:237–247
Burg RW, Snell EE (1969) The bacterial oxidation of vitamin B-6. VI. Pyridoxal dehydrogenase and 4-pyridoxal-lactonase. J Biol Chem 244:2585–2589
Callizot N, Warter JM, Poindron P (2001) Pyridoxine-induced neuropathy in rats: a sensory neuropathy that responds to 4-methylcatechol. Neurobiol Dis 8:626–635
Campanini B, Spyrakis F, Peracchi A et al (2013) Serine racemase: a key player in neuron activity and in neuropathologies. Front Biosci-Landmark 18:1112–1128
Chen JLT, Francis J (2012) Pyridoxamine, advanced glycation inhibition, and diabetic nephropathy. J Am Soc Nephrol 23:6–8
Chick H, El Sadr MM, Worden AN (1940) Occurence of fits of an epileptiform nature in rats maintained for long periods on a diet deprived of vitamin B-6. Biochem J 34:595–600
Chick H, Macrae TF, Martin AJP et al (1938) The water soluble B-vitamins other than aneurin, riboflavin and nicotinic acid required by the pig. Biochem J 32:2207–2224

Chung JY, Choi JH, Hwang CY et al (2008) Pyridoxine induced neuropathy by subcutaneous administration in dogs. J Vet Sci 9:127–131
Coburn SP (1994) A critical review of minimal vitamin B-6 requirements for growth in various species with a proposed method of calculation. Vitam Horm 48:259–300
Coburn SP, Lewis DN, Fink WJ et al (1988a) Human vitamin B-6 pools estimated through muscle biopsies. Am J Clin Nutr 48:291–294
Coburn SP, Mahuren JD (1987) Identification of pyridoxine 3-sulfate, pyridoxal 3-sulfate and N-methylpyridoxine as major urinary metabolites of vitamin B-6 in domestic cats. J Biol Chem 262:2642–2644
Coburn SP, Mahuren JD, Jain M et al (1998) Alkaline phosphatase (EC 3.1.3.1) in serum is inhibited by physiological concentrations of inorganic phosphate. J Clin Endocrinol Metab 83:3951–3957
Coburn SP, Mahuren JD, Kennedy MS et al (1988b) B-6 vitamer content of rat tissues measured by isotope tracer and chromatographic methods. Biofactors 1:307–312
Coburn SP, Mahuren JD, Pauly TA et al (1992) Alkaline phosphatase activity and pyridoxal phosphate concentrations in the milk of various species. J Nutr 122:2348–2353
Coburn SP, Ziegler PJ, Costill DL et al (1991) Response of vitamin-B-6 content of muscle to changes in vitamin B-6 intake in men. Am J Clin Nutr 53:1436–1442
Colombini CE, McCoy EE (1970) Vitamin B-6 metabolism. The utilization of [^{14}C]pyridoxine by the normal mouse. Biochemistry 9:533–538
Coursin DB (1955) Vitamin B-6 deficiency in infants; a follow-up study. Am J Dis Child 90:344–348
Dakshinamurti K, Dakshinamurti S (2014) Neonatal epileptic encephalopathy: role of vitamin B-6 vitamers in diagnosis and therapy. Am J Clin Nutr 100:505–506
Dakshinamurti K, Lal KJ, Ganguly PK (1998) Hypertension, calcium channel and pyridoxine (vitamin B_6). Mol Cell Biochem 188:137–148
Demirbilek H, Alanay Y, Alikasifoglu A et al (2012) Hypophosphatasia presenting with pyridoxine-responsive seizures, hypercalcemia, and pseudotumor cerebri: case report. J Clin Res Pediatr Endocrinol 4:34–38
Di Mauro S, Manes T, Hessle L et al (2002) Kinetic characterization of hypophosphatasia mutations with physiological substrates. J Bone Min Res 17:1383–1391
DiPaolo RV, Spielvogel C, Kanfer JN (1975) Increased lysosomal hydrolase activity in kidney and brain from the vitamin B6-deficient developing rat. Pediatr Res 9:693–697
Douaud G, Refsum H, de Jager CA et al (2013) Preventing Alzheimer's disease-related gray matter atrophy by B-vitamin treatment. Proc Natl Acad Sci USA 110:9523–9528
Ebadi M (1981) Regulation and function of pyridoxal phosphate in CNS. Neurochem Int 3:181–206
Ebadi M, Bifano J (1978) The synthesis of pyridoxal phosphate in rat brain regions. Int J Biochem 9:607–611
Ebadi M, Govitrapong P (1979) Biogenic amine-mediated alteration of pyridoxal phosphate formation in rat brain. J Neurochem 32:845–853
Ebadi MS (1972) Ontogenetic alterations of vitamin B 6 metabolism in rat brain–evidence of acceleration of its turnover rate. Adv Biochem Psychopharmacol 4:141–170
Ebadi MS, Costa E (eds) (1972) Role of Vitamin B-6 in Neurobiology. Advances in Biochemical Psychopharmacology 4
Ebadi MS, McCoy EE, Kugel RB (1970) Interrelationships between pyridoxal phosphate and pyridoxal kinase in rabbit brain. J Neurochem 17:941–948
Ericson KL, Maloney VM, Mahuren JD et al (2008) N-methylpyridoxamine: novel canine vitamin B6 urine metabolite. Bioorg Med Chem Lett 18:1845–1848
Faber CN, Diven WF, Axelrod AE et al (1991) Changes in serum lysosomal enzymes in the vitamin B6-deficient weanling rat. Nutr Res 11:117–128
Fleshood HL, Pitot HC (1970) Metabolism of O-phosphorylethanolamine in animal tissues. I. O-phosphorylethanolamine phospho-lyase: partial purification and characterization. J Biol Chem 245:4414–4420

Fonda ML (1988) Pyridoxamine phosphate oxidase activity in mammalian tissues. Comp Biochem Physiol 90B:731–737

Fonda ML (1992) Purification and characterization of vitamin B-6 phosphate phosphatase from human erythrocytes. J Biol Chem 267:15978–15983

Fonta C, Negyessy L, Renaud L et al (2004) Areal and subcellular localization of the ubiquitous alkaline phosphatase in the primate cerebral cortex: evidence for a role in neurotransmission. Cereb Cortex 14:595–609

Footitt EJ, Clayton PT, Mills K et al (2013) Measurement of plasma B-6 vitamer profiles in children with inborn errors of vitamin B-6 metabolism using an LC-MS/MS method. J Inher Metab Dis 36:139–145

Footitt EJ, Heales SJ, Mills PB et al (2011) Pyridoxal 5′-phosphate in cerebrospinal fluid; factors affecting concentration. J Inher Metab Dis 34:529–538

Fouts PJ, Helmer OM, Lepkovsky S et al (1938) Production of microcytic hypochromic anemia in puppies on synthetic diet deficient in rat antidermatitis factor (vitamin B-6). J Nutr 16:197–207

Furukawa A, Oikawa S, Hasegawa-Ishii S et al (2010) Proteomic analysis of aging brain in SAMP10 mouse: a model of age-related cerebral degeneration. Mech Ageing Dev 131:379–388

Gao G, Fonda ML (1994) Identification of an essential cysteine residue in pyridoxal phosphatase from human erythrocytes. J Biol Chem 269:8234–8239

Gdynia HJ, Muller T, Sperfeld AD et al (2008) Severe sensorimotor neuropathy after intake of highest dosages of vitamin B6. Neuromusc Disord 18:156–158

Ge S, Goh EL, Sailor KA et al (2006) GABA regulates synaptic integration of newly generated neurons in the adult brain. Nature 439:589–593

Geng MY, Saito H, Nishiyama N (1997) Protective effects of pyridoxal phosphate against glucose deprivation-induced damage in cultured hippocampal neurons. J Neurochem 68:2500–2506

Girschick H, Schneider P, Haubitz I et al (2006) Effective NSAID treatment indicates that hyperprostaglandinism is affecting the clinical severity of childhood hypophosphatasia. Orphanet J Rare Dis. doi:10.1186/1750-1172-1-24:

Gohla A, Birkenfeld J, Bokoch GM (2005) Chronophin, a novel HAD-type serine protein phosphatase, regulates cofilin-dependent actin dynamics. Nat Cell Biol 7:21–29

Gregory JF, Park Y, Lamers Y et al. (2013) Metabolomic Analysis Reveals Extended Metabolic Consequences of Marginal Vitamin B-6 Deficiency in Healthy Human Subjects. Plos One 8: e63544

Groziak S, Kirksey A, Hamaker B (1984) Effect of maternal vitamin B-6 restriction on pyridoxal phosphate concentrations in developing regions of the central nervous system in rats. J Nutr 114:727–732

Groziak SM, Kirksey A (1987) Effects of maternal dietary restriction in vitamin B-6 on neocortex development in rats: B-6 vitamer concentrations, volume and cell estimates. J Nutr 117:1045–1052

Guilarte TR (1989) Regional changes in the concentrations of glutamate, glycine, taurine, and GABA in the vitamin B-6 deficient developing rat brain: association with neonatal seizures. Neurochem Res 14:889–897

Hofmann C, Liese J, Schwarz T et al (2013) Compound heterozygosity of two functional null mutations in the ALPL gene associated with deleterious neurological outcome in an infant with hypophosphatasia. Bone 55:150–157

Hwang IK, Yoo KY, Kim DH et al (2007) Time course of changes in pyridoxal 5′-phosphate (vitamin B-6 active form) and its neuroprotection in experimental ischemic damage. Exp Neurol 206:114–125

Ishioka N, Sato J, Nakamura J et al (1995) In vivo modification of $GABA_A$ receptor with a high dose of pyridoxal phosphate induces tonic-clonic convulsion in immature mice. Neurochem Int 26:369–373

Jang YM, Kim DW, Kang TC et al (2003) Human pyridoxal phosphatase: molecular cloning, functional expression, and tissue distribution. J Biol Chem 278:50040–50046

Jansonius JN (1998) Structure, evolution and action of vitamin B_6-dependent enzymes. Curr Opin Struct Biol 8:759–769
Kang JH, Hong ML, Kim DW et al (2004) Genomic organization, tissue distribution and deletion mutation of human pyridoxine 5′-phosphate oxidase. Eur J Biochem 271:2452–2461
Kash SF, Johnson RS, Tecott LH et al (1997) Epilepsy in mice deficient in the 65-kDa isoform of glutamic acid decarboxylase. Proc Natl Acad Sci USA 94:14060–14065
Kerry JA, Rohde M, Kwok F (1986) Brain pyridoxal kinase: purification and characterization. Eur J Biochem 158:581–585
Kim JE, Kim DW, Kwak SE et al (2008) Potential role of pyridoxal-5′-phosphate phosphatase/chronopin in epilepsy. Exp Neurol 211:128–140
Kim JE, Kim DW, Kwak SE et al (2009) Pyridoxal-5′-Phosphate Phosphatase/Chronophin Inhibits Long-Term Potentiation Induction in the rat dentate gyrus. Hippocampus 19:1078–1089
Kim KM, Qin T, Jiang YY et al (2012) Protein domain structure uncovers the origin of aerobic metabolism and the rise of planetary oxygen. Structure 20:67–76
Kim YT, Kwok F, Churchich JE (1988) Interactions of pyridoxal kinase and aspartate aminotransferase emission anisotropy and compartmentation studies. J Biol Chem 263:13712–13717
Koh PO (2010) Proteomic analysis of focal cerebral ischemic injury in male rats. J Vet Med Sci 72:181–185
Kretsch MJ, Sauberlich HE, Newbrun E (1991) Electroencephalographic changes and periodontal status during short-term vitamin-B-6 depletion of young, nonpregnant women. Am J Clin Nutr 53:1266–1274
Kurlemann G, Deufel T, Schuierer G (1997) Pyridoxine-responsive West syndrome and gamma-aminobutyric acid. Eur J Pediatr 156:158–159
Kurlemann G, Ziegler R, Gruneberg M et al (1992) Disturbance of GABA metabolism in pyridoxine-dependent seizures. Neuropediat 23:257–259
Kwak SE, Kim JE, Kim DW et al (2009) Pyridoxine 5′-phosphate oxidase, not pyridoxal kinase, involves in long-term potentiation induction in the rat dentate gyrus. Hippocampus 19:45–56
Kwok F, Churchich JE (1980) Interaction between pyridoxal kinase and pyridoxine-5-P oxidase, two enzymes involved in the metabolism of vitamin B-6. J Biol Chem 255:882–887
Kwon O, Kwok F, Churchich JE (1991) Catalytic and regulatory properties of native and chymotrypsin- treated pyridoxine-5-phosphate oxidase. J Biol Chem 266:22136–22140
Li TK, Lumeng L, Veitch RL (1974) Regulation of pyridoxal 5′-phosphate metabolism in liver. Biochem Biophys Res Commun 61:677–684
Litmanovitz I, Reish O, Dolfin T et al (2002) Glu274Lys/Gly309Arg mutation of the tissue-nonspecific alkaline phosphatase gene in neonatal hypophosphatasia associated with convulsions. J Inherited Metab Dis 25:35–40
Loo YH (1972) Levels of B-6 vitamers and pyridoxal phosphokinase in rat brain during maturation. J Neurochem 19:1835–1837
Loo YH, Mack K (1971) Subcellular distribution of B6 vitamers in cerebral cortex. J Neurochem 18:499–502
Lumeng L, Brashear RE, Li TK (1974) Pyridoxal 5′-phosphate in plasma: source, protein-binding, and cellular transport. J Lab Clin Med 84:334–343
Lumeng L, Li TK (1975) Characterization of the pyridoxal 5′-phosphate and pyridoxamine 5′-phosphate hydrolase activity in rat liver: identity with alkaline phosphatase. J Biol Chem 250:8126–8131
Lyon JB, Bain JA, Williams HL (1962) The distribution of vitamin B-6 in the tissues of the inbred strains of mice fed complete and fed vitamin B-6 deficient rations. J Biol Chem 237:1989–1991
Mahuren JD, Pauly TA, Coburn SP (1991) Identification of 5-pyridoxic acid and 5-pyridoxic acid lactone as metabolites of vitamin B-6 in humans. J Nutr Biochem 2:449–453
Martin DL, Rimvall K (1993) Regulation of gamma-aminobutyric acid synthesis in the brain. J Neurochem 60:395–407

McCormick DB, Gregory ME, Snell EE (1961) Pyridoxal phosphokinases I. Assay, distribution, purification, and properties. J Biol Chem 236:2076–2084

McQuillin A, Rizig M, Gurling HM (2007) A microarray gene expression study of the molecular pharmacology of lithium carbonate on mouse brain mRNA to understand the neurobiology of mood stabilization and treatment of bipolar affective disorder. Pharmacogenet Genomics 17:605–617

Mehansho H, Hamm MW, Henderson LM (1979) Transport and metabolism of pyridoxal and pyridoxal phosphate in the small intestine of the rat. J Nutr 109:1542–1551

Mehta PK, Christen P (2000) The molecular evolution of pyridoxal-5′-phosphate-dependent enzymes. Adv Enzymol Relat Areas Mol Biol 74:129–184

Meisler NT, Thanassi JW (1980) Pyridoxine kinase, pyridoxine phosphate phosphatase and pyridoxine phosphate oxidase activities in control and B-6-deficient rat liver and brain. J Nutr 110:1965–1975

Merrill AH, Henderson JM, Wang E et al (1984) Metabolism of vitamin B-6 by human liver. J Nutr 114:1664–1674

Merrill AH, Horiike K, McCormick DB (1978) Evidence for the regulation of pyridoxal 5′-phosphate formation in liver by pyridoxamine 5′-phosphate oxidase. Biochem Biophys Res Commun 83:984–990

Metzler DE (1981) Regulation and function of pyridoxal phosphate in CNS. Neurochem Int 3:207–209

Miller J, Drost DJ, Jensen E et al (2012) Progressive membrane phospholipid changes in first episode schizophrenia with high field magnetic resonance spectroscopy. Psychiatry Res 201:25–33

Miller LP, Martin DL, Mazumder A et al (1978) Studies on the regulation of GABA synthesis: substrate-promoted dissociation of pyridoxal-5′-phosphate from GAD. J Neurochem 30:361–369

Mizuno N, Kawakami K, Morita E (1980) Competitive inhibition between 4′-substituted pyridoxine analogues and pyridoxal for pyridoxal kinase from mouse brain. J Nutr Sci Vitaminol 26:535–543

Modica-Napolitano JS, Renshaw PF (2004) Ethanolamine and phosphoethanolamine inhibit mitochondrial function in vitro: implications for mitochondrial dysfunction hypothesis in depression and bipolar disorder. Biol Psychiatry 55:273–277

Mooney S, Leuendorf JE, Hendrickson C et al (2009) Vitamin B6: a long known compound of surprising complexity. Molecules 14:329–351

Moriwaki Y, Yamamoto T, Takahashi S et al (2001) Widespread cellular distribution of aldehyde oxidase in human tissues found by immunohistochemistry staining. Histol Histopathol 16:745–753

Nam H, Lewis NE, Lerman JA et al (2012) Network context and selection in the evolution to enzyme specificity. Science 337:1101–1104

Narisawa S, Wennberg C, Millan JL (2001) Abnormal vitamin B6 metabolism in alkaline phosphatase knock-out mice causes multiple abnormalities, but not the impaired bone mineralization. J Pathol 193:125–133

Nishihara Y, Hayashi Y, Fujii T et al (1994) The alkaline phosphatase in human plexus chorioideus. Biochim Biophys Acta 1209:274–278

Nunes ML, Mugnol F, Bica I et al (2002) Pyridoxine-dependent seizures associated with hypophosphatasia in a newborn. J Child Neurol 17:222–224

Ohkawa K, Hirakawa-Sakurai T, Joh K et al (1994) Pyridoxal kinase immunoreactivity in rabbit brain. Neurochem Res 19:1231–1235

Ormazabal A, Oppenheim M, Serrano M et al (2008) Pyridoxal 5′-phosphate values in cerebrospinal fluid: reference values and diagnosis of PNPO deficiency in paediatric patients. Mol Genet Metab 94:173–177

Palm D, Klein HW, Schinzel R et al (1990) The role of pyridoxal 5′-phosphate in glycogen phosphorylase catalysis. Biochem 29:1099–1107

Pandey S, Garg P, Lim KT et al (2013) The efficiency of membrane transport of vitamin B6 coupled to poly(ester amine) gene transporter and transfection in cancer cells. Biomaterials 34:3716–3728

Paul L, Ueland PM, Selhub J (2013) Mechanistic perspective on the relationship between pyridoxal 5′-phosphate and inflammation. Nutr Rev 71:239–244

Percudani R, Peracchi A (2003) A genomic overview of pyridoxal-phosphate-dependent enzymes. EMBO Rep 4:850–854

Percudani R, Peracchi A (2009) The B6 database: a tool for the description and classification of vitamin B6-dependent enzymatic activities and of the corresponding protein families. BMC Bioinform 10:273

Rasmussen KM, Barsa PM, McCormick DB (1979) Pyridoxamine 5′-phosphate oxidase activity in rat tissues during development of riboflavin or pyridoxine deficiency. Proc Soc Exp Biol Med 161:527–530

Reginaldo CD, Selhub J, Paul L et al (2013) Anthranilic acid and 3-hydroxyanthranilic acid, but not kynurenic acid, are associated with plasma pyridoxal-5 phosphate levels. FASEB J 27:1077.21

Rott WH, Williams HL, Bain JA (1960) Intracellular distribution of pyridoxine congeners and related enzymes in mouse brain. Pharmacologist 2:68

Ruth RE, Morgan DG (1986) Dietary pyridoxine and the susceptibility to limbic motor seizures in rats. Exp Neurol 94:441–448

Said HM, Ortiz A, Ma TY (2003) A carrier-mediated mechanism for pyridoxine uptake by human intestinal epithelial Caco-2 cells: regulation by a PKA-mediated pathway. Am J Physiol Cell Physiol 285:C1219–C1225

Said HM (2011) Intestinal absorption of water-soluble vitamins in health and disease. Biochem J 437:357–372

Sakurai T, Asakura T, Matsuda M (1987) Transport and metabolism of pyridoxine and pyridoxal in mice. J Nutr Sci Vitaminol 33:11–19

Sakurai T, Asakura T, Matsuda M (1988) Transport and metabolism of pyridoxine in the intestine of the mouse. J Nutr Sci Vitaminol 43:179–187

Sakurai T, Kuwahara T, Matsuda M (1980a) Subcellular distribution of B-6 vitamers in mouse brain. Jikeikai Med J 27:151–158

Sakurai T, Kuwahara T, Matsuda M (1980b) Vitamin B6 vitamers in mouse brain and their relationship to convulsions. Jikeikai Med J 27:13–21

Salazar P, Tapia R (2001) Seizures induced by intracerebral administration of pyridoxal-5′-phosphate: effect of GABAergic drugs and glutamate receptor antagonists. Neuropharmacology 41:546–553

Sampson DA, O'Connor DK (1989) Response of B-6 vitamers in plasma, erythrocytes and tissues to vitamin B-6 depletion and repletion in the rat. J Nutr 119:1940–1948

Saraswathi S, Bachhawat BK (1963) Phosphatases from human brain I. Purification and properties of pyridoxal phosphate phosphatase. J Neurochem 10:127–133

Schiroli D, Cirrincione S, Donini S et al (2013) Strict reaction and substrate specificity of AGXT2L1, the human O-phosphoethanolamine phospho-lyase. IUBMB Life 65:645–650

Seargeant LE, Stinson RA (1979) Phosphoester specificity of purified human liver alkaline phosphatase. Can J Biochem 57:2000–2007

Seligmann B, Miller LP, Brockman DE et al (1978) Studies on the regulation of GABA synthesis: the interaction of adenine nucleotides and glutamate with brain glutamate decarboxylase. J Neurochem 30:371–376

Shao L, Vawter MP (2008) Shared gene expression alterations in schizophrenia and bipolar disorder. Biol Psychiatry 64:89–97

Sharma SK, Dakshinamurti K (1992) Seizure activity in pyridoxine-deficient adult rats. Epilepsia 33:235–247

Shin YS, Rasshofer R, Endres W (1984) Pyridoxal 5′-phosphate concentrations as marker for vitamin B-6 dependent seizures in the newborn. Lancet 2:870–871

Sia C, Wapnir R, Sokal M et al (1975) Effects of pyridoxine on neonatal hypophosphatasia. Pediatr Res 9:355
Snell EE (1964) Summary of session I and some notes on the metabolism of vitamin B-6. Vitam Horm 22:485–494
Snell EE, Haskell BE (1971) The metabolism of vitamin B-6. Comp Biochem 21:47–71
South M (1999a) Neonatal seizures after pyridoxine use - reply. Lancet 354:2083
South M (1999b) Neonatal seizures after use of pyridoxine in pregnancy. Lancet 353:1940–1941
Spater HW, Novikoff AB, Spater SH et al (1978) Pyridoxal phosphatase: cytochemical localization in GERL and other organelles of rat neurons. J Histochem Cytochem 26:809–821
Spector R (1978a) Vitamin B-6 transport in the central nervous system. In vivo studies. J Neurochem 30:881–887
Spector R (1978b) Vitamin B-6 transport in the central nervous system. In vitro studies. J Neurochem 30:889–897
Spector R, Greenwald L (1978) Transport and metabolism of vitamin B-6 in rabbit brain and choroid plexus. J Biol Chem 253:2373–2379
Spector R, Shikuma SN (1978) Stability of vitamin B-6 accumulation and pyridoxal kinase activity in rabbit brain and choroid plexus. J Neurochem 31:1403–1410
Spector R, Johanson CE (2007) Vitamin transport and homeostasis in mammalian brain: focus on vitamins B and E. J Neurochem 103:425–438
Stanulovic M, Jeremic V, Leskovac V et al (1976) New pathway of conversion of pyridoxal to 4-pyridoxic acid. Enzyme 21:357–369
Stephens MC, Dakshinamurti K (1975) Brain lipids in pyridoxine-deficient young rats. Neurobiol 5:262–269
Stolz J, Vielreicher M (2003) Tpn1p, the plasma membrane vitamin B-6 transporter of Saccharomyces cerevisiae. J Biol Chem 278:18990–18996
Stolz J, Wohrmann HJ, Vogl C (2005) Amiloride uptake and toxicity in fission yeast are caused by the pyridoxine transporter encoded by bsu1+ (car1+). Eukaryot Cell 4:319–326
Suarez I, Bodega G, Fernandez B (2002) Glutamine synthetase in brain: effect of ammonia. Neurochem Int 41:123–142
Surtees R, Mills P, Clayton P (2006) Inborn errors affecting vitamin B6 metabolism. Future Neurol 1:615–620
Tanaka T, Scheet P, Giusti B et al (2009) Genome-wide association study of vitamin B6, vitamin B12, folate, and homocysteine blood concentrations. Am J Hum Genet 84:477–482
Tardif JC, Carrier M, Kandzari DE et al (2007) Effects of pyridoxal-5′-phosphate (MC-1) in patients undergoing high-risk coronary artery bypass surgery: results of the MEND-CABG randomized study. J Thorac Cardiovasc Surg 133:1604–1611
Theriault O, Poulin H, Thomas GR et al (2014) Pyridoxal-5′-phosphate (MC-1), a vitamin B6 derivative, inhibits expressed P2X receptors. Can J Physiol Pharmacol 92:189–196
Thompson MD, Nezarati MM, Gillessen-Kaesbach G et al (2010) Hyperphosphatasia with seizures, neurologic deficit, and characteristic facial features: five new patients with Mabry syndrome. Am J Med Genet A 152A:1661–1669
Thompson MD, Roscioli T, Marcelis C et al (2012) Phenotypic variability in hyperphosphatasia with seizures and neurologic deficit (Mabry syndrome). Am J Med Genet A 158A:553–558
Trezise DJ, Bell NJ, Khakh BS et al (1994) P2 purinoceptor antagonist properties of pyridoxal-5-phosphate. Eur J Pharmacol 259:295–300
van der Ham M, Albersen M, De Koning TJ et al (2012) Quantification of vitamin B6 vitamers in human cerebrospinal fluid by ultra performance liquid chromatography-tandem mass spectrometry. Anal Chim Acta 712:108–114
Veiga-da-Cunha M, Hadi F, Balligand T et al (2012) Molecular identification of hydroxylysine kinase and of ammoniophospholyases acting on 5-phosphohydroxy-L-lysine and phosphoethanolamine. J Biol Chem 287:7246–7255
Verrier ED (2007) Translational research: is there a future? J Thorac Cardiovasc Surg 133:1409–1411

Wada H, Snell EE (1961) The enzymatic oxidation of pyridoxine and pyridoxamine phosphates. J Biol Chem 236:2089–2095

Wang X, Dakshinamurti K, Musat S et al (1999) Pyridoxal 5′-phosphate is an ATP-receptor antagonist in freshly isolated rat cardiomyocytes. J Mol Cell Cardiol 31:1063–1072

Waymire KG, Mahuren JD, Jaje JM et al (1995) Mice lacking tissue non-specific alkaline phosphatase die from seizures associated with defective metabolism of vitamin B-6. Nature Genet 11:45–51

Wei I-L (1999) The influence of dietary restriction on vitamin B-6 vitamer distribution and on vitamin B-6 metabolizing enzymes in rats. J Am Coll Nutr 18:144–151

Whyte MP, Mahuren JD, Fedde KN et al (1988) Perinatal hypophosphatasia: tissue levels of vitamin B-6 are unremarkable despite markedly increased circulating concentrations of pyridoxal 5′-phosphate. J Clin Invest 81:1234–1239

Whyte MP, Mahuren JD, Vrabel LA et al (1985) Markedly increased circulating pyridoxal 5′-phosphate levels in hypophosphatasia. J Clin Invest 76:752–756

Yamamoto H, Sasamoto Y, Miyamoto Y et al (2004) A successful treatment with pyridoxal phosphate for West syndrome in hypophosphatasia. Pediatr Neurol 30:216–218

Zisapel N, Levi M (1980) A calcium stimulated alkaline phosphatase associated with synaptic vesicles. Brain Res 201:385–398

Chapter 12
Tetramisole and Levamisole Suppress Neuronal Activity Independently from Their Inhibitory Action on Tissue Non-specific Alkaline Phosphatase in Mouse Cortex

Lionel G. Nowak, Benoît Rosay, Dávid Czégé and Caroline Fonta

Abstract Tissue non-specific alkaline phosphatase (TNAP) may be involved in the synthesis of GABA and adenosine, which are the main inhibitory neurotransmitters in cortex. We explored this putative TNAP function through electrophysiological recording (local field potential) in slices of mouse somatosensory cortex maintained in vitro. We used tetramisole, a well documented TNAP inhibitor, to block TNAP activity. We expected that inhibiting TNAP with tetramisole would lead to an increase of neuronal response amplitude, owing to a diminished availability of GABA and/or adenosine. Instead, we found that tetramisole *reduced* neuronal response amplitude in a dose-dependent manner. Tetramisole also decreased axonal conduction velocity. Levamisole had identical effects. Several control experiments demonstrated that these actions of tetramisole were independent from this compound acting on TNAP. In particular, tetramisole effects were not stereo-specific and they were not mimicked by another inhibitor of TNAP, MLS-0038949. The decrease of axonal conduction velocity and preliminary intracellular data suggest that tetramisole blocks voltage-dependent sodium channels. Our results imply that levamisole or tetramisole should not be used with the sole purpose of inhibiting TNAP in living excitable cells as it will also block all processes that are activity-dependent. Our data

L.G. Nowak (✉) · B. Rosay · C. Fonta (✉)
Centre de Recherche Cerveau et Cognition (CerCo), Université de Toulouse UPS; CNRS UMR 5549, Toulouse, France
e-mail: lionel.nowak@cerco.ups-tlse.fr

C. Fonta
e-mail: caroline.fonta@cerco.ups-tlse.fr

L.G. Nowak
CNRS, Cerco, Toulouse, France

D. Czégé
Department of Physiology and Neurobiology, Eötvös Loránd University, Budapest, Hungary

C. Fonta and L. Négyessy (eds.), *Neuronal Tissue-Nonspecific Alkaline Phosphatase (TNAP)*, Subcellular Biochemistry 76,
DOI 10.1007/978-94-017-7197-9_12

and a review of the literature indicate that tetramisole may have at least four different targets in the nervous system. We discuss these results with respect to the neurological side effects that were observed when levamisole and tetramisole were used for medical purposes, and that may recur nowadays due to the recent use of levamisole and tetramisole as cocaine adulterants.

Keywords Vitamin B6 · Pyridoxal · Ectonucleotidase · Epilepsy · Inflammatory leukoencephalopathy

12.1 Introduction

Alkaline phosphatases (APs) are a group of enzymes that release inorganic phosphate from phosphomonoesters. APs are ecto-enzymes, thus located on the extracellular side of the cells. Four types of APs have been identified in human. Three have restricted regional distributions and have been named accordingly as intestinal AP, germinal AP and placental AP. The fourth type, the tissue non-specific alkaline phosphatase (TNAP), has been found in all mammalian genera examined. As its name implies, TNAP shows a widespread distribution, being present in various tissues such as blood vessels, bone and cartilage, kidney, liver, lung, etc. (e.g., Borgers 1973; de Bernard et al. 1986; Morris et al. 1992; Hoshi et al. 1997).

In human, mutations of the TNAP gene lead to hypophosphatasia (Rathbun 1948), a rare disease characterized by low level of plasmatic AP and by abnormal bone mineralization (reviewed in: Fraser 1957; Millán 2006; Mornet 2007; Whyte 2010; Taketani et al. 2014; see also Chaps. 1 and 2 in this book). Importantly, the most severe forms of hypophosphatasia (perinatal and, to a lesser extent, infantile hypophosphatasia) are often associated with epileptic seizures (e.g., Rathbun 1948; Fraser 1957; Béthenod et al. 1967; Baumgartner-Sigl et al. 2007; Balasubramaniam et al. 2010; Taketani et al. 2014; see also Chaps. 14 and 15 in this book). In the mouse, inactivation of the TNAP gene produces an animal model of severe human hypophosphatasia, with bone mineralization defect and epileptic seizures as well (Waymire et al. 1995; Narisawa et al. 1997; see also Chap. 3).

Occurrence of epileptic seizures in TNAP KO mice and in the severe forms of human hypophosphatasia suggests that TNAP plays a fundamental role in the control of nerve cell activity—although indirect effects cannot be ruled out. This implies that TNAP should be present in the nervous tissue. Indeed, although early studies (Landow et al. 1942; Bourne 1943) did not report AP activity at the neuronal level in mammals—except in the spinal cord and medulla –, later studies demonstrated significant AP activity in multiple brain structures of various mammalian species. The pharmacological profile of this neuronal AP activity suggested it resulted from TNAP activity (Fonta et al. 2004; Langer et al. 2008) and it has recently been formally ascribed to TNAP gene expression (Ermonval et al. 2009; Brun-Heath et al. 2011).

AP activity in the brain is not homogenous; some structures show strong activity —hypothalamus, superior colliculus, subtancia nigra, spinal cord for example— while other—hippocampus, cerebellum, caudate nucleus, putamen for example— show much weaker activity (Shimizu 1950; Nandy and Bourne 1963; Friede 1966; Sugimura and Mizutani 1979; Langer et al. 2008; Brun-Heath et al. 2011; Street et al. 2013). AP activity is also observed in the cerebral cortex. In primate its level is particularly high in layer 4 of primary visual, auditory and somatosensory cortices (Friede 1966; Fonta and Imbert 2002; Fonta et al. 2004, 2005; Négyessy et al. 2011). In area 17 of primates, the thalamorecipient layers 4Cα and 4Cβ are exquisitely delineated by TNAP activity (Fonta and Imbert 2002; Fonta et al. 2004). In visual cortex, AP activity depends on neuronal activity, as revealed by monocular deprivation experiments (Fonta et al. 2004). TNAP is also highly expressed in the prefrontal cortex of primate (Fonta et al. 2004) and is found preferentially in layer 5 in non-sensory cortices of non-human primates (Friede 1966) and in human (Négyessy et al. 2011). AP activity is less patterned in rodent cortex but clear differences between cortical areas are nevertheless noticeable. For example, layer Ia of the piriform cortex and layer 4 of the somatosensory cortex show stronger AP activity than neighboring layers and surrounding cortical areas (Fonta et al. 2004; Langer et al. 2008; Brun-Heath et al. 2011).

In both cortical and subcortical structures, the suggestion that AP is located on synapses (Nandy and Bourne 1963) has been confirmed by electron microscopy studies, which revealed AP in the synaptic clefts of both excitatory and inhibitory synapses (Sugimura and Mizutani 1979; Mori and Nagano 1985; Fonta et al. 2004, 2005; Hanics et al. 2012; see also Chap. 5, this book). Strong AP activity has also been revealed on the nodes of Ranvier of myelinated axons (Pinner et al. 1964; Mori and Nagano 1985; Fonta et al. 2005; Hanics et al. 2012; see Chap. 5). Altogether, these data suggest that TNAP may be involved in the control of action potential propagation in axons as well as in the control of synaptic transmission.

In this study we used an electrophysiological approach with the aim to determine whether and how TNAP controls neuronal activity in the cerebral cortex, in particular through its putative involvement in the synthesis of two major neurotransmitters, GABA and adenosine. In cortex, both GABA and adenosine are inhibitory neurotransmitters: GABA acts postsynaptically through $GABA_A$ and $GABA_B$ receptors and presynaptically through $GABA_B$ receptors (e.g., Howe et al. 1987a, b; Connors et al. 1988; Deisz and Prince 1989; McCormick 1989). Adenosine mostly acts presynaptically through adenosine A1 receptors (e.g., Collins and Anson 1985; Fontanez and Porter 2006).

GABA is synthesized from glutamate by the glutamic acid decarboxylase (GAD) (e.g., Martin and Rimvall 1993). GAD uses the active form of vitamin B6, pyridoxal phosphate (PLP), as cofactor.[1] PLP cannot cross membranes; only the

[1]Note that GAD is one among multiple PLP-dependent enzymes. Approximately 60 genes encoding for enzymes using PLP as cofactor have been identified in mammals (Percudani and Peracchi 2009). In addition to GAD, some of these enzymes are directly involved in the synthesis of other neurotransmitters such as dopamine and serotonin (Ermonval et al. 2009).

nonphosphorylated form, pyridoxal (PL), can diffuse passively through membranes (e.g., Rifkin et al. 1972; Mehansho and Henderson 1980; see also Chap. 11, this book). TNAP plays a key role at this level as it is responsible for the dephosphorylation of extracellular PLP. In support for the involvement of TNAP in this process, the PLP concentration is markedly elevated in the serum of hypophosphatasic patients (Whyte et al. 1985, 1988). Likewise, PLP concentration in the serum of TNAP KO mice appears to be 20 time higher than in wild type mice (Waymire et al. 1995). In parallel, GABA levels appear to be strongly reduced in the brain of TNAP KO mice (Waymire et al. 1995; Fonta et al. 2012), confirming the importance of TNAP in the control of GABA synthesis. Yet the presence of AP on excitatory synapses suggests that its role is not solely limited to the control of GABA synthesis.

Adenosine synthesis in the extracellular space results from the degradation of ATP. ATP is released by neurons and glial cells (Fields and Burnstock 2006; Abbracchio et al. 2009; Butt 2011). ATP and intermediate nucleotides are rapidly hydrolyzed by a variety of ectonucleotidases. Four different ectonucleotidase families have been identified in the brain (e.g., Zimmermann et al. 2012). The first three families are quite specific: ectonucleoside triphosphate diphosphohydrolase degrades ATP in ADP and ADP in AMP; ectonucleotide pyrophosphatase/phosphodiesterase degrades ATP in AMP; and ecto-5′-nucleotidase degrades AMP in adenosine. In contrast, TNAP shows broader substrate specificity and is capable of degrading ATP to ADP, ADP to AMP, and AMP to adenosine. The ectonucleotidase activity of TNAP has been demonstrated in various tissues such as bone (e.g., Ciancaglini et al. 2010; Simão et al. 2013) and airways (Picher et al. 2003). Supports in favor of a role of TNAP in adenosine synthesis in the brain come from a cultured neuronal cell line study (Ohkubo et al. 2000) and from a recent spinal cord study (Street et al. 2013; see Chap. 13). Yet another study failed to reveal a significant role for TNAP in extracellular adenosine synthesis in the hippocampus (Zhang et al. 2012).

In order to examine the role of TNAP in the control of inhibitory synaptic transmission, we performed electrophysiological experiments on slices of mouse somatosensory cortex maintained in vitro. We used tetramisole to inhibit TNAP activity. Tetramisole is the racemic mixture of two stereoisomers: a levorotatory enantiomer, “levamisole”, and a dextrorotary enantiomer, “dexamisole”. Studies in the seventies showed that levamisole is highly effective at inhibiting TNAP (Van Belle 1972, 1976a, b) while dexamisole has no significant effect on this enzyme (Van Belle 1972, 1976b; Borgers 1973). Levamisole and tetramisole have since been used as TNAP inhibitors in a countless number of studies.

Our hypothesis was that suppressing TNAP activity with tetramisole would result in an *increase* of neuronal response amplitude owing to a reduction of pre- and/or postsynaptic inhibition mediated by GABA and/or adenosine. The results we obtained were opposite to this expectation: tetramisole, at concentration that fully suppress TNAP activity in biochemical assays (1–5 mM), also strongly *suppressed* neuronal activity. Levamisole had a similar effect. This suppression largely resulted from a reduction of action potential transmission along the axons. These results might have been interpreted as revealing a hitherto non-described control of axonal excitability by TNAP present on the nodes of Ranvier, yet several control experiments revealed that neuronal activity suppression by tetramisole and levamisole was *not due to TNAP inhibition* by tetramisole and levamisole. Instead, we propose that tetramisole and levamisole block voltage-dependent sodium channels.

Our results therefore indicate that, in cortex, TNAP is not the unique target of tetramisole and levamisole. In addition, other neuronal processes are likely to be affected by levamisole and tetramisole: indeed, studies showed that, in the peripheral nervous system of mammals, levamisole and tetramisole may also interfere with adrenergic and cholinergic synaptic transmission.

Outside the nervous system, studies suggested that tetramisole and levamisole possess "immunostimulating" properties. For this reason, tetramisole and levamisole have received numerous clinical and pharmaceutical applications in a variety of diseases. Nevertheless, multiple side effects have been reported, including devastating neurological adverse effects such as epilepsy and inflammatory leukoencephalopathy, that have led to discontinue chronic tetramisole and levamisole usage in most countries. Yet these side effects have recently regained the attention of public health specialists because most of the illegal cocaine on the market nowadays appears to be adulterated with levamisole or tetramisole. We discuss the possible links between the neurological side effects of tetramisole and levamisole and the different targets of these compounds in the CNS.

12.2 Methods

12.2.1 Brain Slice Preparation

All procedures were conducted in accordance with the guidelines from the French Ministry of Agriculture (décret 87/848) and from the European Community (directive 86/609) and was approved by the local ethical committee (MP/06/79/11/12, comité d'éthique Midi-Pyrénées pour l'expérimentation animale). Adult (>2 month-old) wild type female mice were used for these experiments. Mouse was anesthetized with isoflurane. Once deeply anesthetized, the mouse was killed by decapitation. The scalp was removed, the skull was drilled, the upper part of the skull was lifted off, and the whole brain was carefully removed. These operations were performed in cold (3–4 °C) *modified* artificial cerebrospinal fluid (mACSF), whose composition was (in mM): NaCl 124, $NaHCO_3$ 26, KCl 3.5, $MgSO_4$ 1, $MgCl_2$ 9, NaH_2PO_4 1.25, and glucose 10. Note that Ca^{++} was omitted while Mg^{++} concentration was 10 mM. The rationale for using such a modified ACSF has been presented elsewhere (Nowak and Bullier 1996). The mACSF was oxygenated for 1 h before the beginning of the surgery with a mixture of 95 % O_2 and 5 % CO_2. Four hundred-micrometer-thick coronal brain slices were cut on a vibratome (752 M vibroslice, Campden Instrument, UK), whose chamber was filled with cold oxygenated mACSF. Once obtained, the slices were kept at room temperature for at least one hour in a storage chamber filled with an in vivo-like (Brumberg et al. 2000; Sanchez-Vives and McCormick 2000) artificial cerebrospinal fluid (ACSF) of the following composition (in mM): NaCl 124, $NaHCO_3$ 26, KCl 3.5, $MgSO_4$ 1, NaH_2PO_4 1.25, $CaCl_2$ 1.2, and glucose 10. This ACSF was continuously bubbled with a 95 % O_2–5 % CO_2 mixture (pH 7.4). Recordings were performed in a submersion type chamber (Scientific System Design, Mississauga, Ontario, Canada) where the temperature was thermostatically held at 33–34 °C. The ACSF was gravity fed at a flow rate of 2.5–3.75 ml/min.

12.2.2 Recording and Stimulation

Neuronal signals examined in this study consisted mostly in local field potentials (LFPs) recorded in layer 4 and in the supragranular layers of the whisker, trunk and hindlimb representations of the primary somatosensory cortex (S1). Intracellular recordings were also attempted in these regions. In both cases, neuronal responses were evoked by extracellular electrical stimulation applied in the white matter or at the white matter-layer 6 border. Some extracellular recordings were also performed in the corpus callosum. In this later case, stimulation was applied in the corpus callosum too.

Tungsten-in-glass microelectrodes, with glass removed from the tip over a length of 45–110 μm, were used for extracellular electrical stimulation. Electrical stimulation was delivered through a stimulation isolation unit (A365 Stimulus Isolated, WPI) and consisted in monopolar, cathodal pulses of 0.2 or 0.3 ms duration. Pulses were delivered either in isolation at a frequency of 0.5 Hz, or as pair of pulses with an interstimulus interval of 20 ms, the pairs being repeated every 10 s (0.1 Hz). Stimulation intensity was between 50 and 180 μA.

LFPs were recorded through tungsten-in-glass microelectrodes and glass micropipettes. Glass micropipettes were pulled on a P97 Flaming Brown micropipette puller from 1.2 mm OD medium walled capillaries with filament (GC120F, Harvard Apparatus). The micropipette tip was broken to a 15–20 μm diameter opening and filled with ACSF (resistance 2–7 MΩ). The signal was amplified with an AxoClamp 2B amplifier (Axon Instrument, Foster City, CA) and further amplified with a Neurolog post-amplifier (final gain: × 1000). The signal was low-pass filtered at 10 kHz. Signals recorded through tungsten-in-glass microelectrodes (5–15 μm exposed tip) were amplified (×1000) and bandpass filtered (0.1 Hz–10 kHz) with the Neurolog recording system. Micropipettes for "sharp" intracellular recording were pulled on the P97 Flaming Brown micropipette puller from 1.2 mm OD medium walled capillaries with filament and filled with K-Acetate 3 M (DC resistance: 60–100 MΩ). The AxoClamp 2B was used for amplification (gain: × 10). The signal was low pass filtered (10 kHz). Signals were digitized with a 1401plus interface (CED, Cambridge, UK) with a digitization rate of 20 kHz.

12.2.3 Sample

The present study is based on a total of 27 successful experiments. By "experiment" we designate an ensemble of data obtained with one pair of stimulation and recording sites in one slice. Given the duration of the protocols used in this study (pharmacological manipulation requiring controls, several concentrations of various pharmacological compounds and recoveries), there was usually only one experiment per slice and per day, and by extension one experiment per mouse (19 experiments). More rarely two experiments were performed the same day on two distinct brain slices obtained from the same mouse, with one of the two experiments being dedicated to corpus callosum recording (8 experiments total).

12.2.4 Analysis

Signals were analyzed offline using spike2 software (CED) and custom scripts. The bulk of the data are issued from extracellular LFP recording evoked by electrical stimulation. The first pass of the LFP analysis consisted in averaging 6 or 30 consecutive sweeps, depending on stimulation frequency (0.1 or 0.5 Hz). This resulted in series of averaged LFPs for each consecutive minute of recording (Fig. 12.1d). These time series allowed following the time-course of the drug effects. They were visualized using Origin software (OriginLab Corporation, USA).

The averaged LFPs obtained in the gray matter (23 experiments) consisted in fast and slow components. The slow component, or field PSP (noted "fPSP" in Figures), reflects postsynaptic responses. Depending on where the recording electrode was located with respect to current sources and sinks (e.g., Mitzdorf 1985), the slow component appeared either as a slow positivity (Figs. 12.1e, 12.6a and 12.7a) or as a slow negativity (Fig. 12.5a). The fast components—always negative—correspond to *population spikes*, that is, to the synchronous discharge of action potentials by a population of neurons located in the vicinity of the recording electrode. Two types of population spikes were identified in our experiments: antidromic and orthodromic. *Antidromic* population spikes (indicated by "APS" in Figures) result from the (unnatural) retrograde propagation of action potentials elicited in the axons at the stimulation site toward the cell body, where spike firing is initiated *without* intercalated synapses. *Orthodromic* population spikes ("OPS" in Figures) are elicited when the (regular) propagation of action potentials along the axons results in postsynaptic responses that are large enough to trigger action potentials in the postsynaptic neurons (Fig. 12.1e, h). Since they are initiated by excitatory synaptic inputs, their latency is longer than that of antidromic responses. In addition, they disappear when synaptic transmission is suppressed. We systematically tested whether the population spikes we were recording from were orthodromic and/or antidromic by suppressing synaptic responses with ACSF solutions without Ca^{++} and containing either 6–10 mM Mg^{++} or 1.2–2 mM Mn^{++}. Antidromic population spikes were present in all but one of the

23 experiments involving gray matter recording, whereas measurable orthodromic responses occurred in 16 of these experiments.

The amplitude of antidromic population spikes was measured as the difference between the peak of the population spike and the pre-stimulus baseline. The latency corresponds to the latency of the peak of the population spikes relative to stimulus onset. As orthodromic population spike followed antidromic responses and were eventually partially merged with them, their amplitude was measured as the difference between the peak and the positivity corresponding to the repolarization of the preceding antidromic population spike. The amplitude of the slow component was measured relative to pre-stimulus baseline. When the orthodromic response showed both a slow component and a population spike, the amplitude of the response usually corresponds to that of the population spike. Population data in text corresponds to the mean ± SEM.

12.2.5 Chemicals

Tetramisole (0.1–5 mM), levamisole (0.5 and 1 mM), adenosine (100 μm), ATP (1 mM) and pyridoxal phosphate (10 μM) were dissolved in ACSF. MLS-0038949 (10 μM final concentration) was first dissolved in DMSO (final DMSO concentration: 0.1 % in ACSF). All chemicals were purchased from Sigma except MLS-0038949 purchased from Merck. When several concentrations of tetramisole or levamisole were used, their order of application was randomized from one experiment to the next.

12.3 Results

12.3.1 Tetramisole Reduces Both Orthodromic and Antidromic Responses in a Dose-Dependent Manner

Figure 12.1a–c illustrates the effects of levamisole (1 mM) on an antidromic population spike. Each trace corresponds to the average of 30 sweeps, representing 1 min of recording in control ACSF (Fig. 12.1a), 1 min of recording after 20 min of perfusion with ACSF containing 1 mM tetramisole (Fig. 12.1b) and 1 min of recording after 20 min of reperfusion with regular ACSF (recovery, Fig. 12.1c). The response was identified as antidromic as it persisted when Ca^{2+} was replaced by Mn^{2+} 2 mM (not shown). The white matter stimulation did not elicit a measurable orthodromic response in this experiment. In the presence of tetramisole (1 mM, Fig. 12.1b), the response amplitude was decreased to 49 % of the amplitude measured in control while the peak latency was increased from 2.6 to 3.5 ms. The amplitude and latency returned to near control values after tetramisole washout (Fig. 12.1c).

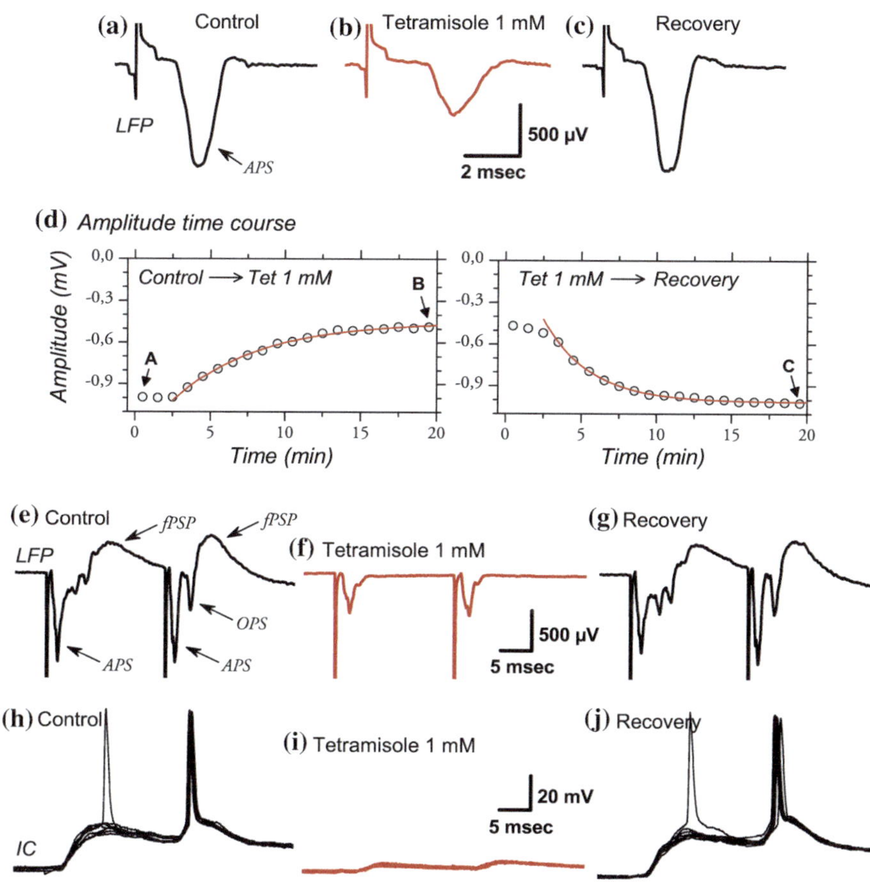

The time-course of the effect of tetramisole for the same experiment is depicted in Fig. 12.1d. The amplitude of the antidromic population spike is represented as a function of time when the control ACSF was replaced by ACSF + tetramisole 1 mM (left panel), and when ACSF + tetramisole 1 mM was replaced by control ACSF (right panel). Each data point represents the amplitude of the response measured in LFP traces averaged over 1 min. The curves are exponentials fitted to the data. The amplitude decreased upon perfusion of tetramisole 1 mM up to a plateau value that was reached after approximately 20 min. Recovery was complete in about 20 min as well.

The effect of tetramisole on orthodromic responses is illustrated with data from another experiment in Fig. 12.1e–g. Electrical stimulation was delivered in the form of pairs of pulses with an interstimulus interval of 20 ms. The LFP (Fig. 12.1e) consisted in an antidromic population spike (first population spike on traces, APS) and in orthodromic responses composed of a slow positivity (fPSP) and of an orthodromic population spike (second population spike on traces, OPS). The

◀ **Fig. 12.1** Both orthodromic and antidromic neuronal responses are reduced by tetramisole. **a–c** Effect of tetramisole 1 mM on an antidromic population spike (*APS*) recorded extracellularly (LFP) in the supragranular layers of mouse somatosensory cortex. Electrical stimulation in white matter (50 μA, 300 μsec, 0.5 Hz). Each trace is the average of 30 sweeps (1 min recording) obtained at the time of switching the perfusion from control ACSF to ACSF with tetramisole 1 mM (**a**), at the end of 20 min perfusion of ACSF with tetramisole 1 mM (**b**), and after 20 min perfusion with control *ACSF* (recovery, **c**). **a, b** Tetramisole reduced peak response amplitude from –1 to –0.49 mV and increased peak response latency from 2.6 to 3.5 ms. **c**: Effects of tetramisole 1 mM were reversible (amplitude: –1.03 mV, latency: 2.7 ms). **d** Time course of the effect of tetramisole on the amplitude of the *APS*. Same experiment as in **a–c**. Symbols correspond to the amplitude of the *APS* measured in *LFP* traces averaged over 1 min. Symbols indicated by letters **a–c** refers to the *LFP* traces displayed in **a–c**. *Solid lines* are exponential curves fitted to the data. The first 3 data points were not included in the fit as they approximately corresponded to the 3 min required for the replacement of one perfusion solution by the other in the recording chamber. *Left panel*: the time constant of amplitude decrease in tetramisole 1 mM was 5 min and 41 s. *Right panel*: the time constant of recovery in regular ACSF was 3 min and 4 s. The R^2 of fit was 0.99 for both fits. **e–g**: Effect of tetramisole 1 mM on APS, orthodromic population spike (OPS) and slow field PSP (fPSP) recorded extracellularly (LFP) in the supragranular layers of mouse somatosensory cortex. Electrical stimulation (80 μA, 200 μsec) was delivered in the white matter as pairs of pulses with an interval of 20 ms. Pairs of pulses delivered every 10 s (0.1 Hz). Traces in **e–g** correspond to the average of 6 sweeps (1 min of recording). **e** The first stimulation of the pair elicited an antidromic population spike (*APS*, –1.09 mV) followed by barely visible orthodromic spikes (not indicated) and by a slow positive component (fPSP). The second stimulation of the pair induced an *APS* whose amplitude (–1.10 mV) was identical to the one obtained after the first stimulation, followed by a well developed orthodromic population spike (OPS) and by a larger *fPSP*. The increase in amplitude of the orthodromic responses was due to paired-pulse facilitation. **f** Orthodromic responses (both OPS and fEPSP) were completely suppressed by tetramisole 1 mM and the antidromic response was reduced to –0.49 (first APS) or –0.51 (second APS) mV. **g** Both antidromic and orthodromic responses recovered upon return to control ACSF. APS amplitudes were –0.97 and –1.10 mV (first and second APS respectively). **h–j** Effect of tetramisole in an intracellularly recorded neuron. The intracellular recording (IC) was performed simultaneously with the LFP recording whose traces are displayed in **e–g**. Each panel shows 8 superimposed sweeps. The stimulation artifact was digitally removed. The white matter stimulation did not elicit an antidromic response in this neuron but it induced an EPSP whose amplitude, in control condition (**h**), was large enough to trigger orthodromic spikes, especially after the second pulse of the stimulus pair as a result of paired-pulse facilitation. **i** EPSP amplitude was drastically reduced by tetramisole (1 mM) and was too small to bring the membrane potential to firing threshold. **j** After tetramisole washout, EPSP amplitude came back to control value and reached spike firing threshold

orthodromic responses were more prominent after the second stimulation as a result of paired-pulse facilitation. In the presence of tetramisole (1 mM), the orthodromic responses were completely suppressed while the antidromic response was reduced to 45 % (first APS) or 46 % (second APS) of the control amplitude (Fig. 12.1f). The effect of tetramisole was reversible (Fig. 12.1g).

One intracellular recording was performed simultaneously with the extracellular LFP recording of Fig. 12.1e–g. In control ACSF, the intracellularly recorded response (Fig. 12.1h) consisted in an excitatory postsynaptic potential (EPSP) of 22 mV amplitude. The EPSP eventually triggered orthodromic action potentials after the first stimulation, and systematically did so after the second stimulation (Fig. 12.1h). The slow EPSP and fast action potentials offer a mirror image of the orthodromic responses recorded extracellularly (Fig. 12.1e). The postsynaptic response was nearly completely suppressed by tetramisole 1 mM (Fig. 12.1i) and the peak of the remnant depolarization (4 mV) remained far below action potential threshold. Both responses recovered upon return to control ACSF (Fig. 12.1j). Since the resting membrane potential did not depolarize in the presence of tetramisole, the reduction of ortho- and antidromic response amplitude cannot be attributed to a depolarization block.

Figure 12.2a represents the effect of different concentrations of tetramisole on the antidromic population spike amplitude at the population level. Response amplitudes were normalized by the control response amplitude and are expressed as percentage of the control response amplitude. The reduction of the antidromic response amplitude was dose-dependent (ANOVA, $P < 0.0001$) and reversible. In comparison to control and recovery, response amplitude was significantly reduced with tetramisole at 0.5, 1, 2 and 5 mM (PLSD Fischer test, $P < 0.0001$ for all comparisons). Response suppression was nearly complete with 2 mM tetramisole (4.4 % ± 0.04 of control response amplitude).

Figure 12.2C shows the dose-response relationship for antidromic response reduction fitted with Hill's equation:

$$R = 100 - 100 \times \frac{C^n}{IC_{50}^n + C^n}$$

where R represent the normalized response amplitude (% of control), C the tetramisole concentration, IC_{50} the concentration producing half the maximal effect, and n the Hill coefficient determining the slope of the curve. The IC_{50} returned by the fit was 0.73 ± 0.08 (SE) mM and the Hill coefficient was 3.04 ± 0.38.

Tetramisole also reduced the amplitude of the orthodromic responses in a dose-dependent manner (ANOVA, $P < 0.0001$, Fig. 12.2b). The amplitude of orthodromic response was significantly less than control and recovery with tetramisole at 0.5 and 1 mM (PLSD Fischer test, $P < 0.0001$ for all comparisons).[2]

Comparing Fig. 12.2a, b, it can be seen that, with tetramisole at 0.5 and 1 mM, orthodromic responses were significantly more reduced than antidromic responses. With 0.5 mM the antidromic response amplitude represented 73 ± 4 % of the control whereas the orthodromic response was reduced to 35 ± 12 % of the control ($P = 0.01$, paired t-test); likewise, with 1 mM tetramisole the antidromic response amplitude (42 ± 4 % of control) was less reduced that the orthodromic response (9 ± 6 % of control) ($P = 0.0002$). The stronger effect of tetramisole on orthodromic response amplitude is confirmed by the Hill's equation fitted to the dose-response relationship, which returned an IC_{50} of 0.41 ± 0.15 mM (Fig. 12.2d). The Hill coefficient was 2.97 ± 2.27.

[2]Orthodromic responses could not be tested with 2 mM tetramisole: among the 4 experiments in which we used this concentration, 2 showed no measurable orthodromic response in the control condition, while in the other 2 experiments the orthodromic response obtained in the control condition failed to recover.

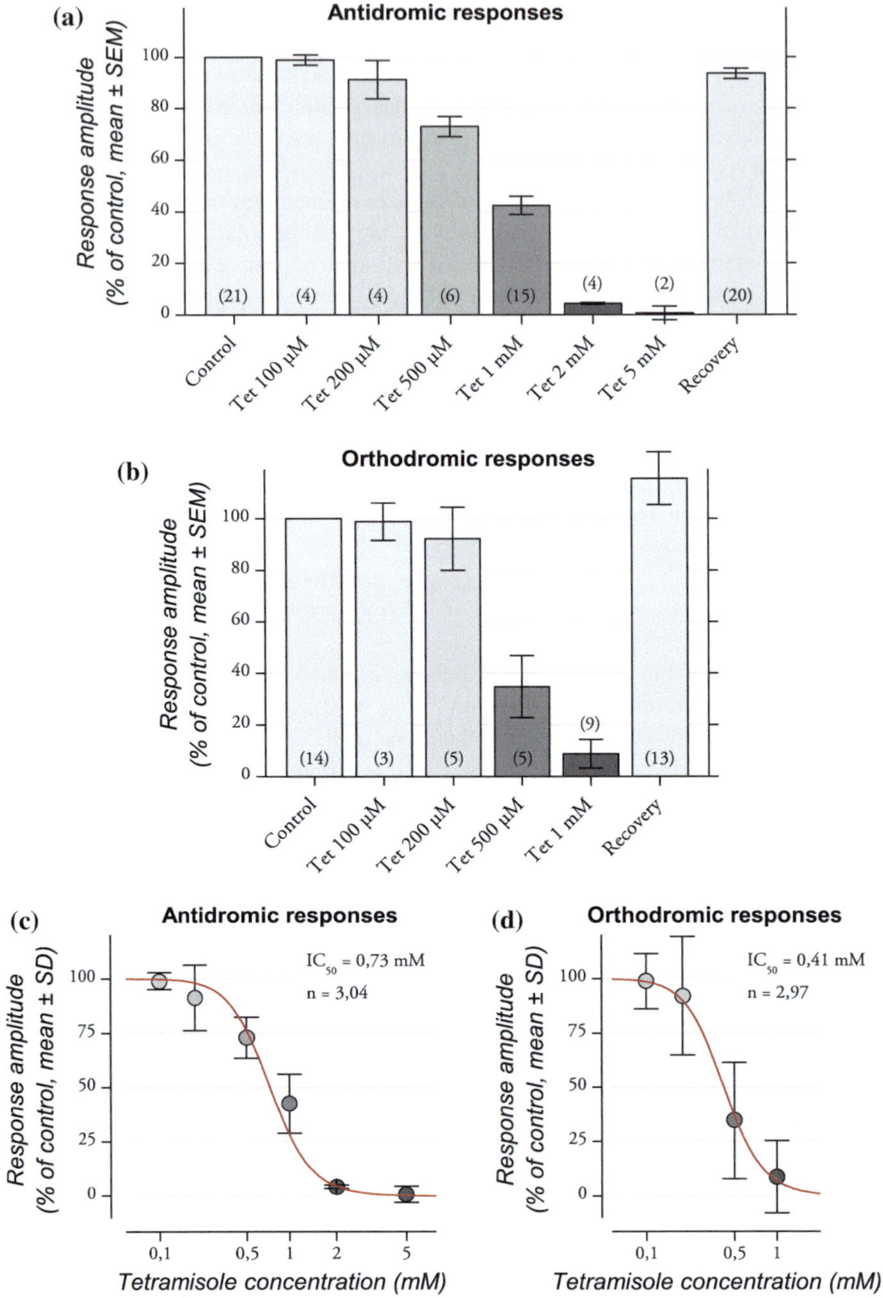
(a)
Antidromic responses
Response amplitude (% of control, mean ± SEM)
100
80
60
40
20
0
(21)
(4)
(4)
(6)
(15)
(4)
(2)
(20)
Control
Tet 100 μM
Tet 200 μM
Tet 500 μM
Tet 1 mM
Tet 2 mM
Tet 5 mM
Recovery
(b)
Orthodromic responses
Response amplitude (% of control, mean ± SEM)
100
80
60
40
20
0
(14)
(3)
(5)
(5)
(9)
(13)
Control
Tet 100 μM
Tet 200 μM
Tet 500 μM
Tet 1 mM
Recovery
(c)
Antidromic responses
Response amplitude (% of control, mean ± SD)
IC50 = 0,73 mM
n = 3,04
100
75
50
25
0
0,1
0,5
1
2
5
Tetramisole concentration (mM)
(d)
Orthodromic responses
Response amplitude (% of control, mean ± SD)
IC50 = 0,41 mM
n = 2,97
100
75
50
25
0
0,1
0,5
1
Tetramisole concentration (mM)

◀ **Fig. 12.2** Reduction of antidromic and orthodromic response amplitude by tetramisole is dose-dependent. **a–b** Bar graphs illustrating effect of various tetramisole ("Tet" on x-axis) concentrations (100 μM–5 mM) on antidromic (**a**) and orthodromic (**b**) response amplitude. Before pooling amplitudes were normalized by the control amplitudes and are expressed as a percentage of control response amplitude. Bar height represents the mean and error bars represent the SEM. Note near complete suppression of antidromic responses with tetramisole at 2 mM, and of orthodromic responses with tetramisole at 1 mM. In **a** and **b** the total number of tested concentrations is larger than the number of controls and recoveries because in some experiments several tetramisole concentrations were successively tested between one control and one recovery. For the discrepancy in the number of controls and recoveries in **a** and **b** see legend of Fig. 12.5. **c–d** Dose-response relationships for antidromic (**c**) and orthodromic (**d**) responses. The dots show mean normalized amplitudes, as in **a–b**, but the error bars represent 1 SD. The continuous lines correspond to the Hill equation (see Results) fitted to the data. The fits were weighted by the variance. The R^2 of the fits was 0.99 for both antidromic and orthodromic responses. The free parameters values (IC_{50} and n) are displayed in the figure

12.3.2 Tetramisole Increases Antidromic Response Latency

In addition to decreasing their amplitudes, tetramisole also increased the latency of the antidromic population spikes. This is illustrated with one example on Fig. 12.3a. In this experiment we applied 3 different tetramisole concentrations. It can be seen that both the onset latency and the latency to peak of the antidromic response increased with increasing concentration of tetramisole. Increase in latency is even visible with a tetramisole concentration of 0.5 mM although this concentration had a marginal effect on response amplitude in this experiment.

For population level analysis we examined and quantified the effect of tetramisole on the latency to peak of the antidromic population spikes. As shown in Fig. 12.3b, tetramisole had a clear, dose-dependent effect on peak latency (ANOVA, $P < 0.0001$). Relative to control (100 %), the latency was significantly increased (+11 ± 1.5 %) with 200 μM tetramisole (PLSD Fischer test, $P = 0.03$), although this concentration had no significant effect on response amplitude (Fig. 12.2a). Increases in latency were also significant with 0.5 mM (+17 ± 2 %, $P = 0.0004$), 1 mM (+41 ± 5 %, $P < 0.0001$) and 2 mM tetramisole; with 2 mM tetramisole, the latency was more than doubled (+110 ± 0.3 %, $P < 0.0001$). Thus, tetramisole not only reduced the amplitude of the responses, but also increased their latencies.

12.3.3 Tetramisole Reduces Amplitude and Increases Latency of Axonal *Population Spikes*

The increase of antidromic response latency could have two origins. The first is that tetramisole affected the somatic membrane potential, leading to an impairment of the coupling between the axon and the cell body. The second is that the increase in response latency originated from an action of tetramisole on the axons themselves.

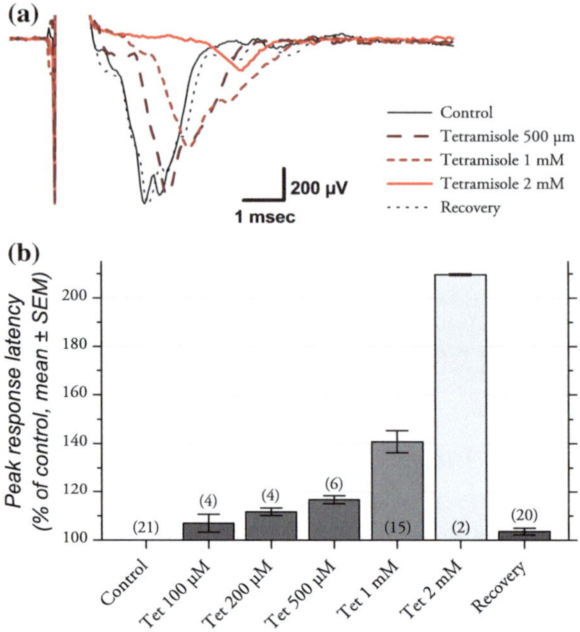

Fig. 12.3 Tetramisole increases antidromic response latency. **a** Effect of tetramisole at different concentrations on antidromic population spikes in one experiment. LFP recordings in the supragranular layers of the barrel cortex. Stimulation in the white matter (60 µA, 200 µsec). Each trace corresponds to the average of 30 sweeps (1 min recording) in control ACSF and in the presence of tetramisole at 0.5, 1 and 2 mM. Tetramisole decreased response amplitude and increased peak response latency (control: 2.33 ms; 0.5 mM: 2.86 ms; 1 mM: 3.47 ms; 2 mM: 4.78 ms; recovery: 2.42 ms). Onset latency also clearly increased in this experiment, indicating that increased peak latency was not simply due to increased spike duration. **b** Bar graph summarizing the effect of tetramisole ("Tet", 0.1–2 mM) on the peak latency of antidromic population spike at the population level. Effect of tetramisole at 2 mM has been examined in 4 experiments but the response was completely suppressed in two of these so that latency could not be measured. Before pooling latencies were normalized by the latency measured in control ACSF and are expressed as a percentage of the control latency. Bar height: mean; error bars: SEM

In order to distinguish between these two possibilities, we made experiments in which both stimulation and recording were performed in the corpus callosum. The corpus callosum is the major axonal tract of the central nervous system. It mostly contains myelinated axons. The electrical stimulation initiated *axonal* population spikes that traveled along the axons. The axonal population spikes were recorded at approximately 1 mm from the stimulating electrode. The example in Fig. 12.4a illustrates the effect of tetramisole at 1 mM: as for antidromic population spikes recorded in the cortical gray matter, the axonal population spike recorded in the white matter showed both reduced amplitude and increased latency. The effects of tetramisole (1 mM) averaged over 4 experiments are summarized in Fig. 12.4b: tetramisole significantly reduced the peak response amplitude to a value representing 56 ± 7 % of the control amplitude (paired t-test, $P = 0.008$) and the peak

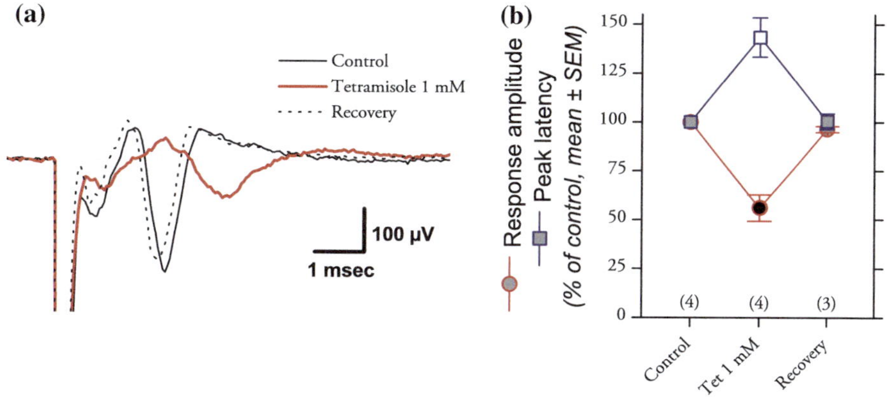

Fig. 12.4 Tetramisole reduces axonal population spike amplitude and reduces axonal conduction velocity. For these experiments both stimulation and recording were performed in the corpus callosum. **a** Example of the effect of tetramisole (1 mM) on axonal population spike amplitude and latency. Each trace corresponds to the average of 6 sweeps (1 min recording). Stimulation in the midline of the corpus callosum; recording approximately 1 mm lateral to the midline, below motor cortex. The amplitude of the axonal population spike was reduced in tetramisole 1 mM (control: –0.33 mV; tetramisole 1 mM: –0.14 mV; recovery: –0.32 mV). The peak latency was increased in tetramisole (control: 2.25 ms; tetramisole 1 mM: 3.54 ms; recovery: 2.10 ms), implying reduced conduction velocity in the axons. **b** Effects of tetramisole 1 mM ("Tet 1 mM") on the amplitude and latency of axonal population spikes, group data. Amplitudes and latencies normalized by control values. The squares represent the mean peak latency and the circles represent the mean amplitude; the error bars represent the SEM

latency was significantly increased by +43 ± 10 % relative to the control latency ($P = 0.02$, Fig. 12.4b). These data indicate that action potential transmission along the axons, hence axonal conduction velocity, was altered by tetramisole.

12.3.4 *Effect of Tetramisole on Antidromic Population Spikes Is not Compensated by Pyridoxal*

The data we have presented thus far suggest that orthodromic and antidromic response suppression by tetramisole results in large part from an impaired transmission of action potentials along the axons. Action potential electrogenesis in myelinated axons takes place at the node of Ranvier. Coincidentally, TNAP appears to be highly concentrated at the nodes of Ranvier. Since one of the main substrate of TNAP is PLP, we hypothesized that tetramisole, by preventing extracellular PLP to PL conversion, perturbed an (unknown to us) intracellular PL/PLP-dependent mechanism involved in action potential initiation and/or propagation. To test this possibility, we recorded antidromic population spikes and compared the effect of tetramisole 1 mM with the effect of tetramisole 1 mM to which we added PL, expecting that exogenous PL would compensate for the inhibition of TNAP by tetramisole. We used a PL concentration of 10 μM, a value much larger than the concentration measured in the CSF (≤0.1 μM: Spector 1978, van der Ham et al. 2012). An example is presented in Fig. 12.5a. The effect of tetramisole alone is identical to that illustrated previously: a reduction of the antidromic population spike amplitude associated with an increased latency. Yet the trace obtained with tetramisole 1 mM + PL 10 μM is nearly identical to that obtained with tetramisole 1 mM alone, indicating that PL did not palliate the action of tetramisole. When examined at the population level (Fig. 12.5b), the mean population spike amplitude measured in the presence of tetramisole did not differ significantly from that

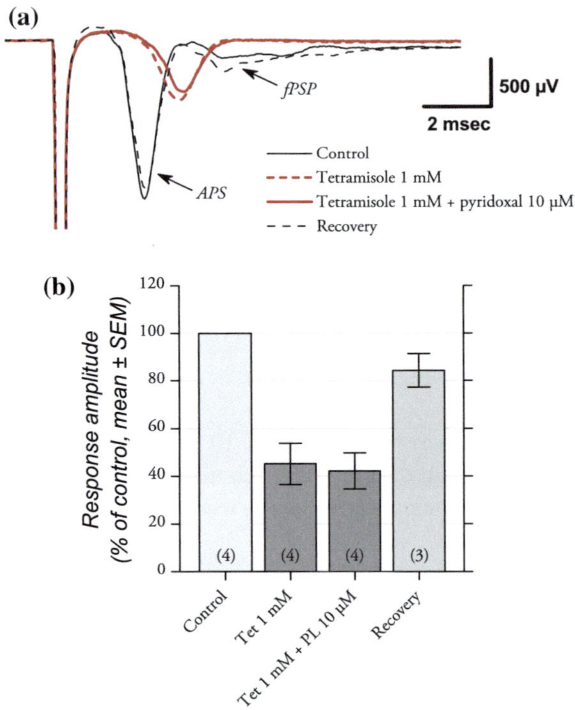

Fig. 12.5 Extracellular PL does not compensate for the effect of tetramisole. **a** Comparison of the effects of tetramisole at 1 mM and tetramisole 1 mM + pyridoxal 10 μM in one experiment. Recording in the supragranular layers of the somatosensory cortex, stimulation in the white matter. Each trace is the average of 30 sweeps (1 min recording) in control ACSF (control and recovery), in ASCF + tetramisole 1 mM, and in ACSF + tetramisole 1 mM + PL 10 μm. Tetramisole reduced the antidromic population spike amplitude and increased its latency, and completely suppressed the slow postsynaptic response (fPSP). PL did not compensate for the effect of tetramisole. **b** Bar graph summarizing the effect of tetramisole 1 mM or tetramisole 1 mM + PL 10 μM on antidromic population spike amplitude. As previously data were normalized by control response amplitude and are expressed as a percentage of control response amplitude. Bar height: mean; error bars: SEM. The discrepancy between the number of controls and recoveries comes from one experiment in which the order of drug application and tests was: control ACSF → tetramisole 1 mM → tetramisole 1 mM + PL 10 μM → tetramisole 1 mM. Response amplitude during the first and second application of tetramisole 1 mM were perfectly identical, and therefore fitted with our stationarity criteria although recovery was not tested in regular ACSF in this experiment

measured in the presence of tetramisole + PL (paired t-test, P = 0.11, n = 4 experiments). On the other hand, both were significantly less than the control response amplitude (tetramisole 1 mM alone: 45 ± 9 % of control, P = 0.008; tetramisole 1 mM + PL 10 μM: 42 ± 8 % of control, P = 0.005). In short, the decreased amplitude and increased latency of the antidromic responses cannot be explained by the alteration of a PL/PLP-dependent mechanism consecutive to TNAP inhibition by tetramisole.

12.3.5 Effect of Tetramisole Is not Compensated by Adenosine and Is not Mimicked by ATP

As outlined in the introduction, TNAP could be one of the ectonucleotidases involved in extracellular adenosine synthesis. Noticeably, a study (Irnich et al. 2002) suggested that adenosine receptors might control the excitability of peripheral nerves axons, although we are not aware of any study demonstrating that this also applies to cortical axons. Therefore, another possibility we examined was that the reduction of antidromic and axonal population spike amplitudes could be consecutive to a decrease in extracellular adenosine concentration due to TNAP inhibition by tetramisole. To explore this possibility, we compared the effect of tetramisole 1 mM alone with those of tetramisole 1 mM to which 100 μM of adenosine was added. This concentration of adenosine corresponds to the one that has maximal or near maximal effect on A1 receptors in cortex (Fontanez and Porter 2006). However, exogenous adenosine did not compensate for the effect of tetramisole. The example shown in Fig. 12.6a shows that the antidromic population spike in the presence of 1 mM tetramisole + 100 μM adenosine was barely distinguishable from that obtained in the presence of tetramisole alone. Both were strongly decreased in amplitude in comparison to control and recovery traces. On average (3 experiments, Fig. 12.6b), the antidromic population spike reduction with tetramisole 1 mM + adenosine 100 μM did not differ significantly from that observed with tetramisole 1 mM alone (P = 0.09, paired t-test, 52 ± 11 % vs. 40 ± 9 % of control response amplitude respectively). Both were significantly reduced in comparison to control (P = 0.04 and P = 0.02 respectively). Therefore, reduction of response amplitude in the presence of tetramisole does not seem to be explained by a reduced extracellular concentration of adenosine following TNAP inhibition by tetramisole.

Alternatively, blocking the ectonucleotidase activity of TNAP may also have led to an increased concentration of extracellular ATP. We therefore examined whether increasing extracellular ATP concentration mimicked the effect of tetramisole on antidromic population spikes (n = 2). As illustrated in Fig. 12.6c, extracellular ATP (1 mM) did not modify the amplitude or the latency of the antidromic responses. This suggests that the effect of tetramisole on antidromic population spikes is not the consequence of a heightened extracellular concentration of ATP due to TNAP inhibition.

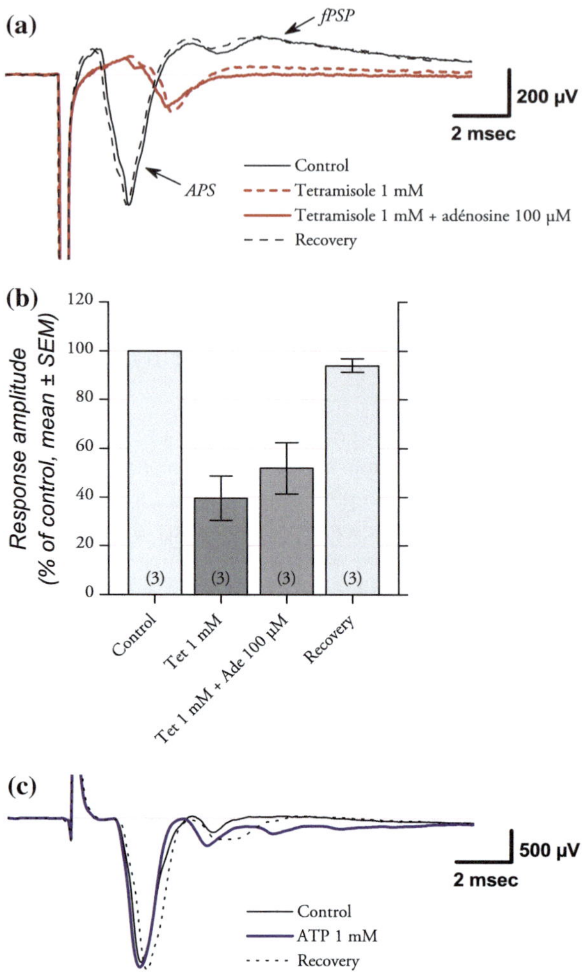

Fig. 12.6 High extracellular adenosine concentration does not compensate for the effect of tetramisole and the effect of tetramisole is not mimicked by high extracellular ATP concentration. **a** Comparison of the effects of tetramisole at 1 mM and tetramisole 1 mM + adenosine 100 μM. Recording in the supragranular layers of the somatosensory cortex, stimulation (100 μA, 300 μs) at the white matter-layer 6 border. Each trace is the average of 30 sweeps (1 min recording) in control ACSF (control and recovery), in ASCF + tetramisole 1 mM, and in ACSF + tetramisole 1 mM + adenosine 100 μm. As previously, tetramisole reduced the amplitude and increased the latency of the APS and eliminated the fPSP. Adenosine did not reverse the effect of tetramisole. **b** Bar graph summarizing the effect of tetramisole 1 mM or tetramisole 1 mM + adenosine (Ade) 100 μM on antidromic population spike amplitude. Amplitudes normalized by control values and expressed as a percentage of control response amplitude. Bar height: mean; error bars: SEM. **c** ATP (1 mM) does not mimic the effect of tetramisole. Recording in the supragranular layers of the somatosensory cortex, stimulation (100 μA, 300 μsec) in the white matter. Each trace is the average of 30 sweeps (1 min recording) in control ACSF (control and recovery) and in ASCF + ATP 1 mM

12.3.6 The Effect of Tetramisole Is not Stereo-Specific

Although they constitute two of the identified metabolites of TNAP, exogenously applied adenosine or PL did not compensate for the effect of tetramisole on neuronal response amplitude. In addition, the IC_{50} for neuronal response suppression by tetramisole (Fig. 12.2c, d) was much higher than that expected given the IC_{50} values reported for TNAP inhibition by levamisole or tetramisole in biochemical assays on cell extracts or cell cultures (Van Belle 1972, 1976a, 1976b; Goldstein et al. 1980; Anagnostou et al. 1996; Calhau et al. 2000; Picher et al. 2003; Sergienko and Millán 2010; Debray et al. 2013): the IC_{50} that has been reported in these studies is between 10 and 70 μM. In addition, the slope of the Hill equation fitted to our data was close to 3, while a Hill coefficient value close to 1 has been reported for the inhibition of TNAP by levamisole (Suzuki et al. 1994). Altogether, these negative results and discrepancies led us to suspect that tetramisole may have acted on a target *other than TNAP*.

Studies showed that levamisole (the levorotatory enantiomer of tetramisole), but not dexamisole (the dextrorotary enantiomer of tetramisole), is effective at inhibiting TNAP (Van Belle 1972, 1976b; Borgers 1973). If inhibition of TNAP activity was responsible for the effects we observed, then it should also be stereo-specific. It follows that the effect observed with a given concentration of tetramisole should be mimicked by levamisole at half that concentration. We therefore compared the effect of levamisole and tetramisole on antidromic response amplitude in 7 experiments. In each of these experiments 1 or 2 different concentrations of levamisole (0.5 or 1 mM) and identical concentrations of tetramisole were used. The prediction was that levamisole at 0.5 mM should have the same effect as tetramisole at 1 mM.

The results we obtained did not fit with this prediction. As illustrated in Fig. 12.7a, levamisole at 0.5 and 1 mM reduced both antidromic and orthodromic response amplitude and increased response latency, in a way comparable to tetramisole. Summary data in Fig. 12.7b show a highly significant reduction of antidromic response amplitude by both tetramisole and levamisole at 0.5 and 1 mM (ANOVA, $P < 0.0001$; PLSD Fisher test: $P < 0.0001$ for levamisole at 0.5 and 1 mM compared to control). However, the reduction of antidromic population spike amplitude by levamisole at 0.5 mM (67 ± 5 % of control response amplitude) was not significantly different ($P = 0.9$) from that of tetramisole at 0.5 mM (68 ± 3 % of control response amplitude). Likewise, the effect of levamisole at 1 mM did not differ significantly ($P = 0.3$) from that of tetramisole at 1 mM (46 ± 6 % and 40 ± 6 % of control response amplitude, respectively). On the other hand, levamisole at 0.5 mM was significantly less potent than tetramisole at 1 mM ($P < 0.0001$). These results show that the effect of tetramisole is not stereo-specific. This strongly suggests that the reduction of response amplitude induced by tetramisole was not mediated by TNAP.

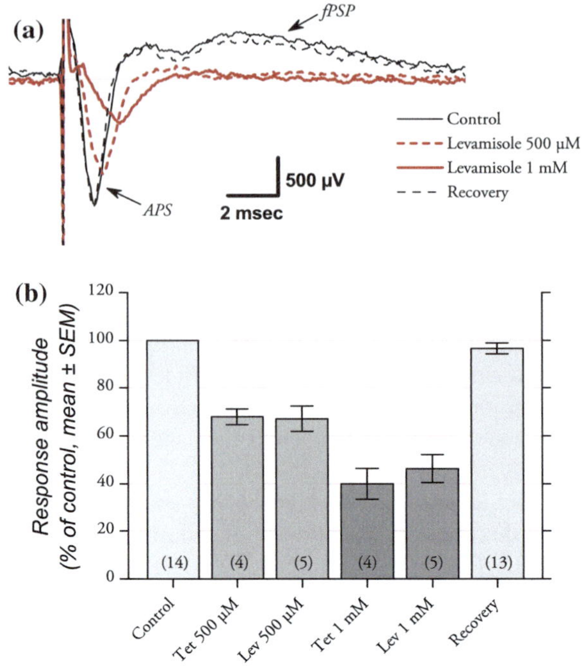

Fig. 12.7 Comparison of the effectiveness of tetramisole and levamisole on antidromic population spike amplitude reduction indicates lack of stereo-specificity in the action of tetramisole. **a** Example of the effects of levamisole at 0.5 and 1 mM. Recording in the supragranular layers of the barrel cortex, stimulation (65 µA, 200 µs) at the white matter-layer 6 border. Each trace is the average of 6 sweeps (1 min recording) in control ACSF (control and recovery) and in ASCF containing either 0.5 or 1 mM levamisole. As with tetramisole, levamisole reduced the amplitude and increased the latency of the antidromic population spike (APS). As with tetramisole, levamisole also suppressed the slow postsynaptic response (fPSP). **b** group data. "Tet" and "Lev" on x-axis refer to tetramisole and levamisole, respectively. Bar height represents the mean percent change in APS amplitude, error bars represent the SEM. Before averaging the amplitudes were normalized by the amplitude obtained in the control condition. The data presented here come from 7 experiments in which at least one concentration of levamisole and one concentration of tetramisole were tested in the same experiment. The total number of controls is 14 because control measurements were performed before both levamisole and tetramisole applications. One recovery not tested. Total number of measurements in levamisole and tetramisole sums to 18 because several concentrations of tetramisole or levamisole were eventually tested between one control and one recovery. The mean amplitude obtained in tetramisole 0.5 mM does not differ significantly from that obtained in levamisole 0.5 mM, and similarly when comparing tetramisole and levamisole at 1 mM

12.3.7 The Effect of Tetramisole Is not Mimicked by MLS-0038949, Another TNAP Inhibitor

An additional proof that tetramisole reduced neuronal response amplitude independently from TNAP inhibition was obtained by using another inhibitor of TNAP. This inhibitor—MLS-0038949—has been recently isolated and shows a very high

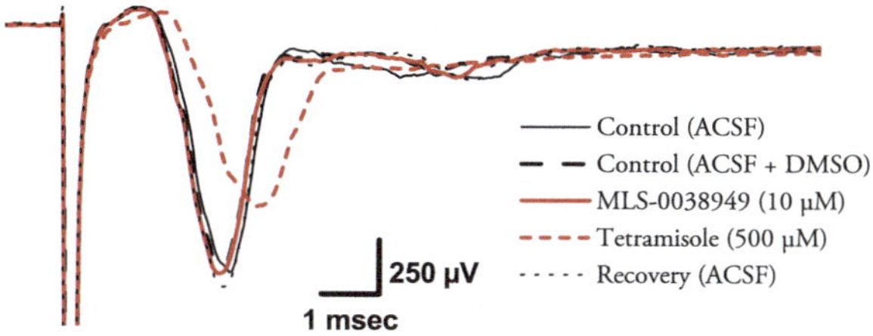

Fig. 12.8 Antidromic population spikes are not affected by MLS-0038949, another TNAP inhibitor. In this experiment, recording was performed in the supragranular layers of the barrel cortex and stimulation (100 μA, 200 μs) was applied in the white matter. Each trace is the average of 6 sweeps (1 min recording). Since MLS-0038949 was dissolved in DMSO, two controls were performed, one in regular ACSF and the other in ACSF + DMSO 0.1 %. DMSO had no effect the antidromic population spike. As in previous experiments, tetramisole (here at 0.5 mM) reduced the amplitude of the antidromic population spike and increased its latency. In contrast, MLS-0038949 (10 μM) had no effect on the amplitude and latency of the antidromic population spike

specificity for TNAP (Dahl et al. 2009). In Dahl et al. (2009) study, the IC_{50} for MLS-0038949 was about 0.2 μM and a complete inactivation of TNAP activity was achieved with 10 μM of MLS-0038949. In control histochemical experiments, we found that MLS-0038949 at 10 μM largely inhibited TNAP activity in mouse brain slices as assessed by histochemisty (NBT-BCIP method, as in Fonta and Imbert 2002) (not illustrated). We therefore examined the effect of MLS-0038949 at 10 μM in 3 experiments. Results of one of these experiments are presented in Fig. 12.8. Since MLS-0038949 was dissolved in DMSO to a final concentration of 0.1 % DMSO in ACSF (Methods), we first checked that DMSO 0.1 % per se had no effect on the response; the response obtained in ACSF + DMSO 0.1 % (Fig. 12.8, dashed line) is indistinguishable from that obtained in regular ACSF. We also checked that tetramisole had its usual action on the antidromic population spike; here again tetramisole (500 μM in this experiment) reduced the response amplitude and increased the peak response latency (Fig. 12.8, short dash). On the other hand, the antidromic population spikes was completely unaffected by MLS-0038949 at 10 μM (Fig. 12.8, thick solid line). The two other experiments failed to show any change on antidromic response amplitude and latency as well. These results confirm that the effects of tetramisole and levamisole on antidromic responses were not due to these compounds inhibiting TNAP activity.

12.3.8 Tetramisole Modifies Action Potential Shape and DV/Dt

In order to untangle how tetramisole reduced antidromic population spike amplitude, we next sought to examine its action on action potential electrogenesis using

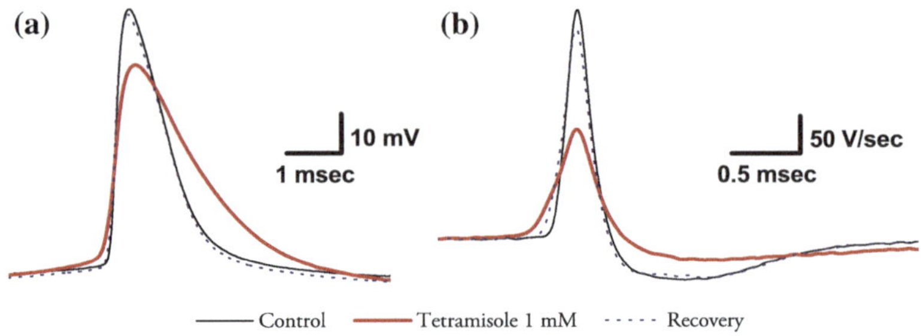

Fig. 12.9 Tetramisole prolongs action potential duration and reduces action potential rate-of-rise. **a** Action potentials were induced by intracellular current injection in an intracellularly recorded neuron (same cell as in Fig. 12.1h–j). Each trace corresponds to the average of 30 action potentials for each condition. Note reduced amplitude and prolonged duration in tetramisole 1 mM. **b** Each trace corresponds to the temporal derivative (dV/dt) of the action potential traces shown in **a**. The maximum rate of rise of the action potential (positive peak in the dV/dt) was reduced to 48 % of the control value in tetramisole 1 mM (control: 316 V/s; tetramisole 1 mM: 152 V/s; recovery: 302 V/s). These data suggest that the voltage-dependent sodium channels responsible for action potential upstroke were partially blocked by tetramisole 1 mM

intracellular recording in somatosensory cortex neurons. Although several cells have been recorded, only one has been held for long enough to obtain a control baseline, to examine the effect of tetramisole (1 mM), and to achieve a complete recovery. The results presented here are therefore very preliminary.

We observed that tetramisole (1 mM) modified the shape of the action potentials. This is illustrated in Fig. 12.9a, where each trace corresponds to the average of 30 action potential traces; these action potentials were induced by intracellular current injection; the mean firing rate induced by these current injections was around 40 spikes/sec in the three conditions. In comparison to the control trace, the action potential in the presence of 1 mM tetramisole showed a decrease in amplitude together with an increased duration. This effect of tetramisole was reversible.

We next computed the first derivative of the action potential (dV/dt). The positive peak in the dV/dt corresponds to the maximum rate-of-rise of the action potential. The maximum rate-of-rise, in turn, is a measure of the inward current underlying the rise of the action potential (Hodgkin and Katz 1949). This inward current is largely, if not entirely, determined by the voltage-dependent sodium conductance (Hodgkin and Katz 1949; Cohen and Strichartz 1977; Hondeghem 1978; Carter and Bean 2009). It can be seen in Fig. 12.9b that tetramisole (1 mM) reduced the maximum rate of rise of the action potential to half the control value. This effect was completely reversible. This result suggests that voltage-dependent sodium channels were partially blocked by tetramisole at 1 mM.

12.4 Discussion

Our study disclosed a new action of tetramisole and levamisole on neuronal signal transmission, which appears to be independent of these compounds acting on TNAP. We showed that tetramisole and levamisole reduced both antidromic and orthodromic response amplitude in mouse somatosensory cortex in a dose-dependent manner. This is a hitherto undocumented effect of tetramisole and levamisole. Yet our control experiments showed that these effects of tetramisole were not explained by inhibition of TNAP. Instead, the decrease of axonal conduction velocity and preliminary intracellular data suggest that tetramisole and levamisole may block voltage-dependent sodium channels. These results imply that, in addition to inhibiting TNAP, levamisole or tetramisole at concentration ≥1 or ≥2 mM completely inhibit all orthodromic and antidromic neuronal activity, respectively, and are therefore likely to suppress all processes that are activity-dependent in living brain cells, such as, among others, axonal growth, myelination or synaptic plasticity.

12.4.1 Suppression of Neuronal Activity by Tetramisole and Levamisole is not Due to TNAP Inhibition

Neuronal responses, both orthodromic and antidromic, were reduced by tetramisole and levamisole (Figs. 12.1, 12.2 and 12.7). These effects are unlikely to be due to TNAP inhibition, for several reasons:

First, the dose-response relationships (Fig. 12.2) appear to be very different from those reported for TNAP inhibition by tetramisole or levamisole. The IC_{50} we report here were 730 μM for antidromic response and 410 μM for orthodromic responses. Both values are much larger than those reported for TNAP inhibition by levamisole in biochemical assays on cell extracts or on cells cultures—in the range of 10–70 μM (Van Belle 1972, 1976a, b; Goldstein et al. 1980; Anagnostou et al. 1996; Picher et al. 2003; Sergienko and Millán 2010; Debray et al. 2013), including brain cell extracts (IC_{50} = 44 μM in Calhau et al. 2000). In addition the exponent of the Hill equation fitted to our data was close to 3, whereas TNAP inhibition has a Hill coefficient close to 1 (Suzuki et al. 1994).

Second, neuronal activity block was not rescued by the application of exogenous adenosine or PL (Figs. 12.5 and 12.6), although these were the metabolites whose extracellular concentrations were presumed to be reduced by TNAP inhibition in our experimental conditions (see Introduction).

Third, in our experiments equivalent concentrations of levamisole and tetramisole reduced neuronal responses by the same amount (Fig. 12.7). Given that tetramisole contains both levamisole and dexamisole in identical proportions, this implies that dexamisole had the same potency at inhibiting neuronal responses as levamisole. This further rules out TNAP inhibition as being responsible for the effects reported here, given that only levamisole if effective at inhibiting TNAP (Van Belle 1972, 1976b; Borgers 1973).

Fourth and finally, blockage of neuronal activity by levamisole and tetramisole were not mimicked by MLS-0038949, a new and highly selective inhibitor of TNAP (Dahl et al. 2009) (Fig. 12.8).

Our results should not be interpreted as meaning that tetramisole or levamisole did not inhibit TNAP in our experimental conditions. Indeed, we performed control histochemical experiments (same method as Fonta and Imbert 2002) using brain slices prepared as for the electrophysiological experiments. These experiments confirmed that AP activity was effectively inhibited by tetramisole (not illustrated). Therefore, TNAP inhibition likely took place in our electrophysiological experiments but its consequences on synaptic transmission, if they occurred, were masked by the suppression of neuronal activity by tetramisole and levamisole.

12.4.2 Tetramisole and Levamisole May Suppress Neuronal Activity by Blocking Voltage-Dependent Sodium Channels

If TNAP inhibition was not responsible for the decrease in neuronal response amplitude produced by levamisole and tetramisole, then what other targets could explain the effects we observed? Two observations suggest that voltage-dependent sodium channels were blocked by tetramisole and levamisole.

First, with concentrations of tetramisole that did not completely suppress the responses, we observed that the latency of the antidromic population spikes increased (Fig. 12.3). The same effect was noticed when recording from axons in the corpus callosum (Fig. 12.4). These results indicate a slowing down of action potential propagation along the axons. Slowing down of action potential propagation speed in axons is not observed with compounds that block voltage-dependent potassium channels (Bostock et al. 1981; Fox and Ruan 1989), but it is typically observed with compounds that block voltage-dependent sodium channels such as TTX (Pinto et al. 2008), phenytoin (Le Quesne et al. 1976; Marcus et al. 1981) and several local anesthetics such as procaine (Franz and Perry 1974) and lidocaine (Raymond 1992; Yokota et al. 1994; De Col et al. 2008). Interestingly, a pharmacological study reported that levamisole possesses local anesthetic properties with a potency representing half that of lidocaine (Onuaguluchi and Igbo 1987); the ED_{50} obtained in this study was quite close to the IC_{50} we report here for the reduction of antidromic population spike amplitude.

Second, studies showed that the compounds that reduce axonal conduction velocity also reduce the peak height of the action potential dV/dt (TTX: Kao and Walker 1982; phenytoin: Selzer 1979, Hershkowitz and Ayala 1981; procaine: Ibusuki et al. 1998; lidocaine: Schwarz and Puil 1998). Since the peak height in the dV/dt is a measure of the sodium current underlying action potential electrogenesis, the reduction of the dV/dt_{max} observed in the present study (Fig. 12.9) further supports the possibility that tetramisole blocks voltage-dependent sodium channels, although additional studies are required for a definitive confirmation. It is also presently unclear whether tetramisole directly blocks voltage-dependent sodium channels, or whether this effect is secondary to tetramisole acting on intracellular signaling pathways that control sodium channel gating properties.

Complete suppression of antidromic responses required a tetramisole concentration of about 2 mM whereas orthodromic responses were nearly completely suppressed with 1 mM tetramisole. Likewise, the IC_{50} for the two types of responses differed by a factor close to 2. We can only speculate on the origin of this discrepancy. The postsynaptic response is proportional to the amount of neurotransmitter released, which is determined by calcium influx inside the presynaptic terminals, which is itself the result of the activation of high threshold voltage-dependent calcium channels by sodium spikes in the axon terminals. Reduction of sodium spike amplitude may then results in a decreased neurotransmitter release but, to fit with our data, a nonlinearity in the presynaptic spike amplitude to postsynaptic response transform would be required. Alternatively, high threshold voltage-dependent calcium channels, which show some structural homologies with voltage-dependent sodium channels (e.g., Zakon 2012), may also be blocked by tetramisole.

12.4.3 Other Targets of Tetramisole and Levamisole in the Nervous System

It has long been shown that tetramisole and levamisole inhibit TNAP. The results we report here further suggest that tetramisole and levamisole may also block voltage-dependent sodium channels. In addition to these two targets, tetramisole and levamisole have been reported to have other actions in the nervous system. We shall first review studies that examined the effect of tetramisole and levamisole using electrophysiological approaches. These studies have been very few and concentrated mostly on nematode neuromuscular junction and mammalian peripheral nervous system.

12.4.3.1 Tetramisole, Levamisole and Acetylcholine Receptors

Tetramisole has originally been isolated for its anthelmintic properties (Thienpont et al. 1966; Raeymaekers et al. 1966). Initially based on screening tests in chicken, tetramisole was proven to be efficient against both intestinal and pulmonary nematode infections in a dozen mammalian species including human and tiger. Levamisole was shown to be the effective compound while dexamisole has no anthelmintic properties (Bullock et al. 1968). The action of tetramisole and levamisole consists in a paralysis of the nematodes (Thienpont et al. 1966; Aceves et al. 1970; Atchison et al. 1992). Nematode paralysis results from a maintained depolarization of their muscle cells (Aceves et al. 1970; Harrow and Gration 1985; Atchison et al. 1992). Muscle cell depolarization is the consequence of levamisole acting as an agonist on neuromuscular acetylcholine (ACh) receptors of the nematode (Lewis et al. 1980; Harrow and Gration 1985; Robertson and Martin 1993).

Fortunately in mammals levamisole does not appear to be an agonist of the neuromuscular ACh receptor (Atchison et al. 1992; Rayes et al. 2004). Instead, on the basis of their pharmacological profiles (Lewis et al. 1980) and of genetic homologies (Fleming et al. 1997), it has been proposed that the neuromuscular ACh receptors that are activated by levamisole in the nematodes are homologous to ganglionic ACh receptors of mammals.

Nevertheless, the effect of levamisole on the ganglionic ACh receptors of mammals ($\alpha3\beta2$ and $\alpha3\beta4$ receptors) is unlike that observed in nematodes (Levandoski et al. 2003): levamisole applied alone has virtually no effect and therefore does not behave as an agonist of the ganglionic ACh receptors. On the other hand, levamisole modulates the response to exogenously applied ACh. Yet this modulation appears to be complex, as it depends on both levamisole and ACh concentrations. Thus the modulation exerted by levamisole can be either facilitatory or inhibitory.

Beside autonomic ganglia, the $\alpha3\beta2$ and $\alpha3\beta4$ receptors can be found in different brain regions but not in cortex (Perry et al. 2002). A large fraction of the nicotinic receptors found in cortex corresponds to the $\alpha4\beta2$ subunits composition, for which, to our knowledge, levamisole action has not been examined. The $\alpha7^*$ is another ACh receptor family that is largely represented in cortex (Paterson and Nordberg 2000) but levamisole does not seem to have any effect on this receptor family (Bartos et al. 2006).

12.4.3.2 Inhibition of Noradrenaline Reuptake by Dexamisole, Levamisole and Tetramisole in Peripheral Nervous System

In addition to cholinergic transmission, noradrenergic transmission in the peripheral nervous system also appears to be affected by tetramisole and levamisole (Vanhoutte et al. 1977; Pires et al. 1979; Gulati et al. 1985). In the various preparations studied (heart, smooth muscle, vas deferens), levamisole or tetramisole had no effect when applied alone and therefore do not appear to be agonists of noradrenaline (NA) receptors. Nonetheless, levamisole and tetramisole produced an enhancement of the response to endogenously released or to exogenously applied NA (Vanhoutte et al. 1977; Pires et al. 1979; Gulati et al. 1985). Response enhancement appeared to result from an inhibition of NA reuptake. The concentrations of levamisole or tetramisole required for near complete NA uptake inhibition were ≤ 40 μM. Dexamisole was more potent than levamisole (Vanhoutte et al. 1977).

The potency of levamisole represented half that of cocaine (Pires et al. 1979) and when NA uptake was already blocked by cocaine, levamisole had no further effect (Gulati et al. 1985), suggesting that levamisole and cocaine were affecting and competing for the same uptake mechanism. Several NA uptake inhibitors, such as cocaine and amphetamines, also inhibit dopamine uptake. Whether this is also the case for levamisole is not known to us but this could provide an explanation as to why illegal cocaine is now so often adulterated with levamisole.

In summary, tetramisole and levamisole may have—at least—4 different effects in the nervous system of mammals:

- Inhibition of TNAP, that may have multiple consequences resulting from interference with numerous PLP-dependent enzymes, including GABA synthesizing enzymes, or from modifications of the extracellular concentrations of ATP and adenosine.
- Modulation of responses mediated by ganglionic ACh receptors.
- Blockage of NA uptake mechanism.
- Blockage of voltage-dependent sodium channels, as reported in the present study.

In addition to these 4 targets, studies suggested that levamisole and tetramisole may affect additional targets in the nervous system, although these are less firmly established. Hence it has been proposed that tetramisole and levamisole could inhibit acetylcholinesterase (Eyre 1970) and monoamine oxidase (Vanhoutte et al. 1977). One study suggested that levamisole interferes with opiate receptors (Spector et al. 1998) but the levamisole doses that have been used in this study have been shown to be deadly in other studies (Mohammad et al. 2006; Rehni and Singh 2010).

12.4.4 Side Effects of Tetramisole and Levamisole Therapies

Levamisole and tetramisole received multiple medical applications: first as an anthelmintic, later on as an immunostimulant for the treatment of a number of diseases.

When used as an anthelmintic, levamisole and tetramisole are used in a single dose and, unless overdosing (Joly et al. 1998), this regimen has not been reported to have serious side effects.[3] On the other hand, serious side effects have been noticed with the chronic (several weeks or months) use of levamisole.

The chronic use of levamisole was largely based on studies reporting that tetramisole and levamisole possess immunostimulating properties (e.g., Renoux and Renoux 1972; Brugmans et al. 1973; Pabst and Crawford 1975; Spreafico et al. 1975;

[3]Adverse reactions after single levamisole or tetramisole doses have been consistently reported in veterinary medicine, and were usually attributed to their actions on ganglionic ACh receptors (reviewed in Hsu 1980).

Renoux et al. 1976; Hadden et al. 1977). Levamisole has thus been used to boost the immune system of patients suffering diseases that were supposed to be associated with, or that received treatments leading to, decreased immunity. Hence levamisole has been used for the treatment of rheumatoid arthritis (Schuermans 1975), of pediatric nephrotic syndrome (Tanphaichitr et al. 1980; Niaudet et al. 1984; Mongeau et al. 1988; British Association for Paediatric Nephrology 1991) and against various skin infections (reviewed in Hadden et al. 1977). Nevertheless, more recent studies have somehow questioned the efficacy of levamisole as an immunostimulant (e.g.: Toivanen et al. 1981; Webster et al. 1982; Aymard et al. 1984; Schiller et al. 1991; Ahmed et al. 1996). Likewise, studies questioned the efficacy of levamisole in the treatment of rheumatoid arthritis (Dinai and Pras 1975) or skin infections (Chang and Fiumara 1978; Seidlin and Straus 1984; Sanchez 2000).

Levamisole has also been used as an immunostimulant in cancer therapy. In this context, levamisole, when used alone, proved to be of limited efficacy (e.g., Toivanen et al. 1981; Treurniet-Donker et al. 1987; Arnaud et al. 1989; Barth and Morton 1995; Moertel et al. 1995) and has even been reported to be worse than placebo (e.g., Chlebowski et al. 1994). On the other hand, quite favorable outcomes have been reported when levamisole was combined with 5-fluorouracil in colon carcinoma treatment (e.g., Moertel et al. 1995).[4]

Thus, with few exceptions (colon carcinoma and pediatric nephrotic syndrome treatment), chronic levamisole treatments proved to be of limited efficacy. More problematical, it appeared that chronic levamisole treatment was associated with multiple side effects, some relatively mild and short-lived such as nausea and diarrhea, other much more severe and eventually life threatening such as dermatitis, cutaneous necrotizing vasculitis, leukopenia and agranulocytosis (e.g., Ruuskanen et al. 1976; Parkinson et al. 1977; MacFarlane and Bacon 1978; Scheinberg et al. 1978; Chang and Fiumara 1978; Toivanen et al. 1981; Niaudet et al. 1984; Moertel et al. 1990; reviewed in Symoens et al. 1978; Larocque and Hoffman 2012; Lee et al. 2012).

Chronic levamisole treatment also led to serious neurological side effects, in particular epileptic seizures and inflammatory leukoencephalopathy, that we shall discuss below.

Beforehand, we shall remind that, although levamisole and tetramisole have several putative targets in the CNS, the involvement of these targets in pathogenic processes depends on their sensitivity to levamisole, hence to the concentration of levamisole in the tissues. With respect to this issue, it is worth mentioning that the plasmatic levamisole concentration is <5 μM in most individuals receiving single, therapeutically relevant levamisole doses (Woestenborghs et al. 1981; Luyckx et al.

[4]In the context of cancer therapy, levamisole has also been reported to possess antiproliferative action in vitro (Kovach et al. 1992; Artwohl et al. 2000). However, the levamisole concentrations required to achieve significant effects in vitro appear to be much higher than the plasmatic concentration of levamisole in clinical therapy. Levamisole concentration in a range similar to the plasmatic concentration measured in patients fails to prevent the proliferation of cancer cell lines in vitro (Grem and Allegra 1989; Wiebke et al. 2003).

1983; Kouassi et al. 1986; Gwilt et al. 2000; Hess et al. 2014), although higher plasmatic concentration (up to 8 μM in Luyckx et al. 1983) have been observed in a minority of individuals. One study further suggested that levamisole may accumulate following repeated daily doses (Reid et al. 1998). In any event, plasmatic levamisole concentration was probably not much higher than 10–15 μM even in the most intensive cures. This rules out the new action of levamisole that we describe here since reduction of neuronal response amplitude and reduction of axonal conduction velocity required tetramisole concentration >100–200 μM. On the other hand, TNAP activity may be significantly affected by levamisole at around 10 μM.

12.4.4.1 Levamisole and Epileptic Seizures

Epileptic seizures have been observed, in all cases in children, during levamisole therapy for the treatment of pediatric nephrotic syndrome (Ruuskanen et al. 1976; Prieur et al. 1978; Palcoux et al. 1994). Seizures occurred after a delay of a few days or weeks after the beginning of the treatment. In Prieur et al. (1978) study, epileptic seizures were observed in 3/50 children. An action of levamisole on nicotinic receptors has been evoked (Palcoux et al. 1994) but alternative mechanisms are quite possible, in particular those involving TNAP inhibition by levamisole.

There are multiple types of epilepsy, each with its own etiology. Quite commonly, impairments of inhibitory mechanisms are involved. In this respect, it is well established that dysfunctions of both GABA- (reviewed in Cossart et al. 2005; Macdonald et al. 2010) and adenosine- (Pagonopoulou et al. 2006; Boison 2012) mediated signaling can lead to epilepsy. As outlined in Introduction, the synthesis of GABA involves TNAP through vitamin B6 metabolism, while the synthesis of extracellular adenosine partially depends, in some structures, on the ectonucleotidase activity of TNAP. That TNAP deficiencies result in epileptic seizures is well documented: in human, epileptic seizures are often observed in perinatal and infantile hypophosphatasia (e.g., Rathbun 1948; Fraser 1957; Béthenod et al. 1967; Baumgartner-Sigl et al. 2007; Balasubramaniam et al. 2010; Taketani et al. 2014; see also Chaps. 14 and 15) and TNAP KO mice, the murine model of severe hypophosphatasia, also present with epileptic seizures (Waymire et al. 1995; Narisawa et al. 1997). In parallel these mice show decreased level of GABA (Waymire et al. 1995; Fonta et al. 2012). In these mice, seizure incidence was reduced, though not definitely suppressed, by administration of PL (Waymire et al. 1995; Narisawa et al. 2001). Altogether, these studies point toward a PLP metabolism deficiency, and its consequence on, at least, GABA synthesis, as the origin of epileptic seizures in TNAP KO mice. In addition to reduced GABA level, we recently observed that adenosine concentration is reduced in the brain of 7 day-old $TNAP^{+/-}$ mice (Fonta et al. 2014). This suggests that reduced adenosine level may also participate in the generation of epileptic seizures in TNAP KO mice.

Since levamisole inhibits TNAP (among others), it is tempting do draw a parallel between the epileptogenic action of levamisole (Ruuskanen et al. 1976; Prieur et al. 1978; Palcoux et al. 1994) and the epileptic phenotype that has been observed in both TNAP KO mice and in severe forms of hypophosphatasia. It is therefore possible that levamisole, by inhibiting TNAP, impairs the balance of excitation and inhibition by reducing the availability of GABA and/or adenosine, leading to epileptic seizures in susceptible individuals.[5]

12.4.4.2 Levamisole and Multifocal Inflammatory Leukoencephalopathy

The first cases of multifocal inflammatory leukoencephalopathy were reported in a small proportion (<1 %) of patients treated for colon carcinoma with 5-fluorouracil and levamisole (Hook et al. 1992; Kimmel and Schutt 1993; Leichman et al. 1993; Neu 1993; Bozik and Gilbert 1994; Critchley et al. 1994; Fassas et al. 1994; Ferroir et al. 1994; additional references in Wu et al. 2006). The patients presented neurological disorders such as diplopia, dysarthria, paresthesia, ataxia, short-term memory impairments and/or loss of consciousness. In all cases MRI examination revealed leukoencephalopathy with multiple white matter lesions. Histological examinations performed from biopsies typically revealed extensive myelin loss and gliosis, but no or only marginal axonal loss. Perivascular lymphocytes and macrophage infiltration were consistently observed, demonstrating an inflammatory process in the leukoencephalopathy (Hook et al. 1992; Neu 1993; Fassas et al. 1994; Bozik and Gilbert 1994; Chen et al. 1994; Savarese et al. 1996; Murray et al. 1997; Israel et al. 2000; Liu et al. 2006; Wu et al. 2006). It was initially unclear whether the lesions were due to levamisole or to 5-fluorouracil but further studies and case reports indicated that levamisole alone could be responsible (Chen et al. 1994; Kimmel et al. 1995; Lucia et al. 1996; Liu et al. 2006; Wu et al. 2006). Years earlier, an experimental study performed in dogs had revealed that chronic levamisole intake resulted in mononuclear cell infiltration in the brain parenchyma, eventually accompanied by demyelination, in the vast majority of tested animals (Vandevelde et al. 1978).

The clinical features of the inflammatory multifocal leukoencephalopathy induced by levamisole are reminiscent of those observed in multiple sclerosis. This is further supported by studies reporting interaction between multiple sclerosis and levamisole: in a clinical trial, levamisole was used with the purpose to alleviate the symptoms of multiple sclerosis, yet levamisole had the opposite effect: it worsened the clinical picture in 5 of the 7 patients (Dau et al. 1976); this led to the premature interruption of the trial. In studies using rat or mouse models of multiple sclerosis (experimental allergic encephalomyelitis, EAE), levamisole increased both the incidence and the severity of the disease, and inflammation and demyelination were exacerbated in animals receiving levamisole in comparison to non-treated animals (Spreafico et al. 1975; Hertz and Deghenghi 1985; Lucchinetti et al. 1997).

---

[5]The epileptic seizures that have been observed in the course of chronic levamisole treatments have all been reported in children but not in adults. This offers another interesting parallel with vitamin B6 metabolism alteration: experimental studies showed that vitamin B6 deficiency results in epileptic crisis in young animals but not in the adults (Daniel et al. 1942; Guilarte 1989).

It is generally admitted that the etiology of multiple sclerosis involves an autoimmune component (for review: Goverman 2009; Cieślak et al. 2011; Stys 2013). In particular, T cells are suspected to play a prominent role in multiple sclerosis. In parallel, it has been reported that the increase in cell-mediated immunity by levamisole largely rests on the recruitment and activation of T cells (e.g., Renoux et al. 1976). Thus the general belief was that levamisole-induced multifocal leukoencephalopathy was the consequence of the immunostimulating action of levamisole in patients with subclinical multiple sclerosis.

Yet, to our knowledge none of the studies that examined the etiology of levamisole-induced multifocal leukoencephalopathy considered the possibility that inhibition of TNAP by levamisole could be involved. A support for this possibility is that TNAP is strategically located on Ranvier's node (Pinner et al. 1964; Mori and Nagano 1985; Fonta et al. 2005; Hanics et al. 2012). Here we propose two, non exclusive hypothetical mechanisms, that both involve TNAP inhibition in levamisole-induced multifocal leukoencephalopathy.

The first mechanism concerns the inflammatory reaction observed in levamisole-induced leukoencephalopathy and a possible role for TNAP in this process through its ectonucleotidase activity. There is now ample evidence that inflammation is tightly associated with purinergic signaling (reviewed in: Khakh and North 2006; Di Virgilio et al. 2009; Trautmann 2009; Cieślak et al. 2011; Sperlágh and Illes 2014). To make a long story short, it has been shown that rises in the extracellular concentration of ATP is pro-inflammatory, whereas extracellular adenosine acts as an anti-inflammatory. It is also worth mentioning that purinergic signaling is likely to control the severity of the deficits and the extent of the lesions in EAE: the inactivation of both adenosine A1 and $A2_A$ receptors aggravates inflammation and demyelination (Tsutsui et al. 2004; Yao et al. 2012) while blockade or inactivation of ATP $P2X_7$ receptors has the opposite effect (Matute et al. 2007; Sharp et al. 2008).

ATP is released by axons, in proportion to action potential firing rate (reviewed in Fields, 2011). In the extracellular space, ATP is rapidly degraded by ectonucleotidases and TNAP, through its strategic location on Ranvier's node, may play an active role in setting the ATP/adenosine balance at this location. Inhibition of TNAP by levamisole may thus lead to an increased extracellular ATP concentration and/or a decreased extracellular adenosine concentration in the close vicinity of myelinated structures. This could trigger a series of reactions, in particular secretion of a variety of pro-inflammatory cytokines, that would ultimately result in monocyte infiltration and myelin loss. This does not rule out the involvement of an autoimmune component in levamisole-induced leukoencephalopathy. If this was the case, levamisole would rather amplify the autoimmune response. In this sense TNAP inhibition by levamisole could be considered as "immunostimulating". However this immunostimulation would not be due to levamisole acting directly on immune cells (Schiller et al. 1991); instead, by altering the extracellular ATP/adenosine balance, levamisole may have an action upstream from the immune system.

The second mechanism where TNAP could play a role concerns the remyelination that may be impaired in levamisole-induced leukoencephalopathy. TNAP is likely to intervene in myelin synthesis. This is supported by the observation that TNAP is highly expressed on axons during development, that is, during myelination (Fonta and Imbert 2002; Fonta et al. 2005) and by the fact that myelination is altered in TNAP KO mice (Hanics et al. 2012). The way TNAP participates in myelination may again involve its function as an ectonucleotidase.

Studies showed that myelination is in large part promoted by action potential firing in the axons (Demerens et al. 1996). Spike firing in the axons leads to ATP release (Fields 2011). Early stages of myelination, which consist in reducing oligodendrocyte precursor cells proliferation and in their differentiation in mature oligodendrocytes, requires adenosine signaling (Stevens et al. 2002). On the other hand, myelination by mature oligodendrocyte is promoted by extracellular ATP, which induces the release of a cytokine, leukemia inhibitory factor, by astrocytes (Ishibashi et al. 2006). Yet leukemia inhibitory factor has a dual action: it promotes myelination at low concentration but inhibits it when present at high concentration. Here again, inhibition of TNAP by levamisole may alter the extracellular ATP/adenosine balance, which would result in compromising (re)myelination: first, TNAP inhibition by levamisole may reduce the availability of adenosine required in the early stages of myelination; secondly, by reducing ATP degradation, TNAP inhibition may lead to an excess of leukemia inhibitory factor in the later stages of myelination.[6]

12.4.4.3 Levamisole and Tetramisole as Cocaine Adulterants

Between the sixties and the nineties, tetramisole and levamisole have been used in a considerable number of medical treatments. Yet, as briefly summarized above, tetramisole and levamisole appeared to be of limited efficacy in most of their applications and, more problematically, their chronic use resulted in serious side-effects in a non-negligible proportion of the patients. For these reasons, tetramisole and levamisole have been discontinued in most countries at the turn of the century, or their uses have been drastically restricted.[7]

---

[6]In addition to its involvement as an ectonucleotidase, TNAP could also intervene in myelination through vitamin B6 metabolism. In support for a role of PLP in myelination, it has been reported that vitamin B6 deficiency during development results in altered myelination (Lott et al. 1978; Kirksey et al. 1990). Indeed, PLP is a cofactor for two enzymes (serine palmitoyltransferase and sphingosine-1-phosphatelyase) involved in sphingolipid synthesis (Bourquin et al. 2011). Among sphingolipids, sphingomyelin, as expected from its name, happens to be particularly abundant in myelin. On the other hand, if myelination impairments observed in levamisole-induced multifocal leukoencephalopathy were consecutive to altered vitamin B6 metabolism, another PLP-dependent process, GABA synthesis, should also be impaired. This should result in occurrences of epileptic seizures in patients with levamisole-induced leukoencephalopathy, which, to our knowledge, have not been reported in case studies.

[7]In France for example, levamisole can still be used for the treatment of pediatric nephrotic syndrome, but under very restrictive conditions.

Yet, near the middle of the last decade, levamisole and tetramisole made an unexpected come-back as it turns out that they are now used as illegal cocaine adulterants (Morley et al. 2006; Fucci 2007; LeGatt 2009). Sixty to eighty percent of the cocaine seized around 2010 in North America or in Europe appeared to contain levamisole (Schneider and Meys 2011; Casale et al. 2012; Lahaie 2012), or, more rarely, tetramisole (Casale et al. 2012). The reasons why traffickers adopted such adulterants have not been trumpeted far and wide. One possibility is that levamisole may be degraded to aminorex, an amphetamine-like compound. Aminorex has been detected in horses that were treated with levamisole (see Larocque and Hoffman 2012) but aminorex is barely detectable in human after levamisole intake (Hess et al. 2014). A more likely reason is that, as inhibitors of noradrenaline uptake, levamisole and dexamisole may have cocaine-like action (see above).

As a result, the undesirable consequences that formerly were observed during chronic levamisole treatment are now observed in a number of cocaine users (reviewed in: Buchanan and Lavonas 2012; Larocque and Hoffman 2012; Lee et al. 2012). This includes thrombocytopenia, agranulocytosis and leukopenia, dermatitis and vasculitis. Occurrence of inflammatory multifocal leukoencephalopathy has also been reported in illegal cocaine users; although not definitively demonstrated, levamisole was suspected to be responsible for these rare instances of leukoencephalopathy (Gilbert 2011; Blanc et al. 2012; González-Duarte and Williams 2013).

Levamisole and tetramisole side-effects, including those that affect the nervous system, might not be something of the past. Understanding the mechanisms responsible for these side-effects may become crucially important in the near future. As we hypothesized above, TNAP may be a key player in these mechanisms, which certainly deserve further study.

Acknowledgment We would like to thank the SARA service and particularly Gregory Marsal for animal care. We acknowledge Pr José Luis Millán for his useful comments on this work while it was in progress. This work was supported by CNRS (PICS 4331), by Hypophosphatasie Europe (grant No. 88461) and by PHC Egide (Balaton 17341 UE).

Conflict of Interest The authors declare no conflict of interest.

References

Abbracchio MP, Burnstock G, Verkhratsky A, Zimmermann H (2009) Purinergic signalling in the nervous system: an overview. Trends Neurosci 32:19–29

Aceves J, Erlij D, Martínez-Marañón R (1970) The mechanism of the paralysing action of tetramisole on Ascaris somatic muscle. Br J Pharmacol 38:602–607

Ahmed FY, Leonard GA, A'Hern R, Taylor AE, Lorentzos A, Atkinson H, Moore J, Nicolson MC, Riches PG, Gore ME (1996) A randomised dose escalation study of subcutaneous interleukin 2 with and without levamisole in patients with metastatic renal cell carcinoma or malignant melanoma. Br J Cancer 74:1109–1113

Anagnostou F, Plas C, Forest N (1996) Ecto-alkaline phosphatase considered as levamisole-sensitive phosphohydrolase at physiological pH range during mineralization in cultured fetal calvaria cells. J Cell Biochem 60:484–494

Arnaud JP, Buyse M, Nordlinger B, Martin F, Pector JC, Zeitoun P, Adloff A, Duez N (1989) Adjuvant therapy of poor prognosis colon cancer with levamisole: results of an EORTC double-blind randomized clinical trial. Br J Surg 76:284–289

Artwohl M, Hölzenbein T, Wagner L, Freudenthaler A, Waldhäusl W, Baumgartner-Parzer SM (2000) Levamisole induced apoptosis in cultured vascular endothelial cells. Br J Pharmacol 131:1577–1583

Atchison WD, Geary TG, Manning B, VandeWaa EA, Thompson DP (1992) Comparative neuromuscular blocking actions of levamisole and pyrantel-type anthelmintics on rat and gastrointestinal nematode somatic muscle. Toxicol Appl Pharmacol 112:133–143

Aymard JP, Janot C, Thibaut G, Bertrand F, Legras B, Lederlin P, Streiff F (1984) Levamisole in chronic lymphocytic leukaemia: a prospective study of 15 patients. Acta Haematol 71:316–321

Balasubramaniam S, Bowling F, Carpenter K, Earl J, Chaitow J, Pitt J, Mornet E, Sillence D, Ellaway C (2010) Perinatal hypophosphatasia presenting as neonatal epileptic encephalopathy with abnormal neurotransmitter metabolism secondary to reduced co-factor pyridoxal-5′-phosphate availability. J Inherit Metab Dis Suppl 3:S25–S33

Barth A, Morton DL (1995) The role of adjuvant therapy in melanoma management. Cancer 75 (S2):726–734

Bartos M, Rayes D, Bouzat C (2006) Molecular determinants of pyrantel selectivity in nicotinic receptors. Mol Pharmacol 70:1307–1318

Baumgartner-Sigl S, Haberlandt E, Mumm S, Scholl-Bürgi S, Sergi C, Ryan L, Ericson KL, Whyte MP, Högler W (2007) Pyridoxine-responsive seizures as the first symptom of infantile hypophosphatasia caused by two novel missense mutations (c.677T > C, p. M226T; c.1112C > T, p.T371I) of the tissue-nonspecific alkaline phosphatase gene. Bone 40:1655–1661

Béthenod M, Cotte MF, Collombel C, Fréderich A, Cotte J (1967) Hypophosphatasie à révélation néo-natale. Amélioration osseuse. Encéphalopathie convulsivante fatale. Ann Pediatr (Paris) 14:835–841

Blanc PD, Chin C, Lynch KL (2012) Multifocal inflammatory leukoencephalopathy associated with cocaine abuse: is levamisole responsible? Clin Toxicol 50:534–535

Boison D (2012) Adenosine dysfunction in epilepsy. Glia 60:1234–1243

Borgers M (1973) The cytochemical application of new potent inhibitors of alkaline phosphatases. J Histochem Cytochem 21:812–824

Bostock H, Sears TA, Sherratt RM (1981) The effects of 4-aminopyryridine and tetraethylamonium ions on normal and demyelinated mammalian nerve fibres. J Physiol 313:301–315

Bourne G (1943) The distribution of alkaline phosphatase in various tissues. Quart J exp physiol 32:1–20

Bourquin F, Capitani G, Grütter MG (2011) PLP-dependent enzymes as entry and exit gates of sphingolipid metabolism. Protein Sci 20:1492–1508

Bozik ME, Gilbert MR (1994) The role of levamisole in 5-fluorouracil-levamisole-associated multifocal leukoencephalopathy. Ann Neurol 36:295

British Association for Paediatric Nephrology (1991) Levamisole for corticosteroid-dependent nephrotic syndrome in childhood. Lancet 337: 1555–1557

Brugmans J, Schuermans V, De Cock W, Thienpont D, Janssen P, Verhaegen H, Van Nimmen L, Louwagie AC, Stevens E (1973) Restoration of host defense mechanisms in man by levamisole. Life Sci 13:1499–1504

Brumberg JC, Nowak LG, McCormick DA (2000) Ionic mechanisms underlying repetitive high-frequency burst firing in supragranular cortical neurons. J Neurosci 20:4829–4843

Brun-Heath I, Ermonval M, Chabrol E, Xiao J, Palkovits M, Lyck R, Miller F, Couraud PO, Mornet E, Fonta C (2011) Differential expression of the bone and the liver tissue non-specific alkaline phosphatase isoforms in brain tissues. Cell Tissue Res 343:521–536

Buchanan JA, Lavonas EJ (2012) Agranulocytosis and other consequences due to use of illicit cocaine contaminated with levamisole. Curr Opin Hematol 19:27–31

Bullock MW, Hand JJ, Waletzky E (1968) Resolution and racemization of dl-tetramisole, dl-6-phenyl-2,3,5,6-tetrahydroimidazo-[2,1-b]thiazole. J Med Chem 11:169–171

Butt AM (2011) ATP: a ubiquitous gliotransmitter integrating neuron-glial networks. Semin Cell Dev Biol 22:205–213

Calhau C, Martel F, Hipólito-Reis C, Azevedo I (2000) Differences between duodenal and jejunal rat alkaline phosphatase. Clin Biochem 33:571–577

Carter BC, Bean BP (2009) Sodium entry during action potentials of mammalian neurons: incomplete inactivation and reduced metabolic efficiency in fast-spiking neurons. Neuron 64:898–909

Casale JF, Colley VL, Legatt DF (2012) Determination of phenyltetrahydroimidazothiazole enantiomers (Levamisole/Dexamisole) in illicit cocaine seizures and in the urine of cocaine abusers via chiral capillary gas chromatography-flame-ionization detection: clinical and forensic perspectives. J Anal Toxicol 36:130–135

Chang TW, Fiumara N (1978) Treatment with levamisole of recurrent herpes genitalis. Antimicrob Agents Chemother 13:809–812

Chen TC, Hinton DR, Leichman L, Atkinson RD, Apuzzo ML, Couldwell WT (1994) Multifocal inflammatory leukoencephalopathy associated with levamisole and 5-fluorouracil: case report. Neurosurgery 35:1138–1142

Chlebowski RTL, Lillington L, Nystrom JS, Sayre J (1994) Late mortality and levamisole adjuvant therapy in colorectal cancer. Br J Cancer 69:1094–1097

Ciancaglini P, Yadav MC, Simão AM, Narisawa S, Pizauro JM, Farquharson C, Hoylaerts MF, Millán JL (2010) Kinetic analysis of substrate utilization by native and TNAP-, NPP1-, or PHOSPHO1-deficient matrix vesicles. J Bone Miner Res 25:716–723

Cieślak M, Kukulski F, Komoszyński M (2011) Emerging role of extracellular nucleotides and adenosine in multiple sclerosis. Purinergic Signal 7:393–402

Cohen IS, Strichartz GR (1977) On the voltage-dependent action of tetrodotoxin. Biophys J 17:275–279

Collins GG, Anson J (1985) Adenosine A1 receptors mediate the inhibitory effects of exogenous adenosine in the rat olfactory cortex slice. Neuropharmacology 24:1077–1084

Connors BW, Malenka RC, Silva LR (1988) Two inhibitory postsynaptic potentials, and $GABA_A$ and $GABA_B$ receptor-mediated responses in neocortex of rat and cat. J Physiol 406:443–468

Cossart R, Bernard C, Ben-Ari Y (2005) Multiple facets of GABAergic neurons and synapses: multiple fates of GABA signalling in epilepsies. Trends Neurosci 28:108–115

Critchley P, Abbott R, Madden FJ (1994) Multifocal inflammatory leukoencephalopathy developing in a patient receiving 5-fluorouracil and levamisole. Clin Oncol (R Coll Radiol) 6:406

Dahl R, Sergienko EA, Su Y, Mostofi YS, Yang L, Simao AM, Narisawa S, Brown B, Mangravita-Novo A, Vicchiarelli M, Smith LH, O'Neill WC, Millán JL, Cosford ND (2009) Discovery and validation of a series of aryl sulfonamides as selective inhibitors of tissue-nonspecific alkaline phosphatase (TNAP). J Med Chem 52:6919–6925

Daniel EP, Kline OD, Tolle CD (1942) A convulsive syndrome in young rats associated with pyridoxine deficiency. J Nutr 23:205–216

Dau PC, Johnson KP, Spitler LE (1976) The effect of levamisole on cellular immunity in multiple sclerosis. Clin Exp Immunol 26:302–309

de Bernard B, Bianco P, Bonucci E, Costantini M, Lunazzi GC, Martinuzzi P, Modricky C, Moro L, Panfili E, Pollesello P, Stagni N, Vittur F (1986) Biochemical and immunohistochemical evidence that in cartilage an alkaline phosphatase is a Ca^{2+}-binding glycoprotein. J Cell Biol 103:1615–1623

De Col R, Messlinger K, Carr RW (2008) Conduction velocity is regulated by sodium channel inactivation in unmyelinated axons innervating the rat cranial meninges. J Physiol 586:1089–1103

Debray J, Chang L, Marquès S, Pellet-Rostaing S, Le Duy D, Mebarek S, Buchet R, Magne D, Popowycz F, Lemaire M (2013) Inhibitors of tissue-nonspecific alkaline phosphatase: design, synthesis, kinetics, biomineralization and cellular tests. Bioorg Med Chem 21:7981–7987

Deisz RA, Prince DA (1989) Frequency-dependent depression of inhibition in guinea-pig neocortex in vitro by $GABA_B$ receptor feed-back on GABA release. J Physiol 412:513–541

Demerens C, Stankoff B, Logak M, Anglade P, Allinquant B, Couraud F, Zalc B, Lubetzki C (1996) Induction of myelination in the central nervous system by electrical activity. Proc Natl Acad Sci USA 93:9887–9892

Di Virgilio F, Ceruti S, Bramanti P, Abbracchio MP (2009) Purinergic signalling in inflammation of the central nervous system. Trends Neurosci 32:79–87

Dinai Y, Pras M (1975) Levamisole in rheumatoid arthritis. Lancet 2(7934):556

Ermonval M, Baudry A, Baychelier F, Pradines E, Pietri M, Oda K, Schneider B, Mouillet-Richard S, Launay JM, Kellermann O (2009) The cellular prion protein interacts with the tissue non-specific alkaline phosphatase in membrane microdomains of bioaminergic neuronal cells. PLoS ONE 4:e6497

Eyre P (1970) Some pharmacodynamic effects of the nematocides: methyridine, tetramisole and pyrantel. J Pharm Pharmacol 22:26–36

Fassas AB, Gattani AM, Morgello S (1994) Cerebral demyelination with 5-fluorouracil and levamisole. Cancer Invest 12:379–383

Ferroir JP, Fenelon G, Beaugerie L, Avenin Recoing D (1994) Leucoencéphalopathie multifocale inflammatoire, complication de la chimiothérapie par 5 fluorouracil et levamisole. Rev Neurol (Paris) 150:471–474

Fields RD, Burnstock G (2006) Purinergic signalling in neuron-glia interactions. Nat Rev Neurosci 7:423–436

Fields RD (2011) Nonsynaptic and nonvesicular ATP release from neurons and relevance to neuron-glia signaling. Semin Cell Dev Biol 22:214–219

Fleming JT, Squire MD, Barnes TM, Tornoe C, Matsuda K, Ahnn J, Fire A, Sulston JE, Barnard EA, Sattelle DB, Lewis JA (1997) Caenorhabditis elegans levamisole resistance genes lev-1, unc-29, and unc-38 encode functional nicotinic acetylcholine receptor subunits. J Neurosci 17:5843–5857

Fonta C, Imbert M (2002) Vascularization in the primate visual cortex during development. Cereb Cortex 12:199–211

Fonta C, Négyessy L, Renaud L, Barone P (2004) Areal and subcellular localization of the ubiquitous alkaline phosphatase in the primate cerebral cortex: evidence for a role in neurotransmission. Cereb Cortex 14:595–609

Fonta C, Negyessy L, Renaud L, Barone P (2005) Postnatal development of alkaline phosphatase activity correlates with the maturation of neurotransmission in the cerebral cortex. J Comp Neurol 486:179–196

Fonta C, Négyessy L, Brun-Heath I, Ermonval M, Czege D, Nowak LG, Frances B, Xiao JS, Mornet E, Millán JL (2012) Bone TNAP in the brain: functions in neurotransmission. Bull Group Int Rech Sci Stomatol Odontol 51:27

Fonta C, Cruz T, Rosay B, Millán JL, Balayssac S, Gilard V, Malet-Martino M, Nowak LG (2014) A dual approach to explore the functions of the tissue nonspecific alkaline phosphatase (TNAP) in the brain. Ninth FENS Forum of Neuroscience, Milan, Italy. Abstract Number: FENS-1232

Fontanez DE, Porter JT (2006) Adenosine A1 receptors decrease thalamic excitation of inhibitory and excitatory neurons in the barrel cortex. Neuroscience 137:1177–1184

Fox DA, Ruan DY (1989) Time- and frequency-dependent effects of potassium channel blockers on large and medium diameter optic tract axons. Brain Res 498:229–242

Franz DN, Perry RS (1974) Mechanisms for differential block among single myelinated and non-myelinated axons by procaine. J Physiol 236:193–210

Fraser D (1957) Hypophosphatasia. Am J Med 22:730–746

Friede RL (1966) A quantitative mapping of alkaline phosphatase in the brain of the rhesus monkey. J Neurochem 13:197–203

Fucci N (2007) Unusual adulterants in cocaine seized on Italian clandestine market. Forensic Sci Int 172:e1

Gilbert JW (2011) J Neuroimaging 21:e188

Goldstein DJ, Rogers CE, Harris H (1980) Expression of alkaline phosphatase loci in mammalian tissues. Proc Natl Acad Sci USA 77:2857–2860

González-Duarte A, Williams R (2013) Cocaine-induced recurrent leukoencephalopathy. Neuroradiol J 26:511–513

Goverman J (2009) Autoimmune T cell responses in the central nervous system. Nat Rev Immunol 9:393–407

Grem JL, Allegra CJ (1989) Toxicity of levamisole and 5-fluorouracil in human colon carcinoma cells. J Natl Cancer Inst 81:1413–1417

Guilarte TR (1989) Regional changes in the concentrations of glutamate, glycine, taurine, and GABA in the vitamin B-6 deficient developing rat brain: association with neonatal seizures. Neurochem Res 14:889–897

Gulati OD, Hemavathi KG, Joshi DP (1985) Interactions of levamisole with some autonomic drugs on guinea-pig vas deferens. J Auton Pharmacol 5:19–23

Gwilt P, Tempero M, Kremer A, Connolly M, Ding C (2000) Pharmacokinetics of levamisole in cancer patients treated with 5-fluorouracil. Cancer Chemother Pharmacol 45:247–251

Hadden JW, Lopez C, O'Reilly RJ, Hadden EM (1977) Levamisole and inosiplex: antiviral agents with immunopotentiating action. Ann NY Acad Sci 284:139–152

Hanics J, Barna J, Xiao J, Millán JL, Fonta C, Négyessy L (2012) Ablation of TNAP function compromises myelination and synaptogenesis in the mouse brain. Cell Tissue Res 349:459–471

Harrow ID, Gration KAF (1985) Mode of action of the anthelmintics morantel, pyrantel and levamisole on muscle cell membrane of the nematode Ascaris suum. Pestic Sci 16:662–672

Hershkowitz N, Ayala GF (1981) Effects of phenytoin on pyramidal neurons of the rat hippocampus. Brain Res 208:487–492

Hertz F, Deghenghi R (1985) Effect of rat and beta-human interferons on hyperacute experimental allergic encephalomyelitis in rats. Agents Actions 16:397–403

Hess C, Ritke N, Sydow K, Mehling LM, Ruehs H, Madea B, Musshoff F (2014) Determination of levamisole, aminorex, and pemoline in plasma by means of liquid chromatography-mass spectrometry and application to a pharmacokinetic study of levamisole. Drug Test Anal 6:1049–1054

Hodgkin AL, Katz B (1949) The effect of sodium ions on the electrical activity of giant axon of the squid. J Physiol 108:37–77

Hondeghem LM (1978) Validity of V_{max} as a measure of the sodium current in cardiac and nervous tissues. Biophys J 23:147–152

Hook CC, Kimmel DW, Kvols LK, Scheithauer BW, Forsyth PA, Rubin J, Moertel CG, Rodriguez M (1992) Multifocal inflammatory leukoencephalopathy with 5-fluorouracil and levamisole. Ann Neurol 3:262–267

Hoshi K, Amizuka N, Oda K, Ikehara Y, Ozawa H (1997) Immunolocalization of tissue non-specific alkaline phosphatase in mice. Histochem Cell Biol 107:183–191

Howe JR, Sutor B, Zieglgänsberger W (1987a) Characteristics of long-duration inhibitory postsynaptic potentials in rat neocortical neurons in vitro. Cell Mol Neurobiol 7:1–18

Howe JR, Sutor B, Zieglgänsberger W (1987b) Baclofen reduces postsynaptic potentials of rat cortical neurones by an action other than its hyperpolarizing action. J Physiol 384:539–569

Hsu WH (1980) Toxicity and drug interactions of levamisole. J Am Vet Med Assoc 176:1166–1169

Ibusuki S, Katsuki H, Takasaki M (1998) The effects of extracellular pH with and without bicarbonate on intracellular procaine concentrations and anesthetic effects in crayfish giant axons. Anesthesiology 88:1549–1557

Irnich D, Tracey DJ, Polten J, Burgstahler R, Grafe P (2002) ATP stimulates peripheral axons in human, rat and mouse—differential involvement of A(2B) adenosine and P2X purinergic receptors. Neuroscience 110:123–129

Ishibashi T, Dakin KA, Stevens B, Lee PR, Kozlov SV, Stewart CL, Fields RD (2006) Astrocytes promote myelination in response to electrical impulses. Neuron 49:823–832

Israel ZH, Lossos A, Barak V, Soffer D, Siegal T (2000) Multifocal demyelinative leukoencephalopathy associated with 5-fluorouracil and levamisole. Acta Oncol 39:117–120

Joly C, Palisse M, Ribbe D, De Calmes O, Genevey P (1998) Intoxication aiguë au levamisole "à dose de cheval". La revue des SAMU 1998:237–240
Kao CY, Walker SE (1982) Active groups of saxitoxin and tetrodotoxin as deduced from actions of saxitoxin analogues on frog muscle and squid axon. J Physiol 323:619–637
Khakh BS, North RA (2006) P2X receptors as cell-surface ATP sensors in health and disease. Nature 442:527–532
Kimmel DW, Schutt AJ (1993) Multifocal leukoencephalopathy: occurrence during 5-fluorouracil and levamisole therapy and resolution after discontinuation of chemotherapy. Mayo Clin Proc 68:363–365
Kimmel DW, Wijdicks EF, Rodriguez M (1995) Multifocal inflammatory leukoencephalopathy associated with levamisole therapy. Neurology 45:374–376
Kirksey A, Morré DM, Wasynczuk AZ (1990) Neuronal development in vitamin B6 deficiency. Ann NY Acad Sci 585:202–218
Kouassi E, Caillé G, Léry L, Larivière L, Vézina M (1986) Novel assay and pharmacokinetics of levamisole and p-hydroxylevamisole in human plasma and urine. Biopharm Drug Dispos 7:71–89
Kovach JS, Svingen PA, Schaid DJ (1992) Levamisole potentiation of fluorouracil antiproliferative activity mimicked by orthovanadate, an inhibitor of tyrosine phosphatase. J Natl Cancer Inst 84:515–519
Lahaie E (2012) Quel est le produit qui circule? In: Pousset M (ed) Observatoire Français des Drogues et des Toxicomanes; cocaïne, données essentielles, pp 35–41
Landow H, Kabat EA, Newman W (1942) Distribution of alkaline phosphatase in normal and in neoplastic tissues of the nervous system. A histochemical study. Arch Neurol Psychiatry 48:518–530
Langer D, Hammer K, Koszalka P, Schrader J, Robson S, Zimmermann H (2008) Distribution of ectonucleotidases in the rodent brain revisited. Cell Tissue Res 334:199–217
Larocque A, Hoffman RS (2012) Levamisole in cocaine: unexpected news from an old acquaintance. Clin Toxicol 50:231–241
Le Quesne PM, Goldberg V, Vajda F (1976) Acute conduction velocity changes in guinea-pigs after administration of diphenylhydantoin. J Neurol Neurosurg Psychiatry 39:995–1000
Lee KC, Ladizinski B, Federman DG (2012) Complications associated with use of levamisole-contaminated cocaine: an emerging public health challenge. Mayo Clin Proc 87:581–586
LeGatt DF (2009) New cocaine cutting agent poses greater risk to users. Clinical & Forensic Toxicology News, Sep 2009
Leichman L, Brown T, Poplin B (1993) Symptomatic, radiologic and pathologic changes in the central nervous system (CNS) associated with 5-fluorouracil (5-FU) and levamisole (LEV) therapy. Proc Am Soc Clin Oncol 12:198
Levandoski MM, Piket B, Chang J (2003) The anthelmintic levamisole is an allosteric modulator of human neuronal nicotinic acetylcholine receptors. Eur J Pharmacol 471:9–20
Lewis JA, Wu CH, Levine JH, Berg H (1980) Levamisole-resistant mutants of the nematode Caenorhabditis elegans appear to lack pharmacological acetylcholine receptors. Neuroscience 5:967–989
Liu HM, Hsieh WJ, Yang CC, Wu VC, Wu KD (2006) Leukoencephalopathy induced by levamisole alone for the treatment of recurrent aphthous ulcers. Neurology 67:1065–1067
Lott IT, Coulombe T, Di Paolo RV, Richardson EP Jr, Levy HL (1978) Vitamin B6-dependent seizures: pathology and chemical findings in brain. Neurology 28:47–54
Lucchinetti CF, Kimmel DW, Pavelko K, Rodriguez M (1997) 5-Fluorouracil and levamisole exacerbate demyelination in susceptible mice infected with Theiler's virus. Exp Neurol 147:123–129
Lucia P, Pocek M, Passacantando A, Sebastiani ML, De Martinis C (1996) Multifocal leucoencephalopathy induced by levamisole. Lancet 348:1450

Luyckx M, Cazin M, Brunet C, Cazin JC, Devulder B, Gosselin P, Adenis L, Cappelaere P, Granier AM, Demaille A (1983) Lymphocytes T and pharmacokinetics after a single oral dose of levamisole in healthy and cancer subjects. Methods Find Exp Clin Pharmacol 5:467–469

Macdonald RL, Kang JQ, Gallagher MJ (2010) Mutations in $GABA_A$ receptor subunits associated with genetic epilepsies. J Physiol 588:1861–1869

Macfarlane DG, Bacon PA (1978) Levamisole-induced vasculitis due to circulating immune complexes. Br Med J 1(6110):407–408

Marcus DJ, Swift TR, McDonald TF (1981) Acute effects of phenytoin on peripheral nerve function in the rat. Muscle Nerve 4:48–50

Martin DL, Rimvall K (1993) Regulation of gamma-aminobutyric acid synthesis in the brain. J Neurochem 60:395–407

Matute C, Torre I, Pérez-Cerdá F, Pérez-Samartín A, Alberdi E, Etxebarria E, Arranz AM, Ravid R, Rodríguez-Antigüedad A, Sánchez-Gómez M, Domercq M (2007) P2X(7) receptor blockade prevents ATP excitotoxicity in oligodendrocytes and ameliorates experimental autoimmune encephalomyelitis. J Neurosci 27:9525–9533

McCormick DA (1989) GABA as an inhibitory neurotransmitter in human cerebral cortex. J Neurophysiol 62:1018–1027

Mehansho H, Henderson LM (1980) Transport and accumulation of pyridoxine and pyridoxal by erythrocytes. J Biol Chem 255:11901–11907

Millán JL (2006). Mammalian alkaline phosphatases. From biology to applications in medicine and biotechnology. Wiley-VCH Verlag GmbH & Co, Weinheim, Germany

Mitzdorf U (1985) Current source-density method and application in cat cerebral cortex: investigation of evoked potentials and EEG phenomena. Physiol Rev 65:37–100

Moertel CG, Fleming TR, Macdonald JS, Haller DG, Laurie JA, Goodman PJ, Ungerleider JS, Emerson WA, Tormey DC, Glick JH, Veerder MH, Mailliard JA (1990) Levamisole and fluorouracil for adjuvant therapy of resected colon carcinoma. N Engl J Med 322:352–358

Moertel CG, Fleming TR, Macdonald JS, Haller DG, Laurie JA, Tangen CM, Ungerleider JS, Emerson WA, Tormey DC, Glick JH, Veeder MH, Mailliard JA (1995) Fluorouracil plus levamisole as effective adjuvant therapy after resection of stage III colon carcinoma: a final report. Ann Int Med 122:321–326

Mohammad FK, Faris GA, Rhayma MS, Ahmed K (2006) Neurobehavioral effects of tetramisole in mice. Neurotoxicology 27:153–157

Mongeau JG, Robitaille PO, Roy F (1988) Clinical efficacy of levamisole in the treatment of primary nephrosis in children. Pediatr Nephrol 2:398–401

Mori S, Nagano M (1985) Electron microscopic cytochemistry of alkaline phosphatase in neurons of rats. Arch Histol Jpn 48:389–397

Morley SR, Forrest ARW, Galloway JH (2006) Levamisole as a contaminant of illicit Cocaine. Int Assoc Forensic Toxicologists 44:6

Mornet E (2007) Hypophosphatasia. Orphanet J Rare Dis 2:40

Morris DC, Masuhara K, Takaoka K, Ono K, Anderson HC (1992) Immunolocalization of alkaline phosphatase in osteoblasts and matrix vesicles of human fetal bone. Bone Miner 19:287–298

Murray CL, Ford WJ, Swenson KK, Heros D, Sperduto PW (1997) Multifocal inflammatory leukoencephalopathy after fluorouracil and levamisole therapy for colon cancer. AJNR Am J Neuroradiol 18:1591–1592

Nandy K, Bourne GH (1963) Alkaline phosphatases in brain and spinal cord. Nature 200:1216–1217

Narisawa S, Fröhlander N, Millán JL (1997) Inactivation of two mouse alkaline phosphatase genes and establishment of a model of infantile hypophosphatasia. Dev Dyn 208:432–446

Narisawa S, Wennberg C, Millán JL (2001) Abnormal vitamin B6 metabolism in alkaline phosphatase knock-out mice causes multiple abnormalities, but not the impaired bone mineralization. J Pathol 193:125–133

Négyessy L, Xiao J, Kántor O, Kovács GG, Palkovits M, Dóczi TP, Renaud L, Baksa G, Glasz T, Ashaber M, Barone P, Fonta C (2011) Layer-specific activity of tissue non-specific alkaline phosphatase in the human neocortex. Neuroscience 172:406–418

Neu IS (1993) Multifocal inflammatory leukencephalopathy caused by adjuvant therapy with 5-fluorouracil and levamisole after resection for an adenocarcinoma of the colon. Acta Neurol Scand 87:70

Niaudet P, Drachman R, Gagnadoux MF, Broyer M (1984) Treatment of idiopathic nephrotic syndrome with levamisole. Acta Paediatr Scand 73:637–641

Nowak LG, Bullier J (1996) Spread of stimulating current in the cortical grey matter of rat visual cortex studied on a new in vitro slice preparation. J Neurosci Methods 67:237–248

Ohkubo S, Kimura J, Matsuoka I (2000) Ecto-alkaline phosphatase in NG108-15 cells : a key enzyme mediating P1 antagonist-sensitive ATP response. Br J Pharmacol 131:1667–1672

Onuaguluchi G, Igbo IN (1987) Comparative local anaesthetic and antiarrhythmic actions of levamisole hydrochloride and lignocaine hydrochloride. Arch Int Pharmacodyn Ther 289:278–289

Pabst HF, Crawford J (1975) L-tetramisole. Enhancement of human lymphocyte response to antigen. Clin Exp Immunol 21:468–473

Pagonopoulou O, Efthimiadou A, Asimakopoulos B, Nikolettos NK (2006) Modulatory role of adenosine and its receptors in epilepsy: possible therapeutic approaches. Neurosci Res 56:14–20

Palcoux JB, Niaudet P, Goumy P (1994) Side effects of levamisole in children with nephrosis. Pediatr Nephrol 8:263–264

Parkinson DR, Cano PO, Jerry LM, Capek A, Shibata HR, Mansell PW, Lewis MG, Marquis G (1977) Complications of cancer immunotherapy with levamisole. Lancet 1(8022):1129–1132

Paterson D, Nordberg A (2000) Neuronal nicotinic receptors in the human brain. Prog Neurobiol 61:75–111

Percudani R, Peracchi A (2009) The B6 database: a tool for the description and classification of vitamin B6-dependent enzymatic activities and of the corresponding protein families. BMC Bioinformatics 10:273

Perry DC, Xiao Y, Nguyen HN, Musachio JL, Dávila-García MI, Kellar KJ (2002) Measuring nicotinic receptors with characteristics of alpha4beta2, alpha3beta2 and alpha3beta4 subtypes in rat tissues by autoradiography. J Neurochem 82:468–481

Picher M, Burch LH, Hirsh AJ, Spychala J, Boucher RC (2003) Ecto 5'-nucleotidase and nonspecific alkaline phosphatase. Two AMP-hydrolyzing ectoenzymes with distinct roles in human airways. J Biol Chem 278:13468–13479

Pinner B, Davison JF, Campbell JB (1964) Alkaline Phosphatase in Peripheral Nerves. Science 145:936–938

Pinto V, Derkach VA, Safronov BV (2008) Role of TTX-sensitive and TTX-resistant sodium channels in Adelta- and C-fiber conduction and synaptic transmission. J Neurophysiol 99:617–628

Pires JG, Futuro-Neto HA, Oliveira AM, Cabral AM (1979) The effect of levamisole on the cardiac responses to sympathomimetics and on the amine uptake process. J Pharm Pharmacol 31:257–258

Prieur AM, Buriot D, Lefur JM, Griscelli C (1978) Possible toxicity of levamisole in children with rheumatoid arthritis. J Pediatr 93:304–305

Raeymaekers AH, Allewijn FT, Vandenberk J, Demoen PJ, Van Offenwert TT, Janssen PA (1966) Novel broad-spectrum anthelmintics. Tetramisole and related derivatives of 6-arylimidazo [2,1-b]thiazole. J Med Chem 9:545–551

Rathbun JC (1948) Hypophosphatasia; a new developmental anomaly. Am J Dis Child 75:822–831

Rayes D, De Rosa MJ, Bartos M, Bouzat C (2004) Molecular basis of the differential sensitivity of nematode and mammalian muscle to the anthelmintic agent levamisole. J Biol Chem 279:36372–36381

Raymond SA (1992) Subblocking concentrations of local anesthetics: effects on impulse generation and conduction in single myelinated sciatic nerve axons in frog. Anesth Analg 75:906–921

Rehni AK, Singh TG (2010) Levamisole-induced reduction in seizure threshold: a possible role of nicotinic acetylcholine receptor-mediated pathway. Naunyn Schmiedebergs Arch Pharmacol 382:279–285
Reid JM, Kovach JS, O'Connell MJ, Bagniewski PG, Moertel CG (1998) Clinical and pharmacokinetic studies of high-dose levamisole in combination with 5-fluorouracil in patients with advanced cancer. Cancer Chemother Pharmacol 41:477–484
Renoux G, Renoux M (1972) Antigenic competition and nonspecific immunity after a rickettsial infection in mice: restoration of antibacterial immunity by phenyl-imidothiazole treatment. J Immunol 109:761–765
Renoux G, Renoux M, Teller MN, McMahon S, Guillaumin JM (1976) Potentiation of T-cell mediated immunity by levamisole. Clin Exp Immunol 25:288–296
Rifkin DB, Compans RW, Reich E (1972) A specific labeling procedure for proteins on the outer surface of membranes. J Biol Chem 247:6432–64327
Robertson SJ, Martin RJ (1993) Levamisole-activated single-channel currents from muscle of the nematode parasite Ascaris suum. Br J Pharmacol 108:170–178
Ruuskanen O, Remes M, Mäkela AL, Isomäki H, Toivanen A (1976) Levamisole and agranulocytosis. Lancet 308:958–959
Sanchez MR (2000) Miscellaneous treatments: thalidomide, potassium iodide, levamisole, clofazimine, colchicine, and D-penicillamine. Clin Dermatol 18:131–145
Sanchez-Vives MV, McCormick DA (2000) Cellular and network mechanisms of rhythmic recurrent activity in neocortex. Nat Neurosci 3:1027–1034
Savarese DM, Gordon J, Smith TW, Litofsky NS, Licho R, Ragland R, Recht L (1996) Cerebral demyelination syndrome in a patient treated with 5-fluorouracil and levamisole. The use of thallium SPECT imaging to assist in noninvasive diagnosis—a case report. Cancer 77:387–394
Scheinberg MA, Bezerra JB, Almeida FA, Silveira LA (1978) Cutaneous necrotising vasculitis induced by levamisole. Br Med J 1(6110):408
Schiller JH, Lindstrom M, Witt PL, Hank JA, Mahvi D, Wagner RJ, Sondel P, Borden EC (1991) Immunological effects of levamisole in vitro. J Immunother 10:297–306
Schneider S, Meys F (2011) Analysis of illicit cocaine and heroin samples seized in Luxembourg from 2005 to 2010. Forensic Sci Int 212:242–246
Schuermans Y (1975) Levamisole in rheumatoid arthritis. Lancet 1(7898):111
Schwarz SK, Puil E (1998) Analgesic and sedative concentrations of lignocaine shunt tonic and burst firing in thalamocortical neurones. Br J Pharmacol 124:1633–1642
Seidlin M, Straus SE (1984) Treatment of mucocutaneous herpes simplex infections. Clin Dermatol 2:100–116
Selzer ME (1979) The effect of phenytoin on the action potential of a vertebrate spinal neuron. Brain Res 171:511–521
Sergienko EA, Millán JL (2010) High-throughput screening of tissue-nonspecific alkaline phosphatase for identification of effectors with diverse modes of action. Nat Protoc 5:1431–1439
Sharp AJ, Polak PE, Simonini V, Lin SX, Richardson JC, Bongarzone ER, Feinstein DL (2008) P2x7 deficiency suppresses development of experimental autoimmune encephalomyelitis. J Neuroinflammation 5:33
Shimizu N (1950) Histochemical studies on the phosphatase of the nervous system. J Comp Neurol 93:201–217
Simão AM, Bolean M, Hoylaerts MF, Millán JL, Ciancaglini P (2013) Effects of pH on the production of phosphate and pyrophosphate by matrix vesicles' biomimetics. Calcif Tissue Int 93:222–232
Spector R (1978) Vitamin B6 transport in the central nervous system: in vivo studies. J Neurochem 30:881–887
Spector S, Munjal I, Schmidt DE (1998) Effects of the immunostimulant, levamisole, on opiate withdrawal and levels of endogenous opiate alkaloids and monoamine neurotransmitters in rat brain. Neuropsychopharmacology 19:417–427

Sperlágh B, Illes P (2014) P2X7 receptor: an emerging target in central nervous system diseases. Trends Pharmacol Sci 35:537–547
Spreafico F, Vecchi A, Mantovani A, Poggi A, Franchi G, Anaclerio A, Garattini S (1975) Characterization of the immunostimulants levamisole and tetramisole. Eur J Cancer 11:555–563
Stevens B, Porta S, Haak LL, Gallo V, Fields RD (2002) Adenosine: a neuron-glial transmitter promoting myelination in the CNS in response to action potentials. Neuron 36:855–868
Street SE, Kramer NJ, Walsh PL, Taylor-Blake B, Yadav MC, King IF, Vihko P, Wightman RM, Millán JL, Zylka MJ (2013) Tissue-nonspecific alkaline phosphatase acts redundantly with PAP and NT5E to generate adenosine in the dorsal spinal cord. J Neurosci 33:11314–11322
Stys PK (2013) Pathoetiology of multiple sclerosis: are we barking up the wrong tree? F1000Prime Rep 5:20
Sugimura K, Mizutani A (1979) Histochemical and cytochemical studies of alkaline phosphatase activity in the synapses of rat brain. Histochemistry 61:123–129
Suzuki K, Yoshimura Y, Hisada Y, Matsumoto A (1994) Sensitivity of intestinal alkaline phosphatase to L-homoarginine and its regulation by subunit-subunit interaction. Jpn J Pharmacol 64:97–102
Symoens J, Veys E, Mielants M, Pinals R (1978) Adverse reactions to levamisole. Cancer Treat Rep 62:1721–1730
Taketani T, Onigata K, Kobayashi H, Mushimoto Y, Fukuda S, Yamaguchi S (2014) Clinical and genetic aspects of hypophosphatasia in Japanese patients. Arch Dis Child 99:211–215
Tanphaichitr P, Tanphaichitr D, Sureeratanan J, Chatasingh S (1980) Treatment of nephrotic syndrome with levamisole. J Pediatr 96:490–493
Thienpont D, Vanparijs OF, Raeymaekers AH, Vandenberk J, Demoen JA, Allewijn FT, Marsboom RP, Niemegeers CJ, Schellekens KH, Janssen PA (1966) Tetramisole (R 8299), a new, potent broad spectrum anthelmintic. Nature 209:1084–1086
Toivanen A, Lassila O, Nordman E (1981) Lack of effect of levamisole on the immune function in melanoma patients. Cancer Immunol Immunother 10:191–195
Trautmann A (2009) Extracellular ATP in the immune system: more than just a "danger signal". Sci Sig 2:pe6
Treurniet-Donker AD, Meischke-de Jongh ML, van Putten WL (1987) Levamisole as adjuvant immunotherapy in breast cancer. Cancer 59:1590–1593
Tsutsui S, Schnermann J, Noorbakhsh F, Henry S, Yong VW, Winston BW, Warren K, Power C (2004) A1 adenosine receptor upregulation and activation attenuates neuroinflammation and demyelination in a model of multiple sclerosis. J Neurosci 24:1521–1529
van Belle H (1972) Kinetics and inhibition of alkaline phosphatases from canine tissues. Biochim Biophys Acta 289:158–168
van Belle H (1976a) Kinetics and inhibition of rat and avian alkaline phosphatases. Gen Pharmacol 7:53–58
van Belle H (1976b) Alkaline phosphatase. I. Kinetics and inhibition by levamisole of purified isoenzymes from humans. Clin Chem 22:972–976
van der Ham M, Albersen M, de Koning TJ, Visser G, Middendorp A, Bosma M, Verhoeven-Duif NM, de Sain-van der Velden MG (2012) Quantification of vitamin B6 vitamers in human cerebrospinal fluid by ultra performance liquid chromatography-tandem mass spectrometry. Anal Chim Acta 712:108–114
Vandevelde M, Boring JG, Hoff EJ, Gingerich DA (1978) The effect of levamisole on the canine central nervous system. J Neuropathol Exp Neurol 37:165–173
Vanhoutte PM, Van Nueten JM, Verbeuren TJ, Laduron PM (1977) Differential effects of the isomers of tetramisole on adrenergic neurotransmission in cutaneous veins of dog. J Pharmacol Exp Ther 200:127–140
Waymire KG, Mahuren JD, Jaje JM, Guilarte TR, Coburn SP, MacGregor GR (1995) Mice lacking tissue non-specific alkaline phosphatase die from seizures due to defective metabolism of vitamin B-6. Nat Genet 11:45–51

Webster DJ, Whitehead RH, Richardson G, Hughes LE (1982) Levamisole: a double blind immunological study. Anticancer Res 2:29–32
Whyte MP (2010) Physiological role of alkaline phosphatase explored in hypophosphatasia. Ann NY Acad Sci 1192:190–200
Whyte MP, Mahuren JD, Fedde KN, Cole FS, McCabe ER, Coburn SP (1988) Perinatal hypophosphatasia: tissue levels of vitamin B6 are unremarkable despite markedly increased circulating concentrations of pyridoxal-5′-phosphate. Evidence for an ectoenzyme role for tissue-nonspecific alkaline phosphatase. J Clin Invest 81:1234–1239
Whyte MP, Mahuren JD, Vrabel LA, Coburn SP (1985) Markedly increased circulating pyridoxal-5′-phosphate levels in hypophosphatasia. Alkaline phosphatase acts in vitamin B6 metabolism. J Clin Invest 76:752–756
Wiebke EA, Grieshop NA, Loehrer PJ, Eckert GJ, Sidner RA (2003) Antitumor effects of 5-fluorouracil on human colon cancer cell lines: antagonism by levamisole. J Surg Res 111:63–69
Woestenborghs R, Michielsen L, Heykants J (1981) Determination of levamisole in plasma and animal tissues by gas chromatography with thermionic specific detection. J Chromatogr 224:25–32
Wu VC, Huang JW, Lien HC, Hsieh ST, Liu HM, Yang CC, Lin YH, Hwang JJ, Wu KD (2006) Levamisole-induced multifocal inflammatory leukoencephalopathy: clinical characteristics, outcome, and impact of treatment in 31 patients. Medicine 85:203–213
Yao SQ, Li ZZ, Huang QY, Li F, Wang ZW, Augusto E, He JC, Wang XT, Chen JF, Zheng RY (2012) Genetic inactivation of the adenosine A(2A) receptor exacerbates brain damage in mice with experimental autoimmune encephalomyelitis. J Neurochem 123:100–112
Yokota T, Saito Y, Miyatake T (1994) Conduction slowing without conduction block of compound muscle and nerve action potentials due to sodium channel block. J Neurol Sci 124:220–224
Zakon HH (2012) Adaptive evolution of voltage-gated sodium channels: the first 800 million years. Proc Natl Acad Sci USA 109(Suppl 1):10619–10625
Zhang D, Xiong W, Chu S, Sun C, Albensi BC, Parkinson FE (2012) Inhibition of hippocampal synaptic activity by ATP, hypoxia or oxygen-glucose deprivation does not require CD73. PLoS ONE 7:e39772
Zimmermann H, Zebisch M, Sträter N (2012) Cellular function and molecular structure of ecto-nucleotidases. Purinergic Signal 8:437–502

Chapter 13
TNAP and Pain Control

Sarah E. Street and Nathaniel A. Sowa

Abstract Chronic pain is one of the most debilitating and expensive diseases, yet current therapies are often insufficient in bringing about long-term relief. Further, many treatments for chronic pain also carry significant side effects. The molecule adenosine has long been identified as a potent inhibitor of nociceptive circuits in the spinal cord; however, the widespread expression of adenosine receptors in many organ systems has limited its use as an analgesic. Recently several 5′ ectonucleotidases, including tissue non-specific alkaline phosphatase (TNAP), have been characterized for their ability to generate endogenous adenosine in nociceptive circuitry of the dorsal spinal cord. These ectonucleotidases have the ability to hydrolyze the endogenous pronociceptive nucleotides like adenosine triphosphate (ATP) into the antinociceptive nucleoside adenosine. This chapter discusses the role of TNAP and other ectonucleotidases in nociceptive circuits, and their potential as future targets of new therapeutics to treat chronic pain.

Keywords Tissue non-specific alkaline phosphatase (TNAP) · Nociception · Dorsal root ganglia · Adenosine · Ectonucleotidase

13.1 The Burden of Chronic Pain

Pain is one of the most common medical complaints in the United States. An estimated 100 million American adults suffer from chronic pain (Relieving Pain in America: A blueprint for transforming prevention, care, education and research

S.E. Street (✉)
Department of Cell Biology and Physiology,
University of North Carolina at Chapel Hill, Chapel Hill, NC, USA
e-mail: Sarah_street@med.unc.edu

N.A. Sowa
Department of Psychiatry, University of North Carolina at Chapel Hill and UNC Hospitals,
Chapel Hill, NC, USA
e-mail: Nathaniel_sowa@med.unc.edu

C. Fonta and L. Négyessy (eds.), *Neuronal Tissue-Nonspecific Alkaline Phosphatase (TNAP)*, Subcellular Biochemistry 76,
DOI 10.1007/978-94-017-7197-9_13

2011). In one study, over one-fourth of all adults in the United States reported that they suffered from pain that persisted for 24 h, and 42 % of those individuals reported that their pain had lasted greater than one year (Health, United States 2006 With Chartbook of Trends in the Health of Americans 2006). Chronic pain can be extremely debilitating, affecting physical and emotional well-being, and patients who suffer from moderate to severe chronic pain report difficulty performing everyday activities, including exercising, sleeping, doing chores, participating in social activities and carrying out familial responsibilities (Chronic Pain in America: Roadblocks to Relief 1999). The total societal cost of chronic pain has been estimated to lie close to 635 billion dollars, which includes money for treatment and lost productivity (Relieving Pain in America: A blueprint for transforming prevention, care, education and research 2011).

Treatments for pain are not only expensive, but they often fall short of alleviating pain without significant side effects. Pharmaceutical treatments are the most common, and include non-steroidal anti-inflammatory drugs (NSAIDS), cyclooxygenase inhibitors, opioids, antidepressants, anticonvulsants and ion-channel blockers. While they are some of the most commonly prescribed drugs on the market, their concomitant side effects such as nausea, drowsiness, dependency, and tolerance make their long-term use problematic for most patients. Other modes of pain relief have subsequently gained popularity, such as injections of local anesthetics, nerve blocks, electrical stimulation, psychological therapy, surgery and even acupuncture. Even despite these treatments, many patients still suffer, necessitating further advances in pain therapy.

Over the past several years, enzymes collectively known as ectonucleotidases have emerged as potential new targets for pain treatment. Ectonucleotidases are extracellular enzymes that break down nucleotides, such as adenosine triphosphate (ATP) into nucleosides like adenosine (Hilaire et al. 2011; Zimmermann 2006). Tissue non-specific alkaline phosphate (TNAP) is the most recent of three ectonucleotidases that have been characterized in nociceptive (pain-sensing) circuits of the dorsal horn in the spinal cord of mice (Street et al. 2013). TNAP recently joined prostatic acid phosphatase (PAP), and ecto-5′-nucleotidase (NT5E) as members of the family of ectonucleotidases whose function is necessary for normal nociceptive signaling in spinal cord circuits (Zylka 2011). This chapter will discuss how recent findings suggest TNAP's ectonucleotidase activity, along with those of PAP and NT5E, is essential in maintaining proper purinergic tone in nociceptive circuits. Further, it will highlight the potential ability to reduce pain that each of these ectonucleotidases harbor.

13.2 Nociceptive Circuitry in the DRG and Spinal Cord

Highly specialized neurons in the periphery, known as primary sensory neurons, provide information about the environment, including the presence of noxious stimuli, to the central nervous system through axonal fibers. Neurons that are specialized to detect noxious stimuli are called nociceptors (Sherrington 1906).

Some nociceptors can detect several forms of noxious stimuli, while others are specialized for one type of input (Julius and Basbaum 2001; Meyer et al. 2006). Fibers arising from these neurons have been shown, using electrophysiological techniques, to be excited by noxious heat and cold, intense pressure, and chemical irritants, but not by innocuous warming or light touch (Burgess and Perl 1967). These neurons house their pseudo-unipolar cell bodies in the trigeminal and dorsal root ganglia (DRG). They then send their central processes into the spinal cord or brainstem where they synapses with second-order sensory neurons. These neurons then either synapse on other neurons in the spinal cord and brainstem, or directly relay information to the thalamus (Meyer et al. 2006).

There are two broad classes of primary nociceptive neurons that encode nociceptive information. First, thinly myelinated Aδ fibers arise from medium diameter cell bodies in the DRG or trigeminal ganglia. These sensory neurons detect noxious heat and mechanical stimuli, and play a role in the sensitization process following burn or chemical injury (Campbell et al. 1979; Dubner and Hu 1977; Meyer et al. 2006; Ringkamp et al. 2001; Treede et al. 1998). Second, small-diameter, unmyelinated C-fibers respond to the same noxious stimuli, but propagate action potentials more slowly, display slowly adapting responses to stimuli, and are heavily involved in the sensitization process following tissue injury (Meyer et al. 2006; Peng et al. 2003).

These small nociceptive DRG neurons are further subdivided into classification based on cellular markers (Woolf and Ma 2007). Peptidergic neurons contain neuropeptides such as calcitonin gene-related protein (CGRP), substance P, and somatostatin, and also express TrkA, the tyrosine kinase receptors that bind nerve growth factor (Fitzgerald 2005; Julius and Basbaum 2001). These neurons predominantly project to lamina I and the outer segment of lamina II in the dorsal horn (Hunt and Mantyh 2001; Zylka et al. 2005). Nonpeptidergic neurons express proteins such as PAP, the G-protein coupled receptors MrgprD, and the ATP-gated ion channel $P2X_3$, bind the plant lectin IB4 (Meyer et al. 2006; Silverman and Kruger 1988a, b; Snider and McMahon 1998; Vulchanova et al. 1998; Zylka et al. 2005, 2008), and project to the inner layer of lamina II (Perry and Lawson 1998; Zylka et al. 2005). These gene expression patterns also loosely translate to the segregation of nociceptive sensory input. For instance, the heat–and capsacin-sensitive channel TRPV1 is mostly expressed in peptidergic neurons (Cavanaugh et al. 2009; McCoy et al. 2012), whereas nonpeptidergic neurons that express MrgprD seem to encode information regarding mechanical stimuli (Cavanaugh et al. 2009). Thus, these categories also have some functional implication.

13.3 ATP and Adenosine's Role in Nociception

ATP and adenosine both play important roles in nociception. ATP is released by nearly all cells, including neurons and glia in response to both physiological and pathological stimuli (Arcuino et al. 2002; Gourine et al. 2010; Matsuka et al. 2008;

Nakamura and Strittmatter 1996). ATP generally exerts its effects through purinergic P2 receptors, of which there are two families—P2X and P2Y. P2X receptors are ligand-gated, cation-selective channels that lead to excitation and pro-nociceptive effects, while P2Y receptors are G-protein coupled receptors (GPCRs) that have been reported to have both pro- and anti-nociceptive effects (Burnstock 2009; Gerevich et al. 2005, 2007; Moriyama et al. 2003; Okada et al. 2002; Sawynok et al. 2006; Tominaga et al. 2001). Of the seven cloned mammalian P2X receptors, $P2X_3$ is the most abundant in DRG neurons. $P2X_3$ receptors colocalize with markers of nonpeptidergic sensory neurons such as MrgprD and PAP (Zylka et al. 2005, 2008). Further, administration of ATP or other $P2X_3$ receptors agonists result in excitation of C-fibers and pain-avoiding behaviors such as thermal hyperalgesia and mechanical allodynia (Dowd et al. 1998; Hamilton et al. 2001, 1999; Hilliges et al. 2002; Sawynok and Reid 1997; Tsuda et al. 2000). This is not surprising since inflammation and nerve injury lead to upregulation of $P2X_3$ expression in DRG and spinal cord neurons, and sensitize the receptor to ATP though phosphorylation (Chen et al. 2001; Dai et al. 2004; Novakovic et al. 1999; Paukert et al. 2001; Zhou et al. 2001; Xu and Huang 2002). In addition to $P2X_3$ receptors, $P2X_{2/3}$ and $P2X_4$ receptors also play a pro-nociceptive role in spinal sensory circuits (Nakagawa et al. 2007; Tsuda et al. 2003). Thus, ATP, in general, has a pro-nociceptive effect through its activation of P2X receptors.

On the other hand, adenosine, a product of ATP hydrolysis, exerts a potent antinociceptive effect. Adenosine activates four subtypes of P1 receptors—A_1, A_{2A}, A_{2B}, and A_3 (Abbracchio et al. 2009; Burnstock 2009). Each of these P1 receptors is a GPCR that couple to different types of G proteins. The A_1 adenosine receptor (A_1R) is expressed by small and medium diameter DRG and trigeminal ganglion neurons, as well as postsynaptic sites in the dorsal horn of the spinal cord and brainstem (Carruthers et al. 2001; Schulte et al. 2003). Activation of A_1Rs results in reduced Ca^{2+} entry, decreased substance P and CGRP release from DRG neurons, as well as inhibition of neurotransmission through both pre- and post-synaptic mechanisms in the spinal cord (Carruthers et al. 2001; Haas and Selbach 2000; Li and Perl 1994; Patel et al. 2001; Santicioli et al. 1993; Sjolund et al. 1997).

Activation of A_1R by adenosine or other agonists administered both peripherally and centrally causes antinociceptive effects. For instance, local administration of A_1R agonists into the rat hind paw leads to a decrease in nociceptive behavior in several different models of inflammatory and neuropathic pain (Aley et al. 1995; Aumeerally et al. 2004; Karlsten et al. 1992; Liu et al. 2000; Taiwo and Levine 1990). Also, spinal application of adenosine, A_1R agonists, or other agents that increase the endogenous adenosine levels have antinociceptive effects in animal models of chronic pain (Dickenson et al. 2000; Sawynok and Liu 2003; Zylka 2011). Conversely, mice lacking A_1R ($A_1R^{-/-}$) show enhanced nociception including thermal sensitivity and thermal hyperalgesia, but have normal responses to mechanical stimuli (Johansson et al. 2001; Wu et al. 2005). Finally, the antinociceptive effect of acupuncture is dependent upon adenosine generation and activation of A_1R (Goldman et al. 2010). Taken together, these data suggest great potential for targeting of A_1R for treatment of chronic pain. However, studies using high doses of A_1R-selective agonists have shown side effects including motor paralysis and autonomic dysfunction (reviewed in Zylka 2011). Thus, in order to harness the potential antinociceptive action of A_1R, a more indirect mode of activation must be determined.

13.4 Ectonucleotidases and Extracellular Adenosine Generation

One source of extracellular adenosine results from the breakdown of nucleotides such as ATP, ADP and AMP. This extracellular production of adenosine relies on the activity of several different enzymes that degrade amine-containing nucleotides in a step-wise manner. While some enyzmes hydrolyze ATP, ADP and AMP, others are specific for a particular adenine-containing nucleotide. These enzymes, known generally as ectonucleotidases, play important roles in the regulation of both P1 receptors (through the production of adenosine) and P2 receptors (through the degradation of ATP) activity (Sawynok 2007; Zimmermann 2006). Since their activity can decrease activation of the pronociceptive P2X receptors, and increase the activation of antinociceptive A_1R receptors, ectonucleotidases sit at a pivotal position in the regulation of nociceptive circuits.

Prior to the discovery of TNAP as an important ectonucleotidase in nociceptive signaling, two other ectonucleotidases were previously identified and studied in the context of pain control in mice. First, a long-standing question in neuroscience was answered when the molecular identity of the fluoride-resistant acid phosphatase, used to identify a sub-population of small-diameter DRG neurons, was shown to be the ectonucleotidase PAP (Zylka et al. 2008). PAP is an acid phosphatase that hydrolyzes AMP into adenosine. In nociceptive circuits, PAP is predominantly expressed in nonpeptidergic DRG neurons, and in their central terminals in lamina II. What was perhaps the most surprising finding of this study however, was spinal injection of PAP led to potent antinociceptive effects in animal models of neuropathic and chronic inflammatory pain (Sowa et al. 2009; Zylka et al. 2008). Further, these antinociceptive effects lasted three-times longer than morphine without any of the apparent side effects. Interestingly, mice lacking A_1R did not show any antinociceptive effects after spinal injection of PAP, suggesting that this antinociceptive activity of PAP was dependent upon the generation of adenosine. Additionally, mice lacking PAP ($Pap^{-/-}$) showed hyperalgesia and mechanical allodynia (sensitization of mechanical sensory neurons) in models of chronic inflammatory pain (Zylka et al. 2008). However, AMP hydrolytic activity was only partly reduced in $Pap^{-/-}$ mice, suggesting that there was at least one other ectonucleotidase responsible for breaking down AMP.

Shortly after PAP was demonstrated to be an important player in nociceptive circuitry, NT5E, an ectonucleotidase that also hydrolyzes AMP into adenosine, was shown to be expressed in DRG neurons (Sowa et al. 2010b). While PAP colocalizes predominantly with nonpeptidergic DRG neurons, NT5E is more broadly expressed across all classes of nociceptive neurons. Similar to PAP, mice lacking NT5E (*Nt5e*$^{-/-}$) showed enhanced nociceptive sensitivity in several behavioral models of pain, and spinal injection of NT5E protein caused anti-nociceptive behavior (Sowa et al. 2010b, c). However, when NT5E was injected into mice lacking A_1R, the anti-nociceptive effect was abolished (Sowa et al. 2010c). Further, when AMP and an adenosine transporter inhibitor were injected into the spinal cords of mice, *Nt5e*$^{-/-}$ mice demonstrated a 50 % reduction in antinociceptive behavior compared to the control (Sowa et al. 2010b). These data suggested that, like PAP, NT5E also was not solely responsible for adenosine generation in nociceptive circuits. Indeed, AMP hydrolysis was decreased, but not abolished in DRG and spinal cord sections from *Nt5e*$^{-/-}$ mice, suggesting that both PAP and NT5E hydrolyze AMP and generate adenosine in the DRG and dorsal horn (Sowa et al. 2010b).

Double knockout (dKO) mice that lack both PAP and NT5E were subsequently bred to determine the extent to which these ectonucleotidases contribute to adenosine generation in the dorsal horn. While AMP hydrolysis was mostly reduced in the dKO mice at both physiological and acidic pH, behavioral studies showed that some AMP hydrolytic activity remained even after PAP and NT5E activity was eliminated (Street et al. 2011). Both *Pap*$^{-/-}$ and *Nt5e*$^{-/-}$ single knockout mice showed an approximately 50 % decrease in antinociceptive behavior when injected with AMP and an inhibitor of the adenosine transporter. However, when the same injection was made in the dKO mouse, a small but significant antinociceptive effect remained, suggesting that there was at least one additional remaining ectonucleotidase in spinal nociceptive circuits.

In an attempt to determine the remaining source of adenosine production in the dKO mice, fast-scan cyclic voltametry (FSCV) was used. FSCV enables the direct measurement of subsecond changes in extracellular adenosine concentration and can be done in mouse spinal cord slices (Swamy and Venton 2007). In this technique a carbon electrode measures the current generated when adenosine changes oxidation states when presented with voltage ramps from −0.5 to 1.5 mV. By measuring the current level generated at 1.0 mV, the voltage at which adenosine is oxidized, the concentration of adenosine can be deduced. Surprisingly, experiments using FSCV showed that even in spinal cord slices from dKO mice, about 30 % of maximum adenosine generation remained (Street et al. 2011). Further, spontaneous transient increases in adenosine concentration were decreased, but not eliminated in the dorsal horn of spinal cord slices from dKO mice (Street et al. 2011). These data strongly suggested that while PAP and NT5E are the predominate contributors to adenosine generation in the dorsal horn, there is at least one remaining ectonucleotidase that is important for the antinociceptive effects of adenosine generation.

13.5 The Role of TNAP in Nociceptive Circuits

FSCV experiments done in dKO animals at neutral pH suggested that AMP hydrolytic activity remained even after genetic deletion of both PAP and NT5E. However, when these experiments were carried out at pH 5.6, virtually no adenosine was detected, suggesting that the remaining ectonucleotidase was not active in acidic environments (Street et al. 2011). These findings led to the hypothesis that perhaps TNAP could be the remaining ectonucleotidase in the dorsal horn of the spinal cord. TNAP can hydrolyze the nucleotides ATP, ADP and AMP to generate adenosine at both physiological and alkaline pH (Ciancaglini et al. 2010; Scheibe et al. 2000; Zimmermann 2006). Further, TNAP has been shown to be widely expressed in the central nervous system, including the developing spinal cord (Fonta et al. 2004; Langer et al. 2008; MacGregor et al. 1995; Narisawa et al. 1994). However, TNAP had not previously been shown to be expressed in, or have a role in adult nociceptive circuits. Thus, the elucidation of TNAP's potential role in nociception was essential to further address the role of ectonucleotidases in the DRG and dorsal horn.

13.6 TNAP Gene Expression in DRG and Spinal Cord

First, to determine whether the TNAP gene was expressed in adult mice, in situ hybridization was performed on adult mouse DRG and spinal cord tissue. TNAP mRNA was detected in nearly all DRG neurons and in neurons throughout the dorsal horn of the spinal cord (Street et al. 2013). This expression pattern slightly differed from PAP and NT5E, since these ectonucleotidases were expressed in subsets of nociceptive and non-nociceptive DRG neurons (Sowa et al. 2010b; Zylka et al. 2008). Further, whereas PAP and NT5E protein expression in the dorsal horn comes primarily from DRG nerve terminals, the TNAP in situ hybridization results suggested that TNAP was expressed in neuronal cell bodies in lamina I and II. These results show that the broad expression of TNAP reported in the developing spinal cord continues into adulthood (MacGregor et al. 1995; Narisawa et al. 1994)

Since TNAP is expressed broadly in many subtypes of DRG neurons, it is likely co-expressed in small- and medium-diameter nociceptive sensory neurons, suggesting that TNAP, NT5E and PAP serve redundant roles in their modulation of nociception. Further, TNAP's localization pattern also suggests that it might serve as an ectonucleotidase in other sensory circuits beyond those activated by noxious stimuli. The level of TNAP expression did not increase in $Pap^{-/-}$ and $Nt5e^{-/-}$ single knockout mice or dKO mice, demonstrating that TNAP expression is not upregulated in response to decreased expression of other ectonucleotidases. Thus, baseline expression of TNAP is adequate to generate enough adenosine to carry out the remaining antinociceptive effects observed in previous experiments (Street et al. 2011, 2013).

13.7 TNAP and Adenosine Generation in Nociceptive Circuits

TNAP, despite being more well known in the nervous system for its hydrolysis of pyridoxal-5′ phosphate and its role in GABA metabolism, also breaks down extracellular nucleotides into adenosine. This ectonucleotidase activity has been reported in the hippocampus, where it plays a role in axon guidance, and also where it serves as the source of adenosine in the absence of NT5E (Diez-Zaera et al. 2011; Zhang et al. 2012). Unfortunately due to TNAP's role in GABA metabolism, *Tnap*$^{-/-}$ mice develop seizures and die approximately 2 weeks after birth due to their inability to generate GABA, making them difficult study (Narisawa et al. 1997; Waymire et al. 1995). However an inhibitor of TNAP, MLS-0038949, was used to study the role of TNAP in adenosine generation (Dahl et al. 2009; Sergienko et al. 2009). Indeed, MLS-0038949 blocked the activation of A_{2B}R when AMP was applied to HEK cells that expressed TNAP and A_{2B}R, showing that it completely inhibits the ectonucleotidase activity of TNAP (Street et al. 2013).

Since TNAP activity was adequately abolished with MLS-0038949, this compound was used to inhibit TNAP activity in AMP histochemistry studies in both DRG and spinal cord sections. Staining of AMP hydrolytic activity was reduced in both DRG and spinal cord in sections from dKO mice, but a substantial amount of activity remained, especially at alkaline pH (Street et al. 2013). However, at both physiological and alkaline pH, AMP hydrolytic activity in slices from dKO mice was completely abolished when MLS-0038949 was present, suggesting that TNAP was the sole remaining ectonucleotidase present in spinal nociceptive circuits.

In order to ensure that MLS-0038949 was specific to TNAP and had not inhibited another unidentified ectonucleotidase, AMP histochemistry was carried out on tissue from an early postnatal *Tnap*$^{-/-}$ mouse. At pH 8.5, both DRG and spinal cord sections showed reduced staining in tissue from *Tnap*$^{-/-}$ mice. To show that the residual histochemical staining was due to PAP and NT5E, we then treated WT and *Tnap*$^{-/-}$ slices with tartrate and α, β methylene ADP, inhibitors of PAP and NT5E respectively. Again, when activity of all three ectonucleotidases was absent, AMP hydrolytic activity was eliminated (Street et al. 2013).

FSCV was then used to quantify the precise levels of adenosine generated in spinal cord slices that lack all three ectonucleotidases (Street et al. 2011). Similar to previous findings, when 100 mM AMP was applied to lamina I and II, a rapid transient increase in adenosine was detected by the electrode (Fig. 13.1a), and this increase in adenosine concentration was decreased in slices from dKO mice (Fig. 13.1b). However, in the presence of MLS-0028949 adenosine generation was dramatically reduced in dKO slices, but not in WT slices, suggesting that TNAP is most likely the last remaining ectonucleotidase expressed in the dorsal horn (Fig. 13.1c, d). When the concentration of the remaining adenosine was quantified, less than 1 μM of adenosine was produced from a 100 mM bolus of AMP (Fig. 13.1e, f). It is also possible that some of this remaining adenosine was the result of spontaneous AMP degradation in the application solution. Since the EC_{50} for adenosine binding to A_1R is approximately 1 μM,

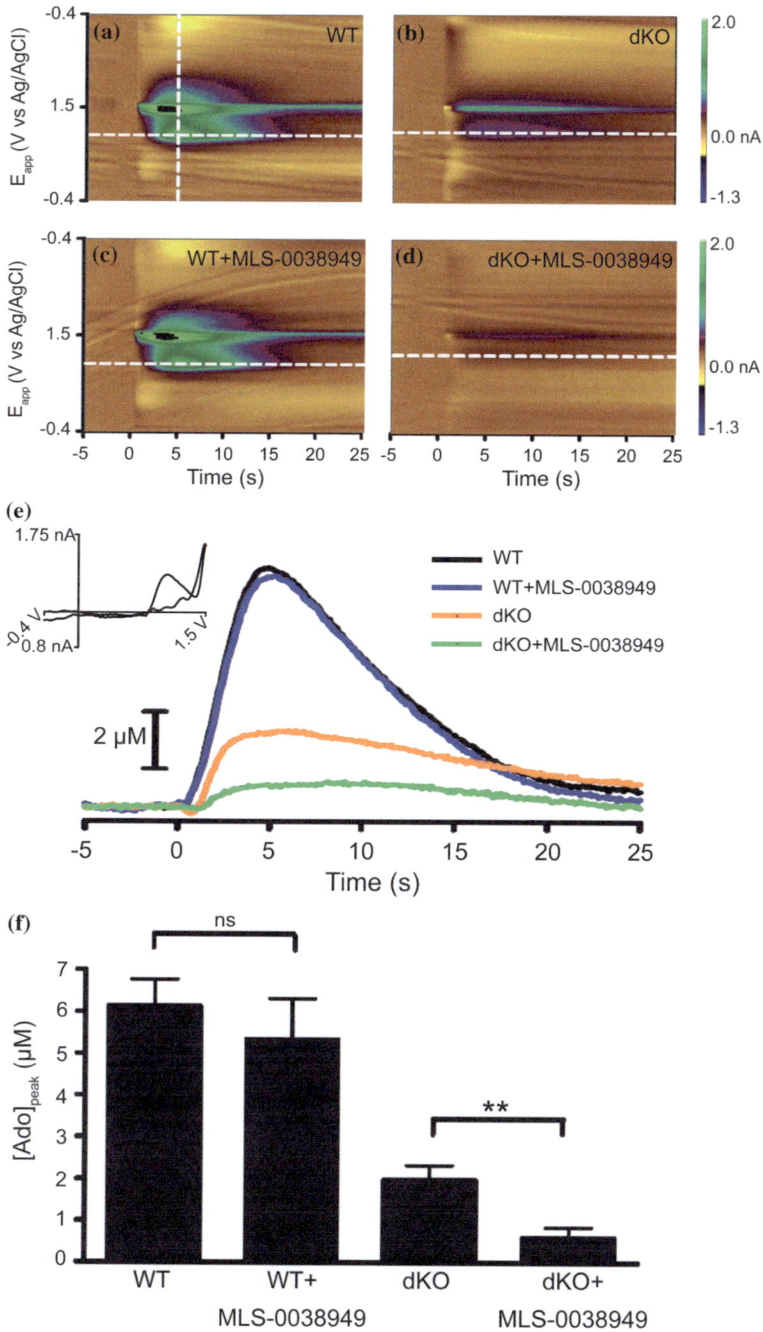
(a) WT
(b) dKO
(c) WT+MLS-0038949
(d) dKO+MLS-0038949
E_app (V vs Ag/AgCl)
-0.4
1.5
-0.4
2.0
0.0 nA
-1.3
-5 0 5 10 15 20 25
Time (s)
(e)
1.75 nA
-0.4 V
0.8 nA
1.5 V
WT
WT+MLS-0038949
dKO
dKO+MLS-0038949
2 μM
(f)
ns
**
[Ado]_peak (μM)
0 1 2 3 4 5 6 7
WT
WT+ MLS-0038949
dKO
dKO+ MLS-0038949

◀ **Fig. 13.1** Inhibition of TNAP, NT5E and PAP abolishes adenosine generation in the dorsal spinal cord. Fast scan cyclic voltametry (FSCV) was used to measure adenosine production at subsecond resolution. (**a–d**) FSCV color plots showing the level of current (color scale) that was generated over time (x-axis) and voltage (y-axis). 100 μM AMP was pressure ejected for 1 s onto lamina II of (**a**, **c**) WT or (**b**, **d**) dKO mice in the absence or presence of MLS-0038949 (50 μM). (**e**) Adenosine concentration, calculated from 1.0 V current (dashed horizontal lines in **a–d**). Inset, cyclic voltammogram confirms adenosine was produced (plotted from *dashed vertical line* in **a**). (**f**) Peak adenosine concentration after pressure ejecting AMP onto lamina II (n = 5 slices for each condition). **P < 0.005. Figure reprinted from Street et al. (2013) with permission

it is unlikely that the concentration of adenosine generated in slices lacking PAP, NT5E, and TNAP is enough to substantially activate A_1Rs (Rittiner et al. 2012). Thus, TNAP is most likely the final ectonucleotidase responsible for the hydrolysis of AMP in the dorsal horn.

13.8 Functional Effects of TNAP Inhibition in Spinal Cord Slices

Since nociceptive DRG neurons synapse onto their target neurons primarily in lamina I and lamina II in the dorsal horn, synaptic neurotransmission between primary and secondary nociceptive neurons can be assessed by measuring the field excitatory post-synaptic potential (fEPSP) generated in lumbar dorsal horn slices after electrical stimulation of a lumbar nerve root (Ruscheweyh and Sandkuhler 2000; Street et al. 2011). AMP and adenosine both inhibit neurotransmission in the dorsal horn through activation of A_1R, although AMP's action has previously been assumed to be indirect through the activity of ectonucleotidase (Dunwiddie et al. 1997; Lao et al. 2001; Li and Perl 1994, 1995; Salter and Henry 1985). Given these previous findings, electrophysiology was then used to determine the functional effects of inhibition of PAP, NT5E, and TNAP (Street et al. 2013).

As previously shown, both AMP and adenosine inhibit the Aδ fEPSP strength in lamina II (Fig. 13.2a). This inhibition is dependent on A_1R since there was no effect of AMP or adenosine in slices from $A_1R^{-/-}$ mice (Fig. 13.2b). AMP and adenosine was then applied to slices from $Pap^{-/-}$, $Nt5e^{-/-}$ and dKO mice. Surprisingly, there was no difference in the inhibitory effect of AMP on fEPSP strength in slices from any of the knockout mice, suggesting that either AMP could activate A_1R without first being hydrolyzed to adenosine or that deletion of all three ectonucleotides was required to sufficiently reduce adenosine generation (Fig. 13.2c). However, when slices from the different strains of mice were incubated in MLS-0038949, the inhibitory effect of AMP was decreased in slices from $Nt5e^{-/-}$ mice, and completely abolished in slices from dKO mice (Fig. 13.2d, e). Since inhibition of all three ectonucleotidases completely reduced AMP's effect on synaptic transmission, these data confirmed that AMP must first be hydrolyzed to adenosine in order to activate A_1R and cause synaptic inhibition.

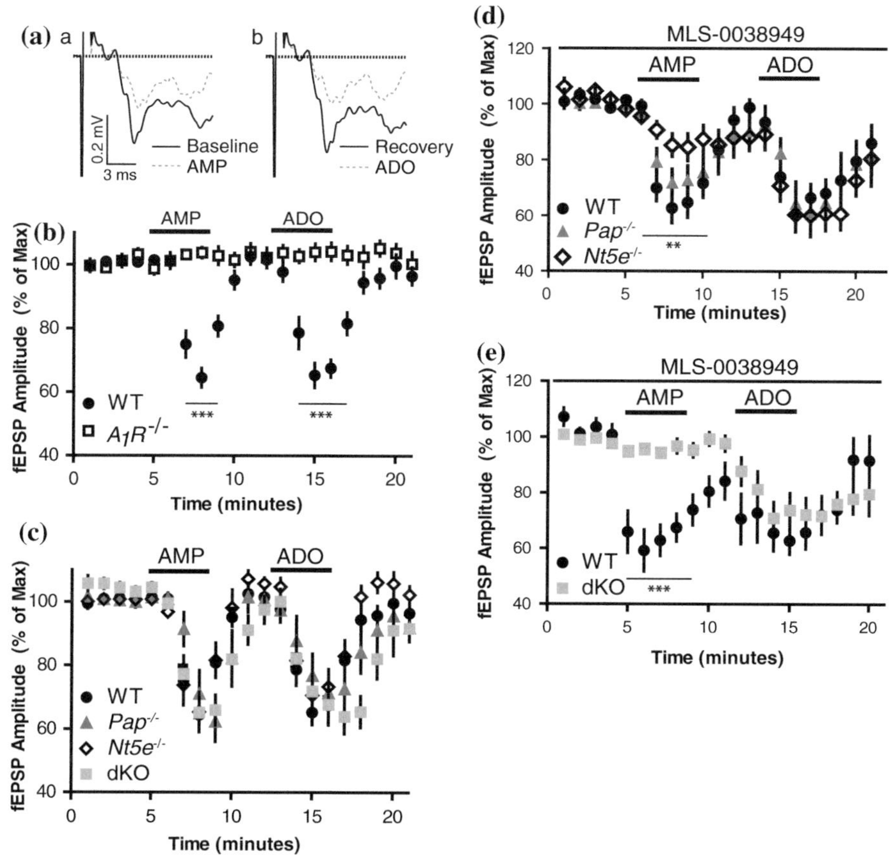

Fig. 13.2 TNAP, NT5E, and PAP are triply redundant in nociceptive circuits of the dorsal spinal cord. **a** Representative Aδ fEPSPs in lamina II of WT spinal cord slices before (*solid line*) and after (*dashed line*) addition of **a** 250 μM AMP or **b** 250 μM adenosine (ADO) to perfusate. **b** Normalized fEPSP amplitude in WT and $A_1R^{-/-}$ spinal cord slices (n = 25 and 11; respectively). ***P < 0.0005. **c** Normalized fEPSP amplitude in WT, $Pap^{-/-}$, $Nt5e^{-/-}$, and dKO spinal cord slices. (n = 25, 10, 20, 16; respectively). No significant differences were found between genotypes. **d** Normalized fEPSP amplitude in WT, $Pap^{-/-}$, and $Nt5e^{-/-}$ slices incubated with TNAP inhibitor MLS-0038949 (50 μM) (n = 11 for all genotypes). There was a significant reduction in the inhibitory effect of AMP on slices from $Nt5e^{-/-}$ mice when compared to WT and $Pap^{-/-}$ (all incubated with MLS-0038949); P < 0.005, **e** Normalized fEPSP amplitude in WT and dKO mice incubated with MLS-0038949 (n = 9 and 12; respectively). ***P < 0.0005. Figure adapted from Street et al. (2013), and reprinted with permission

These results strongly suggest that AMP must first be dephosphorylated into adenosine to have physiological effects downstream of A_1R. Previous studies carried out in heterologous expression systems suggested that AMP might be able to directly activate A_1R in intact organ systems (Rittiner et al. 2012). However, these results demonstrated that adenosine is required to activate G proteins that inhibit cAMP generation (G_i) downstream signal transduction cascades that ultimately result in synaptic inhibition.

More importantly, this experiment strongly suggested that all three enzymes, PAP, NT5E, and TNAP, play a role in generating extracellular adenosine that then inhibits synaptic activity in nociceptive circuits. It also demonstrated that NT5E and TNAP contribute to the majority of adenosine generation since deletion or inhibition of both of these enzymes together resulted in reduced synaptic inhibition. However, removal of activity of all three enzymes was necessary to completely reduce the inhibitory effect of AMP application, showing that these enzymes work redundantly in nociceptive circuits of the spinal cord.

13.9 Proposed Model of Ectonucleotidase Function in Nociceptive Circuits

These studies show that TNAP, along with PAP and NT5E, plays an important role in the modulation of nociception. ATP, released by stimulated sensory neurons, activate P2X and P2Y receptors on both neurons and glia, and is thought to play a role in the transition from acute to chronic pain (Basbaum et al. 2009; Burnstock 2007; Sawynok et al. 2006; Tsuda et al. 2005; Tozaki-Saitoh et al. 2008). Ectonucleotidases then act to terminate this pronociceptive signaling by breaking down ATP, with TNAP, PAP and NT5E contributing to the final step in this breakdown, generating adenosine from AMP. Adenosine then elicits anti-nociceptive effects through A_1R. Thus, TNAP, PAP and NT5E all determine the relative extracellular purinergic tone that is either activating or inhibiting nociceptive neurons. With deletion of these enzymes, less extracellular adenosine is generated, swinging the balance in favor of excitation over inhibition.

Other studies have suggested that alterations in purinergic tone also lead to downstream changes in the levels of membrane phospholipids (Sowa et al. 2010a). For instance, activation of A_1R is thought to inhibit synaptic transmission through the activation of the $G\alpha_i$ subunit. However, other downstream effects of A_1R activation occur through the release and activation of the $G\beta\gamma$, which subsequently activates phospholipase C (PLC) (Jacobson and Gao 2006; Murthy and Makhlouf 1995). PLC hydrolyzes phosphatidylinositol 4, 5-bisphosphate (PIP_2) in the plasma membrane leading to depletion of PIP_2. PIP_2, a membrane phospholipid, interacts with many different transmembrane proteins, including GPCRs and ion channels. Usually this interaction is necessary for physiological activation of ion channels such as the transient receptor potential cation channel subfamily V member 1 (TRPV1) and voltage-gated calcium channels (Caterina et al. 1997, 2000; Davis et al. 2000; Rohacs et al. 2005; Sowa et al. 2010a; Suh and Hille 2005). Decreases in membrane levels of PIP_2 lead to deactivation of excitatory ion channels, resulting in an overall decrease in the excitability of neurons.

In DRG neurons, sustained activation of A_1R decreases nociceptive sensitivity by inhibiting channel activation via PIP_2 depletion (Fig. 13.3) (Sowa et al. 2010a). Since extracellular adenosine concentration is dependent on activity of

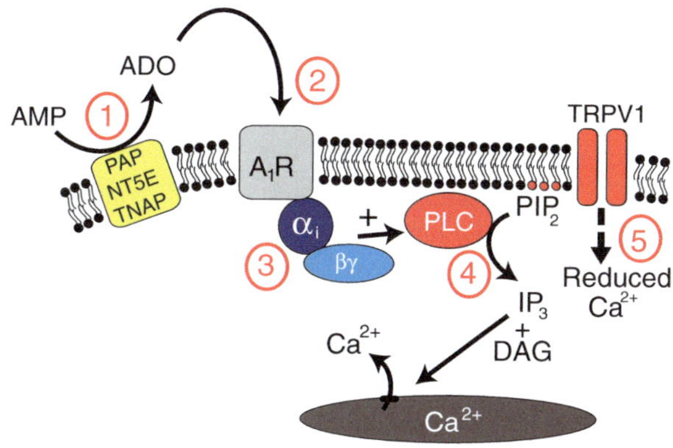

Fig. 13.3 Ectonucleotidases modulate nociceptive sensory neurons by regulating phosphoinositide tone. Proposed model showing the steps involved in ectonucleotidase-catalyzed adenosine generation, and downstream effects. *1* Extracellular AMP is hydrolyzed by TNAP, NT5E, and PAP to generate extracellular adenosine. *2* Adenosine binds and activates A_1R. *3* PLC is activated through the βγ subunit of the G_i G-protein. *4* PLC depletes PIP_2 in the plasma membrane. *5* TRPV1 channels are inhibited following PIP_2 depletion, leading to a decrease in neuronal excitability. Figure adapted from Sowa et al. (2010a) and used with permission

ectonucleotidases, these enzymes influence the levels of PIP_2 in the membrane and hence, the overall excitability of nociceptive neurons. For example, PIP_2 levels are decreased in DRG neurons following injection of PAP, while PIP_2 levels are increased in DRG from *Pap*$^{-/-}$ and *Nt5e*$^{-/-}$ mice (Sowa et al. 2010a). Further, the relative levels of PIP_2 in the membrane of nociceptive neurons sets the degree of sensitization that nociceptive neurons experience following nerve injury or injection of an inflammatory agent. When levels of PIP_2 are low due to increased activity of PAP, both thermal hyperalgesia and mechanical allodynia are reduced (Sowa et al. 2010a). Thus, ectonucleotidases determine activation of A_1R, which in turn determines the levels of PIP_2 in the membrane. PIP_2 then sets the overall excitability of nociceptive neurons through its interaction with ion channels and GPCRs. In fact, by directly increasing the level of PIP_2 in the plasma membrane of DRG neurons through spinal injection of PIP_2, the sensitivity of mice to noxious thermal stimuli is increased.

These findings have led to a theory that emphasizes the interaction of purinergic and cellular "phosphoinositide tone" in nociceptive circuits (Fig. 13.3). With increases in adenosine generation by ectonucleotidases, there is increased A_1R activation and increased depletion of PIP_2. This leads to decreases in synaptic transmission and membrane excitability. However, increases of extracellular ATP and ADP lead to increases in membrane excitability and neuronal sensitization. In this way, the balance of extracellular purinergic tone and membrane PIP_2 levels control the overall excitability and sensitivity of nociceptive circuits. Thus, TNAP, PAP and NT5E play critical roles in determining the overall sensitivity of these circuits.

13.10 Possible Reasons for Redundancy in Ectonucleotidases in the Dorsal Horn

DRG and spinal cord are not the first tissues reported to have redundant expression of ectonucleotidases. For instance, NT5E and TNAP act redundantly to hydrolyze AMP in both the hippocampus and airway epithelial cells (Picher et al. 2003; Zhang et al. 2012). Further TNAP is often co-expressed with NT5E and other ectonucleotidases, suggesting that redundant expression of several different ectonucleotidases is common to many different organ systems and circuits in the central nervous system (Langer et al. 2008; Zimmermann 2006). However, triply redundant expression of acid and alkaline phosphatase, along with another ectonucleotidase, has yet to be reported in any area of the CNS. This raises the question as to why nociceptive circuits display such extensive expression of ectonucleotidases. There are several possibilities that might explain this redundant expression of ectonucleotidases.

13.10.1 Importance of Nucleotide Hydrolysis

The most obvious answer lies in the proposed model linking ectonucleotidases to the control of phosphoinositide tone, which is an important determinant of neuronal sensitivity. If the build-up of extracellular nucleotides leads to sensitization of sensory neurons and the availability of adenosine keeps nociceptive neurons from becoming overly excitable, it follows that the enzymes responsible for this baseline balance have several levels of redundancy. The evidence discussed here clearly shows that loss of just one, or even two of these enzymes does not result in the complete loss of AMP hydrolysis. Therefore, some ability to maintain physiological levels of PIP_2 remains after loss of one or two ectonucleotidases.

Interestingly, some aspects of the knockout mice behavioral data support this conclusion. While both single knockouts (*Pap*$^{-/-}$ or *Nt5e*$^{-/-}$) and the dKO showed increased sensitization in models of chronic inflammatory and neuropathic pain, almost all of the baseline nociceptive behavior was unaffected. Thus, it was not until an inflammatory insult was given to the animal that nociceptive behavior changed (Sowa et al. 2010b; Street et al. 2011; Zylka et al. 2008). This suggests that having at least one remaining enzyme is sufficient to maintain baseline extracellular adenosine levels and phosphoinositide tone. Unfortunately, developing a triple-knockout mouse where TNAP, PAP and NT5E are all deleted would be quite difficult considering that *Tnap*$^{-/-}$ mice die a few weeks after birth. It could be possible to develop a neuron-specific *Tnap*$^{-/-}$ mouse that can then be crossbred with dKO mice, but even this model might not sufficiently reduce TNAP activity since the cellular localization of TNAP in spinal circuits is unknown. However, any *Tnap*$^{-/-}$ mouse model of nociception would also be confounded by the role TNAP plays in somatosensory circuit development. Lumbar nerve roots, used in many mouse models of nociception, are smaller in *Tnap*$^{-/-}$ mice (Narisawa et al. 1997).

13.10.2 pH Specificity

Another possible need for the redundancy of TNAP, PAP and NT5E could be explained by the differences between these enzymes. While all three ectonucleotidases are capable of hydrolyzing AMP, these enzymes have very different functional profiles. NT5E is an exclusive 5′-monophosphatase that selectively hydrolyzes 5′-AMP in the low micromolar range and at optimal neutral pH range. PAP, on the other hand, is a promiscuous phosphomonoesterase that can also degrade 5′-nucleoside monophosphates, and a broader pH range (3–8), while TNAP dephophorylates many extracellular substrates including ATP, ADP, and AMP at neutral and alkaline pH (Ciancaglini et al. 2010; Scheibe et al. 2000).

These differences in activity at varying pH values are especially important in the nervous system, where repeated neuronal stimulation can produce slightly acidic conditions (DeVries 2001; Wemmie et al. 2008). Also, following tissue injury, an "inflammatory soup" arises that contains protons and nucleotides (Julius and Basbaum 2001). In these settings PAP, which is more active at acidic pH may be more capable than NT5E or TNAP at degrading AMP to adenosine. Other studies have also shown that electrical nerve simulation produced acid-alkaline, biphasic, and alkaline-acid-alkaline triphasic swings in extracellular pH in the spinal cord dorsal horn (Sykova and Svoboda 1990). These studies suggest that the changes in extracellular pH at varying levels of sensory input might necessitate the expression of an acid and an alkaline phosphatase, along with the dominant phosphatase that functions at neutral pH.

13.10.3 Substrate Specificity

An additional reason for redundancy might reflect the fact that PAP and TNAP both degrade a wide variety of substrates while NT5E is restricted to only 5′-nucleoside monophosphates. For instance, PAP can degrade another pro-nociceptive compound lysophosphatidic acid (LPA) (Tanaka et al. 2004). LPA is released from platelets and neurons following tissue injury and produces prolonged hyperalgesia and allodynia following activation of central LPA receptors (Inoue et al. 2004; Moolenaar et al. 1997). PAP also has been implicated in acting on cellular substrates that inhibit cell growth and replication (Lin et al. 1992, 1994, 2001; Quintero et al. 2007; Hurt et al. 2012). Therefore PAP might be expressed to perform other roles in addition to its role as an acid phosphatase in DRG and spinal cord.

Neuronal TNAP, in contrast, is most well known for its role in pyridoxal 5′-phosphate metabolism, which is an essential co-factor in GABA synthesis (Narisawa et al. 1997; Waymire et al. 1995). Further, TNAP is not only capable of AMP hydrolysis, but also degradation of ATP and ADP (Ciancaglini et al. 2010). Hence, TNAP might be important in the upstream hydrolysis of extracellular nucleotides as well as AMP hydrolysis. Other ectonucleotidases that degrade ATP and ADP such as nucleoside triphosphate diphosphohydrolases 1–3 (ENTPD1–3) are expressed in DRG neurons, but ENTPD3 activity seems to dominate in nociceptive circuits (Vongtau et al. 2011). Thus, in addition to its primary role in the CNS, TNAP might also serve as a redundant ectonucleotidase that can help out in times of pH shift or large increases in extracellular nucleotides.

13.10.4 Ectonucleotidase Localization

The functional redundancy of TNAP, PAP and NT5E is dependent on the localization of each of these enzymes. While it is known that PAP and NT5E are co-expressed in some DRG neurons, their expression is found mainly in nonpeptidergic neurons and some peptidergic neurons (Sowa et al. 2010b; Zylka et al. 2008). TNAP, on the other hand, is broadly expressed in DRG neurons regardless of size (Street et al. 2013). Thus, there might be different patterns of redundancy depending on the precise primary sensory afferent and its target.

Further, the precise subcellular location of TNAP, PAP and NT5E activity has yet to be determined, leaving many questions unanswered. For instance, are these effects occurring in the peripheral terminals of primary afferents, in DRG cell bodies, in primary afferent terminals in the dorsal spinal cord, or on post-synaptic terminals of dorsal horn neurons? Further, are each of these enzymes expressed in the synapse or are some enzymes also expressed on other membrane areas of neurons and glia to deal with extra-synaptic ATP release? One study placed TNAP localization directly in axonal and dendritic processes, including the synaptic cleft, arguing for a role of TNAP in synaptic transmission (Fonta et al. 2004). A_1R has also been hypothesized to be localized to pre- and post-synaptic membranes in the spinal cord (Lao et al. 2001; Li and Perl 1994; Schulte et al. 2003). It is possible that, depending on the subcellular localization of each of these enzymes and A_1R, they are not redundant at all, but rather all working in their own subcellular microdomains. None of the experiments in the spinal cord to date are able to provide the resolution to identify the precise location where TNAP, PAP and NT5E exert their effects. Until the precise location of these enzymes are known, it is impossible to identify their exact role in the modulation and regulation of nociceptive information.

13.11 Use of TNAP and Other Ectonucleotidases to Treat Pain in Humans

While administration of adenosine and A_1R agonists reduce pain, they also cause paralysis and other dangerous side effects when used to treat pain in humans (Zylka 2011). In contrast, PAP and NT5E, when injected intrathecally in mice, did not produce motor side effects, and caused significant and long-lasting reductions in nociceptive behavior (Sowa et al. 2009, 2010b, c; Zylka et al. 2008). Further, PAP, when injected peripherally into the popliteal fossa of mice, also produced a potent and long-lasting anti-nociception without any visible side effects (Hurt and Zylka 2012). This is likely due to the fact that as enzymes, ectonucleotidases are restricted in their activity by the abundance of their substrate, namely AMP. It seems likely that all three enzymes are producing adenosine throughout the spinal cord, but while that amount made in the dorsal horn is sufficient to mediate antinociception, the amount made in the ventral horn is not enough to inhibit motor function due to the lower endogenous levels of AMP. Thus, this catalytic restriction may limit the amount of A_1R activation when these ectonucleotidases are administered, but this limitation could be advantageous by avoiding unwanted side effects.

Because of the ability to manipulate endogenous levels of nucleotides and thereby limit the level of A_1R activation both in size and in scope, TNAP, PAP and NT5E could prove to be important approaches for the prevention and treatment of chronic pain. Clearly these studies are only the very first steps in the development of new therapies. However, recombinant TNAP has been used to successfully treat rickets in patients with hypophosphatasia, a condition resulting from mutations in the TNAP gene, without adverse side effects (Whyte et al. 2012). This recombinant protein featured a bone-homing fusion tag to target the protein to the correct tissue (Millán et al. 2008). Thus, this study provides evidence of the therapeutic potential of enzyme replacement in humans to treat a chronic disorder (cf Chap. 15). In time, it will be very important to assess the feasibility and safety of using TNAP, PAP, and NT5E recombinant proteins to treat chronic pain in humans. However, in the meantime, the identification of the importance of ectonucleotidases such as TNAP has provided another target for future treatments for chronic pain.

13.12 Summary

Continued research into the role of TNAP in nociceptive circuits will enable further understanding of the exact role of TNAP in pain control. To date, studies have only focused on the expression and function of TNAP in spinal nociceptive circuits. However, there also nuclei in the brainstem, thalamus, insula and cerebral cortex that also play major roles in the perception of pain. TNAP is broadly expressed in many areas of the brain, including the primary somatosensory cortex (Fonta et al. 2004), so it will be important to look at the possible role TNAP could play in higher processing of nociceptive signals and the development of chronic pain. It is also possible that TNAP has additional functional roles in the somatosensory system outside of its ectonucleotidase activity. The identification of TNAP and its important function in spinal nociceptive circuits has opened up many new possibilities for both the study and eventual treatment of pain.

Conflict of Interest The authors declare no conflict of interest.

References

Abbracchio MP, Burnstock G, Verkhratsky A, Zimmermann H (2009) Purinergic signalling in the nervous system: an overview. Trends Neurosci 32(1):19–29

Aley KO, Green PG, Levine JD (1995) Opioid and adenosine peripheral antinociception are subject to tolerance and withdrawal. J Neurosci 15(12):8031–8038

Arcuino G, Lin JH, Takano T, Liu C, Jiang L, Gao Q, Kang J, Nedergaard M (2002) Intercellular calcium signaling mediated by point-source burst release of ATP. Proc Natl Acad Sci USA 99 (15):9840–9845

Aumeerally N, Allen G, Sawynok J (2004) Glutamate-evoked release of adenosine and regulation of peripheral nociception. Neuroscience 127(1):1–11

Basbaum AI, Bautista DM, Scherrer G, Julius D (2009) Cellular and molecular mechanisms of pain. Cell 139(2):267–284

Burgess PR, Perl ER (1967) Myelinated afferent fibres responding specifically to noxious stimulation of the skin. J Physiol 190(3):541–562

Burnstock G (2007) Physiology and pathophysiology of purinergic neurotransmission. Physiol Rev 87(2):659–797

Burnstock G (2009) Purinergic receptors and pain. Curr Pharm Des 15(15):1717–1735

Campbell JN, Meyer RA, LaMotte RH (1979) Sensitization of myelinated nociceptive afferents that innervate monkey hand. J Neurophysiol 42(6):1669–1679

Carruthers AM, Sellers LA, Jenkins DW, Jarvie EM, Feniuk W, Humphrey PP (2001) Adenosine A(1) receptor-mediated inhibition of protein kinase A-induced calcitonin gene-related peptide release from rat trigeminal neurons. Mol Pharmacol 59(6):1533–1541

Caterina MJ, Leffler A, Malmberg AB, Martin WJ, Trafton J, Petersen-Zeitz KR, Koltzenburg M, Basbaum AI, Julius D (2000) Impaired nociception and pain sensation in mice lacking the capsaicin receptor. Science 288(5464):306–313

Caterina MJ, Schumacher MA, Tominaga M, Rosen TA, Levine JD, Julius D (1997) The capsaicin receptor: a heat-activated ion channel in the pain pathway. Nature 389(6653):816–824

Cavanaugh DJ, Lee H, Lo L, Shields SD, Zylka MJ, Basbaum AI, Anderson DJ (2009) Distinct subsets of unmyelinated primary sensory fibers mediate behavioral responses to noxious thermal and mechanical stimuli. Proc Natl Acad Sci USA 106(22):9075–9080

Chen Y, Zhang YH, Zhao ZQ (2001) Novel purinergic sensitivity develops in injured sensory axons following sciatic nerve transection in rat. Brain Res 911(2):168–172

Chrinc Pain in America: Roadblocks to Relief (1999) R.S. Worldwide. American Pain Society and Janssen Pharmeceutica

Ciancaglini P, Yadav MC, Simao AM, Narisawa S, Pizauro JM, Farquharson C, Hoylaerts MF, Millán JL (2010) Kinetic analysis of substrate utilization by native and TNAP-, NPP1-, or PHOSPHO1-deficient matrix vesicles. J Bone Miner Res 25(4):716–723

Dahl R, Sergienko EA, Su Y, Mostofi YS, Yang L, Simao AM, Narisawa S, Brown B, Mangravita-Novo A, Vicchiarelli M, Smith LH, O'Neill WC, Millan JL, Cosford ND (2009) Discovery and validation of a series of aryl sulfonamides as selective inhibitors of tissue-nonspecific alkaline phosphatase (TNAP). J Med Chem 52(21):6919–6925

Dai Y, Fukuoka T, Wang H, Yamanaka H, Obata K, Tokunaga A, Noguchi K (2004) Contribution of sensitized P2X receptors in inflamed tissue to the mechanical hypersensitivity revealed by phosphorylated ERK in DRG neurons. Pain 108(3):258–266

Davis JB, Gray J, Gunthorpe MJ, Hatcher JP, Davey PT, Overend P, Harries MH, Latcham J, Clapham C, Atkinson K, Hughes SA, Rance K, Grau E, Harper AJ, Pugh PL, Rogers DC, Bingham S, Randall A, Sheardown SA (2000) Vanilloid receptor-1 is essential for inflammatory thermal hyperalgesia. Nature 405(6783):183–187

DeVries SH (2001) Exocytosed protons feedback to suppress the Ca2+ current in mammalian cone photoreceptors. Neuron 32(6):1107–1117

Dickenson A, Suzuki R, Reeve AJ (2000) Adenosine as a potential analgesic target in inflammatory adn neuropathic pains. CNS Drugs 13:77–85

Diez-Zaera M, Diaz-Hernandez JI, Hernandez-Alvarez E, Zimmermann H, Diaz-Hernandez M, Miras-Portugal MT (2011) Tissue-nonspecific alkaline phosphatase promotes axonal growth of hippocampal neurons. Mol Biol Cell 22(7):1014–1024

Dowd E, McQueen DS, Chessell IP, Humphrey PP (1998) P2X receptor-mediated excitation of nociceptive afferents in the normal and arthritic rat knee joint. Br J Pharmacol 125(2):341–346

Dubner R, Hu JW (1977) Myelinated (A-delta) nociceptive afferents innervating the monkey's face. J Dental Res 56

Dunwiddie TV, Diao L, Proctor WR (1997) Adenine nucleotides undergo rapid, quantitative conversion to adenosine in the extracellular space in rat hippocampus. J Neurosci 17(20):7673–7682

Fitzgerald M (2005) The development of nociceptive circuits. Nat Rev Neurosci 6(7):507–520

Fonta C, Negyessy L, Renaud L, Barone P (2004) Areal and subcellular localization of the ubiquitous alkaline phosphatase in the primate cerebral cortex: evidence for a role in neurotransmission. Cereb Cortex 14(6):595–609

Gerevich Z, Muller C, Illes P (2005) Metabotropic P2Y1 receptors inhibit P2X3 receptor-channels in rat dorsal root ganglion neurons. Eur J Pharmacol 521(1–3):34–38

Gerevich Z, Zadori Z, Muller C, Wirkner K, Schroder W, Rubini P, Illes P (2007) Metabotropic P2Y receptors inhibit P2X3 receptor-channels via G protein-dependent facilitation of their desensitization. Br J Pharmacol 151(2):226–236

Goldman N, Chen M, Fujita T, Xu Q, Peng W, Liu W, Jensen TK, Pei Y, Wang F, Han X, Chen JF, Schnermann J, Takano T, Bekar L, Tieu K, Nedergaard M (2010) Adenosine A1 receptors mediate local anti-nociceptive effects of acupuncture. Nat Neurosci 13(7):883–888

Gourine AV, Kasymov V, Marina N, Tang F, Figueiredo MF, Lane S, Teschemacher AG, Spyer KM, Deisseroth K, Kasparov S (2010) Astrocytes control breathing through pH-dependent release of ATP. Science 329(5991):571–575

Haas HL, Selbach O (2000) Functions of neuronal adenosine receptors. Naunyn Schmiedebergs Arch Pharmacol 362(4–5):375–381

Hamilton SG, McMahon SB, Lewin GR (2001) Selective activation of nociceptors by P2X receptor agonists in normal and inflamed rat skin. J Physiol 534(Pt. 2):437–445

Hamilton SG, Wade A, McMahon SB (1999) The effects of inflammation and inflammatory mediators on nociceptive behaviour induced by ATP analogues in the rat. Br J Pharmacol 126 (1):326–332

Health, United States 2006 With chartbook of trends in the health of Americans (2006) National Center for Health Statistics, Hyattsville, MD

Hilliges M, Weidner C, Schmelz M, Schmidt R, Orstavik K, Torebjork E, Handwerker H (2002) ATP responses in human C nociceptors. Pain 98(1–2):59–68

Hunt SP, Mantyh PW (2001) The molecular dynamics of pain control. Nat Rev Neurosci 2(2):83–91

Hurt JK, Coleman JL, Fitzpatrick BJ, Taylor-Blake B, Bridges AS, Vihko P, Zylka MJ (2012) Prostatic acid phosphatase is required for the antinociceptive effects of thiamine and benfotiamine. PLoS ONE 7(10):e48562

Hurt JK, Zylka MJ (2012) PAPupuncture has localized and long-lasting antinociceptive effects in mouse models of acute and chronic pain. Mol Pain 8:28

Inoue M, Rashid MH, Fujita R, Contos JJ, Chun J, Ueda H (2004) Initiation of neuropathic pain requires lysophosphatidic acid receptor signaling. Nat Med 10(7):712–718

Jacobson KA, Gao ZG (2006) Adenosine receptors as therapeutic targets. Nat Rev Drug Discov 5 (3):247–264

Johansson B, Halldner L, Dunwiddie TV, Masino SA, Poelchen W, Gimenez-Llort L, Escorihuela RM, Fernandez-Teruel A, Wiesenfeld-Hallin Z, Xu XJ, Hardemark A, Betsholtz C, Herlenius E, Fredholm BB (2001) Hyperalgesia, anxiety, and decreased hypoxic neuroprotection in mice lacking the adenosine A1 receptor. Proc Natl Acad Sci USA 98 (16):9407–9412

Julius D, Basbaum AI (2001) Molecular mechanisms of nociception. Nature 413(6852):203–210

Karlsten R, Gordh T, Post C (1992) Local antinociceptive and hyperalgesic effects in the formalin test after peripheral administration of adenosine analogues in mice. Pharmacol Toxicol 70(6 Pt 1):434–438

Langer D, Hammer K, Koszalka P, Schrader J, Robson S, Zimmermann H (2008) Distribution of ectonucleotidases in the rodent brain revisited. Cell Tissue Res 334(2):199–217

Lao LJ, Kumamoto E, Luo C, Furue H, Yoshimura M (2001) Adenosine inhibits excitatory transmission to substantia gelatinosa neurons of the adult rat spinal cord through the activation of presynaptic A(1) adenosine receptor. Pain 94(3):315–324

Li J, Perl ER (1994) Adenosine inhibition of synaptic transmission in the substantia gelatinosa. J Neurophysiol 72(4):1611–1621

Li J, Perl ER (1995) ATP modulation of synaptic transmission in the spinal substantia gelatinosa. J Neurosci 15(5 Pt 1):3357–3365

Lin MF, DaVolio J, Garcia-Arenas R (1992) Expression of human prostatic acid phosphatase activity and the growth of prostate carcinoma cells. Cancer Res 52(17):4600–4607

Lin MF, Garcia-Arenas R, Xia XZ, Biela B, Lin FF (1994) The cellular level of prostatic acid phosphatase and the growth of human prostate carcinoma cells. Differentiation 57(2):143–149

Lin MF, Lee MS, Zhou XW, Andressen JC, Meng TC, Johansson SL, West WW, Taylor RJ, Anderson JR, Lin FF (2001) Decreased expression of cellular prostatic acid phosphatase increases tumorigenicity of human prostate cancer cells. J Urol 166(5):1943–1950

Liu XJ, White TD, Sawynok J (2000) Potentiation of formalin-evoked adenosine release by an adenosine kinase inhibitor and an adenosine deaminase inhibitor in the rat hind paw: a microdialysis study. Eur J Pharmacol 408(2):143–152

MacGregor GR, Zambrowicz BP, Soriano P (1995) Tissue non-specific alkaline phosphatase is expressed in both embryonic and extraembryonic lineages during mouse embryogenesis but is not required for migration of primordial germ cells. Development 121(5):1487–1496

Matsuka Y, Ono T, Iwase H, Mitrirattanakul S, Omoto KS, Cho T, Lam YY, Snyder B, Spigelman I (2008) Altered ATP release and metabolism in dorsal root ganglia of neuropathic rats. Mol Pain 4:66

McCoy ES, Taylor-Blake B, Zylka MJ (2012) CGRPalpha-expressing sensory neurons respond to stimuli that evoke sensations of pain and itch. PLoS ONE 7(5):e36355

Meyer RA, Ringkamp M, Campbell JN, Raja SN (2006) Peripheral mechanisms of cutaneous nocieption. In: McMahon SB, Koltzenburg M (eds) Wall and melzack's textbook of pain. Elsevier, London, pp 3–29

Millán JL, Narisawa S, Lemire I, Loisel TP, Boileau G, Leonard P, Gramatikova S, Terkeltaub R, Camacho NP, McKee MD, Crine P, Whyte MP (2008) Enzyme replacement therapy for murine hypophosphatasia. J Bone Miner Res 23(6):777–787

Moolenaar WH, Kranenburg O, Postma FR, Zondag GC (1997) Lysophosphatidic acid: G-protein signalling and cellular responses. Curr Opin Cell Biol 9(2):168–173

Moriyama T, Iida T, Kobayashi K, Higashi T, Fukuoka T, Tsumura H, Leon C, Suzuki N, Inoue K, Gachet C, Noguchi K, Tominaga M (2003) Possible involvement of P2Y2 metabotropic receptors in ATP-induced transient receptor potential vanilloid receptor 1-mediated thermal hypersensitivity. J Neurosci 23(14):6058–6062

Murthy KS, Makhlouf GM (1995) Adenosine A1 receptor-mediated activation of phospholipase C-beta 3 in intestinal muscle: dual requirement for alpha and beta gamma subunits of Gi3. Mol Pharmacol 47(6):1172–1179

Nakagawa T, Wakamatsu K, Zhang N, Maeda S, Minami M, Satoh M, Kaneko S (2007) Intrathecal administration of ATP produces long-lasting allodynia in rats: differential mechanisms in the phase of the induction and maintenance. Neuroscience 147(2):445–455

Nakamura F, Strittmatter SM (1996) P2Y1 purinergic receptors in sensory neurons: contribution to touch-induced impulse generation. Proc Natl Acad Sci USA 93(19):10465–10470

Narisawa S, Frohlander N, Millán JL (1997) Inactivation of two mouse alkaline phosphatase genes and establishment of a model of infantile hypophosphatasia. Dev Dyn 208(3):432–446

Narisawa S, Hasegawa H, Watanabe K, Millán JL (1994) Stage-specific expression of alkaline phosphatase during neural development in the mouse. Dev Dyn 201(3):227–235

Novakovic SD, Kassotakis LC, Oglesby IB, Smith JA, Eglen RM, Ford AP, Hunter JC (1999) Immunocytochemical localization of P2X3 purinoceptors in sensory neurons in naive rats and following neuropathic injury. Pain 80(1–2):273–282

Okada M, Nakagawa T, Minami M, Satoh M (2002) Analgesic effects of intrathecal administration of P2Y nucleotide receptor agonists UTP and UDP in normal and neuropathic pain model rats. J Pharmacol Exp Ther 303(1):66–73

Patel MK, Pinnock RD, Lee K (2001) Adenosine exerts multiple effects in dorsal horn neurones of the adult rat spinal cord. Brain Res 920(1–2):19–26

Paukert M, Osteroth R, Geisler HS, Brandle U, Glowatzki E, Ruppersberg JP, Grunder S (2001) Inflammatory mediators potentiate ATP-gated channels through the P2X(3) subunit. J Biol Chem 276(24):21077–21082

Peng YB, Ringkamp M, Meyer RA, Campbell JN (2003) Fatigue and paradoxical enhancement of heat response in C-fiber nociceptors from cross-modal excitation. J Neurosci 23(11):4766–4774

Perry MJ, Lawson SN (1998) Differences in expression of oligosaccharides, neuropeptides, carbonic anhydrase and neurofilament in rat primary afferent neurons retrogradely labelled via skin, muscle or visceral nerves. Neuroscience 85(1):293–310

Picher M, Burch LH, Hirsh AJ, Spychala J, Boucher RC (2003) Ecto 5′-nucleotidase and nonspecific alkaline phosphatase. Two AMP-hydrolyzing ectoenzymes with distinct roles in human airways. J Biol Chem 278(15):13468–13479

Quintero IB, Araujo CL, Pulkka AE, Wirkkala RS, Herrala AM, Eskelinen EL, Jokitalo E, Hellstrom PA, Tuominen HJ, Hirvikoski PP, Vihko PT (2007) Prostatic acid phosphatase is not a prostate specific target. Cancer Res 67(14):6549–6554

Relieving Pain in America: A blueprint for transforming prevention, care, education and research (2011) Institute of Medicine of the National Academies

Ringkamp M, Peng YB, Wu G, Hartke TV, Campbell JN, Meyer RA (2001) Capsaicin responses in heat-sensitive and heat-insensitive A-fiber nociceptors. J Neurosci 21(12):4460–4468

Rittiner JE, Korboukh I, Hull-Ryde EA, Jin J, Janzen WP, Frye SV, Zylka MJ (2012) AMP is an adenosine A1 receptor agonist. J Biol Chem 287(8):5301–5309

Rohacs T, Lopes CM, Michailidis I, Logothetis DE (2005) PI (4, 5) P2 regulates the activation and desensitization of TRPM8 channels through the TRP domain. Nat Neurosci 8(5):626–634

Ruscheweyh R, Sandkuhler J (2000) Differential actions of spinal analgesics on mono-versus polysynaptic Adelta-fibre-evoked field potentials in superficial spinal dorsal horn in vitro. Pain 88(1):97–108

Salter MW, Henry JL (1985) Effects of adenosine 5′-monophosphate and adenosine 5′-triphosphate on functionally identified units in the cat spinal dorsal horn. Evidence for a differential effect of adenosine 5′-triphosphate on nociceptive vs non-nociceptive units. Neuroscience 15(3):815–825

Santicioli P, Del Bianco E, Maggi CA (1993) Adenosine A1 receptors mediate the presynaptic inhibition of calcitonin gene-related peptide release by adenosine in the rat spinal cord. Eur J Pharmacol 231(1):139–142

Sawynok J (2007) Adenosine and ATP receptors. Handb Exp Pharmacol 177:309–328

Sawynok J, Liu XJ (2003) Adenosine in the spinal cord and periphery: release and regulation of pain. Prog Neurobiol 69(5):313–340

Sawynok J, Reid A (1997) Peripheral adenosine 5′-triphosphate enhances nociception in the formalin test via activation of a purinergic p2X receptor. Eur J Pharmacol 330(2–3):115–121

Sawynok J, Reid A, Meisner J (2006) Pain behaviors produced by capsaicin: influence of inflammatory mediators and nerve injury. J Pain 7(2):134–141

Scheibe RJ, Kuehl H, Krautwald S, Meissner JD, Mueller WH (2000) Ecto-alkaline phosphatase activity identified at physiological pH range on intact P19 and HL-60 cells is induced by retinoic acid. J Cell Biochem 76(3):420–436

Schulte G, Robertson B, Fredholm BB, DeLander GE, Shortland P, Molander C (2003) Distribution of antinociceptive adenosine A1 receptors in the spinal cord dorsal horn, and relationship to primary afferents and neuronal subpopulations. Neuroscience 121(4):907–916

Sergienko E, Su Y, Chan X, Brown B, Hurder A, Narisawa S, Millán JL (2009) Identification and characterization of novel tissue-nonspecific alkaline phosphatase inhibitors with diverse modes of action. J Biomol Screen 14(7):824–837

Sherrington CS (1906) The integrative action of the nervous system. Scribner, New York

Silverman JD, Kruger L (1988a) Acid phosphatase as a selective marker for a class of small sensory ganglion cells in several mammals: spinal cord distribution, histochemical properties, and relation to fluoride-resistant acid phosphatase (FRAP) of rodents. Somatosens Res 5 (3):219–246

Silverman JD, Kruger L (1988b) Lectin and neuropeptide labeling of separate populations of dorsal root ganglion neurons and associated "nociceptor" thin axons in rat testis and cornea whole-mount preparations. Somatosens Res 5(3):259–267

Sjolund KF, Sollevi A, Segerdahl M, Lundeberg T (1997) Intrathecal adenosine analog administration reduces substance P in cerebrospinal fluid along with behavioral effects that suggest antinociception in rats. Anesth Analg 85(3):627–632

Snider WD, McMahon SB (1998) Tackling pain at the source: new ideas about nociceptors. Neuron 20(4):629–632

Sowa NA, Street SE, Vihko P, Zylka MJ (2010a) Prostatic acid phosphatase reduces thermal sensitivity and chronic pain sensitization by depleting phosphatidylinositol 4, 5-bisphosphate. J Neurosci 30(31):10282–10293

Sowa NA, Taylor-Blake B, Zylka MJ (2010b) Ecto-5′-nucleotidase (CD73) inhibits nociception by hydrolyzing AMP to adenosine in nociceptive circuits. J Neurosci 30(6):2235–2244

Sowa NA, Vadakkan KI, Zylka MJ (2009) Recombinant mouse PAP has pH-dependent ectonucleotidase activity and acts through A(1)-adenosine receptors to mediate antinociception. PLoS One 4(1):e4248

Sowa NA, Voss MK, Zylka MJ (2010c) Recombinant ecto-5′-nucleotidase (CD73) has long lasting antinociceptive effects that are dependent on adenosine A1 receptor activation. Mol Pain 6:20

St Hilaire C, Ziegler SG, Markello TC, Brusco A, Groden C, Gill F, Carlson-Donohoe H, Lederman RJ, Chen MY, Yang D, Siegenthaler MP, Arduino C, Mancini C, Freudenthal B, Stanescu HC, Zdebik AA, Chaganti RK, Nussbaum RL, Kleta R, Gahl WA, Boehm M (2011) NT5E mutations and arterial calcifications. N Engl J Med 364(5):432–442

Street SE, Kramer NJ, Walsh PL, Taylor-Blake B, Yadav MC, King IF, Vihko P, Wightman RM, Millan JL, Zylka MJ (2013) Tissue-nonspecific alkaline phosphatase acts redundantly with PAP and NT5E to generate adenosine in the dorsal spinal cord. J Neurosci 33(27):11314–11322

Street SE, Walsh PL, Sowa NA, Taylor-Blake B, Guillot TS, Vihko P, Wightman RM, Zylka MJ (2011) PAP and NT5E inhibit nociceptive neurotransmission by rapidly hydrolyzing nucleotides to adenosine. Mol Pain 7:80

Suh BC, Hille B (2005) Regulation of ion channels by phosphatidylinositol 4, 5-bisphosphate. Curr Opin Neurobiol 15(3):370–378

Swamy BE, Venton BJ (2007) Subsecond detection of physiological adenosine concentrations using fast-scan cyclic voltammetry. Anal Chem 79(2):744–750

Sykova E, Svoboda J (1990) Extracellular alkaline-acid-alkaline transients in the rat spinal cord evoked by peripheral stimulation. Brain Res 512(2):181–189

Taiwo YO, Levine JD (1990) Direct cutaneous hyperalgesia induced by adenosine. Neuroscience 38(3):757–762

Tanaka M, Kishi Y, Takanezawa Y, Kakehi Y, Aoki J, Arai H (2004) Prostatic acid phosphatase degrades lysophosphatidic acid in seminal plasma. FEBS Lett 571(1–3):197–204

Tominaga M, Wada M, Masu M (2001) Potentiation of capsaicin receptor activity by metabotropic ATP receptors as a possible mechanism for ATP-evoked pain and hyperalgesia. Proc Natl Acad Sci USA 98(12):6951–6956

Tozaki-Saitoh H, Tsuda M, Miyata H, Ueda K, Kohsaka S, Inoue K (2008) P2Y12 receptors in spinal microglia are required for neuropathic pain after peripheral nerve injury. J Neurosci 28 (19):4949–4956

Treede RD, Meyer RA, Campbell JN (1998) Myelinated mechanically insensitive afferents from monkey hairy skin: heat-response properties. J Neurophysiol 80(3):1082–1093

Tsuda M, Inoue K, Salter MW (2005) Neuropathic pain and spinal microglia: a big problem from molecules in "small" glia. Trends Neurosci 28(2):101–107

Tsuda M, Koizumi S, Kita A, Shigemoto Y, Ueno S, Inoue K (2000) Mechanical allodynia caused by intraplantar injection of P2X receptor agonist in rats: involvement of heteromeric P2X2/3 receptor signaling in capsaicin-insensitive primary afferent neurons. J Neurosci 20 (15):RC90

Tsuda M, Shigemoto-Mogami Y, Koizumi S, Mizokoshi A, Kohsaka S, Salter MW, Inoue K (2003) P2X4 receptors induced in spinal microglia gate tactile allodynia after nerve injury. Nature 424(6950):778–783

Vongtau HO, Lavoie EG, Sevigny J, Molliver DC (2011) Distribution of ecto-nucleotidases in mouse sensory circuits suggests roles for nucleoside triphosphate diphosphohydrolase-3 in nociception and mechanoreception. Neuroscience 193:387–398

Vulchanova L, Riedl MS, Shuster SJ, Stone LS, Hargreaves KM, Buell G, Surprenant A, North RA, Elde R (1998) P2X3 is expressed by DRG neurons that terminate in inner lamina II. Eur J Neurosci 10(11):3470–3478

Waymire KG, Mahuren JD, Jaje JM, Guilarte TR, Coburn SP, MacGregor GR (1995) Mice lacking tissue non-specific alkaline phosphatase die from seizures due to defective metabolism of vitamin B-6. Nat Genet 11(1):45–51

Wemmie JA, Zha XM, Welsh MJ (2008) Acid-sensing ion channels (ASICs) and pH in synapse physiology. In: Hell JW, Ehlers MD (eds) Structural and functional organization of the synapse. Springer, New York, pp 661–681

Whyte MP, Greenberg CR, Salman NJ, Bober MB, McAlister WH, Wenkert D, Van Sickle BJ, Simmons JH, Edgar TS, Bauer ML, Hamdan MA, Bishop N, Lutz RE, McGinn M, Craig S, Moore JN, Taylor JW, Cleveland RH, Cranley WR, Lim R, Thacher TD, Mayhew JE, Downs M, Millán JL, Skrinar AM, Crine P, Landy H (2012) Enzyme-replacement therapy in life-threatening hypophosphatasia. N Engl J Med 366(10):904–913

Woolf CJ, Ma Q (2007) Nociceptors–noxious stimulus detectors. Neuron 55(3):353–364

Wu WP, Hao JX, Halldner L, Lovdahl C, DeLander GE, Wiesenfeld-Hallin Z, Fredholm BB, Xu XJ (2005) Increased nociceptive response in mice lacking the adenosine A1 receptor. Pain 113(3):395–404

Xu GY, Huang LY (2002) Peripheral inflammation sensitizes P2X receptor-mediated responses in rat dorsal root ganglion neurons. J Neurosci 22(1):93–102

Zhang D, Xiong W, Chu S, Sun C, Albensi BC, Parkinson FE (2012) Inhibition of hippocampal synaptic activity by ATP, hypoxia or oxygen-glucose deprivation does not require CD73. PLoS One 7(6):e39772

Zhou J, Chung K, Chung JM (2001) Development of purinergic sensitivity in sensory neurons after peripheral nerve injury in the rat. Brain Res 915(2):161–169

Zimmermann H (2006) Ectonucleotidases in the nervous system. Novartis Found Symp 276:113–128; discussion 128–130, 233–117, 275–181

Zylka MJ (2011) Pain-relieving prospects for adenosine receptors and ectonucleotidases. Trends Mol Med 17(4):188–196

Zylka MJ, Rice FL, Anderson DJ (2005) Topographically distinct epidermal nociceptive circuits revealed by axonal tracers targeted to Mrgprd. Neuron 45(1):17–25

Zylka MJ, Sowa NA, Taylor-Blake B, Twomey MA, Herrala A, Voikar V, Vihko P (2008) Prostatic acid phosphatase is an ectonucleotidase and suppresses pain by generating adenosine. Neuron 60(1):111–122

Part IV
TNAP in Neurological Disorders

Chapter 14
Neurological Symptoms of Hypophosphatasia

Takeshi Taketani

Abstract Hypophosphatasia (HPP) is a bone metabolic disorder caused by mutations in the liver/bone/kidney alkaline phosphatase gene (*ALPL*), which encodes tissue-nonspecific alkaline phosphatase (TNAP). This disease is characterized by disrupted bone and tooth mineralization, and reduced serum AP activity. Along with bone and tooth symptoms, many neurological symptoms, seizure, encephalopathy, intracranial hypertension, mental retardation, deafness, and growth hormone deficiency (GHD), are frequently found in HPP patients. Seizure occurs in severe HPP types soon after birth, and responds to pyridoxine, but is an indicator of lethal prognosis. Encephalopathy rarely presents in severe HPP types, but has severe sequelae. Intracranial hypertension complicated in mild HPP types develops after the age of 1 year and sometimes need neurosurgical intervention. Mental retardation, deafness and GHD are more frequently found in Japanese HPP patients. Mental retardation occurs in all HPP types. Deafness in perinatal lethal type is both conductive and sensorineural. GHD develops in all but perinatal lethal type and the diagnosis tends to delay. The pathogenesis of these neural features of HPP might be due to impairment of both vitamin B6 metabolism and central nervous system development by *ALPL* mutations.

Keywords Seizure • Mental retardation • Deafness • Encephalopathy • Growth hormone deficiency

T. Taketani (✉)
Department of Pediatrics, Shimane University Faculty of Medicine,
89-1, Enya, Izumo, Shimane 693-8501, Japan
e-mail: ttaketani@med.shimane-u.ac.jp

C. Fonta and L. Négyessy (eds.), *Neuronal Tissue-Nonspecific Alkaline Phosphatase (TNAP)*, Subcellular Biochemistry 76,
DOI 10.1007/978-94-017-7197-9_14

14.1 Introduction

Hypophosphatasia (HPP) is a metabolic bone disorder caused by mutations in the liver/bone/kidney alkaline phosphatase (*ALPL*) gene, which encodes tissue-nonspecific alkaline phosphatase (TNAP) (Whyte 2001, 2010; Mornet 2008). This disorder is characterized by defective bone and tooth mineralization, and reduced serum AP activity (Whyte 2001, 2010; Mornet 2008). According to several reports from Western populations, HPP patients exhibit both autosomal dominant (AD) inheritance and autosomal recessive (AR), while almost all HPP patients in the Japanese population are AR (Mornet et al. 2011; Michigami et al. 2005; Ozono and Michigami 2011; Watanabe et al. 2011). Patients with AR inheritance have a severe or mild clinical phenotype, whereas those with AD have a mild phenotype (Whyte 2001, 2010; Mornet 2008; Mornet et al. 2011; Michigami et al. 2005; Wenkert et al. 2011). The clinical severity of HPP often depends on the age of onset (Whyte 2001, 2010; Mornet 2008) (see also Chap. 1). The six clinical types of HPP are: (1) perinatal lethal which is apparent at birth; (2) infantile from 1–6 months; (3) childhood type from the age of 6 months–18 years; (4) odonto type, which is characterized by the premature loss of deciduous teeth by 5 years without apparent bone symptoms; and (5) adult. Interestingly, some patients with prenatal onset, namely the prenatal benign type have ameliorated spontaneous skeletal defects and survive (Ozono et al. 1996; Brun-Heath et al. 2008; Wenkert et al. 2011). The perinatal lethal type usually has poor prognosis because of a profound reduction of bone mineralization; half of patients with the infantile type and all patients with the childhood type survive but experience premature loss of deciduous teeth as well as delayed walking and waddling, which reflect the degree of the skeletal disease (Whyte 2001, 2010; Mornet 2008). Low AP activity contributes to elevated levels of AP substrates, i.e., pyridoxal 5′ phosphate (PLP), phosphoethanolamine (PEA), and inorganic pyrophosphate (Whyte 2001). More than 260 types of *ALPL* mutations have been identified in HPP patients, and 80 % of these are missense mutations according to the *ALPL* mutations database (http://www.sesep.uvsq.fr/03_hypo_mutations.php#mutations). The phenotypes of HPP patients are also closely related to the residual enzyme activity effects of *ALPL* mutations (Zurutuza et al. 1999; Mornet 2000). No curative therapy has been established for HPP. Currently, bone-targeted enzyme-replacement therapy, and cell transplantation from bone marrow and other bone sources are under development (Nishioka et al. 2006; Whyte et al. 2003, 2012; Cahill et al. 2007; Tadokoro et al. 2009 (see Chap. 15).

In addition to bone and tooth symptoms, patients with HPP present with neurological symptoms including seizure, encephalopathy, intracranial hypertension, mental retardation, deafness, and growth hormone deficiency (GHD). This chapter describes these neurological symptoms of HPP in detail.

14.2 Seizure

We summarized several cases of seizures with HPP that have been fully reported (Balasubramaniam et al. 2010; Baumgartner-Sigl et al. 2007; Bethenod et al. 1967; Belachew et al. 2013; Demirbilek et al. 2012; Hofmann et al. 2013; Litmanovitz et al. 2002; Nunes et al. 2002; Sia et al. 1975; Smilari et al. 2005; Yamamoto et al. 2004). Seizures of HPP usually occur from days after birth. Only severe types including perinatal lethal and infantile type have seizures while perinatal benign, childhood, adult and odonto types have not. Therefore, bone symptoms including hypomineralization and shortening or deformity of the extremities and respiratory failure precede appearance of seizures. But, two cases reported by Baumgartner-Sigl et al. (2007) and Belachew et al. (2013) developed refractoryseizures before any sign or symptom of HPP, suggesting that intractable seizures in neonates need to suspect HPP. Tonic or/and clonic seizures, myoclonic convulsion and spasm often occur. Complete blood count and blood biochemical examination such as serum electrolytes, glucose, or ammonia are normal except for low titer of AP. Laboratory findings of bacterial and viral infection in blood, urine, and cerebrospinal fluid (CSF) are negative. Electroencephalogram (EEG) ordinarily reveals mono- or multi-focal epileptic discharge. Hypsarrhythmia, a burst-suppression pattern in EEG, sometimes developed, resulting in West syndrome (Balasubramaniam et al. 2010; Baumgartner-Sigl et al. 2007; Yamamoto et al. 2004). Brain computed tomography and magnetic resonance imaging (MRI) do not indicate abnormalities. Hofmann et al. (2013) first reported that cranial MRI demonstrated progressive cystic and destructive encephalopathy in an infant without hypoxic–ischemic episode. Seizures are refractory to standard anticonvulsants such as diazepam, phenobarbital, phenytoin, clonzepam, or valproic acid. But, pyridoxine (PN, vitamin B6) is the only effective anticonvulsant drug, showing that HPP-related seizures are pyridoxine-responsive seizures (PRS). Initial dosage of PN is 100 mg/day or 30–60 mg/kg/day, and maintenance dosage is 50 mg/day or 5–10 mg/kg/day. The duration of PN administration cannot be clarified because except a single case (Belachew et al. 2013) all studied HPP patients with neonatal seizures died within the age of 18 months, demonstrating that PRS of HPP is an indicator of lethal prognosis compared to ordinary PRS which do not recur after the withdrawal of PN (Basura et al. 2009). In the single case survived as reported by Belachew et al. (2013) the PN supplementation could be stopped without seizure recurrence once enzyme replacement therapy, asfotase alfa (Alexion Pharmaceuticals), a bone-targeted recombinant TNAP, has been started. This enzyme therapy has been reported to improve survival and clinical outcome in HPP (Whyte et al. 2012). Further details on this therapy and long-term neurological outcomes will be reported (see Chap. 15).

PN-dependent seizure was found in a deficiency of a-aminoadipic semialdehyde dehydrogenase (antiquitin), which is encoded by the *ALDH7A1* gene, folinic acid-responsive seizures, which are also caused by antiquitin deficiency, familial hyperphosphatasia (PIGV deficiency) and nutritional vitamin B deficiency (Gospe 2010; Plecko and Stöckler 2010). As for PLP-dependent seizure, there is a deficiency of pridox(am)ine 5′-phosphate oxidase which is encoded by the *PNPO* gene (Gospe 2010; Plecko and Stöckler 2010). It is necessary to confirm the diagnosis of PRS because management of each of these diseases differs.

The pathogenesis of PRS in HPP is not understood. Hypotheses have included cranial deformities, intracerebral hemorrhage, and hypoxia (Whyte et al. 1988), but many of HPP patients without these complications have PRS. Interestingly, neonatal seizure in HPP has a unique metabolic basis due to either accumulation of substrates or diminished products of TNAP-mediated hydrolysis (Whyte 2001). However, PEA excess in HPP does not appear to explain the occurrence of PRS. Instead, PLP accumulation reveals the important role that TNAP in vitamin B6 metabolism: as a plasma membrane-bound ectoenzyme that dephosphorylates PLP to pyridoxal (PL) (Whyte et al. 1988). PL is the form of vitamin B6 that crosses cell plasma membranes to be phosphorylated intracellularly to PLP (Whyte et al. 1988). PLP is the active metabolite of vitamin B6 and is an essential coenzyme for synthesis of various neurotransmitters and biogenic amines. It acts as a co-factor for more than 100 apoenzymes (see Chap. 11). The corresponding holoenzymes catalyse diverse reactions such as transamination, decarboxylation, racemisation, degradation and replacement. In the brain, PLP-dependent enzymes [aromatic amino acid decarboxylase (AADC), branched-chain amino acid 2-oxoglutarate aminotransferase, gamma-aminobutyric acid (GABA) transaminase, glutamate decarboxylase (GAD), glycine cleavage enzyme, kynureninase, kynurenine aminotransferase and L-serine racemase are involved in the metabolism of dopamine, serotonin, glutamate, glycine, GABA, D-serine and taurine (Surtees et al. 2006) (Fig. 14.1). Although low TNAP activity in HPP causes high circulating (extracellular) levels of PLP, plasma PL levels are typically normal (Baumgartner-Sigl

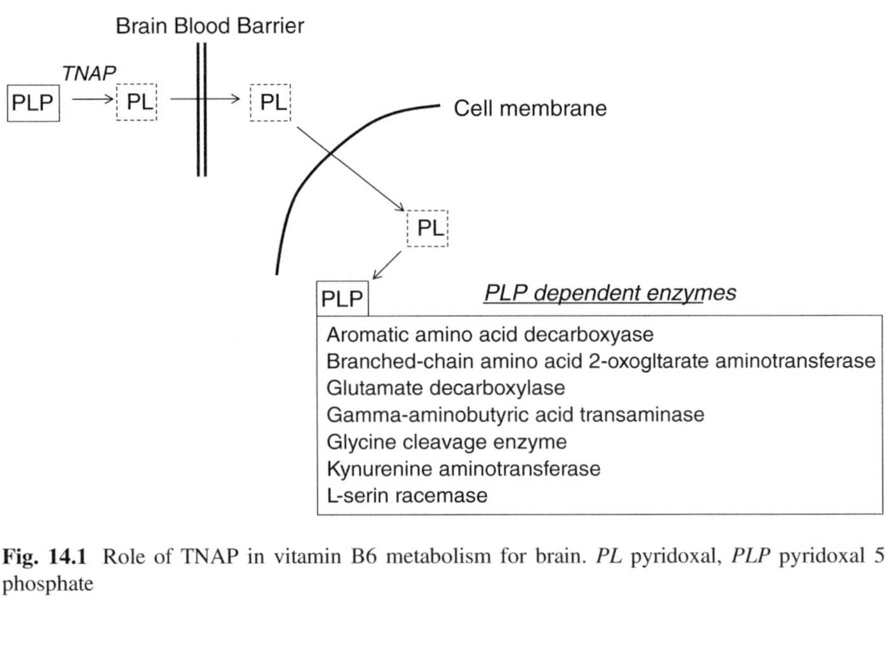

Fig. 14.1 Role of TNAP in vitamin B6 metabolism for brain. *PL* pyridoxal, *PLP* pyridoxal 5′ phosphate

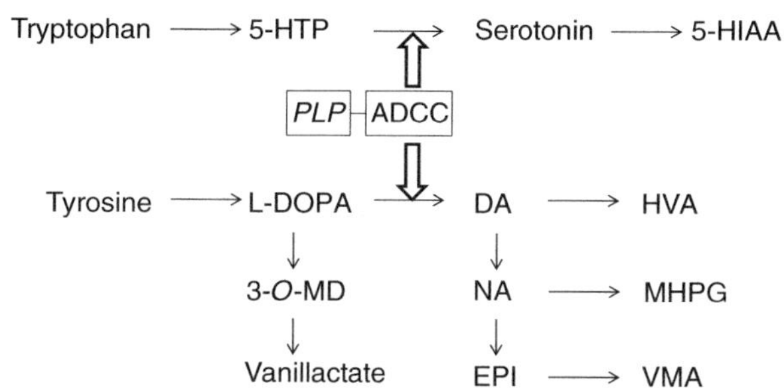

Fig. 14.2 Biochemical pathway related to AADC (modified from Balasubramaniam et al. 2010). *AADC* Aromatic L-amino acid decarboxylase; *5-HTP* hydroxytryptophan, *3-O-MD* 3-*O*-methyldopa, *DA* dopamine, *LDOPA* L-dopamine, *NA* noradrenaline, *EPI* epinephrine, *5-HIAA* 5-hydroxyindole acetic acid, *HVA* homovanillic acid, *MHPG* 3-methoxy-4 hydroxyphenylglycol, *VMA* vanillylmandelic acid, *PL* pyridoxal, *PLP* pyridoxal 5′ phosphate

et al. 2007) (see also Chap. 1). Occasionally, plasma PL levels are low especially in severe HPP cases (Whyte et al. 1988). Low extracellular levels of PL, and resulting low brain PLP levels, may lead to decreased activity of PLP-dependent enzymes (Baumgartner-Sigl et al. 2007). Analysis of biogenic amines in CSF of HPP patients revealed an elevated 3-ortho-methyldopa and 5-hydrotryptophan levels and decreased 5-hydroxyindole acetic acid (Balasubramaniam et al. 2010; Baumgartner-Sigl et al. 2007). Urinary levels of vanillactate were elevated (Balasubramaniam et al. 2010). These metabolites indicate functional deficiency of AADC (Fig. 14.2). Only a few primary AADC deficiency patients were reported with epileptic seizures (Balasubramaniam et al. 2010). Moreover, PL/PLP deficiency in brain cells (despite normal PLP levels in CSF) reduces GABA synthesis because of low GAD activity (Baumgartner-Sigl et al. 2007). PLP is a co-enzyme for GAD, which plays a crucial role in the synthesis of GABA (Baxter 2003). GABA is an inhibitory neurotransmitter and when reduced, the unopposed excitatory neurotransmitters lead to seizure activity (Belachew et al. 2013) (Fig. 14.3). Strong support for this hypothesis comes from null homozygous mouse models null for TNAP gene. These mice have low circulating PL as well as high PLP levels and neonatal seizures that are partially controlled by oral PN/PL (Waymire et al. 1995; Narisawa et al. 2001). In these mice, low PL and PLP levels in brain tissue (before PN treatment) compromised GAD activity and impaired GABA synthesis. Therefore, inadequate levels of PLP in the neurons might lead to PRS.

14.3 Encephalopathy

PRS often presents with epileptic encephalitis (Gospe 2010). Only one HPP patient progresses to encephalopathy (Hofmann et al. 2013). On the patient's seventh day of life, seizures started and rapidly progressed in the following weeks.

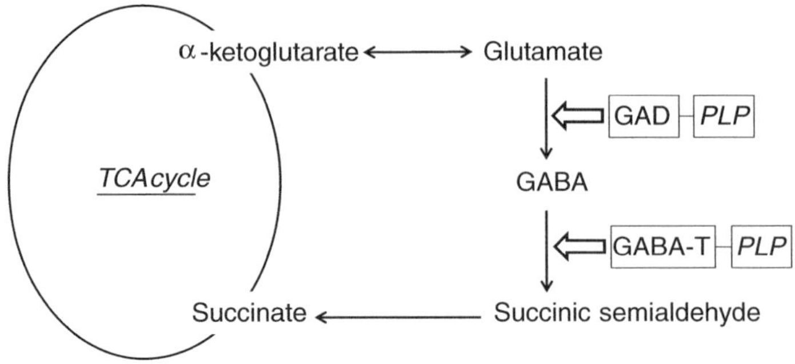

Fig. 14.3 GABA and glutamate metabolic pathway, *GABA* gamma-aminobutyric acid, *GAD* glutamic acid decarboxylase, *GABA-T* GABA transaminase, *PLP* pyridoxal 5′ phosphate, *TCA* tricarboxylic acid

Standard EEG during seizures showed severe generalized hypersynchronous activity but was normal during intermission. Unfortunately, the patient did not show a sustained response to repeated high doses of vitamin B6/pyridoxine, phenobarbital or levetiracetam. The patient continued to have seizures until he was 6 weeks of age when he was deeply sedated. Discontinuation of the sedative medication did not lead to consciousness. EEG-follow up showed discontinuous activity with higher amplitudes than before sedation and hypersynchronous activity was unmasked again. No spontaneous breathing was found and after stimulation the patient opened his eyes without any fixation or spontaneous movements. Particularly synchronous mass movements and exhaustible cloni were obvious and the patient showed a severe muscular hypotonia. Brain MRI revealed a progressive cystic degradation of the cortex and peripheral white matter with nearly complete destruction of the cerebrum. The patients died at the age of 10 weeks.

The patient had PL resistant seizure and obvious brain damage. These findings indicate that the MRI damage pattern of this patient seemed to be due to a hypoxic–ischemic event occurring in conjunction with seizures and was possibly aggravated by the underlying metabolic disease. Considering that there is no clear hypoxic–ischemic episode in this patient's history, the TNAP deficiency is considered to be at least partially involved in this damage pattern leading to neuronal cell death via imbalances or disturbances in phosphate, vitamin B6 or adenosine triphosphate (ATP) metabolism. No similar images have been reported in patients with HPP. TNAP is present in the synapses of the cerebral cortex (Fonta et al. 2004) that are involved in neurotransmitter synthesis, synaptic stabilization, and myelin pattern formation (Fonta et al. 2004, 2005; Négyessy et al. 2011; Hanics et al. 2012). A study in TNAP-KO mice also showed that lack of TNAP compromises myelination and synaptogenesis in the cerebral cortex (Hanics et al. 2012). The Brain Blood Barrier (BBB) is formed by brain capillary endothelial cells (with other cells) which are alkaline phosphatase positive in many animals but lose their marker enzyme in culture (Deracinois et al. 2012) (see Chap. 7). TNAP function in human endothelial cells is not clearly identified. TNAP, however, is known to be involved in the transport of pyridoxine across the BBB subsequently used in neurotransmitter metabolism (Hofmann et al. 2013). TNAP is expressed during early stages of brain vessel development coincident with the appearance of functional BBB. In addition, TNAP seems to be involved in purinergic signaling by producing adenosine through dephosphorylation of ATP. TNAP is localized in growth cones next to ATP-receptors (P2X7) and it seems to be essential for axonal growth of hippocampal neurons by regulating both ligand availability and protein expression of P2X7 receptor (Díez-Zaera et al. 2011). These data suggest that TNAP may play a role in central nervous system (CNS) development and activity-dependent cortical functions (see Chaps. 4, 5 and 18).

14.4 Intracranial Hypertension

The premature closure of sutures is held responsible for intracranial hypertension in patients with HPP (Whyte 2001, 2010; Mornet 2008). Craniosynostosis is a well-known feature of the infantile and childhood HPP (Whyte 2001, 2010; Mornet 2008) while it misses in the adult and odonto-HPP types, thereby suggesting a direct relationship with the degree of enzymatic deficiency (Collmann et al. 2009). Collmann et al. (2009) reported a series of 7 children among 20 cases of infantile and childhood HPP who had craniosynostosis and discussed the functional problems arising from HPP. However, biochemical analysis in these 7 cases did not reveal any overt differences from the 13 other childhood HPP individuals not affected by craniosynostosis. No particular genetic mutational pattern was found in the synostotic individuals. In addition, the age range and the distribution of subtypes did not differ between these two subgroups. Intracranial hypertension developed after the age of 1 year. Neurosurgical intervention was necessary in 4 cases. Secondary ectopia of the cerebellar tonsils were detected in five of the 7 patients and caused hydrosyringomyelia in one of them.

Skull pathology in HPP has been reported as hypomineralization, which, in the most severe perinatal form, results in a "caput membranaceum". In the less severe infantile form, widely diastatic sutures or at least multiple Wormian bones have been reported (Macpherson et al. 1972; Kaplan et al. 1991). They seem to precede sutural obliteration and should, therefore, be considered as functionally closed sutures (Collmann et al. 2009). In fact, MacPherson et al. (1972) found uncalcified osteoid as substrate mimicking this sutural gap. According to Collmann et al. (2009), premature fusion in HPP appears to start with the sagittal or coronal suture. However, synostosis may progressively involve all main cranial sutures during childhood. In these cases, intracranial hypertension putting the optic nerve at risk should be anticipated. Of importance, the lambdoid suture tends to close early, thereby possibly preventing normal expansion of the posterior fossa. This may explain the herniation of the cerebellar tonsils as a secondary phenomenon. Of note, none of patients with intracranial hypertension developed epileptic seizures (Collmann et al. 2009).

Since cranial sutures are frequently involved in infantile and childhood HPP, a multidisciplinary approach for the clinical care is necessary, including long-term neurosurgical surveillance (Collmann et al. 2009). It often progressively involves all cranial sutures and poses significant functional risks to the optic nerves, as well as the spinal cord. The evaluation of HPP in childhood should include plain radiographs and, in case of premature sutural fusion, MR imaging and repeated ophthalmoscopic assessment (Collmann et al. 2009). With regard to the progressive nature of craniosynostosis, careful surveillance is recommended throughout childhood until adolescence and, in the presence of Chiari I malformation or hydrosyringomyelia, probably throughout life (Collmann et al. 2009).

14.5 Neurological Symptoms of Japanese HPP Patients

There are very few group- or country-specific reports on the clinical and genetic features of HPP in pediatric patients. However, the common *ALPL* mutations observed in Japanese patients, i.e., homozygous mutation of c.1559delT and compound heterozygous mutation of c.T979C (p.F327L) have been shown to be associated with relatively lethal and mild types of HPP, respectively (Michigami et al. 2005; Ozono and Michigami 2011; Watanabe et al. 2011). Therefore, we retrospectively examined the clinical and genetic aspects of HPP in total 56 Japanese children including 52 reported by Taketani et al. (2014) Mutations of the *ALPL* gene were analyzed in 35 patients. The frequencies of patients with perinatal lethal, prenatal benign, infantile, childhood, and odonto type HPP were 23, 16, 5, 9, and 3, respectively. There was neither adult type nor mild HPP with AD inheritance. Clinical characteristics by genotype are show in Table 14.1.

About neurological symptoms, seizure was found in 14 patients (11 perinatal lethal, 1 infantile and 2 childhood types). Seizure in perinatal lethal and infantile types was PRS. They occurred in patients within 1 month after birth. Two cases developed seizure before detection of bone symptoms. Computed Tomography and MRI image did not have abnormal findings. Eight patients of them died within the age of 5 years. Two patients with seizure in childhood type were diagnosed as febrile convulsion and epilepsy in each. These two cases did not exhibit seizure in neonatal period. Encephalopathy was found in two patients with perinatal lethal type. Seizure and disturbance of consciousness occurred in both without hypoxic and ischemic episodes. Brain MRI revealed diffuse cerebral edema. All patients had severe sequelae including mental and developmental retardation, incapacitate status and respiratory failure with artificial ventilation. Mental retardation was detected in 11 patients (5 perinatal lethal, 2 prenatal benign, 1 infantile and 3 childhood types). Assessment of mental retardation was performed by: developmental quotient scores, the Enjoji scale of infant analytical development, or intelligence quotient. Four patients among 11 patients with mental retardation (3 perinatal lethal and 1 infantile types) had hypoxic events, the 7 others did not. This suggested that the TNAP deficiency might be partially involved in mental retardation. Deafness

Table 14.1 Clinical characteristics of Japanese patients with hypophosphatasia

	Perinatal lethal (23)	Prenatal benign (16)	Infantile (5)	Childhood (9)
Clinical findings				
Shortening or deformity of the extremities	17	14	0	4
Bone fracture	3	4	1	0
Respiratory failure	23	0	1	0
Seizure	11	0	1	2
Encephalopathy	2	0	0	0
Enlargement of the anterior fontanelle	5	2	4	0
Renal calcification	3	0	0	0
Short stature	5	8	2	6
Failure to thrive	11	4	5	1
Premature loss of deciduous teeth	5	4	0	3
Mental retardation	5	2	1	3
Premature synostosis of the skull	2	0	2	0
Deafness	5	0	0	0
Radiographic findings				
Hypomineralization	22	5	5	0
Loss of bone	7	2	0	0
Deformity of long bones	13	13	0	4
Flared metaphyses	14	8	5	1
Hypolucent mid-metaphyses	5	5	1	0
Osteochondral spurs	0	0	0	0
Narrow thorax	22	1	4	0
Biochemical tests				
AP (IU/L)	19.4	77.3	98	145
Urine PEA (μmol/mg Cr)	7,401	2,119	1,605	873

Numbers indicate the number of patients for each clinical and radiographic criterion. The plasma AP and urine phosphoethanolamine (PEA) values are averages from 56 and 27 patients respectively. In Japan, the normal blood AP range is as follows: median 490 IU/L (range = 59–921) from birth to 1 month; median 617 IU/L (199–1035) during 2–5 months; median 471 IU/L (180–762) from 6 months to 1 year. The lower cut off level during years 1–7 was 170 IU/L. The normal urine PEA range is 31–110 μmol/mg. *Cr* Creatinine. This table was modified from Taketani et al. (2014)

developed in 5 perinatal lethal patients. The auditory tests were normal at birth, suggesting that deafness was acquired after birth and was not congenital. The kind of hearing loss was both conductive and sensorineural pattern by the auditory tests. The degree of deafness varies and two patients with severe deafness needed to wear a hearing device. It is possible that the acquisition of deafness may be exacerbated by hypomineralization of the ear ossicles. However, auditory brainstem response audiometry showed that the brainstem or cerebral cortex was damaged. Approximately half of the patients with a short stature had GHD, rather than mineralization dysfunctions or bone deformities. Short stature was defined as −2 SD of height. The diagnostic criteria of GHD were as follows: (1) height less than −2.5 SD, (2) insulin-like growth factor-1 less than 200 ng/mL, and (3) GH release deficiency in GH secretion test with insulin, arginine, or L-DOPA. GHD in all patients was GH release deficiency in GH secretion test with insulin, arginine, or L-DOPA, demonstrated that GH-secretion by the pituitary gland was decreased. GHD developed in all but perinatal lethal type. Bone symptoms such as shortening or deformity of the extremities occasionally delayed the diagnosis of GHD. Interestingly, there was one patient who was diagnosed with HPP because of the differential diagnosis of short stature, suggesting that we need to consider HPP when further examination of short stature is performed.

ALPL mutations were identified in 35 patients: 14 perinatal lethal, 9 prenatal benign, 5 infantile, 5 childhood, and 2 odonto types (Table 14.2). The remaining 21 patients did not receive genetic testing. Two mutant alleles were identified in 35 patients and no heterozygous mutations were found, suggesting that most Japanese HPP cases are the result of AR inheritance. Mutations in c.1559delT (34 alleles) and p.F327L (12 alleles) were the frequent mutations. The most frequent

Table 14.2 Genetic analysis of the *ALPL* gene in HPP Japanese patients

Clinical types	Genotype	Number of patients
Perinatal lethal type	c.1559delT/c.1559delT	8
	c.1559delT/p.N190del	1
	c.1559delT/p.H324R	1
	c.1559delT/p.G426S	1
	c.1559delT/p.R450C	1
	c.1559delT/p.F327del	1
	p.R223Q/p.R272C	1
Prenatal benign type	c.1559delT/p.F327L	4
	p.F327L/p.R428X	1
	p.F327L/p.G456R	1
	p.F327del/p.R184 W	1
	p.A40 V/p.E191G	2
Infantile type	c.1559delT/p.L299P	2
	c.1559delT/p.F327L	1
	c.1559delT/p.Y436C	1
	p.K224E/p.G426C	1
Childhood type	c.1559delT/pF327L	3
	p.F327L/p.G339R	1
	p.F327L/p.A111T	1
Odonto type	c.1559delT/p.R136H	2

Modified from Taketani et al. (2014)

Japanese HPP genotype was a homozygous mutation (c.1559delT/c.1559delT) and the compound heterozygous mutation of c.1559delT/p.F327L which was observed in 8 patients of each. Interestingly, patients with p.F327L were all HPP types except for the perinatal lethal type. In the Japanese population, the prevalence of the 1559delT homozygous mutation in the *ALPL* gene, which is a common mutation that causes the perinatal lethal form, was estimated to be not less than 1/900,000 (Watanabe et al. 2011). Michigami et al. (2005) reported that c.1559delT represent 40.9 % of severe alleles. This mutation was not found in other countries, suggesting that c.1559delT may be a founder mutation in the Japanese population.

Compared with previous reports of HPP clinical characteristics in Western populations (Whyte 2001, 2010; Mornet 2008), the frequency of mental retardation, deafness, and short stature were more frequent in Japanese HPP patients, while seizure and encephalopathy was approximately the same. Difference in the frequency and the degree of each neurological feature remains unknown, but this difference might depend on the severity of the TNAP activity as well as the differences in the genetic backgrounds of patients. Especially, about the most frequent mutation, c.1559delT, this mutant lost TNAP activity and was localized at the juxtanuclear position, but not on the cell surface (Michigami et al. 2005; Komaru et al. 2005). As referred above, TNAP is likely to have a valuable role in CNS development and functions (Fonta et al. 2004, 2005; Négyessy et al. 2011; Hanics et al. 2012; Díez-Zaera et al. 2011). But patients with the same genotype did not necessarily have the same neurological symptoms. This suggests that cerebral impairment, including seizure, encephalopathy, mental retardation, deafness, and GHD might be the consequence not only of genetic alterations but also of epigenetic changes, ethnic factors, environmental factors and nutritious element.

Conflict of Interest The author declares no conflict of interest.

References

Balasubramaniam S, Bowling F, Carpenter K, Earl J, Chaitow J, Pitt J, Mornet E, Sillence D, Ellaway C (2010) Prenatal hypophosphatasia presenting as neonatal epileptic encephalopathy with abnormal neurotransmitter metabolism secondary to reduced co-factor pyridoxal-5′-phosphate availability. J Inherit Metab Dis 33(Suppl 3):25–33

Basura GJ, Hagland SP, Wiltse AM, Gospe SM Jr (2009) Clinical features and the management of pyridoxine-dependent and pyridoxine-responsive seizures: review of 63 North American cases submitted to a patient registry. Eur J Pediatr 168(6):697–704

Baumgartner-Sigl S, Haberlandt E, Mumm S, Scholl-Bürgi S, Sergi C, Ryan L, Ericson KL, Whyte MP, Högler W (2007) Pyridoxine-responsive seizures as the first symptom of infantile hypophosphatasia caused by two novel missense mutations (c.677T > C, p. M226T; c.1112C > T, p.T371I) of the tissue-nonspecific alkaline phosphatase gene. Bone 40(6):1655–1661

Baxter P (2003) Pyridoxine-dependent seizures: a clinical and biochemical conundrum. Biochim Biophys Acta 1647(1–2):36–41

Bethenod M, Cotte MF, Collombel C, Frederich A, Cotte J. (1967) Neonatal discovery of hypophosphatasia. Bone improvement, fatal convulsant encephalopathy. Ann Pediatr (Paris) 14(12):835–841

Belachew D, Kazmerski T, Libman I, Goldstein AC, Stevens ST, Deward S, VockleyJ Sperling MA, Balest AL (2013) Infantile hypophosphatasia secondary to a novel compound heterozygous mutation presenting with pyridoxine-responsive seizures. JIMD Rep 11:17–24

Brun-Heath I, Chabrol E, Fox M, Drexler K, Petit C, Taillandier A, De azancourt P, Serre JL, Mornet E (2008) A case of lethal hypophosphatasia providing new insights into the perinatal benign form of hypophosphatasia and expression of the ALPL gene. Clin Genet 73(3):245–250

Cahill RA, Wenkert D, Perlman SA, Steele A, Coburn SP, McAlister WH, Mumm S, Whyte MP (2007) Infantile hypophosphatasia: transplantation therapy trial using bone fragments and cultured osteoblasts. J Clin Endocrinol Metab 92(8):2923–2930

Collmann H, Mornet E, Gattenlöhner S, Beck C, Girschick H (2009) Neurosurgical aspects of childhood hypophosphatasia. Childs Nerv Syst 25(2):217–223

Demirbilek H, Alanay Y, Alikaşifoğlu A, Topçu M, Mornet E, Gönç N, Özön A, Kandemir N (2012) Hypophosphatasia presenting with pyridoxine-responsive seizures, hypercalcemia, and pseudotumor cerebri: case report. J Clin Res Pediatr Endocrinol 4(1):34–38

Deracinois B, Duban-Deweer S, Pottiez G, Cecchelli R, Karamanos Y, Flahaut C (2012) TNAP and EHD1 are over-expressed in bovine brain capillary endothelial cells after the re-induction of blood-brain barrier properties. PLoS One 7(10):e48428

Díez-Zaera M, Díaz-Hernández JI, Hernández-Álvarez E, Zimmermann H, Díaz-Hernández M, Miras-Portugal MT (2011) Tissue-nonspecific alkaline phosphatase promotes axonal growth of hippocampal neurons. Mol Biol Cell 22(7):1014–1024

Fonta C, Négyessy L, Renaud L, Barone P (2004) Areal and subcellular localization of the ubiquitous alkaline phosphatase in the primate cerebral cortex: evidence for a role in neurotransmission. Cereb Cortex 14(6):595–609

Fonta C, Negyessy L, Renaud L, Barone P (2005) Postnatal development of alkaline phosphatase activity correlates with the maturation of neurotransmission in the cerebral cortex. J Comp Neurol 486(2):179–196

Gospe SM Jr (2010) Neonatal vitamin-responsive epileptic encephalopathies. Chang Gung Med J 33(1):1–12

Hanics J, Barna J, Xiao J, Millán JL, Fonta C, Négyessy L (2012) Ablation of TNAP function compromises myelination and synaptogenesis in the mouse brain. Cell Tissue Res 349(2):459–471

Hofmann C, Liese J, Schwarz T, Kunzmann S, Wirbelauer J, Nowak J, Hamann J, Girschick H, Graser S, Dietz K, Zeck S, Jakob F, Mentrup B (2013) Compound heterozygosity of two functional null mutations in the ALPL gene associated with deleterious neurological outcome in an infant with hypophosphatasia. Bone 55(1):150–157

Kaplan SB, Kemp SS, Oh KS (1991) Radiographic manifestations of congenital anomalies of the skull. Radiol Clin North Am 29(2):195–218

Komaru K, Ishida Y, Amaya Y, Goseki-Sone M, Orimo H, Oda K (2005) Novel aggregate formation of a frame-shift mutant protein of tissue-nonspecific alkaline phosphatase is ascribed to three cysteine residues in the C-terminal extension. Retarded secretion and proteasomal degradation. FEBS J 272(7):1704–1717

Litmanovitz Reish O, Dolfin T, Arnon S, Regev R, Grinshpan G, Yamazaki M, Ozono K (2002) Glu274Lys/Gly309Arg mutation of the tissue-nonspecific alkaline phosphatase gene in neonatal hypophosphatasia associated with convulsions. J Inherit Metab Dis 25(1):35–40

Macpherson RI, Kroeker M, Houston CS (1972) Hypophosphatasia. J Can Assoc Radiol 23(1):16–26

Michigami T, Uchihashi T, Suzuki A, Tachikawa K, Nakajima S, Ozono K (2005) Common mutations F310L and T1559del in the tissue-nonspecific alkaline phosphatase gene are related to distinct phenotypes in Japanese patients with hypophosphatasia. Eur J Pediatr 164(5):277–282

Mornet E (2000) Hypophosphatasia: the mutations in the tissue-nonspecific alkaline phosphatase gene. Hum Mutat 15(4):309–315

Mornet E (2008) Hypophosphatasia. Best Pract Res Clin Rheumatol 22(1):113–127

Mornet E, Yvard A, Taillandier A, Fauvert D, Simon-Bouy B (2011) A molecular-based estimation of the prevalence of hypophosphatasia in the European population. Ann Hum Genet 75(3):439–445

Narisawa S, Wennberg C, Millán JL (2001) Abnormal vitamin B6 metabolism in alkaline phosphatase knock-out mice causes multiple abnormalities, but not the impaired bone. J Pathol 193(1):125–133

Nishioka T, Tomatsu S, Gutierrez MA, Miyamoto K, Trandafirescu GG, Lopez PL, Grubb JH, Kanai R, Kobayashi H, Yamaguchi S, Gottesman GS, Cahill R, Noguchi A, Sly WS (2006) Enhancement of drug delivery to bone: characterization of human tissue-nonspecific alkaline phosphatase tagged with an acidic oligopeptide. Mol Genet Metab 88(3):244–255

Négyessy L, Xiao J, Kántor O, Kovács GG, Palkovits M, Dóczi TP, Renaud L, Baksa G, Glasz T, Ashaber M, Barone P, Fonta C (2011) Layer-specific activity of tissue non-specific alkaline phosphatase in the human neocortex. Neuroscience 172:406–418

Nunes ML, Mugnol F, Bica I, Fiori RM (2002) Pyridoxine-dependent seizures associated with hypophosphatasia in a newborn. J Child Neurol 17(3):222–224

Ozono K, Yamagata M, Michigami T, Nakajima S, Sakai N, Cai G, Satomura K, Yasui N, Okada S, Nakayama M (1996) Identification of novel missense mutations (Phe310Leu and Gly439Arg) in a neonatal case of hypophosphatasia. J Clin Endocrinol Metab 81(12):4458–4461

Ozono K, Michigami T (2011) Hypophosphatasia now draws more attention of both clinicians and researchers: A Commentary on prevelance of c. 1559delT in ALPL, a common mutation resulting in the perinatal (lethal) form of hypophosphatasias in Japanese and effects of the mutation on heterozygous carriers. J Hum Genet 56(3):174–176

Plecko B, Stöckler S (2010) Vitamin B6 dependent seizures. Can J Neurol Sci 36(Suppl 2):S73–S77

Sia C, Wapnir R, Sokal M, Harper RG, Intizar S, Lifshitz F (1975) Effects of pyridoxine on neonatal hypophosphatasia. Pediatr Res 9:355

Smilari P, Romeo DM, Palazzo P, Meli C, Sorge G (2005) Neonatal hypophosphatasia and seizures. A case report. Minerva Pediatr 57(5):319–323

Surtees R, Mills P, Clayton P (2006) Inborn errors affecting vitamin B6 metabolism. Future Neurol 1:615–620

Tadokoro M, Kanai R, Taketani T, Uchio Y, Yamaguchi S, Ohgushi H (2009) New bone formation by allogeneic mesenchymal stem cell transplantation in a patient with perinatal hypophosphatasia. J Pediatr 154(6):924–930

Taketani T, Onigata K, Kobayashi H, Mushimoto Y, Fukuda S, Yamagichi S (2014) Clinical and genetic aspects of Hypophosphatasia in Japanese patients. Arch Dis Child 99(3):211–215

Watanabe A, Karasugi T, Sawai H, Naing BT, Ikegawa S, Orimo H, Shimada T (2011) Prevalence of c.1559delT in ALPL, a common mutation resulting in the perinatal (lethal) form of hypophosphatasia in Japanese and effects of the mutation on heterozygous carriers. J Hum Genet 56(2):166–168

Waymire KG, Mahuren JD, Jaje JM, Guilarte TR, Coburn SP, MacGregor GR (1995) Mice lacking tissue non-specific alkaline phosphatase die from seizures due to defective metabolism of vitamin B-6. Nat Genet 11(1):45–51

Wenkert D, McAlister WH, Coburn SP, Zerega JA, Ryan LM, Ericson KL, Hersh JH, Mumm S, Whyte MP (2011) Hypophosphatasia: nonlethal disease despite skeletal presentation in utero (17 new cases and literature review). J Bone Miner Res 26(10):2389–2398

Whyte MP, Mahuren JD, Fedde KN, Cole FS, McCabe ER, Coburn SP (1988) Perinatal hypophosphatasia: tissue levels of vitamin B6 are unremarkable despite markedly increased circulating concentrations of pyridoxal-5′-phosphate. Evidence for an ectoenzyme role for tissue-nonspecific alkaline phosphatase. J Clin Invest 81(4):1234–1239

Whyte M (2001) Hypophosphatasia. In: Scriver CR, Beaudet AL, Sly WS, Valle D, Vogelstein B (eds) The metabolic and molecular bases of disease, 8th edn. McGraw-Hill Book Company, New York, pp 5313–5329

Whyte MP, Kurtzberg J, McAlister WH, Mumm S, Podgornik MN, Coburn SP, Ryan LM, Miller CR, Gottesman GS, Smith AK, Douville J, Waters-Pick B, Armstrong RD, Martin PL (2003) Marrow cell transplantation for infantile hypophosphatasia. J Bone Miner Res 18 (4):624–636

Whyte MP (2010) Physiological role of alkaline phosphatase explored in hypophosphatasia. Ann N Y Acad Sci 1192:190–200

Whyte MP, Greenberg CR, Salman NJ, Bober MB, McAlister WH, Wenkert D, Van Sickle BJ, Simmons JH, Edgar TS, Bauer ML, Hamdan MA, Bishop N, Lutz RE, McGinn M, Craig S, Moore JN, Taylor JW, Cleveland RH, Cranley WR, Lim R, Thacher TD, Mayhew JE, Downs M, Millán JL, Skrinar AM, Crine P, Landy H (2012) Enzyme-replacement therapy in life-threatening hypophosphatasia. N Engl J Med 366(10):904–913

Yamamoto H, Sasamoto Y, Miyamoto Y, Murakami H, Kamiyama N (2004) A successful treatment with pyridoxal phosphate for West syndrome in hypophosphatasia. Pediatr Neurol 30 (3):216–218

Zurutuza L, Muller F, Gibrat JF, Taillandier A, Simon-Bouy B, Serre JL, Mornet E (1999) Correlations of genotype and phenotype in hypophosphatasia. Hum Mol Genet 8(6):1039–1046

Chapter 15
Recombinant Enzyme Replacement Therapy in Hypophosphatasia

Christine Hofmann, Franz Jakob, Lothar Seefried, Birgit Mentrup, Stephanie Graser, Horacio Plotkin, Hermann J. Girschick and Johannes Liese

Abstract Hypophosphatasia (HPP) is a rare monogenetic and multisystemic disease with involvement of different organs, including bone, muscle, kidney, lung, gastrointestinal tract and the nervous system. The exact metabolic mechanisms of the effects of TNAP deficiency in different tissues are not understood in detail. There is no approved specific treatment for HPP; therefore symptomatic treatment in order to improve the clinical features is of major interest. Enzyme replacement therapy (ERT) is a relatively new type of treatment based on the principle of administering a medical treatment replacing a defective or absent enzyme. Recently ERT with a bone targeted recombinant human TNAP molecule has been reported to be efficient in ten severely affected patients and improved survival of life

C. Hofmann (✉) · J. Liese
Children's Hospital, University of Würzburg, Josef-Schneider-Str. 2, 97090 Würzburg, Germany
e-mail: hofmann_c5@klinik.uni-wuerzburg.de

J. Liese
e-mail: liese_j@klinik.uni-wuerzburg.de

F. Jakob · L. Seefried · B. Mentrup · S. Graser
Orthopedic Department, Orthopedic Center for Musculoskeletal Research, University of Würzburg, Würzburg, Germany
e-mail: f-jakob.klh@uni-wuerzburg.de

L. Seefried
e-mail: l-seefried.klh@uni-wuerzburg.de

B. Mentrup
e-mail: b-mentrup.klh@uni-wuerzburg.de

S. Graser
e-mail: s-graser.klh@uni-wuerzburg.de

H.J. Girschick
Children's Hospital, Vivantes Hospital im Friedrichshain, Berlin, Germany
e-mail: Hermann.Girschick@mail.uni-wuerzburg.de

H. Plotkin
University of Nebraska Medical Center, Omaha, NE, USA
e-mail: dr_plotkin@yahoo.com

C. Fonta and L. Négyessy (eds.), *Neuronal Tissue-Nonspecific Alkaline Phosphatase (TNAP)*, Subcellular Biochemistry 76,
DOI 10.1007/978-94-017-7197-9_15

threatening forms. These results are very promising especially with regard to the skeletal phenotype but it is unclear whether ERT also has beneficial effects for craniosynostosis and in other affected tissues in HPP such as brain and kidney. Long-term data are not yet available and further systematic clinical trials are needed. It is also necessary to establish therapeutic approaches to help patients who are affected by less severe forms of HPP but also suffer from a significant reduction in quality of life. Further basic research on TNAP function and role in different tissues and on its physiological substrates is critical to gain a better insight in the pathogenesis in HPP. This and further experiences in new therapeutic strategies may improve the prognosis and quality of life of patients with all forms of HPP.

Keywords Hypophosphatasia · Multisystemic disease · TNAP deficiency · TNAP function · Enzyme replacement therapy

15.1 Introduction

Hypophosphatasia (HPP) is a rare metabolic disease with a well characterized molecular cause (see Chap. 2). Due to a genetic defect in the ALPL gene, patients with HPP are deficient in tissue-nonspecific alkaline phosphatase (TNAP) (Mornet et al. 2013). TNAP is a ubiquitous isoenzyme, mostly known for its role in bone mineralization (Millan 2013; Whyte 2010). However, in HPP, TNAP deficiency also results in extra-osseous clinical manifestations. In many case reports HPP has been described as a multisystemic disease with involvement of different organs, including bone, muscle, kidney, lung, gastrointestinal tract and peripheral as well as central nervous system (Hofmann et al. 2013) (see also Chaps. 1, 8 and 14). Its exact biological function especially in different non-osseous tissues is not understood in detail (see Chap. 8). Complete absence of this enzyme appears to be life-threatening. Significant deficiency of TNAP activity and consecutive accumulation of its substrates (inorganic pyrophosphate, PPi; and pyridoxal 5′ phosphate, PLP) may lead to significant damage to multiple vital organs resulting in impaired organ function or even organ failure.

Over the last decades several attempts have been made to symptomatic treatment to improve prognosis and quality of life of patients with HPP of all ages. There is no approved medical therapy for all clinical forms of HPP. However, recently ERT with a bone targeted recombinant human TNAP molecule has been reported to be beneficial in ten severely affected patients and improved survival of forms of HPP considered to be lethal or life-threatening (Whyte et al. 2012). First results with ERT are promising, nevertheless long term data are not yet available and further clinical trials are needed.

15.2 Current Management and Treatment of HPP and Its Limitations

There is no curative treatment of HPP available at the moment. Therefore, symptomatic treatment in order to improve the clinical features, especially with regard to pain and other metabolic phenomena is the current standard of care. Nonsteroidal antiinflammatory drugs (NSAIDs) have been shown to provide significant improvement in patients with significant and recurrent pain. Treatment with teriparatide (recombinant human parathyroid hormone) has been discussed controversially (Camacho et al. 2008; Doshi et al. 2009; Whyte et al. 2007; Sutton et al. 2012). Bisphosphonates, synthetic derivates of PPi, have been reported to be ineffective and are associated with significant side effects (Sutton et al. 2012). Supplementation with magnesium and zinc, the catalytic ions of the TNAP enzyme, seems to have no significant long-term benefit. Dietary phosphate (Pi) restriction and pharmacological binding of dietary Pi is recommended for hyperphosphatemic patients but no long-term studies are available of the long term effects of this treatment. In infants with severe failure to thrive and gastro-intestinal reflux feeding by gastric or jejunal tubes may become necessary.

Regular orodental monitoring by a dentist or orthodontist is very important for patients with all HPP subtypes. Physical exercise and physiotherapy are useful for muscle building in order to stabilize the whole skeleton. However, trauma should be avoided because patients are prone to pathological fractures and fractures may take a long time to heal and might require prolonged casting or stabilization with intra-medullary rods.

Regarding neurological symptoms, different therapeutic options have been reported. Neonatal seizures in severe HPP are caused by reduced vitamin B6 transport in the neural cell and respond to vitamin B6 (pyridoxine) treatment. Neurosurgical interventions may become necessary in patients with neurologic symptoms due to Chiari I malformation or craniosynostosis and elevated intracranial pressure (Collmann et al. 2005, 2009). However, there are different neurological problems which have not been treated successfully so far: e.g. swallowing disorders, persisting nausea and vomiting, irritability in infants, progressive encephalopathy or symptoms like depression and anxiety in adults (Hofmann et al. 2013).

Allogenic stem cell transplantation has been performed with some clinical improvement, but substantial morbidity remained (Whyte et al. 2003; Tadokoro et al. 2009). In infantile HPP, attempts to transplant TNAP-replete bone fragments and cultured osteoblast-like cells obtained from the donor father appeared to gain benefit, but did not cure these patients (Cahill et al. 2007). Treatment with administration of plasma enriched soluble alkaline phosphatase from patients with Paget´s disease have been unsuccessful (Whyte et al. 1982, 1984) suggesting that this therapy fails to adequately target bone, as skeletal mineralization requires the enzymatic activity of TNAP on the plasma membrane of osteoblasts and matrix vesicles and AP activity in the circulation is less important for the bone phenotype.

Given the current limitations in the clinical management and treatment of patients with HPP assessment of enzyme replacement therapy using recombinant human TNAP is of particular interest especially for the frequently fatal perinatal and infantile form of HPP.

15.3 Enzyme Replacement Therapy (ERT) and Lessons Learned from Other Diseases

ERT is a relatively new type of treatment. Since our knowledge about the human genome the genetic cause of different diseases and the cellular function of proteins has increased significantly, new forms of ERT are likely to emerge. The principle of ERT is to administer a medical treatment replacing a defective or absent enzyme making sure that it reaches its target organ. Thus ERT treats the cause, not just the symptoms. However, ERT is not able to correct the underlying genetic defect as it is the case in a gene therapy approach and therefore is not a "cure" for the disease.

Lessons we have learned in ERT in the last decades are based on experience in the treatment of different "enzymopathies", especially of lysosomal storage diseases (LSDs). ERT is currently available for six LSDs and a few non-LSDs (e.g. pancreatitis or digestive enzyme deficiency in cystic fibrosis) (Desnick and Schuchman 2012). The concept of ERT was introduced by Deduve in 1964 (Deduve 1964). However, it took several decades to progress from the concept of ERT to commercial development of the first recombinant enzyme (glucocerebrosidase for Gaucher disease). That delay was mainly due to the time needed for understanding the basic science/cell biology underlying receptor mediated endocytosis and trafficking of the enzyme, the cloning and overexpression of human genes and the generation of animal models.

LSDs is a group of disorders characterized by a deficient activity of a specific lysosomal enzyme and the progressive accumulation of its substrates. ERT has been approved for Gaucher disease, Fabry disease, mucopolysaccharidosis type I, II, and VI, and Pompe disease. The efficacy and safety of ERT for these diseases has been confirmed by extensive clinical trials in the last 20 years.

To be safe and therapeutically useful the recombinant enzyme had to be of human origin, highly purified, available in high quantity and it had to carry the recognition signal permitting an efficient receptor mediated uptake into the target cells. Recombinant glucocerebrosidase secreted by genetically modified Chinese hamster ovary (CHO) cells carrying the ubiquitous mannose-6-phosphate receptor became the first successful pharmaceutical agent for the treatment of LSDs (Grabowski et al. 1995). Subsequently to this proof of concept several human lysosomal enzymes were made in CHO cells (Desnick and Schuchman 2012). CHO cells are relatively easy to grow, perform posttranslational modifications that are almost identical to those in human cells and they facilitate large scale production by selective secretion of the recombinant enzyme into the culture media. Other expression systems, for example human fibroblasts or transgenic animals are also currently being used to produce recombinant enzymes for LSDs.

15.4 Key Principles Learned from the Experience in ERT in These Diseases

Animal models and clinical studies have revealed organ-specific variations in response to ERT due to receptor-mediated biodistribution and tissue delivery (Miranda et al. 2000; Sands et al. 2001). Macromolecular enzymes cannot cross the blood-brain barrier and therefore ERT has no effect on the neurological manifestations of LSDs (Ioannou et al. 2001). Adequate doses are required for delivery to critical sites of pathology. Tissues like spleen, liver, bone marrow or vascular endothelium are easy to reach whereas bone, cartilage, muscle or heart are more difficult to reach and require higher doses. In a normal situation, proteins degrade and are produced again, regulated by the organism itself. In patients requiring ERT recombinant enzyme must be administered regularly, intravenous (i.v.) or subcutaneous (s.c.), at different frequencies (from once per day to once per month) depending on its pharmacokinetics. Presently, continuous and lifelong treatment is recommended in severe LSDs in order to achieve optimal outcomes. Substrate re-accumulation occurs if ERT is interrupted and this may exacerbate clinical manifestations. Years of substrate accumulation could be reversed in certain cells and organs during ERT. The earlier it is initiated, the more effective it is, because some of the irreversible damage of its target organ can be avoided. Experience has documented antibody mediated immune reactions against the recombinant enzymes, more frequently in patients with little or no residual enzymatic activity. This is generally transient with reduction of the non-neutralizing IgG antibody titers against the enzyme and better tolerability in the course of time (Muenzer et al. 2007; Clarke et al. 2009; Harmatz et al. 2006).

However, there are still some remaining challenges to successful ERT that need to be addressed in the upcoming future, such as the development of new techniques for enzyme delivery to difficult-to-reach tissues like bone or brain, or addressing secondary pathological phenomena such as inflammation. In the future appropriate biomarkers need to be identified that reflect therapeutic effectiveness and enable a reliable clinical monitoring of disease activity. Another aim to make future ERT strategies safer and effective in the long-term is the control and management of immunologic reactions due to antibody response. In addition, efforts need to be made to reduce the high costs for development and license of recombinant enzymes for these rare diseases and to optimize accessibility and reimbursement for ERT in general. Furthermore, since long term data are not available for most of the diseases, appropriate studies are needed.

Overall ERT significantly improved the prognosis and quality of life of patients with some LSDs and lessons learned from these experiences contributed to the development of new recombinant enzymes into pharmaceuticals for the treatment of further diseases and their use in clinical trials. A remaining challenge for upcoming new recombinant pharmaceuticals is the problem in tissue and subcellular targeting in ERT which is different for each disease.

15.5 ERT as a Possible Treatment Option in HPP

Similar to LSDs, HPP is an "enzymopathy" and results in an accumulation of non-metabolized substrates. There is no approved medical therapy for HPP. According to case reports of first attempted ERT in HPP (Cahill et al. 2007; Whyte et al. 1982, 1984; Weninger et al. 1989) it became clear that AP activity must be directed to the bone surface, and that levels of TNAP in the circulation are less relevant. Considering this, Asfotase alfa, a human recombinant TNAP protein with a deca-aspartate tail that increases the affinity of the molecule for bone tissue, has been developed by Enobia Pharma (now Alexion) targeting directly sites of skeletal mineralization.

15.6 Asfotase Alfa

Asfotase alfa (Alexion Pharma) is a recombinant, fusion protein comprising the TNAP ectodomain, the constant region of the human IgG1 Fc domain, and a terminal deca-aspartate motif (Fig. 15.1). The soluble glycoprotein consists of 726 amino acids from the catalytic domain of human TNAP. Asfotase alfa is engineered in CHO cells that maintain disulfide bridging and N-linked glycosylation. The TNAP ectodomain is responsible for the enzyme function, the IgG1Fc domain facilitates rapid large-scale drug purification by affinity chromatography and the deca-aspartate motif enables bone targeting. After conjugation with aspartate residues affinity to hydroxyapatite crystals was several times higher than for the unconjugated enzyme (Nishioka et al. 2006) and it showed a significantly greater uptake by bone (Millan et al. 2008).

In its monomeric form the fusion protein has a molecular weight of approximately 84 kDa. It circulates as a homodimer with a molecular weight of around 200 kDa and when analyzed by molecular sieve chromatography it appears to form primarily

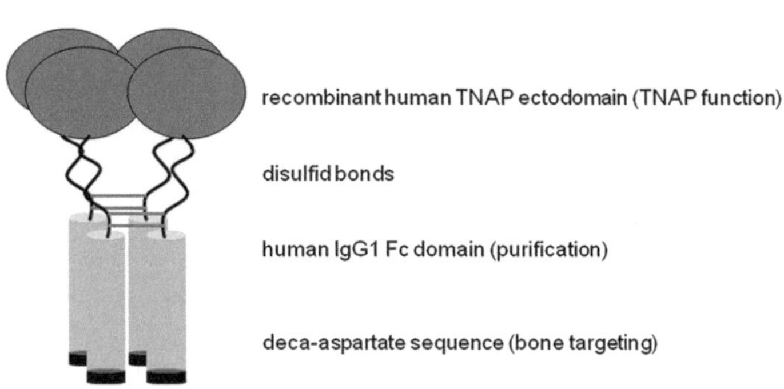

Fig. 15.1 Asfotase alfa. Human recombinant, fusion protein comprising the TNAP ectodomain, the constant region of the human IgG1 Fc domain, and a terminal deca-aspartate motif. Depicted as a homotetramer (from Millan et al. 2008)

homotetramers as it is known from the native, unaltered TNAP (Millan et al. 2008). In adult wild type mice the half-life in blood of Asfostase alfa is approximately 34 h after a single iv bolus of 5 mg/kg with prolonged retention in bone. Skeletal levels of bone-targeted Asfotase alfa seemed stable whereas no sustained accumulation could be seen in muscle. The enzymatic activity in muscle decreased in parallel with its activity in serum (Millan et al. 2008). After subcutaneous administration bioavailability is approximately 43 % in mice (Millan et al. 2008).

15.6.1 Nonclinical Studies

In 2008, Millan et al. successfully used Asfotase alfa for the first time in a TNAP-homozygous null murine ($Akp2^{-/-}$) model that mirrors many of the skeletal and biochemical abnormalities of severe infantile HPP (Millan et al. 2008). These mice are born with normally mineralized skeleton, but develop rickets at day 6 of age and die between day 16–20 suffering severe skeletal hypomineralization, episodes of apnea and seizures attributed to disturbances in PLP (vitamin B6) metabolism (Fedde et al. 1999; Narisawa et al. 1997, 2001; Waymire et al. 1995). Pyridoxine supplementation briefly suppresses the seizures and extends their life-span, but only until post-natal day 18–22 (Narisawa et al. 1997, 2001; Waymire et al. 1995).

Treatment with Asfotase alfa resulted in normal levels of TNAP substrates, normal growth and prevented skeletal and dental manifestations or seizures in $Akp2^{-/-}$ mice compared with mice who received vehicle alone. The study showed that s.c. injection is a suitable means of enzyme administration for newborn mice (Millan et al. 2008). Long term (52 days) high dose (8.2 mg/kg/day) ERT showed long-term survival in contrast with the fatal outcome of the vehicle-treated group (median life span: 18.5 days). Mice sustained their growth rates, normal appearance and apparent wellbeing accompanied by normal plasma pyridoxal, PPi and calcium concentrations. Treatment with daily s.c. injections of Asfotase alfa resulted in serum levels ranging from 650 to 1000 U/l. X-rays showed a normal skeletal phenotype of the feet, thorax and hind limbs. This study is the first report to show that ERT using bone-targeted, recombinant human TNAP prevents biochemical, radiographic and clinical manifestations of infantile HPP in TNAP knockout mice when initiated at birth (Millan et al. 2008). In 2011 McKee et al. (2011) were able to show that daily s.c. injections of Asfotase alfa prevent dentin hypomineralization and restore acellular cementum otherwise absent in $Akp2^{-/-}$ mice. Further investigations showed a clear positive correlation between Asfotase alfa dose and prevention of mineralizing defects. The 80 % effective dose in mice was between 2.8 and 3.2 mg/kg/day for different bones, with long bones showing response to lower daily doses. Median survival, body weight and bone length improved with increasing doses of Asfotase alfa. These dose response relationships strongly supported the pharmacological efficacy of Asfotase alfa for HPP, and provided the experimental basis for the therapeutic range chosen for clinical trials.

15.6.2 Clinical Studies of Asfotase Alfa in HPP

ERT with recombinant TNAP is now being evaluated in different clinical studies (http://clinicaltrials.gov/ct2/show/record/NCT01176266, current enrollement: ENB 010-10). First results of a multinational, open-label phase 2 study in infants and children with life-threatening HPP have recently been published by Whyte et al. (2012). The authors evaluated the safety, tolerability, bioavailability, pharmacokinetics, pharmacodynamics, and efficacy of treatment with Asfotase alfa in 11 patients with severe HPP (perinatal/infantile). The primary endpoint was the amelioration of hypomineralization, as assessed by means of radiography. Further objectives were the effects of Asfotase alfa on respiratory function, growth and gross motor, fine motor, and cognitive development as measured with the Bayley Scales of infants and Toddler Development, third edition (Bayley-III).

15.6.3 Study Design

Eligibility criteria included age of 3 years or less, first signs of HPP before the age of 6 months, serum AP activity below the age-adjusted lower limit of normal, an elevated serum PLP level, radiographic evidence of HPP, failure to thrive, rachitic chest deformity or pyridoxine-responsive seizures, and nontraumatic or poorly healing fractures, hypercalcemia, craniosynostosis, nephrocalcinosis or respiratory compromise. None of the patients had major coexisting disease or a treatable form of rickets and none received bisphosphonates, another investigational drug, or experimental treatment of HPP (e.g. bone marrow transplantation).

The dose of Asfotase alfa used in this study was chosen following efficacy and toxicology preclinical studies and a phase 1 clinical study involving adult patients with HPP. Adults receiving one i.v. dose of 3 mg/kg of body weight and weekly s.c. doses of 2 mg/kg Asfotase alfa showed serum levels in the anticipated therapeutic range (650–1000 U/l). In the published phase 2 study infants and children received Asfotase alfa as single i.v. infusion at a dose of 2 mg/kg, followed by subcutaneous injections three times per week at a dose of 1 mg/kg. The dose was increased up to 3 mg/kg in case of worsening failure to thrive, deteriorating pulmonary function, or no radiographic evidence of skeletal improvement.

15.6.4 Results (Figs. 15.2 and 15.3)

Eleven patients (5 with perinatal, 6 with infantile form of HPP, age: 20 days–36 months) were enrolled in the study, 9 completed 1 year of treatment. One patient was homozygous and the others were compound heterozygous of ALPL missense mutations. One patient died from sepsis after 7.5 months of treatment. The family of one patient withdraw consent when the patient experienced a moderate infusion reaction during the initial iv application of the study drug. The average treatment duration of the 9 participating patients was 18 months (range 12–26).

Treatment with Asfotase alfa (s.c.) was associated with significant radiographic improvement with amelioration of hypomineralization as well as substantial improved respiratory and motor function in these infants with life-threatening HPP.

Although all the study subjects have been classified to have a life-threatening form, disease severity seems to differ significantly in various aspects, like severity of rickets (from no visible bone to moderate severity), respiratory status (4 with ventilation, one with continuous positive airway pressure CPAP, five with oxygen by nasal cannula, one with room air), motor development, gastrointestinal and neurological aspects (one with seizures). Patients' course of disease shows similarities but the clinical course for each patient was different, even in the two with the most severe form of HPP.

15.6.4.1 Skeletal Improvement

An impressive skeletal improvement was seen in all treated patients (Fig. 15.2). Healing of rickets was diagnosed in 9 patients after 6 months of therapy, including increased mineral content, corrected or improved endochondral and membranous bone formation, fracture mending, reduced deformity, resolution of radiolucencies and sclerosis, and extensive modeling and remodeling. Radiographic changes were assessed using two scores: RGI-C (7-point scale of radiographic global impression of changes) developed specifically for this study, and RSS [10-point rickets-severity scale, (Thacher et al. 2000)]. Bone biopsies and densitometry would have been very interesting to follow treatment effects but were precluded for ethical reasons in these severely affected intensive care patients.

15.6.4.2 Craniosynostosis

The occurrence of premature fusion of cranial sutures seems somehow counter-intuitive in view of the absence of skeletal bone mineralization in perinatal and infantile HPP, but it has been recognized that craniosynostosis occurs more frequently in patients with HPP (approximately 40 % of infants with HPP) than in the general population. Interestingly, treatment did not seem to have an influence on the progression of craniosynostosis and skull deformities in these patients. Craniectomies were performed in two patients due to chronic increased intracranial pressure. It is estimated that craniosynostosis affects 1 in 2000 live births and it is part of a skeletal syndrome in 15–40 % (Kimonis et al. 2007). According to current knowledge the dura mater might play an important role in determining closure or patency of the suture and several genetic mutations have been associated with syndromic craniosynostosis, e.g. mutations in different fibroblast growth factor receptor (FGFR) genes and in the transcription factor gene TWIST (Kimonis et al. 2007; Mulliken et al. 2004; Slater et al. 2008). FGF, FGFR and indirectly TWIST

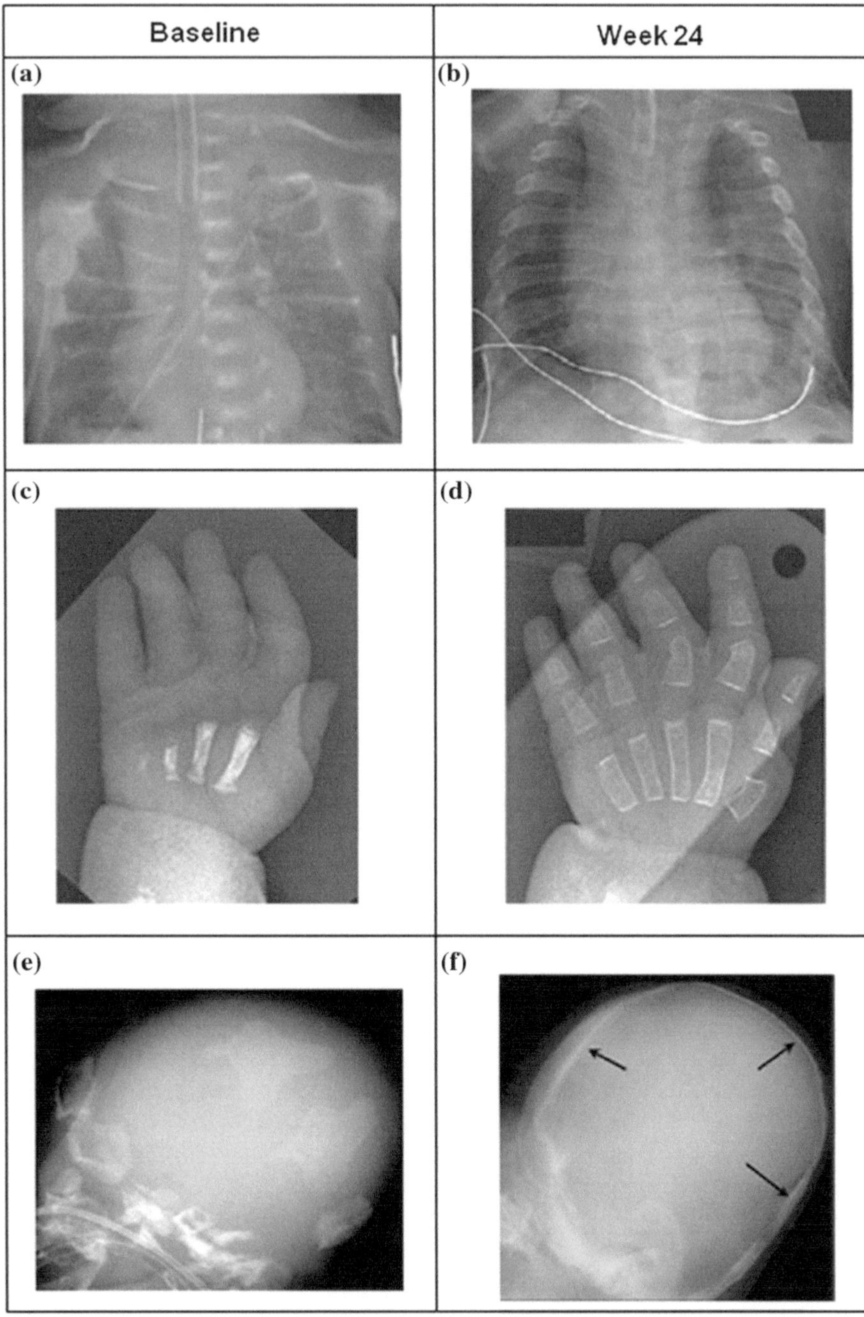
Baseline
Week 24
(a)
(b)
(c)
(d)
(e)
(f)

◀ **Fig. 15.2** X-rays of an infant with HPP before and after 24 weeks of enzyme replacement therapy Baseline radiographs (**a**, **c**, **e**) show extreme hypomineralization of the thorax, left hand and skull. Thin and partially mineralized ribs, clavicles and spine (**a**), multiple non-ossified bones (**c**) and minimal skull mineralization (largely at the base, little in the frontal, parietal, occipital, facial bones and sphenoid). After 24 weeks of ENB-0040 treatment substantial remineralization of all ossified structures (thorax, hand and skull) is apparent (**b**, **d**, **f**). Modified figure from Whyte et al. (2012). Copyright © (2013) Massachusetts Medical Society. Reprinted with permission from Massachusetts Medical Society (MMS)

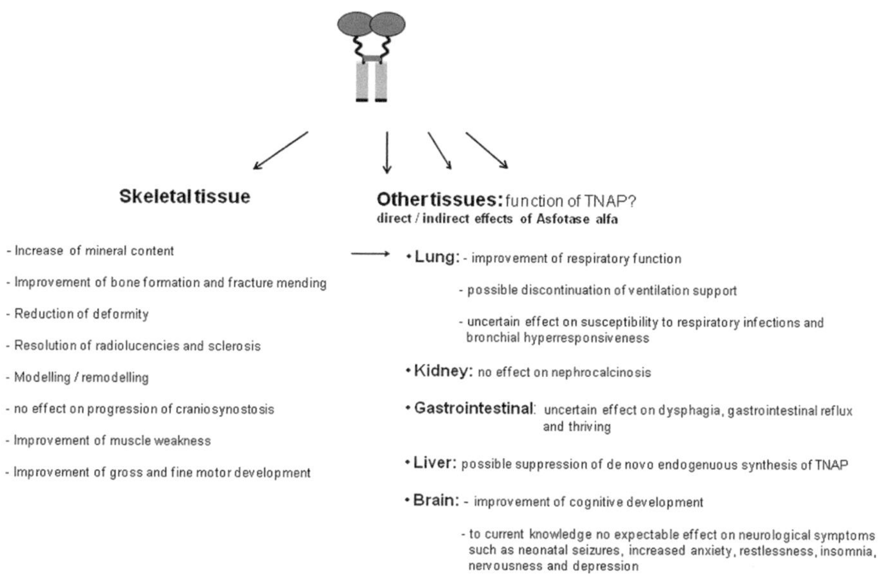

Fig. 15.3 Effects of Asfotase alfa in different tissues and its possible limitations

are known to regulate fetal bone growth and are expressed in cranial sutures during pregnancy (Kimonis et al. 2007). Gain or loss of function mutations might therefore result in disturbances in the fine balance that regulates the multiplication and maturation of the precursor bone cells in the cranial sutures and finally leading to craniosynostosis (Kimonis et al. 2007). Despite this progress, the pathophysiology of sutures is still not understood. The relevance of ALPL deficiency in this context and the exact causative pathways remain unclear. Nevertheless first experiences of enzyme replacement therapy in severe HPP show that long term neurosurgical surveillance is still necessary, since craniosynostosis often results in increased pressure requiring surgical intervention to avoid visual impairment and compromised neurodevelopment. These complex neurosurgical interventions should be performed by a pediatric neurosurgeon or craniofacial surgical team who has experience with treatment of pediatric craniosynostosis and is aware of the special bone situation of ALPL-deficient patients (Collmann et al. 2009).

15.6.4.3 Laboratory Findings

Serum parathyroid hormone PTH is at a low-normal level at baseline in patients with HPP that are often hyercalcemic, and increased during treatment, generally associated with radiographic improvement. Serum calcium and inorganic phosphate levels showed only minor fluctuations. Calcium restriction was stopped, and calcium intake was even increased in some cases to control the rise in PTH and prevent "hungry bone" syndrome. Levels of TNAP substrates (PLP and PPi) decreased.

15.6.4.4 Muscle Function and Motor Development

Improvement in gross and fine motor skills and muscle strength were documented in treated patients. Patients were able to move limbs against gravity, some were taking steps or even walking. Although the pathogenesis of muscular hypotonia in HPP is not understood, the authors suggest a myopathic effect of accumulating TNAP substrates. They speculated that improvement in motor function in Asfotase alfa treated patients might be due to hydrolysis of inorganic pyrophosphate (Whyte et al. 2012; Seshia et al. 1990).

15.6.4.5 Respiratory Situation

All patients showed respiratory compromise at baseline and interestingly respiratory function tended to worsen initially during Asfotase alfa therapy in the infants not receiving ventilatory support. Therefore a close monitoring of these patients is warranted, in particular in the first months after starting Asfotase alfa treatment. Later on respiratory situation improved in all patients, associated with improved mineralization of the thoracic cage. After 12 months of treatment 6 patients were breathing room air (only one at baseline), one received oxygen by nasal cannula, two were on ventilation via a tracheostomy, one of them only at night and went on to be off respiratory support after 2, 5 years of treatment (Four patients were invasively ventilated at baseline, one received CPAP). Rodriguez et al. (2012) showed measurable changes in chest wall dynamics and respiratory mechanics in an infant with thorax deformity and associated hypoplastic lungs after 12 weeks of treatment with Asfotase alfa. A consistent decrease in ventilator requirements and improvement in lung function and chest wall dynamics were observed and correlated with improvement in thoracic cage radiological findings.

15.6.4.6 Gastrointestinal Situation and Thriving

All study patients presented with failure to thrive. Tube feeding (gastric or jejunal tube) was necessary in 7 patients, fundoplication was performed in 3 patients to improve the nutritional status. Dysphagia, gastroesophageal reflux, recurrent emesis and resulting dystrophy represent important additional risk factors for life-threatening complications (e.g. infections), in severe HPP. Therefore active gastrointestinal care and nutritional assistance is mandatory for optimal thriving in patients with HPP, with or without ERT. Long term data of the course of weight and length patients receiving Asfotase alfa has not been published and it is not yet clear what the growth (weight and length) is under ERT treatment for HPP.

15.6.4.7 Kidney

Nine patients had nephrocalcinosis or renal stones at baseline, no new occurrences of renal abnormalities were reported, nephrocalcinosis resolved in one patient and was unchanged in the others. This course suggests no significant improvement of the nephrocalcinosis already present during treatment and no worsening or newly diagnosed renal pathologies. The natural course of renal complication in HPP is not known, although it can be speculated that nephrocalcinosis might have happened due to persisting high levels of PPi and hypercalciuria, as it was present in several patients at baseline, and that Asfotase alfa does not increase the risk of developing this complication.

15.6.4.8 Adverse Events

The most common adverse event was a local injection side reaction that was primarily characterized by mild, localized and transient erythema. Parental consent for treatment was withdrawn for one patient because of mild allergic reaction during iv infusion of Asfotase alfa. There were no clinical signs of hypocalcemia from calcium-avid "hungry" bones nor any treatment-related adverse events suggestive of hypersensitivity or tachyphylaxis. To current knowledge low titers of anti-ENB-0040 antibodies which could be detected in some patients do not influence the clinical outcome. No evidence of ectopic calcifications was found during the treatment period.

Serious adverse events were generally consistent with this seriously ill population and included 58 events during 12 months of treatment, mainly infections, respiratory disorders, nervous-system disorders, musculoskeletal disorders and surgery or medical procedures. Multiple events could occur in a single patient. Three serious adverse events (respiratory distress, craniosynostosis and conductive hearing loss) were considered to be possibly related to the study treatment.

One patient died due to pneumonia and sepsis with multiorgan failure shortly after discharge from long-term hospitalization and 31 weeks of ERT. The 1–1/2-year-old patient who stopped ERT after the first i.v. infusion was later discharged to home ventilation showing remarkable worsening in the x-rays of her thorax at the age of 2–2/3 displaying the natural course of severe HPP.

Since this first phase 2 study was not a placebo-controlled trial and since the manifestations are heterogeneous and differ individually from patient to patient, many questions on the efficacy of ERT with Asfotase alfa still remain open.

Baumgartner-Sigl et al. (2007) reviewed that no patient with severe HPP and pyridoxine-responsive seizures (PRS) survived beyond 18 months of age. But natural history of severe forms of HPP has not been systematically reviewed or published. Leung et al. (2013) recently described a retrospective cohort of Manitoba Mennonites with perinatal HPP with 15 patients who all died by 9 months of age, usually soon after birth, from pulmonary failure. A systemic retrospective study investigating the natural course of disease in patients with severe perinatal and infantile HPP is currently ongoing (http://clinicaltrials.gov/ct2/show/study/NCT01419028).

15.7 Possible Challenges and Limitations of ERT in HPP

Effects of Asfotase alfa on bone and motor function appear promising, but it is unclear whether ERT with Asfotase alfa is effective to prevent damage in other organs like lung, gastrointestinal tract, kidney, central nervous system and skull. For example observed improvement in pulmonary aspects most likely does not result from “TNAP replacement” in the lung but rather from secondary effects due to stabilization of the thoracic cage. Until now the physiological role of TNAP in different extra-osseous tissues is still largely unknown and etiology and the pathogenesis of various clinical manifestations (e.g. gastrointestinal signs, craniosynostosis, neurological or pulmonary symptoms) due to TNAP deficiency in HPP is not understood in detail. Lack or deficiency of TNAP as well as secondary phenomena in each organ may contribute to the complex pathology in this multi-systemic disease. One great challenge is the delivery to difficult-to-reach sites of pathology, such as in brain or lung. Therefore, new combinational therapeutic approaches may be necessary using drugs that enhance delivery to intractable tissues as well as drugs that target alternative and secondary pathological mechanisms, such as substrate accumulation and inflammation. In Asfotase alfa TNAP has been conjugated with aspartate residues to function as a bone anchor (Nishioka et al. 2006). Conjugation with other peptides or molecules might help to address other sites of pathology as well as other forms of administration, such as intra-thecal or as aerosols. The issue of an exact tissue and especially subcellular targeting of recombinant TNAP may have to be addressed more carefully in the future. According to our understanding in TNAP cell biology and function for an efficient ERT in HPP the TNAP protein needs to be enriched in the cell membrane of various cells (osteoblasts, among others) and in the outer surface of the osteoblast matrix vesicle membrane where it is thought to play an important role in the mineralization process and phosphate metabolism as a membrane-bound ectoenzyme. TNAP can be cleaved by a glycosylphosphatidylinositol (GPI)-specific phospholipase D, before it enters the circulation in a soluble anchor free form (Anh et al. 1998, 2001). For complete dimerization and subcellular localization of the GPI-anchored protein, complete posttranslational modification during trans-Golgi network passage is indispensable (Fedde et al. 1988; Hooper 1997). Due to the loss of complete translational modification, various mutated TNAP proteins fail to reach the cell membrane because of rapid premature proteasomal degradation as a kind of quality control (Ishida et al. 2003; Mentrup et al. 2011) (see Chaps. 2 and 9).

Due to the high specificity for hydroxyapatite crystals it will also be interesting to determine if and to what extent other sites at which calcification can occur (e.g. the vasculature) attract Asfotase alfa molecules. Asfotase alfa requires fibrillar collagen to be present for anchoring, which explains why nephrocalcinosis is not worsened by treatment. This should be evaluated in a long-term study, preferably in older patients. In different mouse models it has been illustrated that elevated TNAP favors vascular calcification by hydrolyzing the calcification inhibitor PPi (Hessle et al. 2002; Lomashvili et al. 2008; Narisawa et al. 2007; Villa-Bellosta et al. 2011). Medial vascular calcification follows an active endochondral ossification-like process, involving chondro-osteoblastic conversion of vascular smooth muscle cells with development of osteoblastic features, such as expression of TNAP among others (Zhu et al. 2011). Prevalence of medial vascular calcification corresponds roughly with age and coronary arterial calcification occurring predominantly in the elderly (Towler and Demer 2011). Kiffer-Moreira et al. (2013) were able to show that the combined use of selective PHOSPHO1 and TNAP inhibitors considerably reduce calcification indicating that "phosphatase inhibition" constitutes a viable approach for the prevention and treatment of medial vascular calcification. Based on the knowledge of these preclinical data it seems to be important to monitor administered enzyme doses and serum alkaline phosphate levels very carefully in each patient during a life-long therapy being aware of possible off target effects like medial vascular calcification, especially in the elderly. Furthermore, the need of recombinant TNAP may differ in each patient according to age as natural TNAP levels vary significantly during life (Hofmann et al. 2013).

As ERT in HPP is a very new type of treatment no long term data are available. According to Whyte and colleagues (2012) low titers of antibodies (anti-ENB0040) could be detected in four of the nine patients during the first 6 months but did not appear to influence clinical outcome in the first year of treatment. However, antibody mediated immunologic reactions to the administered enzyme may limit the efficacy of treatment in the future.

Further basic research on TNAP function and role in different tissues and on its physiological substrates has to be done to gain a better understanding of the pathophysiology of HPP. Hopefully this and other new therapeutic strategies may improve the prognosis and quality of life of patients with HPP and may contribute to our understanding of bone metabolism in general.

The availability of potentially effective treatment that improves survival of life-threatening forms of HPP brings up the question for a screening to identify HPP patients as soon as possible to start treatment early without any delay thereby avoiding or at least limiting severe complications as respiratory insufficiency. Diagnosis is often complicated by rareness of the disease. Even severely affected infants have a history of misdiagnosis with different treatment attempts and several hospital transfers before they are correctly diagnosed and ERT could be started in a study centre. In our experience milder forms of HPP often causes even more misdiagnosis problems as symptoms vary in their expression and can resemble other skeletal diseases. Screening programs like a screening of alkaline phosphatase deficiency in serum might decrease the time to diagnosis, optimize the starting point for ERT and early treatment may prevent irreversible damage and improve clinical outcome in HPP. Asfotase alfa has not been licensed in the US or Europe so far. Currently the drug is only available through an open-label study of Asfotase alfa in infants and children ≤5 years of age with HPP (http://clinicaltrials.gov/ct2/show/record/NCT01176266). This phase 2 study is planned to be expanded worldwide and is still recruited in different study centres (Manitoba, Canada; Oakland and Pittsburgh, United States; Würzburg, Germany, Taipei, Taiwan; see also http://clinicaltrials.gov/ct2/show/record/NCT01176266).

HPP patients should be followed by an HPP-experienced multidisciplinary team. Nevertheless the overall number of patients with perinatal HPP is small, and it will therefore be difficult to establish a prospective, randomized placebo-controlled trial. Therefore historical controls might be necessary to define the efficacy of ERT with Asfotase alfa for the various and individually differing manifestations of severe HPP. A systematic retrospective study investigating the natural course of patients with severe perinatal and infantile HPP is currently underway (http://clinicaltrials.gov/ct2/show/record/NCT01419028).

For patients with milder forms controlled trials are necessary to determine the effect of ERT with Asfotase alfa. Eventually patients with milder forms will not qualify for a lifelong and expensive ERT but might profit from an on demand therapy with single injections in somehow challenging situations, like fractures. It is also necessary to establish therapeutic approaches to help patients who are not affected by life-threatening forms but also suffer from a significant reduction in quality of life. Alternatively, a bone anabolic therapy using a anti-sclerostin antibody is being evaluated in a clinical study in Germany (unpublished data Jakob and Seefried, Würzburg).

15.8 Conclusions and Future Directions

In conclusion, HPP is a rare inborn error of bone metabolism, and its diagnosis is often overseen or significantly delayed. As TNAP is a ubiquitous enzyme HPP may present as a multisystemic disease with involvement of different organs, like bone, muscle, kidney, lung, gastrointestinal tract and peripheral as well as central nervous system. The exact metabolic mechanisms of TNAP deficiency and its sequelae in different tissues are not understood in detail. Because there is a considerable variability of clinical manifestations, which may lead to a significant reduction in quality of life, patients should be followed by a HPP-experienced multidisciplinary team. A bone targeted recombinant human TNAP molecule (Asfotase alfa) is currently being evaluated in infants with life-threatening forms of HPP. The preliminary results from clinical studies show significantly improved findings in skeletal radiographs, pulmonary and physical function in infants with life-threatening HPP. These results are very promising, especially with regard to the skeletal phenotype, but it is unclear what the long-term benefits of treatment are for other affected but difficult-to-reach tissues such as brain, kidney or in the case of craniosynostosis (Fig. 15.3). Long-term data are not yet available and further systematic clinical trials are needed.

Conflict of Interest JL and CH received a study grant from Enobia and Alexion Pharma (Cheshire, Connecticut, U.S.) for a phase two study on Asfotase alpha treatment for severe forms of HPP. F. Jakob has received honoraria and travel support for lectures and advice from Eli Lilly, Amgen, Novartis, MSD, Nycomed, Servier, Roche, Enobia and Alexion Pharma, he has received unrestricted research grants from Novartis and is involved in clinical studies related to osteoporosis drugs initiated by Eli Lilly, Amgen, Servier and Novartis. HG served on an advisory board of Enobia/Alexion. HP was a full-time employee of Alexion 2012–1013.

References

Anh DJ, Dimai HP, Hall SL et al (1998) Skeletal alkaline phosphatase activity is primarily released from human osteoblasts in an insoluble form, and the net release is inhibited by calcium and skeletal growth factors. Calcif Tissue Int 62:332–340

Anh DJ, Eden A, Farley JR (2001) Quantitation of soluble and skeletal alkaline phosphatase, and insoluble alkaline phosphatase anchor-hydrolase activities in human serum. Clin Chim Acta 311:137–148

Baumgartner-Sigl S, Haberlandt E, Mumm S et al (2007) Pyridoxine-responsive seizures as the first symptom of infantile hypophosphatasia caused by two novel missense mutations. Bone 40 (6):1655–1661

Cahill RA, Wenkert D, Perlman SA et al (2007) Infantile hypophosphatasia: transplantation therapy trial using bone fragments and cultured osteoblasts. J Clin Endocrinol Metab 92:2923–2930

Camacho PM, Painter S, Kadanoff R (2008) Treatment of adult hypophosphatasia with teriparatide. Endocr Pract 14:204–208

Collmann H, Mornet E, Gattenlohner S et al (2009) Neurosurgical aspects of childhood hypophosphatasia. Childs Nerv Syst 25:217–223

Collmann H, Sorensen N, Krauss J (2005) Hydrocephalus in craniosynostosis: a review. Childs Nerv Syst 21:902–912

Clarke LA, Wraith JE, Beck M et al (2009) Long-term efficacy and safety of laronidase in the treatment of mucopolysaccharidosis I. Pediatrics 123:229–240

Deduve C (1964) From Cytases to Lysosomes. Fed Proc 23:1045–1049

Desnick RJ, Schuchman EH (2012) Enzyme replacement therapy for lysosomal diseases: lessons from 20 years of experience and remaining challenges. Annu Rev Genomics Hum Genet 13:307–335

Doshi KB, Hamrahian AH, Licata AA (2009) Teriparatide treatment in adult hypophosphatasia in a patient exposed to bisphosphonate: a case report. Clin Cases Miner Bone Metab 6:266–269

Fedde KN, Blair L, Silverstein J et al (1999) Alkaline phosphatase knock-out mice recapitulate the metabolic and skeletal defects of infantile hypophosphatasia. J Bone Miner Res 14:2015–2026

Fedde KN, Lane CC, Whyte MP (1988) Alkaline phosphatase is an that acts on micromolar concentrations of natural substrates at physiologic pH in human osteosarcoma (SAOS-2) cells. Arch Biochem Biophys 264:400–409

Grabowski GA, Barton NW, Pastores G et al (1995) Enzyme therapy in type 1 Gaucher disease: comparative efficacy of mannose-terminated glucocerebrosidase from natural and recombinant sources. Ann Intern Med 122:33–39

Harmatz P, Giugliani R, Schwartz I et al (2006) Enzyme replacement therapy for mucopolysaccharidosis VI: a phase 3, randomized, double-blind, placebo- controlled, multinational study of recombinant human N-acetylgalactosamine 4-sulfatase (recombinant human arylsulfatase B or rhASB) and follow-on, open-label extension study. J Pediatr 148:533–539

Hessle L, Johnson KA, Anderson HC et al (2002) Tissue-nonspecific alkaline phosphatase and plasma cell membrane glycoprotein-1 are central antagonistic regulators of bone mineralization. Proc Natl Acad Sci USA 99:9445–9449

Hofmann C, Girschick H, Mentrup B et al (2013) Clinical aspects of hypophosphatasia: an update. Clin Rev Bone Miner Metab 11:60–70

Hooper NM (1997) Glycosyl-phosphatidylinositol anchored membrane enzymes. Clin Chim Acta 266:3–12

Ioannou YA, Zeidner KM, Gordon RE et al (2001) Fabry disease: preclinical studies demonstrate the effectiveness of alpha-galactosidase a replacement in enzyme-deficient mice. Am J Hum Genet 68:14–25

Ishida Y, Komaru K, Ito M et al (2003) Tissue-nonspecific alkaline phosphatase with an Asp (289)->Val mutation fails to reach the cell surface and undergoes proteasome-mediated degradation. J Biochem 134:63–70

Kimonis V, Gold JA, Hoffman TL et al (2007) Genetics of craniosynostosis. Semin Pediatr Neurol 14:150–161

Kiffer-Moreira T, Yadav MC, Zhu D et al (2013) Pharmacological inhibition of PHOSPHO1 suppresses vascular smooth muscle cell calcification. J Bone Miner Res 28:81–91

Leung EC, Mhanni AA, Reed M et al (2013) Outcome of perinatal hypophosphatasia in manitoba mennonites: a retrospective cohort analysis. JIMD Rep 11:73–78

Lomashvili KA, Garg P, Narisawa S et al (2008) Upregulation of alkaline phosphatase and pyrophosphate hydrolysis: potential mechanism for uremic vascular calcification. Kidney Int 73:1024–1030

McKee MD, Nakano Y, Masica DL et al (2011) Enzyme replacement therapy prevents dental defects in a model of hypophosphatasia. J Dent Res 90:470–476

Mentrup B, Marschall C, Barvencik F et al (2011) Functional characterization of a novel mutation localized in the start codon of the tissue-nonspecific alkaline phosphatase gene. Bone 48:1401–1408

Millan JL (2013) The role of phosphatases in the initiation of skeletal mineralization. Calcif Tissue Int 93(4):299–306

Millan JL, Narisawa S, Lemire I et al (2008) Enzyme replacement therapy for murine hypophosphatasia. J Bone Miner Res 23:777–787

Miranda SR, He X, Simonaro CM et al (2000) Infusion of recombinant human acid sphingomyelinase into niemann-pick disease mice leads to visceral, but not neurological, correction of the pathophysiology. FASEB J 14:1988–1995

Mornet E, Hofmann C, Bloch-Zupan A et al (2013) Clinical utility gene card for: hypophosphatasia – update. Eur J Hum Genet. doi:10.1038/ejhg.2013.177

Muenzer J, Gucsavas-Calikoglu M, McCandless SE et al (2007) A phase I/II clinical trial of enzyme replacement therapy in mucopolysaccharidosis II (Hunter syndrome). Mol Genet Metab 90:329–337

Mulliken JB, Gripp KW, Stolle CA et al (2004) Molecular analysis of patients with synostotic frontal plagiocephaly (unilateral coronal synostosis). Plast Reconstr Surg 113:1899–1909

Narisawa S, Frohlander N, Millan JL (1997) Inactivation of two mouse alkaline phosphatase genes and establishment of a model of infantile hypophosphatasia. Dev Dyn 208:432–446

Narisawa S, Wennberg C, Millan JL (2001) Abnormal vitamin B6 metabolism in alkaline phosphatase knock-out mice causes multiple abnormalities, but not the impaired bone mineralization. J Pathol 193:125–133

Nishioka T, Tomatsu S, Gutierrez MA et al (2006) Enhancement of drug delivery to bone: characterization of human tissue-nonspecific alkaline phosphatase tagged with an acidic oligopeptide. Mol Genet Metab 88:244–255

Narisawa S, Harmey D, Yadav MC et al (2007) Novel inhibitors of alkaline phosphatase suppress vascular smooth muscle cell calcification. J Bone Miner Res 22:1700–1710
Rodriguez E, Bober MB, Davey L et al (2012) Respiratory mechanics in an infant with perinatal lethal hypophosphatasia treated with human recombinant enzyme replacement therapy. Pediatr Pulmonol 47:917–922
Sands MS, Vogler CA, Ohlemiller KK et al (2001) Biodistribution, kinetics, and efficacy of highly phosphorylated and non-phosphorylated beta-glucuronidase in the murine model of mucopolysaccharidosis VII. J Biol Chem 276:43160–43165
Seshia SS, Derbyshire G, Haworth JC et al (1990) Myopathy with hypophosphatasia. Arch Dis Child 65:130–131
Slater BJ, Lenton KA, Kwan MD et al (2008) Cranial sutures: a brief review. Plast Reconstr Surg 121:170e–178e
Sutton RA, Mumm S, Coburn SP et al (2012) "Atypical femoral fractures" during bisphosphonate exposure in adult hypophosphatasia. J Bone Miner Res 27:987–994
Tadokoro M, Kanai R, Taketani T et al (2009) New bone formation by allogeneic mesenchymal stem cell transplantation in a patient with perinatal hypophosphatasia. J Pediatr 154:924–930
Thacher TD, Fischer PR, Pettifor JM et al (2000) Radiographic scoring method for the assessment of the severity of nutritional rickets. J Trop Pediatr 46:132–139
Towler DA, Demer LL (2011) Thematic series on the pathobiology of vascular calcification: an introduction. Circ Res 108:1378–1380
Villa-Bellosta R, Wang X, Millan JL et al (2011) Extracellular pyrophosphate metabolism and calcification in vascular smooth muscle. Am J Physiol Heart Circ Physiol 301:H61–H68
Waymire KG, Mahuren JD, Jaje JM et al (1995) Mice lacking tissue non-specific alkaline phosphatase die from seizures due to defective metabolism of vitamin B-6. Nat Genet 11:45–51
Weninger M, Stinson RA, Plenk H et al (1989) Biochemical and morphological effects of human hepatic alkaline phosphatase in a neonate with hypophosphatasia. Acta Paediatr Scand Suppl 360:154–160
Whyte MP (2010) Physiological role of alkaline phosphatase explored in hypophosphatasia. Ann N Y Acad Sci 1192:190–200
Whyte MP, Greenberg CR, Salman NJ et al (2012) Enzyme-replacement therapy in life-threatening hypophosphatasia. N Engl J Med 366:904–913
Whyte MP, Kurtzberg J, McAlister WH et al (2003) Marrow cell transplantation for infantile hypophosphatasia. J Bone Miner Res 18:624–636
Whyte MP, McAlister WH, Patton LS et al (1984) Enzyme replacement therapy for infantile hypophosphatasia attempted by intravenous infusions of alkaline phosphatase-rich Paget plasma: results in three additional patients. J Pediatr 105:926–933
Whyte MP, Mumm S, Deal C (2007) Adult hypophosphatasia treated with teriparatide. J Clin Endocrinol Metab 92:1203–1208
Whyte MP, Valdes R Jr, Ryan LM et al (1982) Infantile hypophosphatasia: enzyme replacement therapy by intravenous infusion of alkaline phosphatase-rich plasma from patients with Paget bone disease. J Pediatr 101:379–386
Zhu D, Mackenzie NC, Millan JL et al (2011) The appearance and modulation of osteocyte marker expression during calcification of vascular smooth muscle cells. PLoS One 6:e19595

Chapter 16
Neurogenetic Aspects of Hyperphosphatasia in Mabry Syndrome

David E.C. Cole and Miles D. Thompson

Abstract An autosomal recessive syndrome of hyperphosphatasia (elevated circulating alkaline phosphatase (AP), seizures and neurologic deficits) was first described by Mabry and colleagues in 1970. Over the ensuing four decades, few cases were reported. In 2010, however, new families were identified and the syndromic nature of the disorder confirmed. Shortly thereafter, next generation sequencing was used to characterize causative defects in the glycosyl phosphatidylinositol (GPI) biosynthetic pathway, based partly on our understanding of how AP is anchored by GPI to the plasma membrane. Whether the seizures and cognitive defects seen in Mabry syndrome patients are attributable in part to the constant hyperphosphatasia is not known, as there are more than 250 other proteins dependent on GPI for their anchoring to the plasma membrane. However, Mabry syndrome may provide a new window on AP function in growth and development.

Keywords Alkaline phosphatase · Glycosylated phosphatidylinositol anchor · Neurodegeneration · Seizures

D.E.C. Cole (✉) · M.D. Thompson
Department of Laboratory Medicine and Pathobiology,
University of Toronto, Toronto, ON, Canada
e-mail: davidec.cole@utoronto.ca

M.D. Thompson
e-mail: miles.thompson@utoronto.ca

D.E.C. Cole
Department of Clinical Pathology, Sunnybrook Health Sciences Center,
Toronto, ON, Canada

C. Fonta and L. Négyessy (eds.), *Neuronal Tissue-Nonspecific Alkaline Phosphatase (TNAP)*, Subcellular Biochemistry 76,
DOI 10.1007/978-94-017-7197-9_16

16.1 Introduction

The term "*hyper*phosphatasia" is used to describe conditions that feature marked persistent elevation of serum alkaline phosphatase (AP) activity as an isolated but sustained biochemical finding, in contrast to the term *hypo*phosphatasia used by most to describe subjects with deleterious mutations of the tissue-nonspecific TNAP gene (*ALPL*). Most often, increased AP activity is not constitutive, but a reflection of increased release from organ damage, be it due to liver cholestasis or increased turnover in mineralizing tissues. Occasionally elevated AP may also result from acute release of other AP isoenzymes, including those from placenta (during pregnancy), intestine (with some forms of severe bowel disease), and sometimes from tumors (Nagao or germ cell isozyme). In these instances, the AP also arises primarily from increased cleavage of dimers or tetramers that are otherwise normally tethered to the plasma membranes of cells expressing AP.

Persistent isolated hyperphosphatasia without organ damage is a rare finding. It may be seen in inherited conditions with constitutively increased skeletal turnover, such as juvenile Paget disease, due to osteoprotegerin deficiency (mutations of the *TNFRSF11B* gene—OMIM *602643), or expansile osteolysis syndrome (*TNFRSF11A* mutations—OMIM *603499). Apart from these disorders, there is a syndrome of persistent hyperphosphatasia associated with a moderate to severe postnatal neurodegenerative disorder with seizures and cognitive deficits (OMIM #239300). This condition was first described by Mabry et al. (1970); hence, the designation Mabry syndrome.

Towards the end of the century, as more cases appeared, the autosomal recessive nature of the syndrome was confirmed. However, by 2003, less than 20 potential cases had been reported and the condition remained poorly characterized (Gomes and Hunter 1970; Kruse et al. 1988; Cole and Whyte 1997). With additional families revealing consanguinity or sibling recurrence, the case for recessive inheritance was more convincing and an intensive search for the causative genetic etiology ensued.

Two reports appeared in 2010 that identified a number of new cases and established additional features of the condition (Thompson et al. 2010; Horn et al. 2010). This was rapidly followed by two seminal publications showing recessive mutations in the genes responsible for synthesis and attachment of the AP membrane anchor, glycosylphosphatidylinositol (GPI) (Krawitz et al. 2010, 2012). With additional evidence for a greatly expanded phenotype and substantial genetic heterogeneity (Thompson et al. 2012), it has become clear that not all cases will have hyperphosphatasia and there appears to be a remarkable degree of variation within and between families with causative mutations (Thompson et al. 2012; Hansen et al. 2013).

We review the early cases along with the prototypical biochemical features, followed by a synthesis of the characteristic features in the more recent studies, with and without identified mutations. Finally, we look more closely at several outstanding issues arising from the variable pathogenetic features, as well as a surprisingly wide spectrum of biochemical and molecular heterogeneity, and the still uncertain role of AP itself in the pathogenesis of the disorder. We conclude with some commentary about potential responsiveness of the seizures to pyridoxine supplements.

16.2 Mabry Syndrome: The First Family

Charlton Mabry and colleagues (1970) reported on four patients from a single consanguineous family (Table 16.1). The family reportedly had ancestral roots in rural Kentucky as far back as the mid-1800s that included African American heritage. Three affected siblings were studied at ages 4 (male), 6 (female), and 7 (male), along with a second cousin, a 14-year-old male. The three cardinal features —hyperphosphatasia, seizures, and cognitive deficit ("mental retardation")—were found in all four affecteds, but not in the parents or 52 other family members.

Although the hyperphosphatasia in the four children was sustained and substantial, there was none of the usual clinical or laboratory evidence for skeletal or hepatic disease. The failure to find any other biochemical anomalies that might point to an inborn error of metabolism led the investigators to explore the properties of the increased circulating AP activity and to look more closely for a storage disorder.

Clinically, a neurometabolic storage disorder seemed likely. While the postnatal period was relatively normal, developmental delay was substantial by 2–3 years of age, and affected children developed seizures that were poorly responsive to the anticonvulsants available at the time. By later childhood, the developmental delay in the affected children was severe, and there was no speech.

It is less clear how the affected children fared physically. The growth parameters (height, weight, and head circumference) in the 6-year-old proband were reported to fall on the 3rd centile, but the affected 14-year-old appears to resemble the unaffected mother, at least in the one available photograph (Mabry et al. 1970; Thompson et al. 2010). Thus, even in this first report, significant intra-familial differences are discernible.

Detailed studies of the enzymatic activity were undertaken, and included a variety of different phosphorylated substrates, buffers, and pH′s. In the four affected sibs, the measured alkaline phosphatase activities ranged from 2.8 to 75 times greater than a quality control normal. Where longitudinal repeat measurements were available, AP activity remained relatively consistent. Efforts to distinguish between tissue-specific isoforms by electrophoretic mobility, biochemical response to inhibitors and heat inactivation, as well as antibody studies and gel filtration analysis all pointed to a 'hepatic' origin of the AP activity in serum. The overall pattern, however, was different from that found with bone and liver disease, consistent with the absence of supportive clinical and laboratory findings for these common AP sources.

Table 16.1 Mabry Syndrome: Summary of 18 Early Cases

	(Mabry et al. 1970)	(Gomes and Hunter 1970)	(Kruse et al. 1988)		(Rabe et al. 1991)	(Cole and Whyte 1997)	
Study characteristics			Severe	Moderate		Severe	Moderate
No. of cases	4	1	3	6	2	1	1
Clinical features							
Birth history	N	Small for dates	N	N	–	Birth asphyxia	N
Consanguinity	+	–	–	–	–	–	–
Siblings affected	+	–	–	+	+	–	–
Seizures	+	+	3/3	3/6		+	+
Blindness/Deafness		Cataracts/-				+/+	+/-
Clinodactyly					++	+	–
Cognitive deficit	Sev	Mod	Sev	Mod	Sev	Sev	Mod
Behavioral problems		–	+	–			–
Esotropia/nystagmus	+	–	–	–	+		+
Facial dysmorphism	?	–			++		±
Funduscopic abnormalities	–	–	–	–		+	–
Bone survey	N	N	N	N	AbN	N	N
Muscle weakness	+	–	++	+		+	+
Growth	3rd %ile	N	N	N	3rd %ile	N	N
Lab findings							
Serum calcium & phosphate	N	N	N	N		N	N
Acid phosphatase	N	N					↑
Osteocalcin			N	N		–	N
Urinary Hydroxyproline		N	N	N			
LFTs	N	N	N	N		N	N
PTH			N	N		N	N
5′NT		N				↑	↑

NOTE In a study by Mabry et al. (1970), only 1 of 4 related patients was described in detail. In studies by Kruse et al. (1988), and Cole and Whyte (1997), cases were stratified according to severity as shown. –, absent; +, present; ++, marked; ↑, increased; 3rd %ile, third centile; 5′NT, 5′-nucleotidase; AbN, Abnormal-; LFTs, liver function tests; Mod, moderate; N, normal; Sev, severe

Tissue studies were similarly ambiguous. No abnormalities were found in a bone marrow, including tests of granulocyte AP content, and liver was histologically normal in both biopsy-derived and wedge-excision tissue samples. In the three older patients, rectal biopsies showed clusters of foam cells (increasing with age) with cytoplasmic inclusions. Periodic-acid Schiff staining for complex carbohydrates showed strongly positive granular aggregates on high-power examination. However, the storage material was not further characterized.

16.3 Other Early Cases

In the single patient reported by Gomes and Hunter (1970), the cognitive deficit was less severe and this small-for-birth infant had cataracts (Table 16.1). Kruse et al. (1988) studied nine patients and suggested they be stratified into two groups. The more severe form in three subjects was associated with behavior problems and profound hypotonia, whereas the other patients were moderately delayed. We studied two patients with different levels of severity (Cole and Whyte 1997) and likewise considered that they could be classified as severe and moderate according to the criteria of Kruse et al. (1988).

At that time, however, convincing evidence for a recessive disorder remained elusive. Rabe et al. (1991) drew attention to the similarity between the hyperphosphatasia with severe neurological deficit, clinodactyly and the Coffin-Siris syndrome. This proved to be an important observation because it drew attention to the presence of variably foreshortened terminal digits (brachytelephalangy) in other cases of Mabry syndrome. Although Coffin-Siris syndrome may feature a distinctive hypoplasia of the fifth phalanx, a review of cases and further analysis of phenotypes led Gillessen-Kaesbach and Meinecke (1999) to consider the brachytelephalangy as a syndrome characteristic, bringing the number of reported cases to 18 by the turn of the century (Table 16.1). A thoughtful and comprehensive review of Coffin-Siris syndrome (Schrier et al. 2012) has since re-classified the two sisters reported by Rabe et al. (1991) as Mabry Syndrome, and considers the minor dysmorphic facial features as consistent with that re-classification. More importantly, updated evaluation by Horn et al. (2010; 2011) and mutation screening by Krawitz et al. (2010) firmly establish these two cases as Mabry syndrome with known phosphatidylinositol glycan anchor biosynthesis, class V (*PIGV*) mutations.

Similarly, phenotypic anomalies in the two sisters born to consanguineous parents that were reported by Marcelis et al. (2007) extended beyond the cardinal features described by Mabry et al. (1970). Hypoplastic terminal phalanges, cleft palate, and an anteriorly displaced anus were also observed. The latter included an anovestibular fistula which required repair.

Consistent with the other cases, however, the liver biopsy was normal and there were none of the usual signs of bone disease that might otherwise explain the increased AP. A muscle biopsy was also unremarkable but MRI showed cortical atrophy. Dysmorphic facial features, observed at 9 years of age in the older sister, included plagiocephaly with hypertelorism and long palpebral fissures, hypoplastic midface with relative prognathism, and small ears (without dysmorphism). The younger sister, at 2 months of age, had arched eyebrows, mild hypertelorism with almond-shaped eyes, and downturned corners of the mouth. Neither the facial features, the other dysmorphic findings, nor the extensive biochemical work-up were felt to be diagnostic, although once again, Coffin-Siris syndrome was suggested, but without conviction (Marcelis et al. 2007). It was thought that the hyperphosphatasia might be due to anticonvulsant therapy, but identification of characteristic (*PIGV*) mutations in these sibs by Krawitz et al. (2010) makes this interpretation much less likely.

In 2010, two groups published their studies of 8 children in 5 families (Horn et al. 2010; Thompson et al. 2010). There were 3 siblings and 2 unrelated singleton cases of German ancestry, one family of two French siblings, and an affected second child of a consanguineous Lebanese couple. All would be considered to have the severe expression of Mabry syndrome described by Kruse et al. (1988) with marked cognitive deficits, seizures that were generally difficult to control, and hyperphosphatasia that varied within and between families. All demonstrated some degree of brachytelephalangy or clinodactyly and a facial dysmorphism.

Additional histologic studies were carried out by Thompson et al. (2010) in an effort to clarify the pathophysiology and confirm some of the findings reported by Mabry et al. (1970). However, maintenance of phenotype in cultured skin fibroblasts has proven to be quite challenging. Storage material, consisting of large refractile cytoplasmic agglomerations could be detected on low-power phase contrast light microscopy, but the accumulation is patchy and unstable over time and with serial passage. On electron microscopy of primary cultured material, membrane-limited granular deposits of low electron density were noted. On light microscopy of the skin obtained at biopsy, it is possible to see Schwann cells with inclusion bodies with granular material around the periphery that stains positive with Periodic-Acid-Schiff (Thompson et al. 2010).

16.4 Initial Genetic Studies

Although it initially seemed reasonable to consider Mabry syndrome as an autosomal recessive disorder with a biochemical phenotype, the small number of cases and the battery of negative laboratory investigations in early cases hampered further pathophysiologic delineation. Marcelis et al. (2007) undertook homozygosity screening using a 10 centimorgan panel of markers and identified a 28-megabase region on chromosome 1 and a 65 megabase region on chromosome 14. No convincing candidate gene was identified.

Krawitz et al. (2010) then applied an exomic search algorithm and identified potential causative mutations using genomic DNA from 8 cases, all previously described. After screening for positives by whole exome sequencing, an identity-by-descent "filter" was applied to the most likely candidates in the three siblings previously described by Horn et al. (2010). This left only two likely mutations—a homozygous single nucleotide polymorphism (SNP) in the SLC9A1 gene and a c.1022C>A SNP in the *PIGV* gene predicting an A341E missense mutation at a site extensively conserved throughout vertebrate evolution.

Given the GPI anchoring of alkaline phosphatase, attention focused on the *PIGV* gene and additional missense mutations were characterized in the remaining five subjects. *PIGV* encodes a mannosyl transferase that constitutes one of more than 18 essential enzymes in the pathway for synthesis and attachment of the GPI anchor to a wide variety of acceptor proteins. These proteins include enzymes (AP and 5′-nucleotidase), as well as cell surface antigens (CD55 and CD59), and receptors (urokinase-type plasminogen receptor) (Murakami et al. 2012).

Apart from opening up a whole new area of biochemical genetic study, the investigations reported by Krawitz et al. (2010) make three key points: (1) that causative *PIGV* mutations are likely to be hypomorphic alleles, not nulls, which are apparently lethal recessive; (2) that the hyperphosphatasia is likely to be only a biomarker for a variety of underlying biochemical disturbances; and (3) that some cases of Mabry will be attributable to mutations in other genes of the GPI-synthetic pathway. This last point was confirmed by the report of *PIGO* mutations in Mabry syndrome subjects, the mutations affecting a downstream step of GPI synthesis that requires ethanolamine phosphate transfer to the nascent glycan link (Krawitz et al. 2012).

16.5 GPI Anchor Biosynthesis—an Outline

The glycosylphosphatidylinositols (GPI) are complex glycolipids synthesized in a wide range of different cell types and anchoring more than 250 proteins (0.5 % of all cell surface proteins) to the plasma membrane (Gamage and Hendrickson 2013; Paulick and Bertozzi 2008). The GPI itself faces outward anchored onto the outer leaflet of the plasma membrane by two long-chain fatty acids linked to a glycerol backbone (Fig. 16.1a). An additional fatty acid linked to the inositol sugar at the 6–O–*myo*inositol position adds a third long chain hydrocarbon and is also embedded in the outer leaflet of the cell membrane. The linker region of the GPI contains carbohydrates and phosphoethanolamine residues, the terminal one of which serves as the point of attachment for the tethered ectoprotein. In the transamidation reaction, a carboxyterminal oligopeptide, recognized by the GPI-specific transpeptidase complex, is cleaved and a new amide bond formed between the terminal phosphoethanolamine of the GPI anchor and the new carboxyterminal amino acid residue, designated omega (ω) (Fig. 16.1b).

Biosynthesis of the GPI anchor itself requires the co-operative action of more than 18 proteins, as summarized and enumerated here (as **{n}**) and in Table 16.2, and depicted in Fig. 16.2. In the first step, a multimeric enzyme complex transfers N-acetylglucosamine (GlcNAc) from its common UDP-GlcNAc substrate to a phosphatidylinositol molecule embedded in the cytoplasmic surface of the endoplasmic reticulum membrane **{1}**. The GlcNAc sugar (in alpha 1→6 linkage) is de-acetylated **{2}.** This is followed by a "flip" of the nascent glucosaminated GPI to the luminal surface (Howard et al. 2014), and the inositol residue is fatty acid acylated **{4}**, generating the third long-chain hydrocarbon anchor embedded in the membrane. Next is the (alpha 1→4) mannosylation by the first of the mannosyltransferases **{5}**, the mannose being supplied by a dolichol phosphate donor.

Another mannosyltransferase, encoded by the *PIGV* gene, catalyzes transfer of an additional mannose in alpha (1→6) linkage **{6}**. Two additional mannosyltransferases add further mannose residues in alpha (1→2) linkage, and

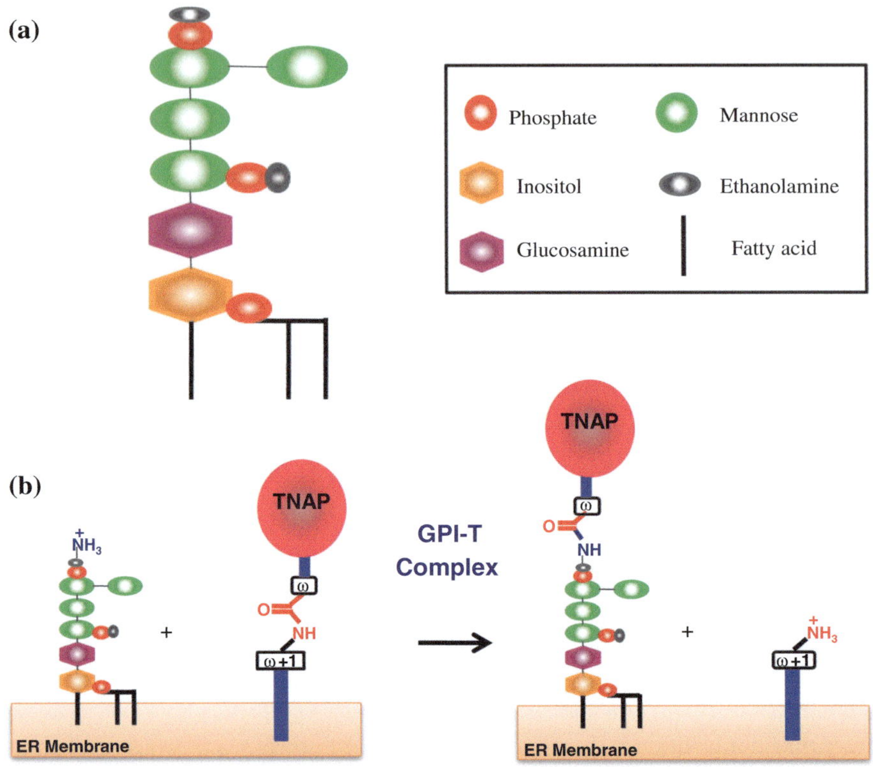

Fig. 16.1 Schematic of the GPI-anchoring step. **a** Structure of the mature GPI anchor, with one fatty acid (*black vertical line*) attached directly to the inositol moiety of phosphatidylinositol and the others attached to the glycerol backbone by alkyl or acyl linkage. **b** Transamidation reaction catalyzed by the GPI-T complex, catalytically supplanting the C-terminal portion of TNAP with a GPI anchor. The new amide bridge links the amine of the terminal phosphoethanolamine residue (attached to the third mannose) with the alpha-carboxyl of the ω amino acid residue in TNAP. The ω + 1 TNAP residue becomes the free amino-terminus of the remnant C-terminal peptide (after Gamage and Hendrickson 2013).

branched alpha (1→2) linkage, respectively **{8, 9}**. A phosphoethanolamine residue, derived from phosphatidyl ethanolamine, is added via 6-O esterification to the terminal alpha (1→2) linked mannose **{10}**. Additional phosphoethanolamines are added to the first and sometimes the second mannose in alpha (1→2) linkage **{7}**. At this point the GPI-anchor is a substrate capable of acting as an intact acceptor for proteins containing the cognate carboxy-terminal transamidation signal motifs. The transamidation reaction itself **{11}** is catalyzed by a complex of five heterologous subunits, encoded by *PIG-K*, *GPAA1*, *PIGS*, *PIGT*, and *PIGU* genes, as reviewed elsewhere (Gamage and Hendrickson 2013).

Once the protein is anchored, the resultant complex undergoes "post-GPI-attachment to proteins" (PGAP) modifications (Maeda and Kinoshita 2011). The fatty acid attached directly to the inositol residue is removed in an

Table 16.2 Genes and proteins in the GPI biosynthesis pathway*

Step	Enzyme	Proteins involved	Gene ID§	Mutant phenotype (OMIM#)
1	GPI-GlcNAc transferase	PIG-A	5277	MCAHS2 (300,868)
		PIG-C	5279	
		PIG-H	5283	
		PIG-Q	9091	
		PIG-Y	84,992	
		DPM2	8818	
2	GlcNAc-PI de-N-acetylase	PIG-L	9487	CHIME syndrome (280,000)
3	Flipase	?		
4	Inositol acyltransferase	PIG-W	284,098	****Mabry**
5	Alpha(1-> 4) mannosyltransferase	PIG-M	93,183	Portal vein thrombosis (610,293)
		PIG-X	54,965	
6	Alpha(1-> 6) mannosyltransferase II	PIG-V	55,650	**Mabry—HPMRS1 (239,300)**
7	EtNP transferase I	PIG-N	23,556	MCAHS1 (614,080)
8	Alpha(1-> 2) mannosyltransferase III	PIG-B	9488	
9	Alpha(1-> 6) mannosyltransferase IV	PIG-Z	80,235	
10	EtNP transferase III	PIG-O	84,720	**Mabry—HPMRS2 (614,749)**
		PIG-F	5281	
11	Transamidase	PIG-K	10,026	
		GPAA1	8733	
		PIG-S	94,005	
		PIG-T	51,604	**MCAHS3 (615,398)**
		PIG-U	128,869	
12	PGAP-1		80,055	
13	PGAP-3		27,315	****Mabry**
14	PGAP-2		93,210	**Mabry—HPMRS3 (614,207)**

*Compiled from Krawitz et al. (2013), Hansen et al. (2013), Maeda and Kinoshita (2011), Gamage and Hendriksen (2013), and Kvarnung et al. (2013). Disorders with abnormalities of alkaline phosphatase metabolism are shown in bold

§ Gene ID and OMIM numbers are tabulated from Genbank

** OMIM numbers to be assigned

enzymatic de-acylation step catalyzed by the PGAP1 protein **{12}**. The phosphoethanolamine attached to the second mannose may also be cleaved at this stage, and the modified molecule trafficked to the Golgi by COPII vesicles. Once in the Golgi, it is subjected to further deacylation by PGAP3, this time removing the polyunsaturated fatty acid at the *sn*2 position of the glycerol backbone, generating a lysoGPI lipid moiety **{13}**. PGAP2, a membrane protein in the Golgi, is required for a final reacylation, largely with stearate, and acquisition of competence to associate with lipid rafts **{14}**. The final secretory step involves transfer of the GPI-anchored TNAP in a lipid raft to the cell surface.

16.6 *PIGV* Mutations—Clinical and Molecular Findings

To date, mutations of *PIGV*, the gene encoding type II GPI-specific mannosyltransferase **{6}**, are the most common found in Mabry syndrome. A series of nine cases (8 families), reviewed in detail by Horn et al. (2014) serves admirably to delineate the range of molecular mutation and the various clinical features of this subgroup. Patients have been of European origin in all but one instance (Pakistani), and were delivered at term to apparently unaffected parents. In all but one instance, missense mutations were identified at conserved sites, and have been shown to generate functional defects in the mannose transfer reaction. In 6 of 9, the mutations were homozygous, but consanguinity was reported in only one.

Where clinical assessment was deemed adequate, cardinal features of Mabry syndrome were present in almost all—that is, global developmental delay (7/7 = 7 of 7 subjects), hyperphosphatasia (9/9), and seizures (8/9). Frequently seen features include: normal neonatal presentation (8/9), normal physical growth in infancy, delayed speech (7/8), facial gestalt (9/9), brachtelephalangy (9/9), and aganglionic megacolon (5/9). Other malformations include anal atresia (3/9), ureteral or renal anomalies (6/9), ventricular septal defect (2/9), cleft palate (2/9). Other anomalies are also noted in singleton cases (Horn et al. 2014). Overall, the age of last assessment is less than 20 years, providing a limited view of long-term natural history. Perhaps screening for hyperphosphatasia in older cohorts with childhood-onset seizures and cognitive deficits will enhance case finding in adulthood and allow better characterization of long-term prognosis.

16.7 *PIGO* and *PGAP2* Mutations—Genetic Heterogeneity

The fact that searches for *PIGV* mutations were negative in about half of the cases diagnosed with Mabry syndrome, and the step-wise nature of GPI-anchor assembly clearly point to the likelihood of genetic heterogenity. With this in mind, Krawitz et al. (2012) used whole-exome sequencing in an English family with affected sisters, and identified compound heterozygosity for mutations of the *PIGO* gene (c.2869C>T predicting p.L957F, and c.2361dup, predicting a downstream frameshift)—a gene encoding the later step of ethanolamine transfer to the glycan linker **{10}**. In an unrelated case—a severely affected 20-month old girl of Caucasian background—another c.2869C>T mutation was accompanied by an intronic c.3069 +5G>A SNP predicted to promote aberrant splicing.

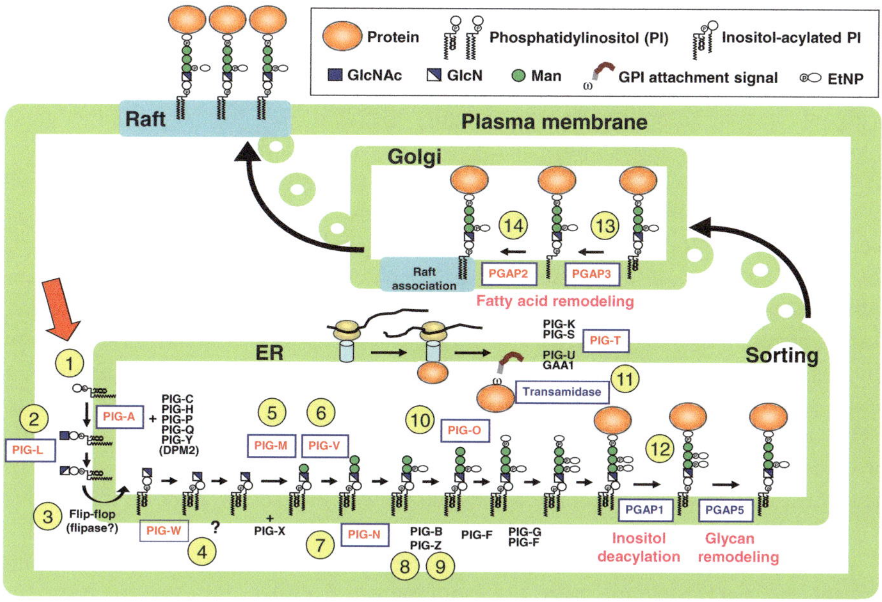

Fig. 16.2 Localization of known biosynthetic defects in the GPI-anchoring pathway. Biosynthetic steps starting with phosphatidylinositol (*large red arrow*) are described in the text and details can be found elsewhere (Maeda and Kinoshita 2011). Numbers correspond to the steps as enumerated in Table 16.2. Steps for which genetic mutations have been clearly linked to an identifiable phenotype (also Table 16.2) are highlighted in bold *red* font (After Maeda and Kinoshita 2011)

To confirm the deleterious nature of the missense mutations, PIGO-deficient Chinese Hamster Ovary (CHO) cells were transiently transfected with wild-type and mutant constructs. Fluorescence-activated cell sorting (FACS) was used to quantitate cell-surface CD59 and endogenous urokinase plasminogen activator receptor (uPAR), both of which are dependent on GPI-anchoring for their expression. Wild-type *PIGO* restored marker levels, but mutants (p.T788Hfs*5 or p.L957F) did not. PIGO content measured by immunoblot analysis was similarly low. In rescued CHO cells, the ratio of cell-derived to secreted (placental) alkaline phosphatase was in the order of ~20 to 1, while it was the opposite proportions in the cell cultures without *PIGO* expression. Similar types of experiments showed that the c.3069 +5G>A mutation leads to excision splicing and loss of the 215bp exon 9.

Clinical findings in the three subjects once again suggest substantial phenotypic variability, but all showed hyperphosphatasia along with psychomotor retardation (delayed walking; delayed speech) and cognitive defects, albeit much more severe in the female infant (not walking, no speech) with compound heterozygosity for a missense and splice mutation. Of note is the one case of marked microcephaly (–5.5 SD) accompanied by other more characteristic features of Mabry syndrome (facial gestalt, brachytelephalangy, megacolon, seizures), and several less frequently recognized anomalies (atrial septal defect, peripheral pulmonary stenosis, left coronal synostosis). As with PIGV, increased severity of the neurologic picture is accompanied by a wider and more severe constellation of major malformations.

Krawitz et al. (2013) carried out further molecular analysis of 13 individuals with intellectual disability and persistently elevated AP, focusing on the genes involved in the GPI biosynthetic pathway. They identified hypomorphic mutations of the *PGAP2* gene **{14}** in a 28-year old Finnish female (compound heterozygous: c.46C>T and c.479C>T predicting p.R16W and p.T160I) and a 41-month old Turkish boy (homozygous: c.380T>C predicting p.L127S). Again, the severity of the disease in one of the cases—here, the eldest boy of three born to consanguineous parents—bears mention. In this patient, there were multiple malformations (megacolon, cleft palate, heart defect, anorectal abnormalies) to accompany the usual features (facial gestalt, seizures, severe developmental delay, hypotonia) and a marked microcephaly (head circumference of –4.5 SD). In PGAP2-deficient CHO cells, rescue of cell-surface expression for GPI-anchored CD59 and CD55 proteins by wild type but not by mutant transfection constituted evidence for causality. The fact that the c.46C>T mutation was less severe may be reflected in the milder patient phenotype.

At the same time, Hansen et al. (2013) describe the identification of overlapping linkage at chr11p15 for two consanguineous families preselected only for non-specific autosomal recessive intellectual disability. Next-generation sequencing and *in silico* analysis were used to identify homozygous mutations in 3 affecteds of Syrian origin (c.296A>G; p.T99C) and in 4 sibs from Pakistan (c.530G>C; p. R177P). Transfection of PGAP2-deficient cells with these variants showed reduced but not absent PGAP2 and partial recovery of cell surface expression, consistent with a hypomorphic phenotype. Hyperphosphatasia was observed in two girls from the first family, but not seizures. In a third child, absence epilepsy was noted, which is consistent with the Mabry syndrome triad, but minor malformations were not detailed. In one child, polygyria and atrophy were noted on MRI, but head circumference in two affected daughters (~ -2.5 SD), not greatly different from the parents (~ -2SD). These results suggest that genetic screening of GPI-anchor genes will reveal mutation carriers who would not have been ascertained as Mabry syndrome on clinical grounds.

16.8 *PIGW* and *PGAP3* Mutations—Extending the Genetic and Clinical Heterogeneity

Most recently, the Krawitz group has identified mutations in a second GPI anchor remodeling gene, *PGAP3* (Howard et al. 2014) **{13}**. Again, the mutations appear to be hypomorphic in nature, although one variant in the compound heterozygous subject from Family B [described before by Thompson et al. (2012)] was a frameshift (c.439dupC). Hyperphosphatasia was seen in all three siblings of one family and the other two index cases. Also present were the marked intellectual deficit, seizures, and the typical facial features. As before, other malformations (microcephaly, cleft palate) were found in some subjects. However, brachytelephalangy was not observed, and the authors suggest that *PGAP3* mutations result in a recognizable subtype of Mabry syndrome. The impaired GPI anchoring was demonstrated by flow cytometry with transfected CHO cells expressing the responsible mutations, and confirmed by PGAP3 protein expression studies.

Finally, Chiyonobu et al. (2014) have described a single case with West syndrome and hyperphosphatasia in which they identified missense mutations of the *PIGW* gene encoding the inositol acyltransferase gene **{4}**. Like the other defects localized to the intermediate steps in GPI anchor biosynthesis, impairment of the inositol acylation step interferes with maturation of the glycan portion and the resultant stimulation of transamidation leads to excessive release of unanchored target protein into the extracellular milieu and causes hyperphosphatasia. If further cases of *PIGW* deficiency demonstrate a high amplitude hypsarrhythmic pattern on interictal EEG, it may be clinically important to define this subtype with the view to delineating whether therapy to correct the associated abnormalities of pyridoxine and gamma aminobutyric acid (GABA) metabolism are amenable to therapy.

16.9 TNAP Abnormalities in GPI-Anchoring Defects

The pathogenesis of hyperphosphatasia in various inherited disorders of GPI-anchor biosynthesis was the subject of a detailed experimental study by Murakami et al. (2012) using knock-out CHO cell lines. They found that transfection of PIGV-deficient CHO cells with the placental alkaline phosphatase (*PLAP*) gene showed very low surface expression and little activity in cell lysates, but greater amounts in cell medium, consistent with enhanced secretion. The secretion, however, requires that the transamidation step cleaves the carboxypeptide tail but does not complete the covalent linkage, so that the mature PLAP is released into the extracellular space with either a defective or absent GPI-anchor.

By contrast, PIGK deficient cells, unable to cleave the propeptide PLAP, showed no secretion at all. Similar studies of other steps in the pathway, suggest that defects in early biosynthetic steps (as far as *PIGM*) will not show excess AP secretion, while those intermediate steps prior to transamidation (*PIGV, PIGB, PIGF*, etc.) will share the hyperphosphatasia phenotype, and the inborn errors of transamidation (e.g. *PIGU*, *PIGK*) may have lowered AP excretion. Consistent with this finding is the presence of hyperphosphatasia in patients with mutations of *PIGO*, as noted above.

The presence of hyperphosphatasia in *PGAP2* requires a somewhat different explanation relating to the fatty acyl moieties maintaining the GPI-anchor in the lipid rafts. In PGAP2-deficient CHO cells, fatty acid transfer is impaired so that the lysoform GPI-anchor accumulates and, although it is transferred to the surface, it may not be incorporated into the lipid rafts, thus predisposing it (and its AP cargo) to premature and/or excessive release from the cell surface (Krawitz et al. 2013).

A more complex picture emerges from an examination of variant CHO cell lines bearing different mutations of the *PGAP3* gene. The p.G92D mutation (family A with three affected sibs) had neglible activity, but protein levels and Golgi localization were normal.

Recently, attention has been drawn to the clinical and biochemical features of an intellectual disability syndrome caused by homozygous mutations in *PIGT*, one of the transamidation genes (Kvarnung et al. 2013) **{11}**. According to the rationale of Murakami et al. (2012), those patients might be expected to manifest a reduction in AP secretion with low circulating levels and perhaps a state resembling *hypo*phosphatasia. Since the *PIGT* mutations are, like other GPI-anchor mutations, hypomorphic, then it might be predicted that a form of mild "secondary hypophosphatasia" might also be present.

However, the investigators found no specific bone abnormalities. Rather, they reported seizures, craniosynostosis, and hypotonia, as well as tooth anomalies and nephrocalcinosis. Apart from one measurement, serum AP in four affected subjects was at or below the stated lower limit of normal in a total of ten separate measurements, while calcium, phosphate and urinary calcium excretion were generally elevated.

Given that these patients had complex, chronic multisystem disease when assessed, the serum AP must be interpreted with caution. Nevertheless, the assay of TNAP-specific metabolites, such as pyridoxal-5′-phosphate or pyrophosphate, would be of considerable interest in *PIGT* mutants. If a state of relative hypophosphatasemia is shown to be part of the PIGT deficiency syndrome, it would become the second rare metabolic disorder to show a secondary hypophosphatasia phenotype (Unger et al. 2002). The other—cleidocranial dysplasia (OMIM #119600—is the result of a mutation in the *RUNX2* gene (OMIM #600211) that regulates TNAP activity (Unger et al. 2002).

16.10 TNAP and Pyridoxine in Hyperphosphatasia

While there is little doubt that most GPI-anchored ectoproteins are affected by disturbances in GPI anchor synthesis, the sentinel finding of elevated circulating TNAP continues to draw considerable attention. This is almost certainly the result of being a well-characterized enzyme activity that is easily and frequently measured in clinical medicine and, in transient form, is a time-tested biomarker for more common diseases. The original studies by Mabry et al. (1970) were very much focused on the enzymatic abnormality. However, we now understand that there is little evidence for a proximal link between the clinical picture and the hyperphosphatasia.

Be that as it may, alkaline phosphatase is a key enzyme for dephosphorylation of circulating pyridoxal-5′-phosphate to pyridoxal, the primary vitamin B_6 derivative that crosses the blood-brain barrier (see Chap. 11). In the absence of enzyme (i.e., primary hypophosphatasia), the seizures observed may respond to vitamin B_6

administration (see Chap. 14). At first glance, then, it is not obvious why pyridoxine might be effective in a state of enzyme excess (*hyper*phosphatasia). In most clinical situations, in fact, enzyme excess is not associated with seizures, despite the decrease in circulating levels of pyridoxal-5′-phosphate (Anderson et al. 1980).

Despite these uncertainties, we assessed the possible benefit of pyridoxine by conducting pyridoxine-challenge electroencephalography in a particularly severe case with very poorly controlled seizures; and what we found was pyridoxine responsiveness of a paradoxical sort (Thompson et al. 2006). In this instance, we had no reason to suspect a deficiency of pyridoxal transport if we considered that the TNAP form of AP found in brain parenchyma may also be elevated. If this kind of increased enzyme activity were intracellular (Murakami et al. 2012), it could contribute to the neurologic deficit and the predisposition to seizures. It would also explain why this form of hyperphosphatasia differs from the much more common conditions that result in increased secretion or release from affected AP-expressing tissues such as bone, liver, and kidney, but no neurologic component. Thus, either elevation or excessive mislocalization of neuronal AP, resulting in focal or more widespread CNS deficiency of intracellular pyridoxal-5′-phosphate, might also be contributing to B_6-responsive seizures in the face of decreased circulating pyridoxal-5-phosphate.

Recent evidence for an expanded role of neuronal AP in primate brain development (see Chap. 5) would be in keeping with the notion that AP is important in vitamin B_6 homeostasis, which in turn modulates GABA synthesis and governs GABAergic neural function, independently of its secretion by liver or bone (Brun-Heath et al. 2011). In this context, it is interesting to note that domoic acid–induced electroencephalographic changes accompanied by GABA depletion are as effectively attenuated by pyridoxine as by valproate or other antiepileptics in an experimental mouse model (Dakshinamurti et al. 2003), consistent with a wider (and perhaps less specific) neuroprotective effect of B vitamins (Rabie et al. 2010).

In our case (Thompson et al. 2006), we have no data to explain why pyridoxine loading in our patient might cause a paradoxical response. Nor can we exclude the possibility that dietary insufficiency of vitamin B_6 may have been an additional predisposing factor in this case. Finally, there remains the question of whether genetic heterogeneity may also play a role in determining whether only some individuals affected with this syndrome might manifest pyridoxine responsiveness but not others.

In contrast to these findings, Heales et al. (2013) looked carefully at pyridoxine and monoamine metabolism in a more conventional case of increased AP. They found a marked lowering of CSF pyridoxal-5′-phosphate but normal CSF concentrations of 5-hydroxyindole acetic acid (5-HIAA) and homovanillic acid—both biomarkers of monoamine status. Although they were able to measure AP in the CSF, the enzyme activity was unstable. Furthermore, there were no data to indicate whether the seizures were responsive to B_6 loading.

A different experience was recently reported by Kuki et al. (2013) in a patient with compound heterozygous mutations in the *PIGO* gene and experimental evidence of defective PIGO expression and GPI anchor synthesis. Their patient, a

9-year-old boy, had the non-specific features of Mabry syndrome (developmental delay, seizures, deafness, facial dysmophism, cardiac defect, megacolon, muscular hypotonia) but also showed persistent hyperphosphatasemia (1200–5960 IU/L), and the characteristic brachytelephalangy. A week after starting a daily regimen of 20 mg/kg body weight pyridoxine, seizures ceased and an EEG showed reduced abnormal activity. Moreover, interruption of the B_6 treatment was associated with recurrent epileptiform activity. Whether this is related in some way to the hyperphosphatasia remains to be seen, but Kuki et al. (2013) point, as we did (Thompson et al. 2006), to the lessons learned with hypophosphatasia. Perhaps the pyridoxal-5′-phosphate PLP deficiency in brain is highly localized in time or space to areas where normal GPI-anchored brain TNAP enzyme activity is required for normal brain growth and development.

16.11 Conclusions

As with hypophosphatasia in the last century, hyperphosphatasia in this century has morphed from an isolated family with a rare but distinctive metabolic finding into an active area of study that involves dozens of investigators in countries around the world. It seems likely that questions of etiology (for the many patients without a molecular diagnosis), and pathogenesis (particularly the role of other defectively anchored proteins) will eventually be answered. The evidence for a beneficial effect of pyridoxine therapy in some, but clearly not all, subjects will undoubtedly encourage clinicians to look carefully at their patients for similar responses, and to turn their attention to other potentially useful therapeutic modalities. If there is one prediction to be made, it is this: There are many more exciting findings to come our way in this fascinating area of TNAP biology.

Acknowledgements We thank Drs. A.Y. Shuen and A. Gozdzik for critical review of the manuscript, and we acknowledge the enthusiastic support proffered by Professors M.P. Whyte and M Michael Cohen Jr. for the very first studies of this condition two decades ago. We also thank the affected families and numerous colleagues around the world for their various contributions.

Conflict of Interest The authors declare no conflict of interest.

References

Anderson BB, O'Brien H, Griffin GE, Mollin DL (1980) Hydrolysis of pyridoxal-5′-phosphate in plasma in conditions with raised alkaline phosphate. Gut 21(3):192–194

Brun-Heath I, Ermonval M, Chabrol E, Xiao J, Palkovits M, Lyck R, Miller F, Couraud PO, Mornet E, Fonta C (2011) Differential expression of the bone and the liver tissue non-specific alkaline phosphatase isoforms in brain tissues. Cell Tissue Res 343(3):521–536. doi:10.1007/s00441-010-1111-4

Chiyonobu T, Inoue N, Morimoto M, Kinoshita T, Murakami Y (2014) Glycosylphosphatidylinositol (GPI) anchor deficiency caused by mutations in PIGW is associated with West syndrome and hyperphosphatasia with mental retardation syndrome. J Med Genet 51(3):203–207. doi:10.1136/jmedgenet-2013-102156

Cole D, Whyte M (1997) Hyperphosphatasia syndromes. In: Cohen M Jr, Baum B (eds) Studies in stomatology and craniofacial biology. IOS Press, Amsterdam, pp 245–272

Dakshinamurti K, Sharma SK, Geiger JD (2003) Neuroprotective actions of pyridoxine. Biochim Biophys Acta 1647(1–2):225–229

Gamage DG, Hendrickson TL (2013) GPI transamidase and GPI anchored proteins: oncogenes and biomarkers for cancer. Crit Rev Biochem Mol Biol 48(5):446–464. doi:10.3109/10409238.2013.831024

Gillessen-Kaesbach G, Meinecke P (1999) Brachytelephalangy, hyperphosphatasia and mental retardation: a newly recognized autosomal recessive condition. Med Genet 11 (117)

Gomes WJ, Hunter JL (1970) Mental retardation, cataracts, and unexplained hyperphosphatasia. Arch Dis Child 45(243):726–727

Hansen L, Tawamie H, Murakami Y, Mang Y, ur Rehman S, Buchert R, Schaffer S, Muhammad S, Bak M, Nothen MM, Bennett EP, Maeda Y, Aigner M, Reis A, Kinoshita T, Tommerup N, Baig SM, Abou Jamra R (2013) Hypomorphic mutations in PGAP2, encoding a GPI-anchor-remodeling protein, cause autosomal-recessive intellectual disability. Am J Hum Genet 92(4):575–583. doi:10.1016/j.ajhg.2013.03.008

Heales S, Neergheen V, Pope S, Oppenheim M, Dadhra J, Eltze C, Kurian M, Gissen P, Mills P, Clayton P (2013) P-518: Hyperphosphatasia associated with markedly low CSF pyridoxal phosphate (PLP) but no impairment of monoamine metabolism [2013 12th International Congress of Inborn Errors of Metabolism (ICIEM) abstracts]. J Inherit Metab Dis 36(2):S248. doi:10.1007/s10545-013-9633-z

Horn D, Krawitz P, Mannhardt A, Korenke GC, Meinecke P (2011) Hyperphosphatasia-mental retardation syndrome due to PIGV mutations: expanded clinical spectrum. Am J Med Genet Part A 155A(8):1917–1922. doi:10.1002/ajmg.a.34102

Horn D, Schottmann G, Meinecke P (2010) Hyperphosphatasia with mental retardation, brachytelephalangy, and a distinct facial gestalt: delineation of a recognizable syndrome. Eur J Med Genet 53(2):85–88. doi:10.1016/j.ejmg.2010.01.002

Horn D, Wieczorek D, Metcalfe K, Baric I, Palezac L, Cuk M, Petkovic Ramadza D, Kruger U, Demuth S, Heinritz W, Linden T, Koenig J, Robinson PN, Krawitz P (2014) Delineation of PIGV mutation spectrum and associated phenotypes in hyperphosphatasia with mental retardation syndrome. Eur J Hum Genet 22(6):762–767. doi:10.1038/ejhg.2013.241

Howard MF, Murakami Y, Pagnamenta AT, Daumer-Haas C, Fischer B, Hecht J, Keays DA, Knight SJ, Kolsch U, Kruger U, Leiz S, Maeda Y, Mitchell D, Mundlos S, Phillips JA 3rd, Robinson PN, Kini U, Taylor JC, Horn D, Kinoshita T, Krawitz PM (2014) Mutations in PGAP3 impair GPI-anchor maturation, causing a subtype of hyperphosphatasia with mental retardation. Am J Hum Genet 94(2):278–287. doi:10.1016/j.ajhg.2013.12.012

Krawitz PM, Murakami Y, Hecht J, Kruger U, Holder SE, Mortier GR, Delle Chiaie B, De Baere E, Thompson MD, Roscioli T, Kielbasa S, Kinoshita T, Mundlos S, Robinson PN, Horn D (2012) Mutations in PIGO, a member of the GPI-anchor-synthesis pathway, cause hyperphosphatasia with mental retardation. Am J Hum Genet 91(1):146–151. doi:10.1016/j.ajhg.2012.05.004

Krawitz PM, Murakami Y, Riess A, Hietala M, Kruger U, Zhu N, Kinoshita T, Mundlos S, Hecht J, Robinson PN, Horn D (2013) PGAP2 mutations, affecting the GPI-anchor-synthesis pathway, cause hyperphosphatasia with mental retardation syndrome. Am J Hum Genet 92(4):584–589. doi:10.1016/j.ajhg.2013.03.011

Krawitz PM, Schweiger MR, Rodelsperger C, Marcelis C, Kolsch U, Meisel C, Stephani F, Kinoshita T, Murakami Y, Bauer S, Isau M, Fischer A, Dahl A, Kerick M, Hecht J, Kohler S, Jager M, Grunhagen J, de Condor BJ, Doelken S, Brunner HG, Meinecke P, Passarge E, Thompson MD, Cole DE, Horn D, Roscioli T, Mundlos S, Robinson PN (2010) Identity-by-descent filtering of exome sequence data identifies PIGV mutations in hyperphosphatasia mental retardation syndrome. Nat Genet 42(10):827–829. doi:10.1038/ng.653

Kruse K, Hanefeld F, Kohlschutter A, Rosskamp R, Gross-Selbeck G (1988) Hyperphosphatasia with mental retardation. J Pediatr 112(3):436–439

Kuki I, Takahashi Y, Okazaki S, Kawawaki H, Ehara E, Inoue N, Kinoshita T, Murakami Y (2013) Vitamin B6-responsive epilepsy due to inherited GPI deficiency. Neurology 81(16): 1467–1469. doi:10.1212/WNL.0b013e3182a8411a

Kvarnung M, Nilsson D, Lindstrand A, Korenke GC, Chiang SC, Blennow E, Bergmann M, Stodberg T, Makitie O, Anderlid BM, Bryceson YT, Nordenskjold M, Nordgren A (2013) A novel intellectual disability syndrome caused by GPI anchor deficiency due to homozygous mutations in PIGT. J Med Genet 50(8):521–528. doi:10.1136/jmedgenet-2013-101654

Mabry CC, Bautista A, Kirk RF, Dubilier LD, Braunstein H, Koepke JA (1970) Familial hyperphosphatase with mental retardation, seizures, and neurologic deficits. J Pediatr 77(1): 74–85

Maeda Y, Kinoshita T (2011) Structural remodeling, trafficking and functions of glycosylphosphatidylinositol-anchored proteins. Prog Lipid Res 50(4):411–424. doi:10.1016/j.plipres.2011.05.002

Marcelis CL, Rieu P, Beemer F, Brunner HG (2007) Severe mental retardation, epilepsy, anal anomalies, and distal phalangeal hypoplasia in siblings. Clin Dysmorphol 16(2):73–76. doi:10.1097/MCD.0b013e3280147130

Murakami Y, Kanzawa N, Saito K, Krawitz PM, Mundlos S, Robinson PN, Karadimitris A, Maeda Y, Kinoshita T (2012) Mechanism for release of alkaline phosphatase caused by glycosylphosphatidylinositol deficiency in patients with hyperphosphatasia mental retardation syndrome. J Biol Chem 287(9):6318–6325. doi:10.1074/jbc.M111.331090

Paulick MG, Bertozzi CR (2008) The glycosylphosphatidylinositol anchor: a complex membrane-anchoring structure for proteins. Biochemistry 47(27):6991–7000. doi:10.1021/bi8006324

Rabe P, Haverkamp F, Emons D, Rosskamp R, Zerres K, Passarge E (1991) Syndrome of developmental retardation, facial and skeletal anomalies, and hyperphosphatasia in two sisters: nosology and genetics of the Coffin-Siris syndrome. Am J Med Genet 41(3):350–354. doi:10.1002/ajmg.1320410317

Rabie T, Muhlhofer W, Bruckner T, Schwab A, Bauer AT, Zimmermann M, Bonke D, Marti HH, Schenkel J (2010) Transient protective effect of B-vitamins in experimental epilepsy in the mouse brain. J Mol Neurosci: MN 41(1):74–79. doi:10.1007/s12031-009-9286-4

Schrier SA, Bodurtha JN, Burton B, Chudley AE, Chiong MA, D'Avanzo MG, Lynch SA, Musio A, Nyazov DM, Sanchez-Lara PA, Shalev SA, Deardorff MA (2012) The Coffin-Siris syndrome: a proposed diagnostic approach and assessment of 15 overlapping cases. Am J Med Genet Part A 158A(8):1865–1876. doi:10.1002/ajmg.a.35415

Thompson MD, Killoran A, Percy ME, Nezarati M, Cole DE, Hwang PA (2006) Hyperphosphatasia with neurologic deficit: a pyridoxine-responsive seizure disorder? Pediatr Neurol 34(4):303–307. doi:10.1016/j.pediatrneurol.2005.08.020

Thompson MD, Nezarati MM, Gillessen-Kaesbach G, Meinecke P, Mendoza-Londono R, Mornet E, Brun-Heath I, Squarcioni CP, Legeai-Mallet L, Munnich A, Cole DE (2010) Hyperphosphatasia with seizures, neurologic deficit, and characteristic facial features: Five new patients with Mabry syndrome. Am J Med Genet Part A 152A(7):1661–1669. doi:10.1002/ajmg.a.33438

Thompson MD, Roscioli T, Marcelis C, Nezarati MM, Stolte-Dijkstra I, Sharom FJ, Lu P, Phillips JA, Sweeney E, Robinson PN, Krawitz P, Yntema HG, Andrade DM, Brunner HG, Cole DE (2012) Phenotypic variability in hyperphosphatasia with seizures and neurologic deficit (Mabry syndrome). Am J Med Genet Part A 158A(3):553–558. doi:10.1002/ajmg.a.35202

Unger S, Mornet E, Mundlos S, Blaser S, Cole DE (2002) Severe cleidocranial dysplasia can mimic hypophosphatasia. Eur J Pediatr 161(11):623–626. doi:10.1007/s00431-002-0978-9

Chapter 17
The Role of Tissue Non-specific Alkaline Phosphatase (TNAP) in Neurodegenerative Diseases: Alzheimer's Disease in the Focus

Katherine A.B. Kellett and Nigel M. Hooper

Abstract Tissue non-specific alkaline phosphatase (TNAP) is present on neuronal membranes and induces neuronal toxicity via tau dephosphorylation; a mechanism which could play a role in the neuronal loss seen in Alzheimer's disease (AD). TNAP increases in the plasma following brain injury and cerebrovascular disease. In this chapter we summarise our previous work which looked at changes in TNAP activity in the brain and plasma of AD individuals and discuss whether these changes may be reflective of neuronal loss. Our data demonstrate that TNAP activity is significantly increased in the brain in both the sporadic and familial forms of AD and that TNAP activity is significantly increased in the plasma in AD patients. In addition, we describe a significant inverse correlation between plasma TNAP activity and cognitive function in AD. Using these data we propose a model for TNAP-induced neurodegeneration in AD resulting from tau dephosphorylation following secretion of tau from neuronal cells.

Keywords Tau · Cognitive function · Neuronal toxicity · Plasma · Human brain

17.1 Introduction

Alzheimer's disease (AD is the most common form of dementia and the impact of this neurodegenerative disease is increasing as the population ages (Burns and Iliffe 2009). AD symptoms progress from mild memory problems (mild cognitive impairment; MCI) through to severe cognitive deficits. Studies of ageing and

K.A.B. Kellett (✉) · N.M. Hooper
Institute of Brain, Behaviour and Mental Health, Faculty of Medical and Human Sciences, University of Manchester, Manchester M13 9PT, UK
e-mail: katherine.kellett@manchester.ac.uk

N.M. Hooper
e-mail: nigel.hooper@manchester.ac.uk

C. Fonta and L. Négyessy (eds.), *Neuronal Tissue-Nonspecific Alkaline Phosphatase (TNAP)*, Subcellular Biochemistry 76,
DOI 10.1007/978-94-017-7197-9_17

cognition suggest that impairment in multiple cognitive domains may be observed several years before a clinical diagnosis of AD is made (Matthews et al. 2007). It has been suggested that the observed cognitive dysfunction is not qualitatively different from that seen in normal ageing, proposing the idea that there is a continuum from normal ageing to preclinical dementia (Brayne 2007). AD is characterised by the formation of extracellular amyloid plaques and intracellular neurofibrillary tangles. Amyloid plaques are composed of amyloid-β peptide formed from the sequential cleavage of the amyloid precursor protein (APP) by the β-site APP cleaving enzyme-1 (BACE1) and the ϒ–secretase complex. Neurofibrillary tangles are composed of hyperphosphorylated and aggregated tau (Iqbal et al. 2005) and initially appear in the entorhinal cortex and hippocampus, before the expansion of tau pathology into other neighbouring areas (Braak and Braak 1991).

The spread of tau pathology in the Alzheimer's brain was characterised by Braak and Braak in 1991 and their method of scoring tau pathology is still used for pathological analysis (Braak and Braak 1991). This characteristic spread of tau pathology has led to the idea of tau'propagation' or 'infectivity', and it has been shown that extracellular tau aggregates can enter cells to induce misfolding of intracellular tau (Frost et al. 2009). Different methods have been used to demonstrate tau propagation: the injection of brain extract from mice expressing a mutated form of human tau demonstrated that tau pathology could spread from the injection site to neighbouring regions (Clavaguera et al. 2009); and the expression of a human tau transgene in specific neurons showed the spread of tau from these neurons to neighbouring, non-expressing neurons, inducing degeneration (de Calignon et al. 2012). The propagation of tau pathology is thought to involve a trans-synaptic mechanism, spreading along anatomically connected networks (Liu et al. 2012) and although the mechanism of tau secretion and tau uptake are still to be confirmed, one study has suggested that tau fibril uptake occurs via heparin sulphate proteoglycans (Holmes et al. 2013). Tau pathology is accompanied by neuronal loss, following which, tau can be found in the extracellular space either in a monomeric or an aggregated form where it is assembled into extracellular ghost tangles (Cras et al. 1995).

While the pathology of AD has been extensively studied, the underlying causes of the disease remain elusive. Less than 5 % of AD cases are inherited (Mullan et al. 1992), caused by mutations in the genes encoding APP or the presenilins in the ϒ-secretase complex. The remaining 95 % of cases are termed sporadic AD usually occurring after age 65 (hence often referred to as late-onset AD). The causes of sporadic AD are less clear; although sporadic AD is not caused by dominant mutations like familial AD (Pimplikar 2009), there may be a genetic component to some cases of sporadic AD as indicated in recent years by genome-wide association studies (GWAS). These studies use large cohorts to detect single nucleotide polymorphisms (SNPs) that are related to sporadic AD. A large number of genes have been identified by the various GWAS but nine have consistently been identified in all studies (CLU (clusterin), BIN1 (bridging integrator 1), PICALM (phosphatidylinositol clathrin associated lymphoid myeloid), CD2AP (CD2 associated protein), EPHA1 (ephrin receptor A1), CR1 (complement receptor 1), ABCA7 (ATP binding cassette transporter 7), CD33 (Myeloid cell surface antigen CD33), and the MS4A (Membrane-spanning 4-domains subfamily A) gene cluster) (Harold et al. 2009; Bertram and Tanzi 2010; Seshadri et al. 2010; Carrasquillo et al. 2011; Hollingworth et al. 2011; Hu et al. 2011; Wijsman et al. 2011); it is estimated that these genes may account for up to 50 % of sporadic AD cases (Morgan 2011) although how the proteins that these genes encode are involved in the initiation and/or progression of AD is still unclear.

Tissue non-specific alkaline phosphatase (TNAP) is widely expressed in the brain including in the occipital-, frontal- and temporal lobe areas of the cerebral cortex (Negyessy et al. 2011) (see also Chaps. 5 and 6) and also in the hippocampus (Diez-Zaera et al. 2011; Street et al. 2013) where tau pathology initially appears. The high TNAP expression in specific brain areas suggests links to the thalamo-cortical connections (Fonta et al. 2004; Negyessy et al. 2011) which have been shown to be affected in AD (Zhou et al. 2013). TNAP has been shown to have a role in both brain development and brain function (see Chaps. 4, 14 and 18), but does it also have a role in neurodegenerative disease? Our studies, published in 2011 (Kellett et al. 2011; Vardy et al. 2012) and reviewed here, examine the changes in TNAP activity that occur as a result of AD.

17.2 Changes in CSF and Plasma TNAP Associated with CNS Injury; Could Changes in Plasma TNAP Reflect Neuronal Loss in the Brain?

Changes in cerebrospinal fluid (CSF) (Lampl et al. 1990) and plasma alkaline phosphatase activity (Yamashita et al. 1989; Meythaler et al. 1998) occur as a result of central nervous system injury, including non-traumatic brain injury (Meythaler et al. 1998). The changes in TNAP in brain-damaged patients was shown to increase as secondary brain damage developed and plasma TNAP concentration was shown to correlate with the functional outcome (Yamashita et al. 1989). These previous studies suggested that changes in plasma TNAP as a result of brain injury were an indirect consequence due to liver function changes; but is this an incorrect assumption based on the misconception that TNAP primarily reflects liver function? Could the increase in TNAP in the plasma and CSF as a result of brain injury, traumatic or non-traumatic, be reflective of neuronal loss? The correlation of plasma TNAP with functional outcome would suggest that it might. Previous to our studies on TNAP in AD, elevated alkaline phosphatase activity had been reported in patients with neurodegenerative disease. A study contrasting AD with vascular dementia (VD) indicated that alkaline phosphatase activity was significantly increased in VD compared to AD but did not, unfortunately, study these groups in relation to age-matched controls or correlate alkaline phosphatase activity with cognitive function (Cacabelos et al. 2004).

17.3 TNAP Activity Is Increased in the Brain and in the Plasma in Alzheimer's Disease

Our studies investigated the changes in TNAP activity associated with AD in both post-mortem brain samples and in plasma samples from AD patients. In both post-mortem brain samples and in plasma we identified the isoform of alkaline phosphatase being measured as TNAP by using the TNAP-specific inhibitor, levamisole.

Our study (Vardy et al. 2012) demonstrated a significant increase in TNAP activity in post-mortem brain samples from the hippocampus of sporadic AD patients (Fig. 17.1a). TNAP has been shown to be present in the hippocampus (Diez-Zaera et al. 2011; Street et al. 2013) and the increase in TNAP activity in the AD patients compared to non-demented controls suggests a role for TNAP in the progression of AD. In addition to analysing TNAP activity in samples from sporadic AD patients, we also examined post-mortem hippocampal brain samples from familial AD patients. As in sporadic AD, TNAP activity was significantly increased in the hippocampus of familial AD patients compared to age-matched controls

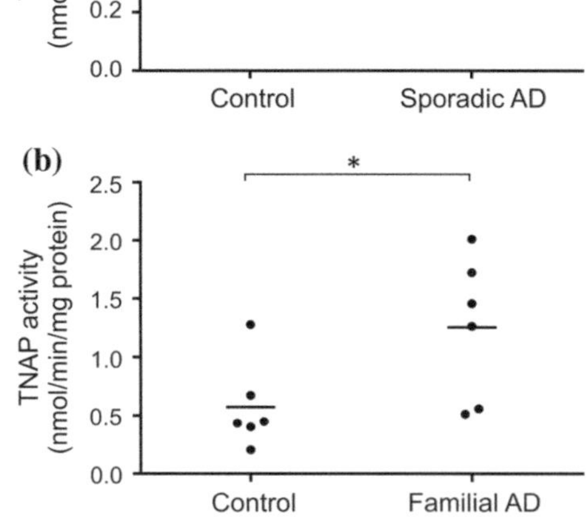

Fig. 17.1 TNAP activity is increased in the hippocampus in AD. TNAP activity in hippocampal brain samples from **a** sporadic AD compared to age-matched controls and **b** from familial AD compared to age-matched controls. Data shown as grouped scatter plot with bar representing the mean. Sporadic AD: n = 5 per group; Familial AD: n = 6 per group. *$p < 0.05$. (Vardy et al. 2012)

(Fig. 17.1b). As the major risk factor for sporadic AD is age we also investigated changes in TNAP activity in samples from an ageing cohort; the results showed that TNAP activity does not correlate with age ($r = 0.063$, $p = 0.853$). The increase in TNAP in both sporadic and familial AD demonstrate that the changes seen are likely to be a consequence of AD-associated changes in the brain rather than a primary mechanism of disease. Our data supported previous work showing that TNAP activity is increased in temporal lobe brain samples in AD (Diaz-Hernandez et al. 2010), demonstrating that TNAP activity is also increased in the hippocampus, both in the sporadic and familial forms of the disease.

In addition to measuring TNAP activity in the brain we also designed 2 independent studies (Kellett et al. 2011; Vardy et al. 2012) with separate cohorts to investigate whether TNAP activity was also altered in the plasma in AD patients. For both of these studies the two main aims were to examine plasma TNAP and to correlate this with cognitive function. Our Leeds cohort consisted of 110 'probable' AD patients, in accordance with international diagnostic criteria (National Institute of Neurological and Communicative Disorders and Stroke-Alzheimer's Disease and Related Disorders Association work group) (McKhann et al. 1984) and 110 age- and gender-matched control subjects without evidence of cognitive impairment assessed using the Mini-Mental State Examination (MMSE) (Folstein et al. 1975). Our OPTIMA (Oxford Project To Investigate Memory and Aging) cohort consisted of 121 'probable' (McKhann et al. 1984) or pathologically confirmed AD patients, 89 mild-cognitively impaired (MCI) patients and 180 cognitively-screened non-demented controls assessed using the Cambridge Examination for Mental Disorders (CAMCOG) (Roth et al. 1986). In both of these cohorts we demonstrated that TNAP activity is significantly increased in the plasma of AD patients compared to controls and that these changes are not a result of other underlying conditions, such as liver, bone or inflammatory diseases (Fig. 17.2a, b). In addition, the inclusion of MCI patients in the study (Kellett et al. 2011) identified that this change in TNAP is evident in the early phase of cognitive impairment with raised TNAP levels in MCI patients compared to controls but at an intermediate level to the increase observed in the AD group (Fig. 17.2b). Plasma TNAP therefore appears to be related to clinical diagnosis and, we therefore predicted, could be related to cognitive function. Using the results from cognitive examinations in both cohorts we were able to correlate TNAP activity in plasma with cognitive function in all subjects, including controls. Interestingly, the results showed that TNAP activity significantly inversely correlated with cognitive function in AD patients from our Leeds cohort ($r_s = -0.211$, $p = 0.027$) and significantly inversely correlated with cognitive function in all groups in our OPTIMA cohort (AD ($z = -3.61$, $p = 0.0003$); MCI ($z = -2.49$, $p = 0.013$); Control ($z = -2.21$, $p = 0.027$)) suggesting that changes in plasma TNAP activity may reflect changes in cognition in any individual. Our final analysis looked at follow-up over 5 years in our OPTIMA cohort. The results of this analysis determined that TNAP activity remained

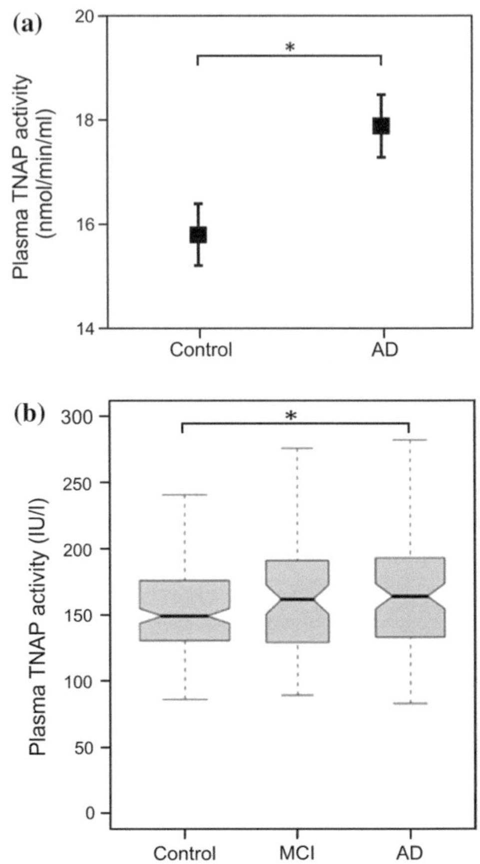

Fig. 17.2 TNAP activity in plasma is significantly increased in AD. **a** In our Leeds cohort TNAP activity was significantly increased in AD compared to age-matched controls (mean ± SEM: 17.88 ± 0.603 and 15.97 ± 0.596 nmol/min/ml, respectively, p = 0.018); data represented in an error bar graph showing adjusted mean ± SEM, n = 110 per group (Vardy et al. 2012). **b** In our OPTIMA cohort, TNAP activity was significantly increase in AD compared to control (median (*lower quartile*, *upper quartile*): 165.5 (139.5, 195.8) and 149.5 (130.0, 178.0), respectively, p = 0.005). TNAP activity in MCI patients was at a level in between the control and AD groups (164.0 (129.0, 196.0)) but was not significantly different from either the control or AD group. Data represented in a box and whisker plot showing data from initial patient assessment, control n = 180, MCI n = 89, AD n = 121 (Kellett et al. 2011)

significantly inversely correlated to cognition over all assessments during the 5 year follow-up in AD patients. The analysis showed that the rate of decline is not dependent on TNAP but that changes in TNAP reflect changes in cognitive function in any given individual.

17.4 TNAP as a Marker of AD?

While TNAP activity is increased in the plasma and correlates with cognitive function in all groups (control, MCI and AD), our observations in post-mortem brain samples indicate that changes in TNAP are not correlated with age and are therefore not a result of the normal ageing process. These changes in the brain are a consequence of the disease process, reflective, therefore, of AD-associated changes. However, the correlation of plasma TNAP activity with cognitive function in all groups would suggest that changes in the plasma may also be reflective of a process occurring in all subjects irrespective of diagnosis. What is unclear is whether the increase in plasma TNAP activity is reflective of increased brain TNAP or whether there is a related systemic change in TNAP activity. While there is an increase in plasma TNAP activity in AD the mean values in the AD and control cohorts remained in the clinically 'normal' range (10-140 IU/l); in contrast, in traumatic brain injury, plasma alkaline phosphatase activity has been shown to increase above 200 IU/l (Yamashita et al. 1989). From these data it could therefore be concluded that TNAP could reflect brain injury and perhaps neuronal loss, with acute increases following traumatic brain injury and long-term low-level changes in AD. Unfortunately, however, as plasma TNAP activity remains within the 'normal' range it could not be used as a predictive or diagnostic biomarker for AD. There is, however, scope to further investigate TNAP as a marker for AD progression and/or effectiveness of therapeutic intervention within an individual patient, or perhaps, to further investigate plasma TNAP measurement in combination with other plasma proteins that have been shown to correlate with cognitive function for a combined AD marker.

17.5 TNAP and Tau in AD

TNAP has been shown to have a specific role in the neurotoxic events causing AD as a result of its action on hyperphosphorylated tau (See Chap. 18). TNAP can dephosphorylate the extracellular hyperphosphorylated tau that surrounds the neurons (Fig. 17.3). This dephosphorylated tau is an agonist for M1 and M3 muscarinic receptors and its action causes a sustained calcium influx into the cell which is neurotoxic and ultimately results in neuronal loss (Gomez-Ramos et al. 2008; Diaz-Hernandez et al. 2010). This mechanism linking TNAP activity to tau dephosphorylation and the resulting neuronal loss provides evidence of a direct role for TNAP in the progression of AD. The neurotoxic effects resulting from tau dephosphorylated by TNAP suggests that TNAP activity in the brain may therefore be reflective of neuronal loss. Interestingly, it was also noted that activation of muscarinic receptors by dephosphorylated tau increases TNAP expression (Diaz-Hernandez et al. 2010) (Fig. 17.3). This suggests a positive feedback loop to drive tau dephosphorylation by increased TNAP expression, possibly driven by the

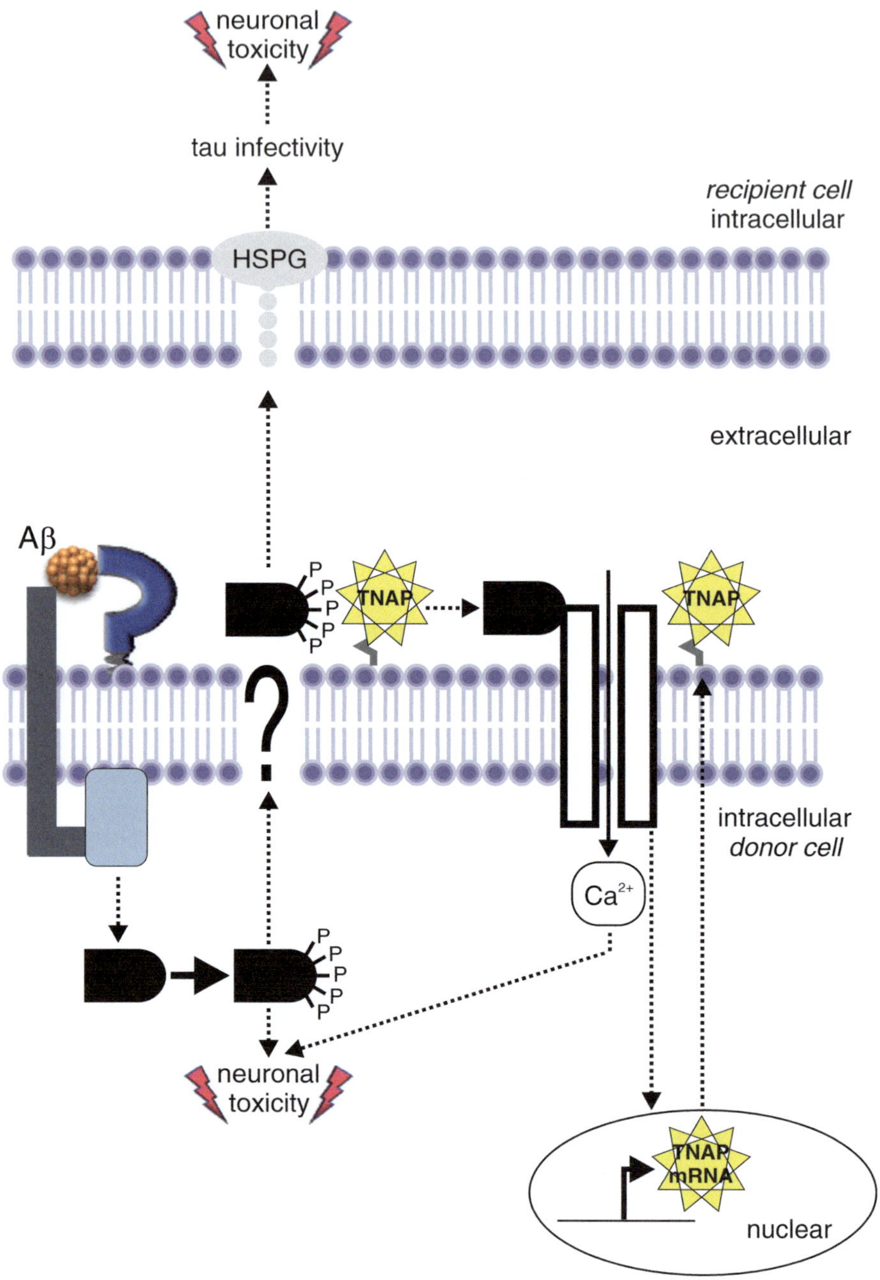
neuronal toxicity
tau infectivity
recipient cell
intracellular
HSPG
extracellular
Aβ
P
P
P
P
P
TNAP
TNAP
?
intracellular
donor cell
Ca2+
P
P
P
P
P
neuronal toxicity
TNAP mRNA
nuclear

◀ **Fig. 17.3** The role of TNAP in AD. Schematic diagram to show how toxicity may be mediated via TNAP in AD through tau dephosphorylation. Aβ binds to a receptor on the cell surface and initiates a signalling cascade resulting in tau hyperphosphorylation. The hyperphosphorylated tau can itself cause toxicity resulting in its release into the extracellular space following neuronal death. Alternatively, the hyperphosphorylated tau can be secreted from the donor cell via an, as yet, unknown mechanism. Following secretion, the tau can be taken up into neighbouring neurons, potentially via heparin sulphate proteoglycans (HSPG) resulting in tau infectivity in the recipient cell. Alternatively, the secreted hyperphosphorylated tau can be dephosphorylated by TNAP and the resulting dephosphorylated tau can act on M1 and M3 muscarinic receptors causing calcium influx and neuronal toxicity. The activation of the M1 and M3 muscarinic receptors also induces increased expression of TNAP thereby exacerbating its effects on extracellular hyperphosphorylated tau

increased formation of hyperphosphorylated tau seen in AD. This increase in tau dephosphorylation, however, promotes neurotoxicity via its action on muscarinic receptors. Together, the evidence from our work (Kellett et al. 2011) and others (Diaz-Hernandez et al. 2010) that TNAP activity is increased in the brain regions affected in AD supports a role for TNAP in AD, suggesting that with AD progression, brain TNAP activity increases alongside tau levels, causing an increase in neuronal loss and a decline in cognitive function.

17.6 Conclusions

In conclusion, our work has demonstrated an increase in brain and plasma TNAP activity in AD and has shown that this change in TNAP activity inversely correlates with cognitive function. A direct link between TNAP activity and tau provides a mechanism for increased neurotoxicity and neuronal loss in the AD brain which would correspond to the resulting decline in cognition seen in AD and inversely correlated to TNAP activity. Taken together these results suggest that changes in TNAP activity may reflect neuronal loss in AD.

Conflict of Interest The authors declare that the research was conducted in the absence of any commercial or financial relationships that could be construed as a potential conflict of interest.

References

Bertram L, Tanzi RE (2010) Alzheimer disease: new light on an old CLU. Nat Rev Neurol 6 (1):11–13

Braak H, Braak E (1991) Neuropathological stageing of Alzheimer-related changes. Acta Neuropathol 82(4):239–259

Brayne C (2007) The elephant in the room—healthy brains in later life, epidemiology and public health. Nat Rev Neurosci 8(3):233–239

Burns A, Iliffe S (2009) Alzheimer's disease. BMJ 338:b158

Cacabelos R, Fernandez-Novoa L, Corzo L, Amado L, Pichel V, Lombardi V, Kubota Y (2004) Phenotypic profiles and functional genomics in Alzheimer's disease and in dementia with a vascular component. Neurol Res 26(5):459–480

Carrasquillo MM, Belbin O, Hunter TA, Ma L, Bisceglio GD, Zou F, Crook JE, Pankratz VS, Sando SB, Aasly JO, Barcikowska M, Wszolek ZK, Dickson DW, Graff-Radford NR, Petersen RC, Morgan K, Younkin SG (2011) Replication of BIN1 association with Alzheimer's disease and evaluation of genetic interactions. JAD 24(4):751–758

Clavaguera F, Bolmont T, Crowther RA, Abramowski D, Frank S, Probst A, Fraser G, Stalder AK, Beibel M, Staufenbiel M, Jucker M, Goedert M, Tolnay M (2009) Transmission and spreading of tauopathy in transgenic mouse brain. Nat Cell Biol 11(7):909–913

Cras P, Smith MA, Richey PL, Siedlak SL, Mulvihill P, Perry G (1995) Extracellular neurofibrillary tangles reflect neuronal loss and provide further evidence of extensive protein cross-linking in Alzheimer disease. Acta Neuropathol 89(4):291–295

de Calignon A, Polydoro M, Suarez-Calvet M, William C, Adamowicz DH, Kopeikina KJ, Pitstick R, Sahara N, Ashe KH, Carlson GA, Spires-Jones TL, Hyman BT (2012) Propagation of tau pathology in a model of early Alzheimer's disease. Neuron 73(4):685–697

Diaz-Hernandez M, Gomez-Ramos A, Rubio A, Gomez-Villafuertes R, Naranjo JR, Miras-Portugal MT, Avila J (2010) Tissue-nonspecific alkaline phosphatase promotes the neurotoxicity effect of extracellular tau. J Biol Chem 285(42):32539–32548

Diez-Zaera M, Diaz-Hernandez JI, Hernandez-Alvarez E, Zimmermann H, Diaz-Hernandez M, Miras-Portugal MT (2011) Tissue-nonspecific alkaline phosphatase promotes axonal growth of hippocampal neurons. Mol Biol Cell 22(7):1014–1024

Folstein MF, Folstein SE, McHugh PR (1975) Mini-mental state. a practical method for grading the cognitive state of patients for the clinician. J Psychiatr Res 12(3):189–198

Fonta C, Negyessy L, Renaud L, Barone P (2004) Areal and subcellular localization of the ubiquitous alkaline phosphatase in the primate cerebral cortex: evidence for a role in neurotransmission. Cereb Cortex 14(6):595–609

Frost B, Jacks RL, Diamond MI (2009) Propagation of tau misfolding from the outside to the inside of a cell. J Biol Chem 284(19):12845–12852

Gomez-Ramos A, Diaz-Hernandez M, Rubio A, Miras-Portugal MT, Avila J (2008) Extracellular tau promotes intracellular calcium increase through M1 and M3 muscarinic receptors in neuronal cells. Mol Cell Neurosci 37(4):673–681

Harold D, Abraham R, Hollingworth P, Sims R, Gerrish A, Hamshere ML, Pahwa JS, Moskvina V, Dowzell K, Williams A, Jones N, Thomas C, Stretton A, Morgan AR, Lovestone S, Powell J, Proitsi P, Lupton MK, Brayne C, Rubinsztein DC, Gill M, Lawlor B, Lynch A, Morgan K, Brown KS, Passmore PA, Craig D, McGuinness B, Todd S, Holmes C, Mann D, Smith AD, Love S, Kehoe PG, Hardy J, Mead S, Fox N, Rossor M, Collinge J, Maier W, Jessen F, Schurmann B, van den Bussche H, Heuser I, Kornhuber J, Wiltfang J, Dichgans M, Frolich L, Hampel H, Hull M, Rujescu D, Goate AM, Kauwe JS, Cruchaga C, Nowotny P, Morris JC, Mayo K, Sleegers K, Bettens K, Engelborghs S, De Deyn PP, Van Broeckhoven C, Livingston G, Bass NJ, Gurling H, McQuillin A, Gwilliam R, Deloukas P, Al-Chalabi A, Shaw CE, Tsolaki M, Singleton AB, Guerreiro R, Muhleisen TW, Nothen MM, Moebus S, Jockel KH, Klopp N, Wichmann HE, Carrasquillo MM, Pankratz VS, Younkin SG, Holmans PA, O'Donovan M, Owen MJ, Williams J (2009) Genome-wide association study identifies variants at CLU and PICALM associated with Alzheimer's disease. Nat Genet 41(10):1088–1093

Hollingworth P, Harold D, Sims R, Gerrish A, Lambert JC, Carrasquillo MM, Abraham R, Hamshere ML, Pahwa JS, Moskvina V, Dowzell K, Jones N, Stretton A, Thomas C, Richards A, Ivanov D, Widdowson C, Chapman J, Lovestone S, Powell J, Proitsi P, Lupton MK, Brayne C, Rubinsztein DC, Gill M, Lawlor B, Lynch A, Brown KS, Passmore PA, Craig D, McGuinness B, Todd S, Holmes C, Mann D, Smith AD, Beaumont H, Warden D, Wilcock G, Love S, Kehoe PG, Hooper NM, Vardy ER, Hardy J, Mead S, Fox NC, Rossor M, Collinge J, Maier W, Jessen F, Ruther E, Schurmann B, Heun R, Kolsch H, van den Bussche H, Heuser I, Kornhuber J, Wiltfang J, Dichgans M, Frolich L,

Hampel H, Gallacher J, Hull M, Rujescu D, Giegling I, Goate AM, Kauwe JS, Cruchaga C, Nowotny P, Morris JC, Mayo K, Sleegers K, Bettens K, Engelborghs S, De Deyn PP, Van Broeckhoven C, Livingston G, Bass NJ, Gurling H, McQuillin A, Gwilliam R, Deloukas P, Al-Chalabi A, Shaw CE, Tsolaki M, Singleton AB, Guerreiro R, Muhleisen TW, Nothen MM, Moebus S, Jockel KH, Klopp N, Wichmann HE, Pankratz VS, Sando SB, Aasly JO, Barcikowska M, Wszolek ZK, Dickson DW, Graff-Radford NR, Petersen RC, van Duijn CM, Breteler MM, Ikram MA, DeStefano AL, Fitzpatrick AL, Lopez O, Launer LJ, Seshadri S, Berr C, Campion D, Epelbaum J, Dartigues JF, Tzourio C, Alperovitch A, Lathrop M, Feulner TM, Friedrich P, Riehle C, Krawczak M, Schreiber S, Mayhaus M, Nicolhaus S, Wagenpfeil S, Steinberg S, Stefansson H, Stefansson K, Snaedal J, Bjornsson S, Jonsson PV, Chouraki V, Genier-Boley B, Hiltunen M, Soininen H, Combarros O, Zelenika D, Delepine M, Bullido MJ, Pasquier F, Mateo I, Frank-Garcia A, Porcellini E, Hanon O, Coto E, Alvarez V, Bosco P, Siciliano G, Mancuso M, Panza F, Solfrizzi V, Nacmias B, Sorbi S, Bossu P, Piccardi P, Arosio B, Annoni G, Seripa D, Pilotto A, Scarpini E, Galimberti D, Brice A, Hannequin D, Licastro F, Jones L, Holmans PA, Jonsson T, Riemenschneider M, Morgan K, Younkin SG, Owen MJ, O'Donovan M, Amouyel P, Williams J (2011) Common variants at ABCA7, MS4A6A/MS4A4E, EPHA1, CD33 and CD2AP are associated with Alzheimer's disease. Nat Genet 43(5):429–435

Holmes BB, Devos SL, Kfoury N, Li M, Jacks R, Yanamandra K, Ouidja MO, Brodsky FM, Marasa J, Bagchi DP, Kotzbauer PT, Miller TM, Papy-Garcia D, Diamond MI (2013) Heparan sulfate proteoglycans mediate internalization and propagation of specific proteopathic seeds. PNAS 110(33):E3138–E3147

Hu X, Pickering EH, Hall SK, Naik S, Liu YC, Soares H, Katz E, Paciga SA, Liu W, Aisen PS, Bales KR, Samad TA,John SL (2011) Genome-wide association study identifies multiple novel loci associated with disease progression in subjects with mild cognitive impairment. Transl Psychiatry 1:e54

Iqbal K, Alonso Adel C, Chen S, Chohan MO, El-Akkad E, Gong CX, Khatoon S, Li B, Liu F, Rahman A, Tanimukai H, Grundke-Iqbal I (2005) Tau pathology in Alzheimer disease and other tauopathies. Biochim Biophys Acta 1739(2–3):198–210

Kellett KA, Williams J, Vardy ER, Smith AD, Hooper NM (2011) Plasma alkaline phosphatase is elevated in Alzheimer's disease and inversely correlates with cognitive function. Int J Mol Epidemiol Genet 2(2):114–121

Lampl Y, Paniri Y, Eshel Y, Sarova-Pinchas I (1990) Alkaline phosphatase level in CSF in various brain tumors and pulmonary carcinomatous meningitis. J Neurooncol 9(1):35–40

Liu L, Drouet V, Wu JW, Witter MP, Small SA, Clelland C, Duff K (2012) Trans-synaptic spread of tau pathology in vivo. PLoS ONE 7(2):e31302

Matthews FE, McKeith I, Bond J, Brayne C (2007) Reaching the population with dementia drugs: what are the challenges? Int J Geriatr Psychiatry 22(7):627–631

McKhann G, Drachman D, Folstein M, Katzman R, Price D, Stadlan EM (1984) Clinical diagnosis of Alzheimer's disease: report of the NINCDS-ADRDA work group under the auspices of department of health and human services task force on Alzheimer's disease. Neurology 34(7):939–944

Meythaler JM, Hazlewood J, DeVivo MJ, Rosner M (1998) Elevated liver enzymes after nontraumatic intracranial hemorrhages. Arch Phys Med Rehabil 79(7):766–771

Morgan K (2011) The three new pathways leading to Alzheimer's disease. Neuropathol Appl Neurobiol 37(4):353–357

Mullan M, Crawford F, Axelman K, Houlden H, Lilius L, Winblad B, Lannfelt L (1992) A pathogenic mutation for probable Alzheimer's disease in the APP gene at the N-terminus of beta-amyloid. Nat Genet 1(5):345–347

Negyessy L, Xiao J, Kantor O, Kovacs GG, Palkovits M, Doczi TP, Renaud L, Baksa G, Glasz T, Ashaber M, Barone P, Fonta C (2011) Layer-specific activity of tissue non-specific alkaline phosphatase in the human neocortex. Neuroscience 172:406–418

Pimplikar SW (2009) Reassessing the amyloid cascade hypothesis of Alzheimer's disease. Int J Biochem Cell Biol 41(6):1261–1268

Roth M, Tym E, Mountjoy CQ, Huppert FA, Hendrie H, Verma S, Goddard R (1986) CAMDEX. A standardised instrument for the diagnosis of mental disorder in the elderly with special reference to the early detection of dementia. British J Psych 149:698–709

Seshadri S, Fitzpatrick AL, Ikram MA, DeStefano AL, Gudnason V, Boada M, Bis JC, Smith AV, Carassquillo MM, Lambert JC, Harold D, Schrijvers EM, Ramirez-Lorca R, Debette S, Longstreth WT Jr, Janssens AC, Pankratz VS, Dartigues JF, Hollingworth P, Aspelund T, Hernandez I, Beiser A, Kuller LH, Koudstaal PJ, Dickson DW, Tzourio C, Abraham R, Antunez C, Du Y, Rotter JI, Aulchenko YS, Harris TB, Petersen RC, Berr C, Owen MJ, Lopez-Arrieta J, Varadarajan BN, Becker JT, Rivadeneira F, Nalls MA, Graff-Radford NR, Campion D, Auerbach S, Rice K, Hofman A, Jonsson PV, Schmidt H, Lathrop M, Mosley TH, Au R, Psaty BM, Uitterlinden AG, Farrer LA, Lumley T, Ruiz A, Williams J, Amouyel P, Younkin SG, Wolf PA, Launer LJ, Lopez OL, van Duijn CM, Breteler MM (2010) Genome-wide analysis of genetic loci associated with Alzheimer disease. JAMA 303(18): 1832–1840

Street SE, Kramer NJ, Walsh PL, Taylor-Blake B, Yadav MC, King IF, Vihko P, Wightman RM, Millan JL, Zylka MJ (2013) Tissue-Nonspecific alkaline phosphatase acts redundantly with PAP and NT5E to generate adenosine in the dorsal spinal cord. J Neurosci 33(27): 11314–11322

Vardy ER, Kellett KA, Cocklin SL, Hooper NM (2012) Alkaline phosphatase is increased in both brain and plasma in Alzheimer's disease. Neurodegener Dis 9(1):31–37

Wijsman EM, Pankratz ND, Choi Y, Rothstein JH, Faber KM, Cheng R, Lee JH, Bird TD, Bennett DA, Diaz-Arrastia R, Goate AM, Farlow M, Ghetti B, Sweet RA, Foroud TM, Mayeux R (2011) Genome-wide association of familial late-onset Alzheimer's disease replicates BIN1 and CLU and nominates CUGBP2 in interaction with APOE. PLoS Genet 7(2):e1001308

Yamashita M, Sasaki M, Mii K, Tsuzuki M, Takakura K, Yoshinoya S, Ohkubo A (1989) Measurement of serum alkaline phosphatase isozyme I in brain-damaged patients. Neurol Med Chir (Tokyo) 29(11):995–998

Zhou B, Liu Y, Zhang Z, An N, Yao H, Wang P, Wang L, Zhang X, Jiang T (2013) Impaired functional connectivity of the thalamus in alzheimer's disease and mild cognitive impairment: a resting-state FMRI study. Curr Alzheimer Res 10(7):754–766

Chapter 18
TNAP Plays a Key Role in Neural Differentiation as well as in Neurodegenerative Disorders

Miguel Diaz-Hernandez, Felix Hernandez, Maria Teresa Miras-Portugal and Jesús Avila

Abstract New evidences have been reported that point to the ecto-enzyme, tissue-nonspecific alkaline phosphatase (TNAP), as a key element at the origin of two opposite phenomena, neuronal differentiation and neuronal degeneration. During brain development, TNAP plays an essential role for establishing neuronal circuits. The pro-neuritic effect induced by TNAP, which results in axonal length increase, is due to its enzymatic hydrolysis of extracellular ATP at the surrounding area of the axonal growth cone. In this way, the activation of P2X7 receptor is prevented and as a consequence there is no inhibition of axonal growth. The existence of a close functional interrelation between both purinergic elements is finally supported by the fact that both elements may control, in a reciprocal way, the expression level of the other. On the opposite stage, recent evidences indicate that TNAP plays a key role in spreading the neurotoxicity effect induced by extracellular hyperphosphorylated tau protein, the main component of intracellular neurofibrillary tangles present in the brain of Alzheimer disease patients. TNAP exhibits a broad substrate specificity and in addition to nucleotides it is able to dephosphorylate extracellular proteins, such as the hyperphosphorylated tau protein once it

M. Diaz-Hernandez (✉) · M.T. Miras-Portugal
Departamento de Bioquímica y Biología Molecular, Facultad de Veterinaria, Universidad Complutense de Madrid, 28040 Madrid, Spain
e-mail: migueldiaz@ucm.es

M.T. Miras-Portugal
e-mail: mtmiras@ucm.es

M. Diaz-Hernandez · M.T. Miras-Portugal
Instituto de Investigación Sanitaria San Carlos IDISSC, Madrid, Spain

F. Hernandez · J. Avila
Centro de Biología Molecular "Severo Ochoa", CSIC-UAM, 28049 Madrid, Spain
e-mail: fhernandez@cbm.csic.es

J. Avila
e-mail: javila@cbm.csic.es

F. Hernandez · J. Avila
Centro de Investigación Biomédica en Red Sobre Enfermedades Neurodegenerativas (CIBERNED), Madrid, Spain

C. Fonta and L. Négyessy (eds.), *Neuronal Tissue-Nonspecific Alkaline Phosphatase (TNAP)*, Subcellular Biochemistry 76,
DOI 10.1007/978-94-017-7197-9_18

is released to the extracellular medium. Dephosphorylated tau protein behaves as an agonist of muscarinic M1 and M3 receptors, provoking a robust and sustained intracellular calcium increase that finally triggering neuronal death. Besides, activation of muscarinic receptors by dephosphorylated tau increases the expression of TNAP, which could explain the increase in TNAP activity and protein levels detected in Alzheimer disease.

Keywords Muscarinic receptor · Tau protein · Axonal growth · Purinergic receptor · Alzheimer disease

18.1 Involvement of TNAP in Brain Development

One of the most important events that take place during embryonic development of brain is the formation and growth of neuronal axons. This process allows neurons to establish their precise connectivity patterns and organize the future neuronal circuits that are essential for the proper functionality of the mature brain.

The axon formation and growth is a process regulated by different extracellular signals, such as neurotrophic factors, neurotransmitters and other signaling molecules. These signals mainly control the axonal growth and branching through the control of polymerization and depolymerization of microtubules and microfilaments (Bradke and Dotti 1999; da Silva and Dotti 2002) Although many of these signals have been widely studied, little is known about the mechanisms by which neurons modulate or integrate all the signals impinging upon them for getting a correct axonal growth. On the first part of this chapter the evidences suggesting that TNAP plays a key role on axonal growth will be exposed.

18.2 TNAP Regulates the Axonal Growth by Controlling the Purinergic Signaling

Purines act as neurotransmitters and modulators in the central and peripheral nervous system, but also extracellular purines can exert a direct trophic influence on the development and maintenance of the nervous system (Rathbone et al. 1999). In fact, it has been reported that ATP is able to reduce neurite extension from motoneuron-containing neural tube explants of rat embryos at embryonic day (E) 12 (Cheung et al. 2005). ATP behaves as neurotransmitter in the central nervous system being able to activate ionotropic P2X receptors or (Burnstock 2007; Edwards et al. 1992; Evans et al. 1992) metabotropic P2Y receptors (Abbracchio et al. 2006). The ionotropic P2X receptors are ligand-gated cationic channels whose activation transiently increases the intracellular Ca^{2+} concentrations by permitting the influx of this ion independently of voltage-dependent Ca^{2+}-channel activation

(Khakh 2001). It's widely known that, while the Ca^{2+} influx or a transient increase on the concentration of this divalent cation inside growth cones negatively regulates the rate of axon outgrowth, however the reduction of the intracellular Ca^{2+} levels on this localization accelerates the axon extension (Gomez and Spitzer 1999). Recently, it has been reported that ATP is able to induce intracellular calcium increase on the axonal growth cone, provoking the generation of a Ca^{2+} wave that remains at the distal part of the axon and does not reach the somatodendritic compartment. These focal calcium influxes also change the growth cone morphology, going from lamellipodial to filopodial morphology. P2X7 receptor was identified as the one responsible for the above mentioned phenomena. Additionally, it was also checked that when hippocampal neurons were cultured in presence of specific antagonist of the P2X7 receptor, axons had a remarkable growth in length (Diaz-Hernandez et al. 2008).

Other interesting fact that was observed with cultured hippocampal neurons was the reduction on extracellular ATP levels that takes place during the first days on culture, precisely when neurons are acquiring their specific morphologies and a neurite that emerge from cell body becomes the axon. It was observed that low levels of extracellular ATP correlates with a significant increase on TNAP levels, especially in the axonal growth cone (Fig. 18.1a, b). In fact, it was observed that the pharmacological inhibition of TNAP prevented the decline of extracellular ATP levels in the culture medium. Moreover, pharmacological inhibition of TNAP or its specific knockdown with short hairpin RNA interference inhibited the growth and branching of principal axons (Fig. 18.1c, d). However, neither activation nor inhibition of adenosine receptors influenced the axonal growth, excluding, in this way, the contribution of adenosine, which is the main product obtained from the extracellular hydrolysis of ATP, on the TNAP-mediated effects. Immunocytochemical studies performed with specific antibodies confirmed that TNAP was mainly expressed at axonal growth cones, in a similar way to that observed for the P2X7 receptor, suggesting the existence of a close functional interrelation between both proteins (Fig. 18.1b). Whereby, TNAP is able to induce axonal growth by hydrolyzing the ATP in the immediate environment of the receptor, preventing in this way the activation of P2X7 receptor (Fig. 18.1d–f). A strong relationship between both proteins was unveiled when it was demonstrated that the inhibition of P2X7 receptor reduced the TNAP expression and, the addition of exogenous TNAP enhanced the P2X7 receptor expression (Diez-Zaera et al. 2011) (see also Chap. 4).

18.3 TNAP Contribution to Embryonic Development

During the early embryonic development, TNAP is strongly expressed in neuronal tube, especially on a subpopulation of neuroectodermal cells, the migrating priomordial germ cells (Chiquoine 1954). In mice, these cells move from the epithelium of hind gut until genital ridges around embryonic day 8 (E8) (Godin et al. 1990).

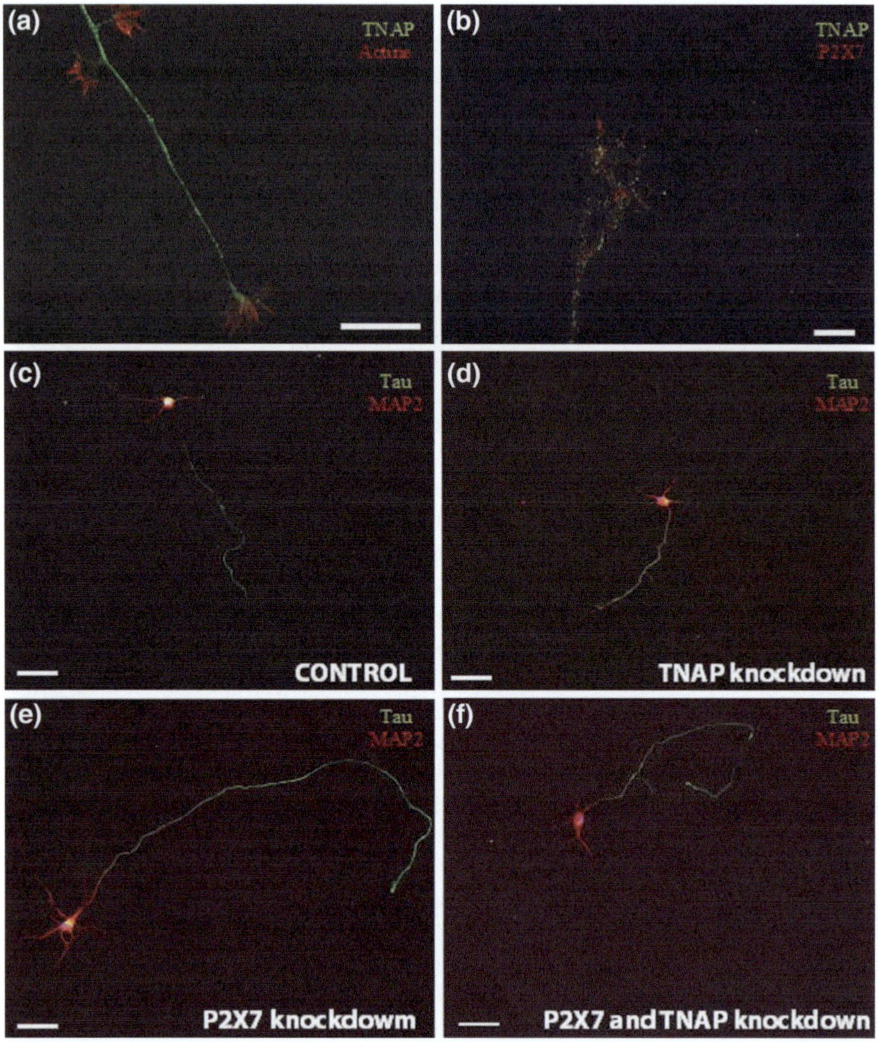

Fig. 18.1 The pair consisting of TNAP ectoenzyme ectoenzyme and the P2X7 receptor controls the axonal growth. **a**, **b** Localization of TNAP in cultured hippocampal neurons. Cultured hippocampal neurons double-stained with antibodies against TNAP (*green*) and phalloidin, which specifically binds to F-actin (*red*) (**a**), and TNAP (*green*) and P2X7 (*red*) (**b**). **c–f** Double immune-fluorescent labeling of hippocampal neurons transfected at 1 day in vitro (DIV) and fixed at 3 DIV. Neurons were stained with antibodies against Tau (*green*) and microtubule-associated protein2 (MAP2) (*red*). Representative fluorescence images of neurons transfected with a plasmid encoding EGFP (**c**), or with shRNA TNAP, a short hairpin RNA interference that specifically knocks down the expression of TNAP (**d**), or with shRNA P2X7, a short hairpin RNA interference that specifically knocks down the expression of P2X7 receptor (**e**), or sh RNA P2X7 plus sh RNA TNAP (**f**). *Scale bars* **a** and **b** 25 μm. *Scale bar* **c–f** 50 μm

In addition, TNAP activity has also been found on pioneer growth cones of nerve cells coming from cranial nerves and dorsal roots. Taking into account the control that TNAP carries over the axonal growth cone, it appears reasonable to postulate that this enzyme is involved on the migration of immature nerve cells emerging from mesencephalon and spinal cord (Narisawa et al. 1994). However, it is necessary to bear in mind that initially the modulatory role of purinergic receptors on axonal growth was unknown. Thus, the first explanation for the involvement of this enzyme on the migratory and elongation processes was suggested to be due to protein-protein interaction domain at its extracellular region (residues 400–430) that allows it to bind to proteins from extracellular matrix (Bossi et al. 1993).

Another explanation arose for the last states of mouse embryonic development, around day E14, when a strong activity of TNAP has been also found in cells from the ventricular and subventricular zones, but its expression in the outer layers is reduced (Langer et al. 2007) (see also Chap. 4). These findings highlighted that these regions have an increased proliferative activity either during development or in adult brain (Altman and Bayer 1990). Perhaps the TNAP expression on membrane of progenitor cells may be responding to the necessity of this kind of cells have to possess an efficient mechanism to control the nucleotide availability in the close environment of purinergic receptors that are expressed in these cells (Langer et al. 2007). In good agreement with this hypothesis it has been reported that the Ca^{2+} waves propagating through radial glial cells in the proliferative cortical ventricular zone are mediated by P2Y1 receptor activation. The fact that the disruption of calcium waves among the embryonic neuronal progenitors induces a strong reduction of the cell proliferation on ventricular zone (Weissman et al. 2004) suggest that control of the purinergic signals that arrive to these progenitor cells along the embryonic development is essential for a correct brain development.

18.4 Role of TNAP in the Synaptic Function

Once established the synaptic contact, TNAP continuous playing an important role on the synaptic functionality, largely due to the fact that is implicated in several phenomena related to the synaptic plasticity, some of which will be commented below.

Opposite to its expression pattern during the embryonic development, in adults TNAP was also found in the synaptic cleft of cortical connections of primate and human (Fonta et al. 2004; Negyessy et al. 2011). Despite, it still remains unknown which are the functions that the ecto-enzyme plays in this localization, it has been related with the metabolism or hydrolysis of several important compounds for the synaptic functionality in the CNS. So, it has been described that TNAP regulates the concentration of pirodoxal 5′-phosphate (PLP) (Fedde and Whyte 1990), a cofactor of GAD65 (Martin and Barke 1998) essential for the GABA synthesis. Since the GABA is one of the main inhibitory neurotransmitters in the CNS, an alteration in its synthesis could induce an unbalance between excitatory and inhibitory impulses in the cortex.

One of the most important functions of TNAP in the CNS is the extracellular hydrolysis of nucleotides whereby it finishes the nucleotide extracellular signaling. However, an important derived consequence from its activity is the generation of extracellular adenosine, the final product from nucleotide hydrolysis (Zimmermann et al. 2012). Adenosine also behaves as a neurotransmitter in the CNS and is involved on several physiological events through the activation of their own receptors (Burnstock et al. 2011). In addition, adenosine is also capable to modulate the neurotransmission mediated by nucleotides. In this way, it has been reported that in nerve endings from rat midbrain, the adenosine generated by TNAP from nucleotide hydrolysis is enough to activate the A1 adenosine receptors, thus increasing the affinity and response of the presynaptic ionotropic receptors sensitive to diadenosine polyphosphates expressed on those nerve endings (Diaz-Hernandez et al. 2000, 2002a). Considering that the activation of these ionotropic receptors can trigger the vesicular release of several neurotransmitters in the CNS, such as ACh, Glutamate or GABA, (Diaz-Hernandez et al. 2002b; Gualix et al. 2003; Gomez-Villafuertes et al. 2003), anything that regulates the activity of these receptors will be significant for synaptic function.

Finally, it is important to take into account that TNAP has broad substrate specificity, being able to dephosphorylate extracellular proteins (see Chaps. 9 and 10). In this way it has been recently described its capability to dephosphorylate the hyperphosphorylated tau protein. Tau proteins are preferably located in the intracellular compartment, associated to microtubules, as one of its main functions is to modulate and stabilize the axonal microtubules. However, in pathological conditions, this protein can be hyperphosphorylated by different enzymes leading to its self-aggregation that finally results in the formation of intracellular tangles by association of helical filaments of hyperphosphorylated tau protein. These aggregates compromise cell viability and after rupture of plasma membrane, these tangles reach the interstitial space where they go through a slowly disassembly, then releasing the monomeric hyperphosphorylated tau, that will be dephosphorylated by TNAP in the extracellular milieu (Diaz-Hernandez et al. 2010).

18.5 Role of TNAP on Alzheirmer Disease Progression

Alzheimer disease (AD) is a disabling neurodegenerative disease affecting around 30 million persons on the whole world. It is expected that by 2050, the percentage of affected patients will reach 1 on 80 humans (Brookmeyer et al. 2007). The characteristic symptoms of this disorder appear at the adult age. Initially these symptoms are characterized by a difficulty to remember newly learned information, but with aging, these symptoms evolve to a total memory loss, behavioral changes, and finally driving to the death of patient. At neuropathologycal level, the disease is characterized by a deep reduction of cortical and hippocampal volume as well as by the presence of two aberrant structures, the extracellular amyloid plaques formed by the amyloid beta peptide, generated by the processing of amyloid precursor protein

(APP), and second, the intracellular neurofibrillary tangles (NFTs), that are mainly built by hyperphosphorylated tau protein (Grundke-Iqbal et al. 1986).

Along the AD progression, the pathology associated with tau protein follows a reproducible pattern, the hyperphosphorylated tau appearing first in the entorhinal cortex and hippocampus, and from there spreading to surrounding areas (Braak and Braak 1991). Since this pathological process is associated with neuronal death, after the initial rupture of cytoplasmic membrane, the intracellular components, including the Tau protein, can reach the interstitial space, either in its soluble or aggregated form coming from intracellular tangles. Taking into account this stage it was suggested that extracellular tau could be contributing to the spreading of the disease. This hypothesis has been confirmed demonstrating that extracellular tau was able to induce a toxic effect on hippocampal cultured neurons. In addition, it was also tested that this toxicity varies according to the degree of protein aggregation, the monomeric tau protein being the most toxic form. The mechanism underling this toxic effect has been linked with an increase in intracellular calcium levels that unbalances the intracellular calcium homeostasis finally inducing the cell death (Gomez-Ramos et al. 2006). Similar mechanisms has been also described in other neurodegenerative disorders associated with other proteins, such as the α-synuclein (Danzer et al. 2007) or prion protein (Demuro et al. 2005). In case of the tau protein it was reported that its toxic effect was mediated by the muscarinic receptor. In fact, it was found that specific antagonists of either muscarinic receptors subtypes M1 and M3 are able to avoid the toxic effect induced by tau protein receptor activation (Gomez-Ramos et al. 2008). Accordingly, the muscarinic receptor binding region was identified in the C-terminal region of the tau protein between the residues from 391 to 407 (Gomez-Ramos et al. 2008).

Once the receptor target of extracellular tau protein identified the question raised was, why tau and no other native muscarinic receptors agonist, like acetylcholine (ACh), is able to induce a neurotoxic effect through the muscarinic receptor activation? To clarify this issue, several kinetic parameters related with the response profile and properties of muscarinic receptors after being activated by tau or ACh were analyzed. The results revealed that the muscarinic receptors have an affinity for tau protein around one order of magnitude higher than for ACh. Furthermore, while the repeated stimulation of muscarinic receptor with ACh induced a desensitization state, the reiterated stimulation by tau did not induce this phenomenon (Gomez-Ramos et al. 2009). In addition, tau protein is much more stable than ACh in the extracellular milieu (Gomez-Ramos et al. 2009), allowing the protein agonist to reach more distant localizations and so contributing to the spreading of cellular destruction in AD. Thus, tau behaves as an agonist. All these data provide valuable information to understand why tau and not ACh is able to induce toxicity on neural cells bearing M1 or M3 muscarinic receptors.

In opposite way to the agonist role of the soluble tau monomer mentioned above, the existence of an inverse correlation between the number of NFTs and the number of living neurons in the hippocampus has been widely described (Fukutani et al. 1995; Cras et al. 1995; Bondareff et al. 1989), suggesting that intracellular hyperphophorylated tau is also playing a key role in the progression of the disease.

But, are both events related? Recently TNAP has been described as a key element that can link both events. Thus, it has been found that extracellular tau could be dephosphorylated by TNAP and that tau behaves as a muscarinic receptors agonist, whose activation causes a robust and sustained intracellular calcium increase that finally triggers the neuronal death (Diaz-Hernandez et al. 2010). In addition, it was also reported that muscarinic receptors activation by the dephosphorylated tau also increases both TNAP expression and phosphorylation levels of intracellular tau. To sum up, several intracellular pathways will be activated that finally result in a robust and sustained increase in intracellular calcium levels as well as an increase in phosphorylation levels of intracellular tau and TNAP expression (Gomez-Ramos et al. 2006; Diaz-Hernandez et al. 2010). Taking into account all these data we can postulate the next scenario which is summarized in Fig. 18.2.

After an initial undetermined injury, or even, as it has been recently suggested, through its own vesicular release (Simon et al. 2012), the tau protein, once dephosphorylated by TNAP, can reach the extracellular space where it can activate muscarinic receptors. As a consequence the intracellular levels of hyper-phosphorylated tau are increased, and eventually will result in the formation of neurofibrillary tangles. In addition, TNAP expression levels will be also increased, which will cause a more efficient de-phosphorylation of extracellular tau, closing in this way a positive feedback loop whose final consequence will be to induce a constant muscarinic receptor activation. These events will disturb cellular homeostasis that will

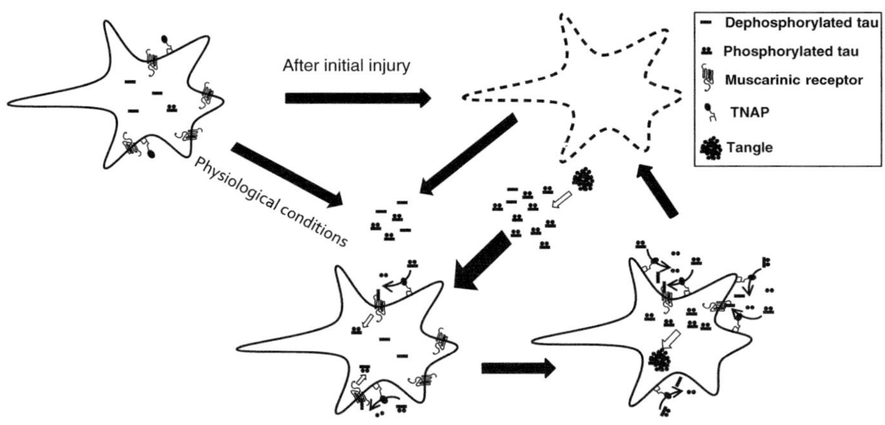

Fig. 18.2 Proposed model that the involvement of TNAP in the progression of Alzheimer disease. As a consequence of an initial undetermined injury or under physiological conditions, tau protein reaches the extracellular space where it is dephosphorylated by TNAP and then activates muscarinic receptors. The activation of muscarinic receptors by tau protein have several consequences: (1) increase of the phosphorylation level of intracellular tau that can trigger the formation of intracellular tangles, (2) increase of TNAP expression levels that subsequently can give rise to cellular death. Cell death also increases the extracellular tau levels on a longer time scale, mainly due the slow disassembly of the extracellular neurofibrillary tangles, one of the main sources of the extracellular tau. All these events allow tau protein to reach distant brain areas, contributing, on this way, to the spreading of disease. *Open arrows* assembling/disassembling Tau

compromise cell viability giving rise among others to the subsequent rupture of plasma membrane and the release of intracellular content to interstitial space. In addition, extracellular levels of tau proteins will be increased due to the formation of new intracellular tangles induced by muscarinic receptor activation, prior the cell death. Once at the extracellular media, the neurofibrillary tangles suffer a slow disassembly and degradation, allowing the proteins to reach distant brain areas and thus contributing to the spreading of this aberrant cycle of stimulation, cell death and release to new surrounding areas (Diaz-Hernandez et al. 2010).

In good agreement with this hypothesis, in a recent preclinical assay performed on more than 100 AD patients it has been reported that TNAP activity is significantly increased in the hippocampus of the AD patients compared with the age-matched controls, independently whether they were diagnosed with sporadic or familial AD. In the same study it was also noticed that TNAP enzymatic activity was increased in the blood of AD patients (Vardy et al. 2012), postulating in this way the possible role of TNAP as a biological marker of the disease progression (see Chap. 17).

Conflict of Interest All authors declare no conflict of interest on this work.

References

Abbracchio MP, Burnstock G, Boeynaems JM, Barnard EA, Boyer JL, Kennedy C, Knight GE, Fumagalli M, Gachet C, Jacobson KA, Weisman GA (2006) International Union of Pharmacology LVIII: update on the P2Y G protein-coupled nucleotide receptors: from molecular mechanisms and pathophysiology to therapy. Pharmacol Rev 58(3):281–341. doi:10.1124/pr.58.3.3

Altman J, Bayer SA (1990) Horizontal compartmentation in the germinal matrices and intermediate zone of the embryonic rat cerebral cortex. Exp Neurol 107(1):36–47

Bondareff W, Mountjoy CQ, Roth M, Hauser DL (1989) Neurofibrillary degeneration and neuronal loss in Alzheimer's disease. Neurobiol Aging 10(6):709–715

Bossi M, Hoylaerts MF, Millan JL (1993) Modifications in a flexible surface loop modulate the isozyme-specific properties of mammalian alkaline phosphatases. J Biol Chem 268(34): 25409–25416

Braak H, Braak E (1991) Neuropathological stageing of Alzheimer-related changes. Acta Neuropathol 82(4):239–259

Bradke F, Dotti CG (1999) The role of local actin instability in axon formation. Science 283(5409):1931–1934

Brookmeyer R, Johnson E, Ziegler-Graham K, Arrighi HM (2007) Forecasting the global burden of Alzheimer's disease. Alzheimers Dement 3(3):186–191. doi:10.1016/j.jalz.2007.04.381

Burnstock G (2007) Physiology and pathophysiology of purinergic neurotransmission. Physiol Rev 87(2):659–797. doi:10.1152/physrev.00043.2006

Burnstock G, Fredholm BB, Verkhratsky A (2011) Adenosine and ATP receptors in the brain. Curr Top Med Chem 11(8):973–1011

Cras P, Smith MA, Richey PL, Siedlak SL, Mulvihill P, Perry G (1995) Extracellular neurofibrillary tangles reflect neuronal loss and provide further evidence of extensive protein cross-linking in Alzheimer disease. Acta Neuropathol 89(4):291–295

Cheung KK, Chan WY, Burnstock G (2005) Expression of P2X purinoceptors during rat brain development and their inhibitory role on motor axon outgrowth in neural tube explant cultures. Neuroscience 133(4):937–945. doi:10.1016/j.neuroscience.2005.03.032

Chiquoine AD (1954) The identification, origin, and migration of the primordial germ cells in the mouse embryo. Anat Rec 118(2):135–146

da Silva JS, Dotti CG (2002) Breaking the neuronal sphere: regulation of the actin cytoskeleton in neuritogenesis. Nat Rev Neurosci 3(9):694–704. doi:10.1038/nrn918

Danzer KM, Haasen D, Karow AR, Moussaud S, Habeck M, Giese A, Kretzschmar H, Hengerer B, Kostka M (2007) Different species of alpha-synuclein oligomers induce calcium influx and seeding. J Neurosci 27(34):9220–9232. doi:10.1523/JNEUROSCI.2617-07.2007

Demuro A, Mina E, Kayed R, Milton SC, Parker I, Glabe CG (2005) Calcium dysregulation and membrane disruption as a ubiquitous neurotoxic mechanism of soluble amyloid oligomers. J Biol Chem 280(17):17294–17300. doi:10.1074/jbc.M500997200

Diaz-Hernandez M, del Puerto A, Diaz-Hernandez JI, Diez-Zaera M, Lucas JJ, Garrido JJ, Miras-Portugal MT (2008) Inhibition of the ATP-gated P2X7 receptor promotes axonal growth and branching in cultured hippocampal neurons. J Cell Sci 121(Pt 22):3717–3728. doi:10.1242/jcs.034082

Diaz-Hernandez M, Gomez-Ramos A, Rubio A, Gomez-Villafuertes R, Naranjo JR, Miras-Portugal MT, Avila J (2010) Tissue-nonspecific alkaline phosphatase promotes the neurotoxicity effect of extracellular tau. J Biol Chem 285(42):32539–32548. doi:10.1074/jbc.M110.145003

Diaz-Hernandez M, Pereira MF, Pintor J, Cunha RA, Ribeiro JA, Miras-Portugal MT (2002a) Modulation of the rat hippocampal dinucleotide receptor by adenosine receptor activation. J Pharmacol Exp Ther 301(2):441–450

Diaz-Hernandez M, Pintor J, Castro E, Miras-Portugal MT (2002b) Co-localisation of functional nicotinic and ionotropic nucleotide receptors in isolated cholinergic synaptic terminals. Neuropharmacology 42(1):20–33

Diaz-Hernandez M, Pintor J, Miras-Portugal MT (2000) Modulation of the dinucleotide receptor present in rat midbrain synaptosomes by adenosine and ATP. Br J Pharmacol 130(2):434–440. doi:10.1038/sj.bjp.0703300

Diez-Zaera M, Diaz-Hernandez JI, Hernandez-Alvarez E, Zimmermann H, Diaz-Hernandez M, Miras-Portugal MT (2011) Tissue-nonspecific alkaline phosphatase promotes axonal growth of hippocampal neurons. Mol Biol Cell 22(7):1014–1024. doi:10.1091/mbc.E10-09-0740

Edwards FA, Gibb AJ, Colquhoun D (1992) ATP receptor-mediated synaptic currents in the central nervous system. Nature 359(6391):144–147. doi:10.1038/359144a0

Evans RJ, Derkach V, Surprenant A (1992) ATP mediates fast synaptic transmission in mammalian neurons. Nature 357(6378):503–505. doi:10.1038/357503a0

Fedde KN, Whyte MP (1990) Alkaline phosphatase (tissue-nonspecific isoenzyme) is a phosphoethanolamine and pyridoxal-5′-phosphate ectophosphatase: normal and hypophosphatasia fibroblast study. Am J Hum Genet 47(5):767–775

Fonta C, Negyessy L, Renaud L, Barone P (2004) Areal and subcellular localization of the ubiquitous alkaline phosphatase in the primate cerebral cortex: evidence for a role in neurotransmission. Cereb Cortex 14(6):595–609. doi:10.1093/cercor/bhh021

Fukutani Y, Kobayashi K, Nakamura I, Watanabe K, Isaki K, Cairns NJ (1995) Neurons, intracellular and extracellular neurofibrillary tangles in subdivisions of the hippocampal cortex in normal ageing and Alzheimer's disease. Neurosci Lett 200(1):57–60

Godin I, Wylie C, Heasman J (1990) Genital ridges exert long-range effects on mouse primordial germ cell numbers and direction of migration in culture. Development 108(2):357–363

Gomez-Ramos A, Diaz-Hernandez M, Cuadros R, Hernandez F, Avila J (2006) Extracellular tau is toxic to neuronal cells. FEBS Lett 580(20):4842–4850. doi:10.1016/j.febslet.2006.07.078

Gomez-Ramos A, Diaz-Hernandez M, Rubio A, Diaz-Hernandez JI, Miras-Portugal MT, Avila J (2009) Characteristics and consequences of muscarinic receptor activation by tau protein. Eur Neuropsychopharmacol 19(10):708–717. doi:10.1016/j.euroneuro.2009.04.006

Gomez-Ramos A, Diaz-Hernandez M, Rubio A, Miras-Portugal MT, Avila J (2008) Extracellular tau promotes intracellular calcium increase through M1 and M3 muscarinic receptors in neuronal cells. Mol Cell Neurosci 37(4):673–681. doi:10.1016/j.mcn.2007.12.010

Gomez-Villafuertes R, Pintor J, Gualix J, Miras-Portugal MT (2003) GABAB receptor-mediated presynaptic potentiation of ATP ionotropic receptors in rat midbrain synaptosomes. Neuropharmacology 44(3):311–323

Gomez TM, Spitzer NC (1999) In vivo regulation of axon extension and pathfinding by growth-cone calcium transients. Nature 397(6717):350–355. doi:10.1038/16927

Grundke-Iqbal I, Iqbal K, Tung YC, Quinlan M, Wisniewski HM, Binder LI (1986) Abnormal phosphorylation of the microtubule-associated protein tau (tau) in Alzheimer cytoskeletal pathology. Proc Natl Acad Sci USA 83(13):4913–4917

Gualix J, Gomez-Villafuertes R, Diaz-Hernandez M, Miras-Portugal MT (2003) Presence of functional ATP and dinucleotide receptors in glutamatergic synaptic terminals from rat midbrain. J Neurochem 87(1):160–171

Khakh BS (2001) Molecular physiology of P2X receptors and ATP signalling at synapses. Nat Rev Neurosci 2(3):165–174. doi:10.1038/35058521

Langer D, Ikehara Y, Takebayashi H, Hawkes R, Zimmermann H (2007) The ectonucleotidases alkaline phosphatase and nucleoside triphosphate diphosphohydrolase 2 are associated with subsets of progenitor cell populations in the mouse embryonic, postnatal and adult neurogenic zones. Neuroscience 150(4):863–879. doi:10.1016/j.neuroscience.2007.07.064

Martin DL, Barke KE (1998) Are GAD65 and GAD67 associated with specific pools of GABA in brain? Perspect Dev Neurobiol 5(2–3):119–129

Narisawa S, Hasegawa H, Watanabe K, Millan JL (1994) Stage-specific expression of alkaline phosphatase during neural development in the mouse. Dev Dyn 201(3):227–235. doi:10.1002/aja.1002010306

Negyessy L, Xiao J, Kantor O, Kovacs GG, Palkovits M, Doczi TP, Renaud L, Baksa G, Glasz T, Ashaber M, Barone P, Fonta C (2011) Layer-specific activity of tissue non-specific alkaline phosphatase in the human neocortex. Neuroscience 172:406–418. doi:10.1016/j.neuroscience.2010.10.049

Rathbone MP, Middlemiss PJ, Gysbers JW, Andrew C, Herman MA, Reed JK, Ciccarelli R, Di Iorio P, Caciagli F (1999) Trophic effects of purines in neurons and glial cells. Prog Neurobiol 59(6):663–690

Simon D, Garcia-Garcia E, Gomez-Ramos A, Falcon-Perez JM, Diaz-Hernandez M, Hernandez F, Avila J (2012) Tau overexpression results in its secretion via membrane vesicles. Neurodegener Dis 10(1–4):73–75. doi:10.1159/000334915

Vardy ER, Kellett KA, Cocklin SL, Hooper NM (2012) Alkaline phosphatase is increased in both brain and plasma in Alzheimer's disease. Neurodegener Dis 9(1):31–37. doi:10.1159/000329722

Weissman TA, Riquelme PA, Ivic L, Flint AC, Kriegstein AR (2004) Calcium waves propagate through radial glial cells and modulate proliferation in the developing neocortex. Neuron 43(5):647–661. doi:10.1016/j.neuron.2004.08.015

Zimmermann H, Zebisch M, Strater N (2012) Cellular function and molecular structure of ecto-nucleotidases. Purinergic Signal 8(3):437–502. doi:10.1007/s11302-012-9309-4

Index

C. Fonta and L. Négyessy (eds.), *Neuronal Tissue-Nonspecific Alkaline Phosphatase (TNAP)*, Subcellular Biochemistry 76,
DOI 10.1007/978-94-017-7197-9

R

MIX
Papier aus verantwortungsvollen Quellen
Paper from responsible sources
FSC® C105338

If you have any concerns about our products,
you can contact us on
ProductSafety@springernature.com

In case Publisher is established outside the EU,
the EU authorized representative is:
Springer Nature Customer Service Center GmbH
Europaplatz 3, 69115 Heidelberg, Germany

Printed by Libri Plureos GmbH
in Hamburg, Germany